WITHDRAWN
WRIGHT STATE UNIVERSITY LIBRARIES

Problems on
Mapping Class Groups
and Related Topics

Proceedings of Symposia in
PURE MATHEMATICS

Volume 74

Problems on
Mapping Class Groups
and Related Topics

Benson Farb
Editor

American Mathematical Society
Providence, Rhode Island

2000 *Mathematics Subject Classification.* Primary 58D29, 20F38, 30F60, 14D22, 57M99, 20F34, 20F36.

Library of Congress Cataloging-in-Publication Data
Problems on mapping class groups and related topics / Benson Farb, editor.
  p. cm. — (Proceedings of symposia in pure mathematics ; v. 74)
  Includes bibliographical references.
  ISBN-13: 978-0-8218-3838-9 (alk. paper)
  ISBN-10: 0-8218-3838-5 (alk. paper)
  1. Mappings (Mathematics)–Congresses. 2. Class groups (Mathematics)–Congresses.
I. Farb, Benson.

QA360.P76 2006
511.3′3–dc22   2006048369

Copying and reprinting. Material in this book may be reproduced by any means for educational and scientific purposes without fee or permission with the exception of reproduction by services that collect fees for delivery of documents and provided that the customary acknowledgment of the source is given. This consent does not extend to other kinds of copying for general distribution, for advertising or promotional purposes, or for resale. Requests for permission for commercial use of material should be addressed to the Acquisitions Department, American Mathematical Society, 201 Charles Street, Providence, Rhode Island 02904-2294, USA. Requests can also be made by e-mail to reprint-permission@ams.org.

Excluded from these provisions is material in articles for which the author holds copyright. In such cases, requests for permission to use or reprint should be addressed directly to the author(s). (Copyright ownership is indicated in the notice in the lower right-hand corner of the first page of each article.)

© 2006 by the American Mathematical Society. All rights reserved.
The American Mathematical Society retains all rights
except those granted to the United States Government.
Copyright of individual articles may revert to the public domain 28 years
after publication. Contact the AMS for copyright status of individual articles.
Printed in the United States of America.

∞ The paper used in this book is acid-free and falls within the guidelines
established to ensure permanence and durability.
Visit the AMS home page at http://www.ams.org/

10 9 8 7 6 5 4 3 2 1   11 10 09 08 07 06

Dye mon, gen mon.

Behind the mountains,
more mountains.

— **Haitian proverb**

Contents

Preface ix

I. Cohomological, Combinatorial and Algebraic Structure

Four Questions about Mapping Class Groups
   M. BESTVINA   3

Some Problems on Mapping Class Groups and Moduli Space
   B. FARB   11

Finiteness and Torelli Spaces
   R. HAIN   57

Fifteen Problems about the Mapping Class Groups
   N. IVANOV   71

Problems on Homomorphisms of Mapping Class Groups
   M. KORKMAZ   81

The Mapping Class Group and Homotopy Theory
   I. MADSEN   91

Probing Mapping Class Groups Using Arcs
   R. PENNER   97

Relations in the Mapping Class Group
   B. WAJNRYB   115

II. Connections with 3-manifolds, Symplectic Geometry and Algebraic Geometry

Mapping Class Group Factorizations and Symplectic 4-manifolds: Some Open Problems
   D. AUROUX   123

The Topology of 3-manifolds, Heegaard Distance and the Mapping Class Group of a 2-manifold
   J. BIRMAN   133

Lefschetz Pencils and Mapping Class Groups
   S.K. DONALDSON   151

Open Problems in Grothendieck-Teichmüller Theory
    P. LOCHAK AND L. SCHNEPS   165

III. Geometric and Dynamical Aspects

Mapping Class Group Dynamics on Surface Group Representations
    W. GOLDMAN   189

Geometric Properties of the Mapping Class Group
    U. HAMENSTÄDT   215

Problems on Billiards, Flat Surfaces and Translation Surfaces
    P. HUBERT, H. MASUR, T. SCHMIDT AND A. ZORICH   233

Problems in the Geometry of Surface Group Extensions
    L. MOSHER   245

Surface Subgroups of Mapping Class Groups
    A.W. REID   257

Weil-Petersson Perspectives
    S. WOLPERT   269

IV. Braid Groups, $\mathrm{Out}(F_n)$ and other Related Groups

Braid Groups and Iwahori-Heck Algebras
    S. BIGELOW   285

Automorphism Groups of Free Groups, Surface Groups and Free Abelian Groups
    M. BRIDSON AND K. VOGTMANN   301

Problems: Braid Groups, Homotopy, Cohomology and Representations
    F.R. COHEN   317

Cohomological Structure of the Mapping Class Group and Beyond
    S. MORITA   329

From Braid Groups to Mapping Class Groups
    L. PARIS   355

Preface

The study of mapping class groups, the moduli space of Riemann surfaces, Teichmüller geometry and related areas has seen a recent influx of young mathematicians. Inspired by this, I had the idea to solicit from some of the senior people in the area papers that would focus primarily on open problems and directions. I proposed that these problems might range in scope from specific computations to broad programs. The idea was then to bring these papers together into one source, most likely a book. This book would then be a convenient location where younger (and indeed all) researchers could go in order to find problems that might inspire them to further work. I was especially interested in having problems formulated explicitly and accessibly. The result is this book.

The appearance of mapping class groups in mathematics is ubiquitous; choosing topics to cover seemed an overwhelming task. In the end I chose to solicit papers which would likely focus on those aspects of the topic most deeply connected with geometric topology, combinatorial group theory, and surrounding areas.

Content. For organizational purposes the papers here are divided into four groups. This division is by necessity somewhat arbitrary, and a number of the papers could just as easily have been grouped differently.

The problems discussed in Part I focus on the combinatorial and (co)homological group-theoretic aspects of mapping class groups, and the way in which these relate to problems in geometry and topology. The most remarkable recent success in this direction has been the proof by Madsen and Weiss of the Morita-Mumford-Miller Conjecture on the stable cohomology of mapping class groups. Further problems arising from this work are described in Madsen's paper. Other cohomological aspects, including those related to various subgroups, most notably the Torelli group, are discussed in the papers of Bestvina and Hain. The combinatorial and geometric group theory of mapping class groups admits a rich and interesting structure. Ideas and problems coming out of this point of view are discussed in the papers of Farb, Ivanov, Korkmaz, Penner and Wajnryb.

Part II concentrates on connections between various classification problems in topology and their combinatorial reduction to (still open) problems about mapping class groups. In dimension three this reduction is classical. It arises from the fact that every 3-manifold is a union of two handlebodies glued along their boundaries. This construction and many of the problems arising from it are described in Birman's paper. The reduction of the classification of 4-dimensional symplectic manifolds to purely combinatorial topological questions about surfaces and mapping class groups is more recent. The general idea is that (by a theorem of Donaldson) each closed symplectic 4-manifold admits a *symplectic Lefschetz pencil*. These are a kind of "fibration with singularities", and the main piece of data that determines a Lefschetz pencil is its monodromy, which is a collection of mapping classes. These ideas and a number of problems arising from them are presented in the papers of Auroux and Donaldson. Finally, connections with algebraic geometry and number theory via Grothendieck-Teichmüller theory are given in the paper of Lochak and Schneps. One can begin to see this connection, for example, in Belyi's theorem that a complex algebraic curve is defined over $\bar{Q}$ if and only if it is a branched cover over S^2 branched only over $\{0, 1, \infty\}$.

A wide variety of problems, from understanding billiard trajectories to the classification of Kleinian groups, can be reduced to differential and synthetic geometry problems about moduli space. Such problems and connections are discussed in Part III in the papers of Hamenstädt, Mosher, Reid and Wolpert. Those with heavily dynamical flavor are discussed in the papers of Goldman and Hubert, Masur, Schmidt and Zorich.

Mapping class groups are related, both concretely and philosophically, to a number of other groups. While braid groups are technically a special example of a type of mapping class group, the study of these groups has its own special flavor, and in most instances much more is known in this case. The papers of Bigelow, Cohen and Paris concentrate on problems related to braid groups. There has also been a long-running analogy between mapping class groups, linear groups, and automorphism groups of free groups. Problems relating to this analogy are explored in the papers of Bridson and Vogtmann and Morita.

Acknowledgements. The entire content of this book is due to the authors of the individual papers. I feel priviledged to have edited a collection of papers from such experts. It is a great pleasure to thank them for their generosity and their time, not to mention their willingness to openly share their ideas and viewpoints. I must admit my surprise at how little nagging I had to do to complete this project; indeed, every single paper in this volume (except, I must admit, mine) was completed in a timely manner. I hope and believe that the visions and problems shared by these authors here will have a significant influence on the development of the field.

I would like to thank Sergei Gelfand for his continued support, and for his prodding, without which this project would not have been completed.

Part I

Cohomological, Combinatorial and Algebraic Structure

Four Questions about Mapping Class Groups

Mladen Bestvina

In this note I will present my four favorite questions about mapping class groups. The first two are particularly dear to my heart and I frequently ponder some of their aspects.

1. Systoles and rational cohomology

Let $\mathcal{T}_g$ denote the Teichmüller space of marked complete hyperbolic structures on the surface S_g of genus g, and let
$$\mathcal{M}_g = \mathcal{T}_g/MCG(S_g)$$
denote the quotient by the mapping class group (thus $\mathcal{M}_g$ is the space of unmarked hyperbolic structures). To each hyperbolic surface $\Sigma \in \mathcal{M}_g$ we can associate the length $L(\Sigma)$ of a shortest nontrivial closed geodesic in Σ. Such a geodesic is called a *systole*. This gives us a continuous function
$$L : \mathcal{M}_g \to (0, \infty)$$

EXERCISE 1. Show that
- L attains a maximum μ_g.
- Every set of the form $L^{-1}([\epsilon, \mu_g])$ ($\epsilon > 0$) is compact.

In other words, the function
$$\Psi := -log\ L : \mathcal{M}_g \to [-log(\mu_g), \infty)$$
is a proper function. We would like to regard Ψ as a kind of a Morse function on $\mathcal{M}_g$ in order to study its topology. Of course, L is not even smooth, but that shouldn't stop us. (To see that L is not smooth imagine a smooth arc in $\mathcal{M}_g$ along which the length of a curve α has positive derivative and the length of a curve β has negative derivative. Suppose that the lengths of α and β are equal at an interior point p of the arc, and that α is a systole on one side of this point and β on the other. Then L restricted to this arc fails to be smooth at p. For $g = 1$ one can see that L is not smooth by an explicit calculation – see below.)

We want to examine the change in topology of the sublevel sets
$$\Psi^{-1}([-log(\mu_g), t])$$
as t passes through the "critical values" of Ψ.

It is a theorem of Akrout [**Akr03**] that $\Psi \circ q$ is a topological Morse function on $\mathcal{T}_g$ where $q : \mathcal{T}_g \to \mathcal{M}_g$ is the quotient map. Recall that a point $x \in X$

©2006 Mladen Bestvina

in a topological n-manifold X is said to be *regular point* of a continuous function $f : X \to \mathbb{R}$ if there is a (topological) chart around $x = 0$ in which $f(z) = const + z_1$. Otherwise, x is a *critical point* of f. Also recall that a point $x \in X$ is said to be a *nondegenerate critical point* of f if in a local (topological) chart around $x = 0$ we have $f(z) = const - z_1^2 - \cdots - z_\lambda^2 + z_{\lambda+1}^2 + \cdots + z_n^2$ for some λ. This λ does not depend on the choice of such a chart and is called the *index* of the critical point x. The nondegenerate critical points are necessarily isolated. The function f is *topologically Morse* if it is proper and all critical points are nondegenerate. The usual results of Morse theory, such as Morse inequalities and the construction of a homotopy model with cells corresponding to critical points, hold in the topological category.

This construction makes sense in the case $g = 1$ as well, even though a torus does not have a hyperbolic metric. In that case we take $\mathcal{T}_1$ to be the space of marked Euclidean (i.e. flat) metrics of area 1. $\mathcal{T}_1$ can be identified with hyperbolic plane $\mathbb{H}^2$ and $MCG(S_1) = SL_2(\mathbb{Z})$. The quotient $\mathcal{M}_1$ is the $(2, 3, \infty)$ hyperbolic orbifold whose underlying space is homeomorphic to $\mathbb{R}^2$. A point in $\mathcal{M}_1$ can be thought of as a lattice Λ in $\mathbb{C} = \mathbb{R}^2$ of area 1 (up to rotations), with the corresponding flat torus $\mathbb{R}^2/\Lambda$.

EXERCISE 2. In the case $g = 1$

- the maximal value of L is $\mu_1 = \sqrt{\frac{2}{\sqrt{3}}}$ and it is attained (only) at the hexagonal lattice $\Lambda = span(1, \tau)$ where τ is a primitive 6th root of unity.
- Let $\Lambda_\square = span(1, i)$ be the square lattice. Let $q : \mathcal{T}_1 \to \mathcal{M}_1$ be the projection. The points of $q^{-1}(\Lambda_\square)$ are (topological) saddle points of $\Psi \circ q : \mathcal{T}_1 \to [-log(\mu_1), \infty)$. Hint: View $\mathcal{T}_1$ as the space of discrete orientation preserving embeddings $\mathbb{Z}^2 \to \mathbb{R}^2$ modulo homothety. Normalizing so that $(1, 0) \mapsto (1, 0)$ let (x, y) denote the image of $(0, 1)$. For the square lattice we have $(x, y) = (0, 1)$, and in a neighborhood of $(0, 1)$ the function $L \circ q$ is given by
$$L \circ q(x, y) = \min(1, \sqrt{x^2 + y^2})/\sqrt{y}$$
(dividing by the square root of the area). Then $(0, 1)$ is a "topological saddle point".
- Show that $\Lambda_\square$ is not a critical point for Ψ.
- If a lattice Λ has a unique (up to sign) nonzero element of smallest norm, then L is smooth near Λ and Λ is not a critical point of L.
- If a lattice Λ has exactly two (up to sign) nonzero elements of smallest norm, but it is not the square lattice, then L is not smooth near Λ but Λ is not a critical point (there is a homeomorphic chart in which L is the height function).

Thus Ψ is the simplest possible proper function on $\mathcal{M}_1 \cong \mathbb{R}^2$: it has a unique minimum and no other critical points.

Schmutz Schaller has studied the nature of some of the critical points of $\Psi \circ q$ and in particular has a complete description in genus 2 [**SS99**].

Now I want to point out that Ψ is "rationally Morse". To explain this, we recall the following fact.

Fact. Let X be an ANR (e.g. a simplicial complex or a topological manifold) and let G be a finite group of homeomorphisms of X.

(1) If X is rationally acyclic then X/G is rationally acyclic.
(2) If X is rationally the n-sphere and G acts by fixing $H_n(X;\mathbb{Q})$ (we say G preserves orientation) then X/G is rationally the n-sphere.
(3) If X is rationally the n-sphere, some elements fix $H_n(X;\mathbb{Q})$ and others act by changing the sign (we say G reverses orientation), then X/G is rationally acyclic.

A similar fact holds for pairs (X, A).

We now apply this fact as follows. Let $\tilde{\Sigma} \in \mathcal{T}_g$ and $\Psi q(\tilde{\Sigma}) = t$. Consider the pair
$$(X, A) = (q^{-1}\Psi^{-1}[0,t], q^{-1}\Psi^{-1}[0,t] \setminus \{\tilde{\Sigma}\})$$
The group of symmetries $G(\tilde{\Sigma})$ of $\tilde{\Sigma}$ (i.e. the stabilizer of $\tilde{\Sigma}$ in $MCG(S_g)$) acts on (X, A). Let $\Sigma = q(\tilde{\Sigma})$. The Fact then implies that the rational cohomology of
$$(\Psi^{-1}[0,t], \Psi^{-1}[0,t] \setminus \{\Sigma\})$$
is either trivial (when $\tilde{\Sigma}$ is regular or if $G(\tilde{\Sigma})$ is orientation reversing) or it is 1-dimensional and concentrated in dimension λ (if $\tilde{\Sigma}$ is critical of index λ and $G(\tilde{\Sigma})$ is orientation preserving). In the latter case we say that Σ is rationally critical of index λ. Morse inequalities hold in this context.

PROBLEM 1.1. *Study the function*
$$\Psi : \mathcal{M}_g \to [-\log(\mu_g), \infty)$$
as a (rational) Morse function. Classify its rational critical points. Deduce properties of the rational cohomology of $\mathcal{M}_g$ (see below for specific statements).

EXERCISE 3. Ψ has only finitely many rationally critical points.

I don't really know what to expect and don't have enough confidence to state any conjectures. However, there are things I hope for. The function Ψ should be a "simplest possible" proper function on $\mathcal{M}_g$. If I may be bold, I will state these Hopes:

(1) The number of rationally critical points of index k does not depend on g as long as $g >> k$.
(2) By A denote the free polynomial algebra
$$A = \mathbb{Q}[\kappa_1, \kappa_2, \kappa_3, \cdots]$$
in generators κ_i of degree $2i$. Let A_k be the degree k homogeneous piece of A. For every n and for sufficiently large $g >> n$ the following holds: if $k \leq n$ then $\Psi : \mathcal{M}_g \to [-\log(\mu_g), \infty)$ has $\dim_{\mathbb{Q}} A_k$ rationally critical points of index k. (This is 0 if k is odd.)

EXERCISE 4. Let $\Sigma \in \mathcal{M}_g$ be a hyperbolic surface in which the set of systoles does not fill. Then Σ is not a rationally critical point.

Finally, let me explain where the Hopes above come from. For any k the (integral and rational) cohomology $H^k(MCG(S_g))$ is independent of g as long as $g >> k$ [**Har85**], [**Iva89**]. This is reflected in Hope (1). It was recently proved by Madsen and Weiss [**MW**] that in the stable range $H^*(MCG(S_g); \mathbb{Q})$ coincides with A, thus settling the conjecture of Mumford [**Mum83**]. Classes κ_i are known as Miller-Morita-Mumford classes [**Mil86**],[**Mor84**],[**Mor87**]. It was known previously [**Mil86**] that A injects in the stable range. Thus Hope (2) would give an alternative proof of Mumford's conjecture.

2. Torelli groups

By $\mathcal{I}_g$ denote the subgroup of $MCG(S_g)$ consisting of mapping classes that act trivially in $H_1(S_g)$. The group $\mathcal{I}_g$ is the *genus g Torelli group*. The following problem was posed by Geoff Mess [**Kir97**, Problem 2.9].

PROBLEM 2.1. *Determine the finiteness properties of $\mathcal{I}_g$. For which k is $H_k(\mathcal{I}_g)$ finitely generated? For which k is there a $K(\mathcal{I}_g, 1)$ with finite k-skeleton (one says $\mathcal{I}_g$ is of type F_k)?*

EXERCISE 5. $\mathcal{I}_1$ is trivial. $\mathcal{I}_g$ is torsion-free for any g.

The following is known:
- $\mathcal{I}_g$ is finitely generated for $g \geq 3$ [**Joh83**].
- $\mathcal{I}_2$ is not finitely generated [**MM86**]; in fact $\mathcal{I}_2$ is a free group of infinite rank with a basis in 1-1 correspondence with orthogonal splittings of $H_1(S_g) = \mathbb{Z}^{2g}$ (with respect to the intersection pairing) [**Mes92**].
- $H_3(\mathcal{I}_3)$ is not finitely generated [**Mes92**].

One might expect that $\mathcal{I}_g$ is of type $F_{k(g)}$ and $H_{k(g)+1}(\mathcal{I}_g)$ is not finitely generated (for $g \geq 3$) for a certain function $k(g) \to \infty$ (one would presumably expect $k(g)$ to be linear with slope 1 or 2). One might also expect that $\mathcal{I}_g$ has dimension[1] $k(g) + 1$. A phenomenon of this sort is well known. Perhaps the simplest example is the class of groups K_g defined as the kernel of the homomorphism f from the g-fold product of rank 2 free groups to $\mathbb{Z}$ that sends all $2g$ standard generators to the generator of $\mathbb{Z}$. It is a theorem of Bieri [**Bie76**] (motivated by an earlier work of Stallings) that K_g has type F_{g-1} and $H_g(K_g)$ is not finitely generated. There is a Morse theoretic proof of this in [**BB97**]. I would propose an approach to the above question via Morse theory, similar in spirit to Problem 1.1.

First, one has a natural $K(\mathcal{I}_g, 1)$, namely the quotient $Y_g = \mathcal{T}_g/\mathcal{I}_g$. Since $\mathcal{I}_g$ is torsion-free, Y_g is a (noncompact) smooth manifold. We would like to define and study a Morse function on Y_g.

A point in Y_g is represented by a hyperbolic surface Σ of genus g together with a "homology marking", i.e. an identification between $H_1(\Sigma)$ and $H_1(S_g) = \mathbb{Z}^{2g}$. Given an element $\gamma \in \mathbb{Z}^{2g}$ we have a corresponding $x_\gamma \in H_1(\Sigma)$ and we denote by $L_\gamma(\Sigma)$ the length of a shortest cycle representing x_γ. This shortest cycle will necessarily be a sum of simple closed geodesics any two of which either coincide or are disjoint. The shortest representative may not be unique and the function

$$L_\gamma : Y_g \to \mathbb{R}$$

is generally not smooth (certainly for $g \geq 2$).

PROBLEM 2.2. *Let $\gamma_1, \cdots, \gamma_{2g}$ be the standard basis of $\mathbb{Z}^{2g}$. Study the function*

$$L = \sum L_{\gamma_i} : Y_g \to [0, \infty)$$

as a Morse function. Find critical sets and deduce properties of the topology of Y_g. For example, estimate from above and below the homotopy dimension[2] of Y_g.

---

[1] the smallest dimension of a $K(\mathcal{I}_g, 1)$, or the cohomological dimension (they should be equal for this group)

[2] the smallest dimension of a space homotopy equivalent to Y_g

EXERCISE 6. For $g = 1$ we have $Y_1 = \mathcal{T}_1 = \mathbb{H}^2$ and L is a smooth proper function with a unique critical point, which is the minimum (corresponding to the square lattice).

There are many technical obstacles when $g \geq 2$. L is not smooth, but we are used to this from the previous problem. A bigger concern is that L is not proper. For example, take a ray given by pinching a separating curve α. When α is small it has a long cyclindrical neighborhood and the minimizing cycles will not cross α. It follows that L is bounded along the ray. To fix this problem we "complete" Y_g by adding points at infinity. In the first approximation, for a separating curve α we will add $Y_{g_1}^1 \times Y_{g_2}^1 \times S^1$ where α separates into surfaces of genus g_1 and $g_2 = g - g_1$, S^1 is the twisting parameter, and Y_g^1 is the analog of Y_g where we allow one puncture. More precisely, a copy of $Y_{g_1}^1 \times Y_{g_2}^2 \times S^1$ is added for every splitting of $H_1(S_g) = \mathbb{Z}^{2g}$ as a direct sum of two orthogonal subspaces with respect to the intersection pairing. We also add similar spaces for pairs of disjoint separating curves etc.

PROBLEM 2.3. *Work out the details of this construction of the completion $\overline{Y}_g$ of Y_g. Show that $L : Y_g \to [0, \infty)$ extends to*
$$\overline{L} : \overline{Y}_g \to [0, \infty)$$
and is a proper map. Ideally, inclusion $Y_g \to \overline{Y}_g$ should be a homotopy equivalence (and $\overline{Y}_g - Y_g$ should be the boundary of the manifold with corners $\overline{Y}_g$). Study $\overline{L}$ as a "Morse" function. Critical points of $\overline{L}$ are most likely not isolated; however, explain the change in the homotopy type of $\overline{L}^{-1}[0, t]$ as t passes through a critical value.

For example, take the case $g = 2$. Spaces added at infinity are $Y_1^1 \times Y_1^1 \times S^1 = \mathbb{H}^2 \times \mathbb{H}^2 \times S^1$ and the restriction of $\overline{L}$ to such a set factors through $\mathbb{H}^2 \times \mathbb{H}^2$ where it has a unique minimum and no other critical points. Thus this set contains a "mincircle".

Hope. A point of $\overline{Y}_2$ is a local minimum of $\overline{L}$ iff it belongs to one of these mincircles. In addition there are isolated critical points of index 1 that connect up these circles in a tree-like pattern. Thus $\overline{Y}_2$ and Y_2 are homotopy equivalent to a wedge of circles, one circle for every splitting, recovering the theorem of Mess mentioned above.

3. Action dimension

The focus of this section is the following.

PROBLEM 3.1. *What is the smallest $n = n(g)$ such that $MCG(S_g)$ admits a properly discontinuous action on $\mathbb{R}^n$? on a contractible n-manifold? (The answers are expected to be the same.)*

Of course, $MCG(S_g)$ acts properly discontinuously on Teichmüller space $\mathcal{T}_g \cong \mathbb{R}^{6g-6}$ so it is a natural guess that $n(g) = 6g - 6$ (for $g \geq 2$; clearly $n(1) = 2$).

The question is motivated by the work in [**BKK02**] where a technique is developed for finding lower bounds on the dimension $n(\Gamma)$ of a contractible manifold where a given group Γ acts properly discontinuously. The optimal $n(\Gamma)$ is called the *action dimension* of Γ. For example, $n(F_2^g) = 2g$ and if Γ is a lattice in a connected semisimple Lie group G then $n(\Gamma) = \dim G/K$ (where K is a maximal compact subgroup) [**BF02a**].

As observed in [**BKK02**], if B_k is the braid group on k strands then $n(B_k) = 2k - 3$ ($k \geq 2$) which equals the dimension of the corresponding Teichmüller space.

The method is this. Find a convenient finite simplicial complex L that does not embed in $\mathbb{R}^m$ "for homological reasons". Such a complex is called an *m-obstructor complex*. Complexes used in all applications are joins of pointed spheres (a pointed sphere S_+^k is a sphere S^k union a disjoint point). E.g. the "utilities graph" is the join of two pointed 0-spheres, it does not embed in the plane for homological reasons, and is a 2-obstructor complex. For example, if
$$L = S_+^{k_1} * S_+^{k_2} * \cdots * S_+^{k_p}$$
then L is an m-obstructor complex[3] for
$$m = k_1 + k_2 + \cdots + k_p + 2p - 2$$
and all obstructor complexes used in [**BKK02**] and [**BF02b**] have this form. Then one wants to find a map
$$F : L \times [0, \infty) \to X$$
which is proper and expanding and X is a proper metric space on which the given group Γ acts isometrically, properly discontinuously and cocompactly. The map F is *expanding* if for any two disjoint simplices $\sigma, \tau \subset L$ we have
$$\lim_{t \to \infty} d(F(\sigma \times [t, \infty)), F(\tau \times [t, \infty))) = \infty$$
For the case $\Gamma = MCG(S_g)$ it is natural to take X to be the subset $\mathcal{T}_g^{\geq \epsilon}$ of Teichmüller space consisting of marked surfaces with no closed geodesics of length $< \epsilon$ for a certain small $\epsilon > 0$. When such a map is found it then follows from [**BKK02**] that $n(\Gamma) \geq m + 2$.

PROBLEM 3.2. *Find a $(6g - 8)$-obstructor complex L and a proper expanding map*
$$F : L \times [0, \infty) \to \mathcal{T}_g^{\geq \epsilon}$$

4. Bounded cohomology

Recall that a *quasihomomorphism* on a group Γ is a function
$$f : \Gamma \to \mathbb{R}$$
such that
$$|f(\gamma_1 \gamma_2) - f(\gamma_1) - f(\gamma_2)|$$
is uniformly bounded over all $\gamma_1, \gamma_2 \in \Gamma$. In [**BF02c**] it is shown that there are many quasihomomorphisms
$$f : MCG(S_g) \to \mathbb{R}$$
that are
(1) bounded on every subgroup of $MCG(S_g)$ consisting of mapping classes fixing a simple closed curve, and
(2) unbounded on the subgroup generated by (necessarily a pseudo-Anosov homeomorphism) $\phi_f \in MCG(S_g)$.

---

[3]For concreteness we triangulate each sphere as the join of 0-spheres and we triangulate L as the join.

In other words, the subgroups in (1) are "small" in the sense that the set
$$G_1 \cdot G_2 \cdot \dots \cdot G_p = \{g_1 g_2 \cdots g_p | g_i \in G_i\}$$
fails to contain high powers of ϕ_f for any p and any subgroups G_i as in (1).

Now I would like to propose that certain other subgroups of $MCG(S_g)$ are small as well. Let $q: S_g \to S'$ be a covering map of degree > 1. Define
$$G(q) = \{\phi \in MCG(S_g) | \phi \text{ is a lift of a mapping class on } S'\}$$

PROBLEM 4.1. *Show that there are many quasihomomorphisms* $f: MCG(S_g) \to \mathbb{R}$ *that satisfy (1) and (2) above plus*

(3) *f is bounded on every $G(q)$.*

I remark that the consequence about product sets of $G(q)$'s was shown to be true by a different method (Bestvina-Feighn, unpublished)

References

[Akr03] H. Akrout, *Singularités topologiques des systoles généralisées*, Topology **42** (2003), no. 2, 291–308. MR1 941 437

[BB97]   Mladen Bestvina and Noel Brady, *Morse theory and finiteness properties of groups*, Invent. Math. **129** (1997), no. 3, 445–470. MR98i:20039

[BF02a] Mladen Bestvina and Mark Feighn, *Proper actions of lattices on contractible manifolds*, Invent. Math. **150** (2002), no. 2, 237–256. MR2004d:57042

[BF02b] _____, *Proper actions of lattices on contractible manifolds*, Invent. Math. **150** (2002), no. 2, 237–256. MR1933585 (2004d:57042)

[BF02c] Mladen Bestvina and Koji Fujiwara, *Bounded cohomology of subgroups of mapping class groups*, Geom. Topol. **6** (2002), 69–89 (electronic). MR2003f:57003

[Bie76] Robert Bieri, *Normal subgroups in duality groups and in groups of cohomological dimension* 2, J. Pure Appl. Algebra **7** (1976), no. 1, 35–51. MR52 #10904

[BKK02] Mladen Bestvina, Michael Kapovich, and Bruce Kleiner, *Van Kampen's embedding obstruction for discrete groups*, Invent. Math. **150** (2002), no. 2, 219–235. MR2004c:57060

[Har85] John L. Harer, *Stability of the homology of the mapping class groups of orientable surfaces*, Ann. of Math. (2) **121** (1985), no. 2, 215–249. MR87f:57009

[Iva89] N. V. Ivanov, *Stabilization of the homology of Teichmüller modular groups*, Algebra i Analiz **1** (1989), no. 3, 110–126. MR91g:57010

[Joh83] Dennis Johnson, *The structure of the Torelli group. I. A finite set of generators for $\mathcal{I}$*, Ann. of Math. (2) **118** (1983), no. 3, 423–442. MR85a:57005

[Kir97] Robion Kirby, *Problems in low-dimensional topology*, Geometric topology (Athens, GA, 1993) (Robion Kirby, ed.), AMS/IP Stud. Adv. Math., vol. 2, Amer. Math. Soc., Providence, RI, 1997, pp. 35–473. MR1 470 751

[Mes92] Geoffrey Mess, *The Torelli groups for genus 2 and 3 surfaces*, Topology **31** (1992), no. 4, 775–790. MR93k:57003

[Mil86] Edward Y. Miller, *The homology of the mapping class group*, J. Differential Geom. **24** (1986), no. 1, 1–14. MR88b:32051

[MM86]   Darryl McCullough and Andy Miller, *The genus 2 Torelli group is not finitely generated*, Topology Appl. **22** (1986), no. 1, 43–49. MR87h:57015

[Mor84] Shigeyuki Morita, *Characteristic classes of surface bundles*, Bull. Amer. Math. Soc. (N.S.) **11** (1984), no. 2, 386–388. MR85j:55032

[Mor87] _____, *Characteristic classes of surface bundles*, Invent. Math. **90** (1987), no. 3, 551–577. MR89e:57022

[Mum83] David Mumford, *Towards an enumerative geometry of the moduli space of curves*, Arithmetic and geometry, Vol. II, Progr. Math., vol. 36, Birkhäuser Boston, Boston, MA, 1983, pp. 271–328. MR85j:14046

[MW]   Ib Madsen and Michael S. Weiss, *The stable moduli space of Riemann surfaces: Mumford's conjecture*, arXiv, math.AT/0212321.

[SS99]   Paul Schmutz Schaller, *Systoles and topological Morse functions for Riemann surfaces*, J. Differential Geom. **52** (1999), no. 3, 407–452. MR2001d:32018

Some Problems on Mapping Class Groups and Moduli Space

Benson Farb

Contents

1. Introduction
2. Subgroups and submanifolds: Existence and classification
3. Combinatorial and geometric group theory of Mod_g
4. Problems on the geometry of $\mathcal{M}_g$
5. The Torelli group
6. Linear and nonlinear representations of Mod_g
7. Pseudo-Anosov theory

References

1. Introduction

This paper contains a biased and personal list of problems on mapping class groups of surfaces. One of the difficulties in this area has been that there have not been so many easy problems. One of my goals here is to formulate a number of problems for which it seems that progress is possible. Another goal is the formulation of problems which will force us to penetrate more deeply into the structure of mapping class groups. Useful topological tools have been developed, for example the Thurston normal form, boundary theory, the reduction theory for subgroups, and the geometry and topology of the complex of curves. On the other hand there are basic problems which seem beyond the reach of these methods. One of my goals here is to pose problems whose solutions might require new methods.

1.1. Universal properties of Mod_g and $\mathcal{M}_g$. Let Σ_g denote a closed, oriented surface of genus g, and let Mod_g denote the group of homotopy classes of orientation-preserving homeomorphisms of Σ_g. The mapping class group Mod_g, along with its variations, derives much of its importance from its universal properties. Let me explain this for $g \geq 2$. In this case, a classical result of Earle-Eells gives that the identity component $\text{Diff}^0(\Sigma_g)$ of $\text{Diff}^+(\Sigma_g)$ is contractible. Since

---

Supported in part by NSF grants DMS-9704640 and DMS-0244542.

©2006 Benson Farb

$\text{Mod}_g = \pi_0 \text{Diff}^+(\Sigma_g)$ by definition, we have a homotopy equivalence of classifying spaces:

$$\text{BDiff}^+(\Sigma_g) \simeq \text{BMod}_g \tag{1}$$

Let $\mathcal{M}_g$ denote the moduli space of Riemann surfaces. The group Mod_g acts properly discontinuously on the Teichmüller space Teich_g of marked, genus g Riemann surfaces. Since Teich_g is contractible it follows that $\mathcal{M}_g$ is a $K(\text{Mod}_g, 1)$ space, i.e. it is homotopy equivalent to the spaces in (1). From these considerations it *morally* follows that, for any topological space B, we have the following bijections:

$$\left\{ \begin{array}{c} \text{Isomorphism classes} \\ \text{of } \Sigma_g\text{-bundles over} \\ B \end{array} \right\} \longleftrightarrow \left\{ \begin{array}{c} \text{Homotopy classes} \\ \text{of} \\ \text{maps } B \to \mathcal{M}_g \end{array} \right\} \longleftrightarrow \left\{ \begin{array}{c} \text{Conjugacy classes} \\ \text{of representations} \\ \rho : \pi_1 B \to \text{Mod}_g \end{array} \right\} \tag{2}$$

I use the term "morally" because (2) is not exactly true as stated. For example, one can have two nonisomorphic Σ_g bundles over S^1 with finite monodromy and with classifying maps $f : S^1 \to \text{Mod}_g$ having the same image, namely a single point. The problem here comes from the torsion in Mod_g. This torsion prevents $\mathcal{M}_g$ from being a manifold; it is instead an orbifold, and so we need to work in the category of orbifolds. This is a nontrivial issue which requires care. There are two basic fixes to this problem. First, one can simply replace $\mathcal{M}_g$ in (2) with the classifying space BMod_g. Another option is to replace Mod_g in (2) with any torsion-free subgroup $\Gamma < \text{Mod}_g$ of finite index. Then Γ acts freely on Teich_g and the corresponding finite cover of $\mathcal{M}_g$ is a manifold. In this case (2) is true as stated. This torsion subtlety will usually not have a major impact on our discussion, so we will for the most part ignore it. This is fine on the level of rational homology since the homotopy equivalences described above induce isomorphisms:

$$H^*(\mathcal{M}_g, \mathbf{Q}) \approx H^*(\text{BDiff}^+(\Sigma_g), \mathbf{Q}) \approx H^*(\text{Mod}_g, \mathbf{Q}) \tag{3}$$

There is a unique complex orbifold structure on $\mathcal{M}_g$ with the property that these bijections carry over to the holomorphic category. This means that the manifolds are complex manifolds, the bundles are non-isotrivial (i.e. the holomorphic structure of the fibers is not locally constant, unless the map $B \to \mathcal{M}_g$ is trivial), and the maps are holomorphic. For the third entry of (2), one must restrict to such conjugacy classes with holomorphic representatives; many conjugacy classes do not have such a representative.

For $g \geq 3$ there is a canonical Σ_g-bundle $\mathcal{U}_g$ over $\mathcal{M}_g$, called the *universal curve* (terminology from algebraic geometry), for which the (generic) fiber over any $X \in \mathcal{M}_g$ is the Riemann surface X. The first bijection of (2) is realized concretely using $\mathcal{U}_g$. For example given any smooth $f : B \to \mathcal{M}_g$, one simply pulls back the bundle $\mathcal{U}_g$ over f to give a Σ_g-bundle over B. Thus $\mathcal{M}_g$ plays the same role for surface bundles as the (infinite) Grassmann manifolds play for vector bundles. Again one needs to be careful about torsion in Mod_g here, for example by passing to a finite cover of $\mathcal{M}_g$. For $g = 2$ there are more serious problems.

An important consequence is the following. Suppose one wants to associate to every Σ_g-bundle a (say integral) cohomology class on the base of that bundle, so that this association is *natural*, that is, it is preserved under pullbacks. Then each such cohomology class must be the pullback of some element of $H^*(\mathcal{M}_g, \mathbf{Z})$. In this sense the classes in $H^*(\mathcal{M}_g, \mathbf{Z})$ are universal. After circle bundles, this is the next simplest nonlinear bundle theory. Unlike circle bundles, this study connects in a fundamental way to algebraic geometry, among other things.

Understanding the sets in (2) is interesting even in the simplest cases.

Example 1.1. (Surface bundles over S^1). Let $B = S^1$. In this case (2) states that the classfication of Σ_g-bundles over S^1, up to bundle isomorphism, is equivalent to the classification of elements of Mod_g up to conjugacy. Now, a fixed 3-manifold may fiber over S^1 in infinitely many different ways, although there are finitely many fiberings with fiber of fixed genus. Since it is possible to compute these fiberings[1], the homeomorphism problem for 3-manifolds fibering over S^1 can easily be reduced to solving the conjugacy problem for Mod_g. This was first done by Hemion [**He**].

Example 1.2. (Arakelov-Parshin finiteness). Now let $B = \Sigma_h$ for some $h \geq 1$, and consider the sets and bijections of (2) in the holomorphic category. The Geometric Shafarevich Conjecture, proved by Arakelov and Parshin, states that these sets are finite, and that the holomorphic map in each nontrivial homotopy class is unique. As beautifully explained in [**Mc1**], from this result one can derive (with a branched cover trick of Parshin) finiteness theorems for rational points on algebraic varieties over function fields.

Remark on universality. I would like to emphasize the following point. While the existence of characteristic classes associated to *every* Σ_g-bundle is clearly the first case to look at, it seems that requiring such a broad form of universality is too constraining. One reflection of this is the paucity of cohomology of $\mathcal{M}_g$, as the Miller-Morita-Mumford conjecture (now theorem due to Madsen-Weiss [**MW**]) shows. One problem is that the requirement of naturality for *all* monodromies simply kills what are otherwise natural and common classes. Perhaps more natural would be to consider the characteristic classes for Σ_g-bundles with torsion-free monodromy. This would lead one to understand the cohomology of various finite index subgroups of Mod_g.

Another simple yet striking example of this phenomenon is Harer's theorem that

$$H^2(\mathcal{M}_g, \mathbf{Q}) = \mathbf{Q}$$

In particular the signature cocycle, which assigns to every bundle $\Sigma_g \to M^4 \to B$ the signature $\sigma(M^4)$, is (up to a rational multiple) the only characteristic class in dimension 2. When the monodromy representation $\pi_1 B \to \mathrm{Mod}_g$ lies in the (infinite index) Torelli subgroup $\mathcal{I}_g < \mathrm{Mod}_g$ (see below), σ is always zero, and so is useless. However, there are infinitely many homotopy types of surface bundles M^4 over surfaces with $\sigma(M^4) = 0$; indeed such families of examples can be taken to have monodromy in $\mathcal{I}_g$. We note that there are no known elements of $H^*(\mathrm{Mod}_g, \mathbf{Q})$ which restrict to a nonzero element of $H^*(\mathcal{I}_g, \mathbf{Q})$.

We can then try to find, for example, characteristic classes for Σ_g-bundles with monodromy lying in $\mathcal{I}_g$, and it is not hard to prove that these are just pullbacks of classes in $H^*(\mathcal{I}_g, \mathbf{Z})$. In dimensions one and two, for example, we obtain a large number of such classes (see [**Jo1**] and [**BFa2**], respectively).

---

[1] This is essentially the fact that the Thurston norm is computable.

I hope I have provided some motivation for understanding the cohomology of subgroups of Mod_g. This topic is wide-open; we will discuss a few aspects of it below.

Three general problems. Understanding the theory of surface bundles thus leads to the following basic general problems.

(1) For various finitely presented groups Γ, classify the representations $\rho : \Gamma \to \text{Mod}_g$ up to conjugacy.
(2) Try to find analytic and geometric structures on $\mathcal{M}_g$ in order to apply complex and Riemannian geometry ideas to constrain possibilities on such ρ.
(3) Understand the cohomology algebra $H^*(\Gamma, K)$ for various subgroups $\Gamma < \text{Mod}_g$ and various modules K, and find topological and geometric interpretations for its elements.

We discuss below some problems in these directions in §2, §4 and §5, respectively. One appealing aspect of such problems is that attempts at understanding them quickly lead to ideas from combinatorial group theory, complex and algebraic geometry, the theory of dynamical systems, low-dimensional topology, symplectic representation theory, and more. In addition to these problems, we will be motivated here by the fact that Mod_g and its subgroups provide a rich and important collection of examples to study in combinatorial and geometric group theory; see §3 below.

Remark on notational conventions. We will usually state conjectures and problems and results for Mod_g, that is, for closed surfaces. This is simply for convenience and simplicity; such conjectures and problems should always be considered as posed for surfaces with boundary and punctures, except perhaps for some sporadic, low-genus cases. Similarly for other subgroups such as the Torelli group $\mathcal{I}_g$. Sometimes the extension to these cases is straight-forward, but sometimes it isn't, and new phenomena actually arise.

1.2. The Torelli group and associated subgroups. One of the recurring objects in this paper will be the Torelli group. We now briefly describe how it fits in to the general picture.

Torelli group. Algebraic intersection number gives a symplectic form on $H_1(\Sigma_g, \mathbf{Z})$. This form is preserved by the natural action of Mod_g. The *Torelli group* $\mathcal{I}_g$ is defined to be the kernel of this action. We then have an exact sequence

$$(4) \qquad 1 \to \mathcal{I}_g \to \text{Mod}_g \to \text{Sp}(2g, \mathbf{Z}) \to 1$$

The genus g *Torelli space* is defined to be the quotient of Teich_g by $\mathcal{I}_g$. Like $\mathcal{M}_g$, this space has the appropriate universal mapping properties. However, the study of maps into Torelli space is precisely *complementary* to the theory of holomorphic maps into $\mathcal{M}_g$, as follows. Any holomorphic map $f : B \to \mathcal{M}_g$ with $f_*(B) \subseteq \mathcal{I}_g$, when composed with the (holomorphic) period mapping $\mathcal{M}_g \to \mathcal{A}_g$ (see §4.4 below), lifts to the universal cover $\widetilde{\mathcal{A}_g}$, which is the Siegel upper half-space (i.e. the symmetric space $\text{Sp}(2g, \mathbf{R})/\text{SU}(g)$). Since the domain is compact, the image of this holomorphic lift is constant. Hence f is constant.

The study of $\mathcal{I}_g$ goes back to Nielsen (1919) and Magnus (1936), although the next big breakthrough came in a series of remarkable papers by Dennis Johnson in the late 1970's (see [**Jo1**] for a summary). Still, many basic questions about $\mathcal{I}_g$ remain open; we add to the list in §5 below.

Group generated by twists about separating curves. The *group generated by twists about separating curves*, denoted $\mathcal{K}_g$, is defined to be the subgroup Mod_g generated by the (infinitely many) Dehn twists about separating (i.e. bounding) curves in Σ_g. The group $\mathcal{K}_g$ is sometimes called the *Johnson kernel* since Johnson proved that $\mathcal{K}_g$ is precisely the kernel of the so-called Johnson homomorphism. This group is a featured player in the study of the Torelli group. Its connection to 3-manifold theory begins with Morita's result that every integral homology 3-sphere comes from removing a handlebody component of some Heegaard embedding $h : \Sigma_g \hookrightarrow S^3$ and gluing it back to the boundary Σ_g by an element of $\mathcal{K}_g$. Morita then proves the beautiful result that for a fixed such h, taking the Casson invariant of the resulting 3-manifold actually gives a *homomorphism* $\mathcal{K}_g \to \mathbf{Z}$. This is a starting point for Morita's analysis of the Casson invariant; see [**Mo1**] for a summary.

Johnson filtration. We now describe a filtration of Mod_g which generalizes (4). This filtration has become a basic object of study.

For a group Γ we inductively define $\Gamma_0 := \Gamma$ and $\Gamma_{i+1} = [\Gamma, \Gamma_i]$. The chain of subgroups $\Gamma \supseteq \Gamma_1 \supseteq \cdots$ is the *lower central series* of Γ. The group Γ/Γ_i is i-step nilpotent; indeed Γ/Γ_i has the universal property that any homomorphism from Γ to any i-step nilpotent group factors through Γ/Γ_i. The sequence $\{\Gamma/\Gamma_i\}$ can be thought of as a kind of Taylor series for Γ.

Now let $\Gamma := \pi_1 \Sigma_g$. It is a classical result of Magnus that $\bigcap_{i=1}^{\infty} \Gamma_i = 1$, that is Γ is *residually nilpotent*. Now Mod_g acts by outer automorphisms on Γ, and each of the subgroups Γ_i is clearly characteristic. We may thus define for each $k \geq 0$:

$$(5) \qquad \mathcal{I}_g(k) := \ker(\text{Mod}_g \to \text{Out}(\Gamma/\Gamma_k))$$

That is, $\mathcal{I}_g(k)$ is just the subgroup of Mod_g acting trivially on the kth nilpotent quotient of $\pi_1 \Sigma_g$. Clearly $\mathcal{I}_g(1) = \mathcal{I}_g$; Johnson proved that $\mathcal{I}_g(2) = \mathcal{K}_g$. The sequence $\mathcal{I}_g = \mathcal{I}_g(1) \supset \mathcal{I}_g(2) \supset \cdots$ is called the *Johnson filtration*; it has also been called the *relative weight filtration*. This sequence forms a (but not *the*) lower central series for $\mathcal{I}_g$. The fact that $\bigcap_{i=1}^{\infty} \Gamma_i = 1$ easily implies that the Aut versions of the groups defined in (5) have trivial intersection. The stronger fact that $\bigcap_{i=1}^{\infty} \mathcal{I}_g(i) = 1$, so that $\mathcal{I}_g$ is residually nilpotent, is also true, but needs some additional argument (see [**BL**]).

Acknowledgments. Thanks to everyone who sent in comments on earlier drafts of this paper. Tara Brendle, Curt McMullen and Andy Putman made a number of useful comments. I am especially grateful to Chris Leininger, Dan Margalit and Ben McReynolds, each of whom passed on to me extensive lists of comments and corrections. Their efforts have greatly improved this paper. Finally, I would like to express my deep appreciation to John Stiefel for his constant friendship and support during the last six months.

2. Subgroups and submanifolds: Existence and classification

It is a natural problem to classify subgroups H of Mod_g. By classify we mean to give an effective list of isomorphism or commensurability types of subgroups, and also to determine all embeddings of a given subgroup, up to conjugacy. While there are some very general structure theorems, one sees rather quickly that the problem as stated is far too ambitous. One thus begins with the problem of finding various invariants associated to subgroups, by which we mean invariants of their isomorphism type, commensurability type, or more extrinsic invariants which depend on the embedding $H \to \mathrm{Mod}_g$. One then tries to determine precisely which values of a particular invariant can occur, and perhaps even classify those subgroups having a given value of the invariant. In this section we present a selection of such problems.

Remark on subvarieties. The classification and existence problem for subgroups and submanifolds of $\mathcal{M}_g$ can be viewed as algebraic and topological analogues of the problem, studied by algebraic geometers, of understanding (complete) subvarieties of $\mathcal{M}_g$. There is an extensive literature on this problem; see, e.g., [**Mor**] for a survey. To give just one example of the type of problem studied, let

$$c_g := \max\{\dim_{\mathbf{C}}(V) : V \text{ is a complete subvariety of } \mathcal{M}_g\}$$

The goal is to compute c_g. This is a kind of measure of where $\mathcal{M}_g$ sits between being affine (in which case c_g would be 0) and projective (in which case c_g would equal $\dim_{\mathbf{C}}(\mathcal{M}_g) = 3g - 3$). While $\mathcal{M}_2$ is affine and so $c_2 = 0$, it is known for $g \geq 3$ that $1 \leq c_g < g-1$; the lower bound is given by construction, the upper bound is a well-known theorem of Diaz.

2.1. Some invariants. The notion of *relative ends* ends(Γ, H) provides a natural way to measure the "codimension" of a subgroup H in a group Γ. To define $e(\Gamma, H)$, consider any proper, connected metric space X on which Γ acts properly and cocompactly by isometries. Then ends(Γ, H) is defined to be the number of ends of the quotient space X/H.

QUESTION 2.1 (Ends spectrum). *What are the possibile values of* ends(Mod_g, H) *for finitely-generated subgroups* $H < \mathrm{Mod}_g$?

It is well-known that the moduli space $\mathcal{M}_g$ has one end. The key point of the proof is that the complex of curves is connected. This proof actually gives more: any cover of $\mathcal{M}_g$ has one end; see, e.g., [**FMa**]. However, I do not see how this fact directly gives information about Question 2.1.

Commensurators. Asking for two subgroups of a group to be conjugate is often too restrictive a question. A more robust notion is that of commensurability. Subgroups Γ_1, Γ_2 of a group H are *commensurable* if there exists $h \in H$ such that $h\Gamma_1 h^{-1} \cap \Gamma_2$ has finite index in both $h\Gamma_1 h^{-1}$ and in Γ_2. One then wants to classify subgroups up to commensurability; this is the natural equivalence relation one studies in order to coarsify the relation of "conjugate" to ignore finite index information. The primary commensurability invariant for subgroups $\Gamma < H$ is the *commensurator of* Γ *in* H, denoted $\mathrm{Comm}_H(\Gamma)$, defined as:

$$\mathrm{Comm}_H(\Gamma) := \{h \in H : h\Gamma h^{-1} \cap \Gamma \text{ has finite index in both } \Gamma \text{ and } h\Gamma h^{-1}\}$$

The commensurator has most commonly been studied for discrete subgroups of Lie groups. One of the most striking results about commensurators, due to Margulis, states that if Γ is an irreducible lattice in a semisimple[2] Lie group H then $[\mathrm{Comm}_H(\Gamma) : \Gamma] = \infty$ if and only if Γ is arithmetic. In other words, it is precisely the arithmetic lattices that have infinitely many "hidden symmetries".

PROBLEM 2.2. *Compute* $\mathrm{Comm}_{\mathrm{Mod}_g}(\Gamma)$ *for various subgroups* $\Gamma < \mathrm{Mod}_g$.

---

[2]By *semisimple* we will always mean linear semisimple with no compact factors.

Paris-Rolfsen and Paris (see, e.g., [**Pa**]) have proven that most subgroups of Mod_g stabilizing a simple closed curve, or coming from the mapping class group of a subsurface of S, are self-commensurating in Mod_g. Self-commensurating subgroups, that is subgroups $\Gamma < H$ with $\mathrm{Comm}_H(\Gamma) = \Gamma$, are particularly important since the finite-dimensional unitary dual of Γ injects into the unitary dual of H; in other words, any unitary representation of H induced from a finite-dimensional irreducible unitary representation of Γ must itself be irreducible.

Volumes of representations. Consider the general problem of classifying, for a fixed finitely generated group Γ, the set

$$\mathcal{X}_g(\Gamma) := \mathrm{Hom}(\Gamma, \mathrm{Mod}_g)/\mathrm{Mod}_g$$

of conjugacy classes of representations $\rho : \Gamma \to \mathrm{Mod}_g$. Here the representations ρ_1 and ρ_2 are conjugate if $\rho_1 = C_h \circ \rho_2$, where $C_h : \mathrm{Mod}_g \to \mathrm{Mod}_g$ is conjugation by some $h \in \mathrm{Mod}_g$. Suppose $\Gamma = \pi_1 X$ where X is, say, a smooth, closed n-manifold. Since $\mathcal{M}_g$ is a classifying space for Mod_g, we know that for each for each $[\rho] \in \mathcal{X}_g(\Gamma)$ there exists a smooth map $f : X \to \mathcal{M}_g$ with $f_* = \rho$, and that f is unique up to homotopy.

Each n-dimensional real cocycle ξ on $\mathcal{M}_g$ then gives a well-defined invariant

$$\nu_\xi : \mathcal{X}_g(\Gamma) \to \mathbf{R}$$

defined by

$$\nu_\xi([\rho]) := \int_X f^*\xi$$

It is clear that $\nu_\xi([\rho])$ does not depend on the choices, and indeed depends only on the cohomology class of ξ. As a case of special interest, let X be a $2k$-dimensional manifold and let ω_{WP} denote the Weil-Petersson symplectic form on $\mathcal{M}_g$. Define the *complex k-volume* of $\rho : \pi_1 X \to \mathrm{Mod}_g$ to be

$$\mathrm{Vol}_k([\rho]) := \int_X f^*\omega_{\mathrm{WP}}^k$$

PROBLEM 2.3 (Volume spectrum). *Determine for each $1 \leq k \leq 3g - 3$ the image of $\mathrm{Vol}_k : \mathcal{X}_g(\Gamma) \to \mathbf{R}$. Determine the union of all such images as Γ ranges over all finitely presented groups.*

It would also be interesting to pose the same problem for representations with special geometric constraints, for example those with holomorphic or totally geodesic (with respect to a fixed metric) representatives. In particular, how do such geometric properties constrain the set of possible volumes? Note that Mirzakhani [**Mir**] has given recursive formulas for the Weil-Petersson volumes of moduli spaces for surfaces with nonempty totally geodesic boundary.

Invariants from linear representations. Each linear representation $\psi : \mathrm{Mod}_g \to \mathrm{GL}_m(\mathbf{C})$ provides us with many invariants for elements of $\mathcal{X}_g(\Gamma)$, simply by composition with ψ followed by taking any fixed class function on $\mathrm{GL}_m(\mathbf{C})$. One can obtain commensurability invariants for subgroups of Mod_g this way as well. While no *faithful* ψ is known for $g \geq 3$ (indeed the existence of such a ψ remains a major open problem), there are many such ψ which give a great deal of information. Some computations using this idea can be found in [**Su2**]. I think further computations would be worthwhile.

2.2. Lattices in semisimple Lie groups.
While Ivanov proved that Mod_g is not isomorphic to a lattice in a semisimple Lie group, a recurring theme has been the comparison of algebraic properties of Mod_g with such lattices and geometric/topological properties of moduli space $\mathcal{M}_g$ with those of locally symmetric orbifolds. The question (probably due to Ivanov) then arose: "Which lattices Γ are subgroups of Mod_g?" This question arises from a slightly different angle under the algebro-geometric guise of studying locally symmetric subvarieties of moduli space; see [**Ha1**]. The possibilities for such Γ are highly constrained; theorems of Kaimanovich-Masur, Farb-Masur and Yeung (see [**FaM, Ye**]), give the following.

THEOREM 2.4. *Let Γ be an irreducible lattice in a semisimple Lie group $G \neq \text{SO}(m,1), \text{SU}(n,1)$ with $m \geq 2, n \geq 1$. Then any homomorphism $\rho : \Gamma \to \text{Mod}_g$ with $g \geq 1$ has finite image.*

Theorem 2.4 does not extend to the cases $G = \text{SO}(m,1)$ and $G = \text{SU}(n,1)$ in general since these groups admit lattices Γ which surject to $\mathbf{Z}$. Now let us restrict to the case of *injective* ρ, so we are asking about which lattices Γ occur as subgroups of Mod_g. As far as I know, here is what is currently known about this question:

(1) (Lattices $\Gamma < \text{SO}(2,1)$): Each such Γ has a finite index subgroup which is free or $\pi_1 \Sigma_h$ for some $h \geq 2$. These groups are plentiful in Mod_g and are discussed in more detail in §2.3 below.

(2) (Lattices $\Gamma < \text{SO}(3,1)$): These exist in $\text{Mod}_{g,1}$ for $g \geq 2$ and in Mod_g for $g \geq 4$, by the following somewhat well-known construction. Consider the Birman exact sequence

(6) $$1 \to \pi_1 \Sigma_g \to \text{Mod}_{g,1} \xrightarrow{\pi} \text{Mod}_g \to 1$$

Let $\phi \in \text{Mod}_g$ be a pseudo-Anosov homeomorphism, and let $\Gamma_\phi < \text{Mod}_{g,1}$ be the pullback under π of the cyclic subgroup of Mod_g generated by ϕ. The group Γ_ϕ is isomorphic to the fundamental group of a Σ_g bundle over S^1, namely the bundle obtained from $\Sigma_g \times [0,1]$ by identifying $(x,0)$ with $(\phi(x),1)$. By a deep theorem of Thurston, such manifolds admit a hyperbolic metric, and so Γ_ϕ is a cocompact lattice in $\text{SO}(3,1) = \text{Isom}^+(\mathbf{H}^3)$. A branched covering trick of Gonzalez-Diaz and Harvey (see [**GH**]) can be used to find Γ_ϕ as a subgroup of Mod_h for appropriate $h \geq 4$. A variation of the above can be used to find nonuniform lattices in $\text{SO}(3,1)$ inside Mod_g for $g \geq 4$.

(3) (Cocompact lattices $\Gamma < \text{SO}(4,1)$): Recently John Crisp and I [**CF**] found one example of a cocompact lattice $\Gamma < \text{SO}(4,1) = \text{Isom}^+(\mathbf{H}^4)$ which embeds in Mod_g for all sufficiently large g. While we only know of one such Γ, it has infinitely many conjugacy classes in Mod_g. The group Γ is a right-angled Artin group, which is commensurable with a group of reflections in the right-angled 120-cell in $\mathbf{H}^4$.

(4) (Noncocompact lattices $\Gamma < \text{SU}(n,1), n \geq 2$): These Γ have nilpotent subgroups which are not virtually abelian. Since every nilpotent, indeed solvable subgroup of Mod_g is virtually abelian, Γ is not isomorphic to any subgroup of Mod_g.

Hence the problem of understanding which lattices in semisimple Lie groups occur as subgroups of Mod_g comes down to the following.

QUESTION 2.5. *Does there exist some* $\mathrm{Mod}_g, g \geq 2$ *that contains a subgroup* Γ *isomorphic to a cocompact (resp. noncocompact) lattice in* $\mathrm{SO}(m,1)$ *with* $m \geq 5$ *(resp.* $m \geq 4$*)? a cocompact lattice in* $\mathrm{SU}(n,1), n \geq 2$*? Must there be only finitely many conjugacy classes of any such fixed* Γ *in* Mod_g*?*

In light of example (3) above, I would like to specifically ask: can Mod_g contain infinitely many isomorphism types of cocompact lattices in $\mathrm{SO}(4,1)$?

Note that when Γ is the fundamental group of a (complex) algebraic variety V, then it is known that there can be at most finitely many representations $\rho : \Gamma \to \mathrm{Mod}_g$ which have *holomorphic* representatives, by which we mean the unique homotopy class of maps $f : V \to \mathcal{M}_g$ with $f_* = \rho$ contains a holomorphic map. This result follows from repeatedly taking hyperplane sections and finally quoting the result for (complex) curves. The result for these is a theorem of Arakelov-Parshin (cf. Example 1.1 above, and §2.3 below.)

For representations which do not *a priori* have a holomorphic representative, one might try to find a harmonic representative and then to prove a Siu-type rigidity result to obtain a holomorphic representative. One difficulty here is that it is not easy to find harmonic representatives, since (among other problems) every loop in $\mathcal{M}_g, g \geq 2$ can be freely homotoped outside every compact set. For recent progress, however, see [**DW**] and the references contained therein.

2.3. Surfaces in moduli space. Motivation for studying representations $\rho : \pi_1 \Sigma_h \to \mathrm{Mod}_g$ for $g, h \geq 2$ comes from many directions. These include: the analogy of Mod_g with Kleinian groups (see, e.g., [**LR**]); the fact that such subgroups are the main source of locally symmetric families of Riemann surfaces (see §2.2 above, and [**Ha1**]) ; and their appearance as a key piece of data in the topological classification of surface bundles over surfaces (cf. (2) above).

Of course understanding such ρ with holomorphic representatives is the Arakelov-Parshin Finiteness Theorem discussed in Example 1.1 above. Holomorphicity is a key feature of this result. For example, one can prove (see, e.g., [**Mc1**]) that there are finitely many such ρ with a holomorphic representative by finding a *Schwarz Lemma for* $\mathcal{M}_g$: any holomorphic map from a compact hyperbolic surface into $\mathcal{M}_g$ endowed with the Teichmüller metric is distance decreasing. The finiteness is also just not true without the holomorphic assumption (see Theorem 2.7 below). We therefore want to recognize when a given representation has a holomorphic representative.

PROBLEM 2.6 (Holomorphic representatives). *Find an algorithm or a group-theoretic invariant which determines or detects whether or not a given representation* $\rho : \pi_1 \Sigma_h \to \mathrm{Mod}_g$ *has a holomorphic representative.*

Note that a necessary, but not sufficient, condition for a representation $\rho : \pi_1 \Sigma_h \to \mathrm{Mod}_g$ to be holomorphic is that it be *irreducible*, i.e. there is no essential isotopy class of simple closed curve α in Σ_g such that $\rho(\pi_1 \Sigma_h)(\alpha) = \alpha$. I believe it is not difficult to give an algorithm to determine whether or not any given ρ is irreducible or not.

We would like to construct and classify (up to conjugacy) such ρ. We would also like to compute their associated invariants, such as

$$\nu(\rho) := \int_{\Sigma_h} f^* \omega_{\mathrm{WP}}$$

where ω_{WP} is the Weil-Petersson 2-form on $\mathcal{M}_g$, and where $f : \Sigma_h \to \mathcal{M}_g$ is any map with $f_* = \rho$. This would give information on the signatures of surface bundles over surfaces, and also on the Gromov co-norm of $[\omega_{\text{WP}}] \in H^*(\mathcal{M}_g, \mathbf{R})$.

The classification question is basically impossible as stated, since e.g. surface groups surject onto free groups, so it is natural to first restrict to injective ρ. Using a technique of Crisp-Wiest, J. Crisp and I show in [**CF**] that irreducible, injective ρ are quite common.

THEOREM 2.7. *For each $g \geq 4$ and each $h \geq 2$, there are infinitely many Mod_g-conjugacy classes of injective, irreducible representations $\rho : \pi_1 \Sigma_h \to \operatorname{Mod}_g$. One can take the images to lie inside the Torelli group $\mathcal{I}_g$. Further, for any $n \geq 1$, one can take the images to lie inside the subgroup of Mod_g generated by n^{th} powers of all Dehn twists.*

One can try to use these representations, as well as those of [**LR**], to give new constructions of surface bundles over surfaces with small genus base and fiber and nonzero signature. The idea is that Meyer's signature cocycle is positively proportional to the Weil-Petersson 2-form ω_{WP}, and one can actually explicitly integrate the pullback of ω_{WP} under various representations, for example those glued together using Teichmüller curves.

Note that Theorem 2.7 provides us with infinitely many topological types of surface bundles, each with irreducible, faithful monodromy, all having the same base and the same fiber, and all with signature zero.

2.4. Normal subgroups. It is a well-known open question to determine whether or not Mod_g contains a normal subgroup H consisting of all pseudo-Anosov homeomorphisms. For genus $g = 2$ Whittlesey [**Wh**] found such an H; it is an infinitely generated free group. As far as I know this problem is still wide open.

Actually, when starting to think about this problem I began to realize that it is not easy to find *finitely generated* normal subgroups of Mod_g which are not commensurable with either $\mathcal{I}_g$ or Mod_g. There are many normal subgroups of Mod_g which are not commensurable to either $\mathcal{I}_g$ or to Mod_g, most notably the terms $\mathcal{I}_g(k)$ of the Johnson filtration for $k \geq 2$ and the terms of the lower central series of $\mathcal{I}_g$. However, the former are infinitely generated and the latter are likely to be infinitely generated; see Theorem 5.4 and Conjecture 5.5 below.

QUESTION 2.8 (Normal subgroups). *Let Γ be a finitely generated normal subgroup of Mod_g, where $g \geq 3$. Must Γ be commensurable with Mod_g or with $\mathcal{I}_g$?*

One way of constructing *infinitely generated* normal subgroups of Mod_g is to take the group generated by the nth powers of all Dehn twists. Another way is to take the normal closure N_ϕ of a single element $\phi \in \operatorname{Mod}_g$. It seems unclear how to determine the algebraic structure of these N_ϕ, in particular to determine whether N_ϕ is finite index in Mod_g, or in one of the $\mathcal{I}_g(k)$. The following is a basic test question.

QUESTION 2.9. *Is it true that, given any pseudo-Anosov $\phi \in \operatorname{Mod}_g$, there exists $n = n(\phi)$ such that the normal closure of ϕ^n is free?*

Gromov discovered the analogous phenomenon for elements of hyperbolic type inside nonelementary word-hyperbolic groups; see [**Gro**], Theorem 5.3.E.

One should compare Question 2.8 to the Margulis Normal Subgroup Theorem (see, e.g. [**Ma**]), which states that if Λ is any irreducible lattice in a real, linear semisimple Lie group with no compact factors and with **R**-rank at least 2, then any (not necessarily finitely generated) normal subgroup of Λ is finite and central or has finite index in Λ. Indeed, we may apply this result to analyzing normal subgroups Γ of Mod_g. For $g \geq 2$ the group $\operatorname{Sp}(2g, \mathbf{Z})$ satisfies the hypotheses of Margulis's theorem, and so the image $\pi(\Gamma)$ under the natural representation $\pi : \operatorname{Mod}_g \to \operatorname{Sp}(2g, \mathbf{Z})$ is normal, hence is finite or finite index. This proves the following.

PROPOSITION 2.10 (Maximality of Torelli). *Any normal subgroup Γ of Mod_g containing $\mathcal{I}_g$ is commensurable either with Mod_g or with $\mathcal{I}_g$.*

Proposition 2.10 is a starting point for trying to understand Question 2.8. Note too that Mess [**Me**] proved that the group $\mathcal{I}_2$ is an infinitely generated free group, and so it has no finitely generated normal subgroups. Thus we know that if Γ is any finitely generated normal subgroup of Mod_2, then $\pi(\Gamma)$ has finite index in $\operatorname{Sp}(4, \mathbf{Z})$. This in turn gives strong information about Γ; see [**Fa3**].

One can go further by considering the Malcev Lie algebra $\mathfrak{t}_g$ of $\mathcal{I}_g$, computed by Hain in [**Ha3**]; cf. §5.4 below. The normal subgroup $\Gamma \cap \mathcal{I}_g$ of $\mathcal{I}_g$ gives an $\operatorname{Sp}(2g, \mathbf{Z})$-invariant subalgebra $\mathfrak{h}$ of $\mathfrak{t}_g$. Let $H = H_1(\Sigma_g, \mathbf{Z})$. The Johnson homomorphism $\tau : \mathcal{I}_g \to \wedge^3 H/H$ is equivariant with respect to the action of $\operatorname{Sp}(2g, \mathbf{Z})$, and $\wedge^3 H/H$ is an irreducible $\operatorname{Sp}(2g, \mathbf{Z})$-module. It follows that the first quotient in the lower central series of $\Gamma \cap \mathcal{I}_g$ is either trivial or is all of $\wedge^3 H/H$. With more work, one can extend the result of Proposition 2.10 from $\mathcal{I}_g$ to $\mathcal{K}_g = \mathcal{I}_g(2)$; see [**Fa3**].

THEOREM 2.11 (Normal subgroups containing $\mathcal{K}_g$). *Any normal subgroup Γ of Mod_g containing $\mathcal{K}_g$ is commensurable to $\mathcal{K}_g$, $\mathcal{I}_g$ or Mod_g.*

One can continue this "all or nothing image" line of reasoning to deeper levels of $\mathfrak{t}_g$. Indeed, I believe one can completely reduce the classification of normal subgroups of Mod_g, at least those that contain some $\mathcal{I}_g(k)$, to some symplectic representation theory problems, such as the following.

PROBLEM 2.12. *For $g \geq 2$, determine the irreducible factors of the graded pieces of the Malcev Lie algebra $\mathfrak{t}_g$ of $\mathcal{I}_g$ as Sp-modules.*

While Hain gives in [**Ha3**] an explicit and reasonably simple presentation for $\mathfrak{t}_g$ when $g > 6$, Problem 2.12 still seems to be an involved problem in classical representation theory.

As Chris Leininger pointed out to me, all of the questions above have natural "virtual versions". For example, one can ask about the classification of normal subgroups of finite index subgroups of Mod_g. Another variation is the classification of *virtually normal* subgroups of Mod_g, that is, subgroups whose normalizers in Mod_g have finite index in Mod_g.

2.5. Numerology of finite subgroups. The Nielsen Realization Theorem, due to Kerckhoff, states that any finite subgroup $F < \operatorname{Mod}_g$ can be realized as a group of automorphisms of some Riemann surface X_F. Here by *automorphism* group we mean group of (orientation-preserving) isometries in some Riemannian metric, or equivalently in the hyperbolic metric, or equivalently the group of biholomorphic automorphisms of X_F. An easy application of the uniformization theorem gives these equivalences. It is classical that the automorphism group $\mathrm{Aut}(X)$ of a closed Riemann surface of genus $g \geq 2$ is finite. Thus the study of finite subgroups of $\mathrm{Mod}_g, g \geq 2$ reduces to the study of automorphism groups of Riemann surfaces.

Let $N(g)$ denote the largest possible order of $\mathrm{Aut}(X)$ as X ranges over all genus g surfaces. Then for $g \geq 2$ it is known that

$$8(g+1) \leq N(g) \leq 84(g-1) \qquad (7)$$

the lower bound due to Accola and Maclachlan (see, e.g., [**Ac**]); the upper due to Hurwitz. It is also known that each of these bounds is achieved for infinitely many g and is not achieved for infinitely many g. There is an extensive literature seeking to compute $N(g)$ and variations of it. The most significant achievement in this direction is the following result of M. Larsen [**La**].

THEOREM 2.13 (Larsen). *Let H denote the set of integers $g \geq 2$ such that there exists at least one compact Riemann surface X_g of genus g with $|\mathrm{Aut}(X_g)| = 84(g-1)$. Then the series $\sum_{g \in H} g^{-s}$ converges absolutely for $\Re(s) > 1/3$ and has a singularity at $s = 1/3$.*

In particular, the g for which $N(g) = 84(g-1)$ occur with the same frequency as perfect cubes. It follows easily from the Riemann-Hurwitz formula that the bound $84(g-1)$ is achieved precisely for those surfaces which isometrically (orbifold) cover the $(2,3,7)$ orbifold. Thus the problem of determining which genera realize the $84(g-1)$ bound comes down to figuring out the possible (finite) indices of subgroups, or what is close to the same thing finite quotients, of the $(2,3,7)$ orbifold group. Larsen's argument uses in a fundamental way the classification of finite simple groups.

PROBLEM 2.14. *Give a proof of Theorem 2.13 which does not depend on the classification of finite simple groups.*

To complete the picture, one would like to understand the frequency of those g for which the lower bound in (7) occurs.

PROBLEM 2.15 (Frequency of low symmetry). *Let H denote the set of integers $g \geq 2$ such that $N(g) = 8(g+1)$. Find the s_0 for which the series $\sum_{g \in H} g^{-s}$ converges absolutely for the real part $\Re(s)$ of s satisfying $\Re(s) > s_0$, and has a singularity at $s = s_0$.*

There are various refinements and variations on Problem 2.15. For example, Accola proves in [**Ac**] that when g is divisible by 3, then $N(g) \geq 8(g+3)$, with the bound attained infinitely often. One can try to build on this for other g, and can also ask for the frequency of this occurence.

One can begin to refine Hurwitz's Theorem by asking for bounds of orders of groups of automorphisms which in addition satisfy various algebraic constraints, such as being nilpotent, being solvable, being a p-group, etc. There already exist a number of theorems of this sort. For example, Zomorrodian [**Zo**] proved that if $\mathrm{Aut}(X_g)$ is nilpotent then it has order at most $16(g-1)$, and if this bound is attained then $g-1$ must be a power of 2. One can also ask for lower bounds in this context. As these kinds of bounds are typically attained and not attained for infinitely many g, one then wants to solve the following.

PROBLEM 2.16 (Automorphism groups with special properties). *Let P be a property of finite groups, for example being nilpotent, solvable, or a p-group. Prove* a version of Larsen's theorem which counts those g for which the upper bound of $|\operatorname{Aut}(X_g)|$ is realized for some X_g with $\operatorname{Aut}(X_g)$ having P. Similarly for lower bounds. Determine the least g for which each given bound is realized.

Many of the surfaces realizing the extremal bounds in all of the above questions are *arithmetic*, that is they are quotients of $\mathbf{H}^2$ by an arithmetic lattice. Such lattices are well-known to have special properties, in particular they have a lot of symmetry. On the other hand arithmetic surfaces are not typical. Thus to understand the "typical" surface with symmetry, the natural problem is the following.

PROBLEM 2.17 (Nonarithmetic extremal surfaces). *Give answers to all of the above problems on automorphisms of Riemann surfaces for the collection of nonarithmetic surfaces. For example find bounds on orders of automorphism groups which are nilpotent, solvable, p-groups, etc. Prove that these bounds are sharp for infinitely many g. Determine the frequency of those g for which such bounds are sharp. Determine the least genus for which the bounds are sharp.*

The model result for these kind of problems is the following.

THEOREM 2.18 (Belolipetsky [**Be**]). *Let X_g be any non-arithmetic Riemann surface of genus $g \geq 2$. Then*

$$|\operatorname{Aut}(X_g)| \leq \frac{156}{7}(g-1)$$

Further, this bound is sharp for infinitely many g; the least such is $g = 50$.

The key idea in the proof of Theorem 2.18 is the following. One considers the quotient orbifold $X_g/\operatorname{Aut}(X_g)$, and computes via the Riemann-Hurwitz formula that it is the quotient of $\mathbf{H}^2$ by a triangle group. Lower bounds on the area of this orbifold give upper bounds on $|\operatorname{Aut}(X_g)|$; for example the universal lower bound of $\pi/21$ for the area of every 2-dimensional hyperbolic orbifold gives the classical Hurwitz $84(g-1)$ theorem. Now Takeuchi classified all arithmetic triangle groups; they are given by a finite list. One can then use this list to refine the usual calculations with the Riemann-Hurwitz formula to give results such as Theorem 2.18; see [**Be**]. This idea should also be applicable to Problem 2.17.

I would like to mention a second instance of the theme of playing off algebraic properties of automorphism groups versus the numerology of their orders. One can prove that the maximal possible order of an automorphism of a Riemann surface X_g is $4g+2$. This bound is easily seen to be achieved for each $g \geq 2$ by considering the rotation of the appropriately regular hyperbolic $(4g+2)$-gon. Kulkarni [**Ku**] proved that there is a *unique* Riemann surface W_g admitting such an automorphism. Further, he proved that $\operatorname{Aut}(W_g)$ is cyclic, and he gave the equation describing W_g as an algebraic curve.

PROBLEM 2.19 (Canonical basepoints for $\mathcal{M}_g$). *Find other properties of automorphisms or automorphism groups that determine a unique point of $\mathcal{M}_g$. For example, is there a unique Riemann surface of genus $g \geq 2$ whose automorphism group is nilpotent, and is the largest possible order among nilpotent automorphism groups of genus g surfaces?*

A *Hurwitz surface* is a hyperbolic surface attaining the bound $84(g-1)$. As the quotient of such a surface by its automorphism group is the $(2, 3, 7)$ orbifold, which has a unique hyperbolic metric, it follows that for each $g \geq 2$ there are finitely many

Hurwitz surfaces. As these are the surfaces of maximal symmetry, it is natural to ask precisely how many there are.

QUESTION 2.20 (Number of Hurwitz surfaces). *Give a formula for the number of Hurwitz surfaces of genus g. What is the frequency of those g for which there is a unique Hurwitz surface?*

3. Combinatorial and geometric group theory of Mod_g

Ever since Dehn, Mod_g has been a central example in combinatorial and geometric group theory. One reason for this is that Mod_g lies at a gateway: on one side are matrix groups and groups naturally equipped with a geometric structure (e.g. hyperbolic geometry); on the other side are groups given purely combinatorially. In this section we pose some problems in this direction.

3.1. Decision problems and almost convexity.

Word and conjugacy problems. Recall that the *word problem* for a finitely presented group Γ asks for an algorithm which takes as input any word w in a fixed generating set for Γ, and as output tells whether or not w is trivial. The *conjugacy problem* for Γ asks for an algorithm which takes as input two words, and as output tells whether or not these words represent conjugate elements of Γ.

There is some history to the word and conjugacy problems for Mod_g, beginning with braid groups. These problems have topological applications; for example the conjugacy problem for Mod_g is one of the two main ingredients one needs to solve the homeomorphism problem for 3-manifolds fibering over the circle[3]. Lee Mosher proved in [**Mos**] that Mod_g is automatic. From this it follows that there is an $O(n^2)$-time algorithm to solve the word problem for Mod_g; indeed there is an $O(n^2)$-time algorithm which puts each word in a fixed generating set into a unique normal form. However, the following is still open.

QUESTION 3.1 (Fast word problem). *Is there a sub-quadratic time algorithm to solve the word problem in Mod_g?*

One might guess that $n \log n$ is possible here, as there is such an algorithm for certain relatively (strongly) hyperbolic groups (see [**Fa3**]), and mapping class groups are at least weakly hyperbolic, as proven by Masur-Minsky (Theorem 1.3 of [**MM1**]).

The conjugacy problem for Mod_g is harder. The original algorithm of Hemion [**He**] seems to give no reasonable (even exponential) time bound. One refinement of the problem would be to prove that Mod_g is *biautomatic*. However, even a biautomatic structure gives only an exponential time algorithm to solve the conjugacy problem. Another approach to solving the conjugacy problem is the following.

PROBLEM 3.2 (Conjugator length bounds). *Prove that there exist constants C, K, depending only on S, so that if $u, v \in \text{Mod}_g$ are conjugate, then there exists $g \in \text{Mod}_g$ with $||g|| \leq K \max\{||u||, ||v||\} + C$ so that $u = gvg^{-1}$.*

Masur-Minsky ([**MM2**], Theorem 7.2) solved this problem in the case where u and v are pseudo-Anosov; their method of hierarchies seems quite applicable to solving Problem 3.2 in the general case. While interesting in its own right, even the solution to Problem 3.2 would not answer the following basic problem.

---

[3]The second ingredient, crucial but ignored by some authors, is the computability of the Thurston norm.

PROBLEM 3.3 (Fast conjugacy problem). *Find a polynomial time algorithm to solve the conjugacy problem in* Mod_g. *Is there a quadratic time algorithm, as for the word problem?*

As explained in the example on page 13, a solution to Problem 3.3 would be a major step in finding a polynomial time algorithm to solve the homeomorphism problem for 3-manifolds that fiber over the circle.

Almost convexity. In [**Ca**] Cannon initiated the beautiful theory of almost convex groups. A group Γ with generating set S is *almost convex* if there exists $C > 0$ so that for each $r > 0$, and for any two points $x, y \in \Gamma$ on the sphere of radius r in Γ with $d(x, y) = 2$, there exists a path γ of length at most C connecting x to y and lying completely inside the ball of radius r in Γ. There is an obvious generalization of this concept from groups to spaces. One strong consequence of the almost convexity of Γ is that for such groups one can recursively build the Cayley graph of Γ near any point x in the n-sphere of Γ by only doing a local computation involving elements of Γ lying (universally) close to x; see [**Ca**]. In particular one can build each n-ball, $n \geq 0$, and so solve the word problem in an efficient way.

QUESTION 3.4 (Almost convexity). *Does there exist a finite generating set for* Mod_g *for which it is almost convex?*

One would also like to know the answer to this question for various subgroups of Mod_g. Here is a related, but different, basic question about the geometry of Teichmüller space.

QUESTION 3.5. *Is* $\text{Teich}(\Sigma_g)$, *endowed with the Teichmüller metric, almost convex?*

Note that Cannon proves in [**Ca**] that fundamental groups of closed, negatively curved manifolds are almost convex with respect to any generating set. He conjectures that this should generalize both to the finite volume case and to the nonpositively curved case.

3.2. The generalized word problem and distortion. In this subsection we pose some problems relating to the ways in which subgroups embed in Mod_g.

Generalized word problem. Let Γ be a finitely presented group with finite generating set S, and let H be a finitely generated subgroup of Γ. The *generalized word problem*, or GWP, for H in Γ, asks for an algorithm which takes as input an element of the free group on S, and as output tells whether or not this element represents an element of H. When H is the trivial subgroup then this is simply the word problem for Γ. The group Γ is said to have *solvable generalized word problem* if the GWP is solvable for every finitely generated subgroup H in Γ.

The generalized word problem, also called the *membership, occurrence* or *Magnus problem*, was formulated by K. Mihailova [**Mi**] in 1958, but special cases had already been studied by Nielsen [**Ni**] and Magnus [**Ma**]. Mihailova [**Mi**] found a finitely generated subgroup of a product $F_m \times F_m$ of free groups which has unsolvable generalized word problem.

Now when $g \geq 2$, the group Mod_g contains a product of free groups $F_m \times F_m$ (for any $m > 0$). Further, one can find copies of $F_m \times F_m$ such that the generalized word problem for these subgroups is solvable inside Mod_g. To give one concrete example, simply divide Σ_g into two subsurfaces S_1 and S_2 which intersect in a (possibly empty) collection of curves. Then take, for each $i = 1, 2$, a pair f_i, g_i of independent pseudo-Anosov homeomorphisms. After perhaps taking powers of f_i and g_i if necessary, the group generated by $\{f_1, g_1, f_2, g_2\}$ will be the group we require; one can pass to finite index subgroups if one wants $m > 2$. The point here is that such subgroups are not distorted (see below), as they are convex cocompact in the subgroups $\text{Mod}(S_i)$ of Mod_g, and these in turn are not distorted in Mod_g. It follows that the generalized word problem for Mihailova's subgroup $G < F_m \times F_m$ is not solvable in Mod_g.

PROBLEM 3.6 (Generalized word problem). *Determine the subgroups H in Mod_g for which the generalized word problem is solvable. Give efficient algorithms to solve the generalized word problem for these subgroups. Find the optimal time bounds for such algorithms.*

Of course this problem is too broad to solve in complete generality; results even in special cases might be interesting. Some solutions to Problem 3.6 are given by Leininger-McReynolds in [**LM**]. We also note that Bridson-Miller (personal communication), extending an old result of Baumslag-Roseblade, have proven that any product of finite rank free groups has solvable generalized word problem with respect to any *finitely presented* subgroup. In light of this result, it would be interesting to determine whether or not there is a finitely presented subgroup of Mod_g with respect to which the generalized word problem is not solvable.

Distortion and quasiconvexity. There is a refinement of Problem 3.6. Let H be a finitely generated subgroup of a finitely generated group Γ. Fix finite generating sets on both H and Γ. This choice gives a *word metric* on both H and Γ, where $d_\Gamma(g, h)$ is defined to be the minimal number of generators of Γ needed to represent gh^{-1}. Let $\mathbf{N}$ denote the natural numbers. We say that a function $f : \mathbf{N} \longrightarrow \mathbf{N}$ is a *distortion function* for H in Γ if for every word w in the generators of Γ, if w represents an element $\overline{w} \in H$ then

$$d_H(1, \overline{w}) \leq f(d_\Gamma(1, \overline{w}))$$

In this case we also say that "H has distortion $f(n)$ in Γ." It is easy to see that the growth type of f, i.e. polynomial, exponential, etc., does not depend on the choice of generators for either H or Γ. It is also easy to see that $f(n)$ is constant if and only if H is finite; otherwise f is at least linear. It is proved in [**Fa1**] that, for a group Γ with solvable word problem, the distortion of H in Γ is recursive if and only if H has solvable generalized word problem in Γ. For some concrete examples, we note that the center of the 3-dimensional integral Heisenberg group has quadratic distortion; and the cyclic group generated by b in the group $\langle a, b : aba^{-1} = b^2 \rangle$ has exponential distortion since $a^n b a^{-n} = b^{2^n}$.

PROBLEM 3.7 (Distortion). *Find the possible distortions of subgroups in Mod_g. In particular, compute the distortions of $\mathcal{I}_g$. Determine the asymptotics of the distortion of $\mathcal{I}_g(k)$ as $k \to \infty$. Is there a subgroup $H < \text{Mod}_g$ that has precisely polynomial distortion of degree $d > 1$?*

There are some known results on distortion of subgroups in Mod_g. Convex cocompact subgroups (in the sense of [**FMo**]) have linear distortion in Mod_g; there are many such examples where H is a free group. Abelian subgroups of Mod_g have linear distortion (see [**FLMi**]), as do subgroups corresponding to mapping class groups of subsurfaces (see [**MM2, Ham**]).

I would guess that $\mathcal{I}_g$ has exponential distortion in Mod_g. A first step to the question of how the distortion of $\mathcal{I}_g(k)$ in Mod_g behaves as $k \to \infty$ would be to determine the distortion of $\mathcal{I}_g(k+1)$ in $\mathcal{I}_g(k)$. The "higher Johnson homomorphisms" (see, e.g., [**Mo3**]) might be useful here.

A stronger notion than linear distortion is that of quasiconvexity. Let $S = S^{-1}$ be a fixed generating set for a group Γ, and let $\pi : S^* \to \Gamma$ be the natural surjective homomorphism from the free monoid on S to Γ sending a word to the group element it represents. Let $\sigma : \Gamma \to S^*$ be a (perhaps multi-valued) section of π; that is, σ is just a choice of paths in Γ from the origin to each $g \in \Gamma$. We say that a subgroup $H < \Gamma$ is *quasiconvex* (with respect to σ) if there exists $K > 0$ so that for each $h \in H$, each path $\sigma(h)$ lies in the K-neighborhood of H in Γ. Quasiconvexity is a well-known and basic notion in geometric group theory. It is easy to see that if H is quasiconvex with respect to some collection of quasigeodesics then H has linear distortion in Mod_g (see [**Fa1**]).

PROBLEM 3.8. *Determine which subgroups of Mod_g are quasiconvex with respect to some collection of geodesics.*

This question is closely related to, but different than, the question of *convex cocompactness* of subgroups of Mod_g, as defined in [**FMo**], since the embedding of Mod_g in Teich_g via any orbit is exponentially distorted, by [**FLMi**].

3.3. Decision problems for subgroups. As a collection of groups, how rich and varied can the set of subgroups of Mod_g be? One instance of this general question is the following.

QUESTION 3.9. *Does every finitely presented subgroup $H < \text{Mod}_g$ have solvable conjugacy problem? is it combable? automatic?*

Note that every finitely-generated subgroup of a group with solvable word problem has solvable word problem. The same is not true for the conjugacy problem: there are subgroups of $\text{GL}(n, \mathbf{Z})$ with unsolvable conjugacy problem; see [**Mi**].

It is not hard to see that Mod_g, like $\text{GL}(n, \mathbf{Z})$, has finitely many conjugacy classes of finite subgroups. However, we pose the following.

PROBLEM 3.10. *Find a finitely presented subgroup $H < \text{Mod}_g$ for which there are infinitely many conjugacy classes of finite subgroups in H.*

The motivation for this problem comes from a corresponding example, due to Bridson [**Br**], of such an H in $\text{GL}(n, \mathbf{Z})$. One might to solve Problem 3.10 by extending Bridson's construction to $\text{Sp}(2g, \mathbf{Z})$, pulling back such an H, and also noting that the natural map $\text{Mod}_g \to \text{Sp}(2g, \mathbf{Z})$ is injective on torsion.

Another determination of the variety of subgroups of Mod_g is the following. Recall that the *isomorphism problem* for a collection $\mathcal{S}$ of finitely presented groups asks for an algorithm which takes as input two presentations for two elements of $\mathcal{S}$ and as output tells whether or not those groups are isomorphic.

QUESTION 3.11 (Isomorphism problem for subgroups). *Is the isomorphism problem for the collection of finitely presented subgroups of Mod_g solvable?*

Note that the isomorphism problem is not solvable for the collection of all finitely generated linear groups, nor is it even solvable for the collection of finitely generated subgroups of $\text{GL}(n, \mathbf{Z})$.

There are many other algorithmic questions one can ask; we mention just one more.

QUESTION 3.12. *Is there an algorithm to decide whether or not a given subgroup $H < \text{Mod}_g$ is freely indecomposable? Whether or not H splits over $\mathbf{Z}$?*

3.4. Growth and counting questions. Recall that the *growth series* of a group Γ with respect to a finite generating set S is defined to be the power series

$$(8) \qquad f(z) = \sum_{i=0}^{\infty} c_i z^i$$

where c_i denotes the cardinality $\#B_\Gamma(i)$ of the ball of radius i in Γ with respect to the word metric induced by S. We say that Γ has *rational growth* (with respect to S) if f is a rational function, that is the quotient of two polynomials. This is equivalent to the existence of a linear recurrence relation among the c_i; that is, there exist $m > 0$ real numbers $a_1, \ldots, a_m \geq 0$ so that for each r:

$$c_r = a_1 c_{r-1} + \cdots + a_m c_{r-m}$$

Many groups have rational growth with respect to various (sometimes every) generating sets. Examples include word-hyperbolic groups, abelian groups, and Coxeter groups. See, e.g., [**Harp2**] for an introduction to the theory of growth series.

In [**Mos**], Mosher constructed an automatic structure for Mod_g. This result suggests that the following might have a positive answer.

QUESTION 3.13 (Rational growth). *Does Mod_g have rational growth function with respect to some set of generators? with respect to every set of generators?*

Of course one can also ask the same question for any finitely generated subgroup of Mod_g, for example $\mathcal{I}_g$. Note that the existence of an automatic structure is not known to imply rationality of growth (even for one generating set); one needs in addition the property that the automatic structure consist of geodesics. Unfortunately Mosher's automatic structure does not satisfy this stronger condition.

It is natural to ask for other recursive patterns in the Cayley graph of Mod_g. To be more precise, let P denote a property that elements of Mod_g might or might not have. For example, P might be the property of being finite order, of lying in a fixed subgroup $H < \text{Mod}_g$, or of being pseudo-Anosov. Now let

$$c_P(r) = \#\{B_{\text{Mod}_g}(r) \cap \{x \in \text{Mod}_g : x \text{ has P}\}\}$$

We now define the *growth series for the property P*, with respect to a fixed generating set for Γ, to be the power series

$$f_P(z) = \sum_{i=0}^{\infty} c_P(i) z^i$$

QUESTION 3.14 (Rational growth for properties). *For which properties P is the function f_P is rational?*

Densities. For any subset $S \subset \text{Mod}_g$, it is natural to ask how common elements of S are in Mod_g. There are various ways to interpret this question, and the answer likely depends in a strong way on the choice of interpretation[4]. One way to formalize this is via the *density* $d(S)$ of S in Mod_g, where

$$d(S) = \lim_{r \to \infty} \frac{\#[B(r) \cap S]}{\#B(r)} \tag{9}$$

where $B(r)$ is the number of elements of Mod_g in the ball of radius r, with respect to a fixed set of generators. While for subgroups $H < \mathrm{Mod}_g$ the number $d(H)$ itself may depend on the choice of generating sets for H and Mod_g, it is not hard to see that the (non)positivity of $d(H)$ does not depend on the choices of generating sets.

As the denominator and (typically) the numerator in (9) are exponential, one expects that $d(S) = 0$ for most S. Thus is it natural to replace both the numerator and denominator of (9) with their logarithms; we denote the corresponding limit as in (9) by $d_{\log}(S)$, and we call this the *logarithmic density* of S in Mod_g.

The following is one interpretation of a folklore conjecture.

CONJECTURE 3.15 (Density of pseudo-Anosovs). *Let $\mathcal{P}$ denote the set of pseudo-Anosov elements of Mod_g. Then $d(\mathcal{P}) = 1$.*

J. Maher [**Mah**] has recently proven that a random walk on Mod_g lands on a pseudo-Anosov element with probability tending to one as the length of the walk tends to infinity. I. Rivin [**Ri**] has proven that a random (in a certain specific sense) element of Mod_g is pseudo-Anosov by proving a corresponding result for $\mathrm{Sp}(2g, \mathbf{Z})$. While the methods in [**Mah**] and [**Ri**] may be relevant, Conjecture 3.15 does not seem to follow directly from these results. As another test we pose the following.

CONJECTURE 3.16. $d(\mathcal{I}_g) = 0$.

Even better would be to determine $d_{\log}(\mathcal{I}_g)$. Conjecture 3.16 would imply that $d(\mathcal{I}_g(m)) = 0$ for each $m \geq 2$. It is not hard to see that $\mathcal{I}_g(m)$ has exponential growth for each $g \geq 2, m \geq 1$, and one wants to understand how the various exponential growth rates compare to each other. In other words, one wants to know how common an occurence it is, as a function of k, for an element of Mod_g to act trivially on the first k terms of the lower central series of $\pi_1 \Sigma_g$.

PROBLEM 3.17 (Logarithmic densities of the Johnson filtration). *Determine the asymptotics of $d_{\log}(\mathcal{I}_g(m))$ both as $g \to \infty$ and as $m \to \infty$.*

Indeed, as far as I know, even the asymptotics of the (logarithmic) density of the kth term of the lower central series of $\pi_1 \Sigma_g$ in $\pi_1 \Sigma_g$ as $k \to \infty$ has not been determined.

Entropy. The *exponential growth rate* of a group Γ with respect to a finite generating set S is defined as

$$w(\Gamma, S) := \lim_{r \to \infty} (B_r(\Gamma, S))^{1/r}$$

where $B_r(\Gamma, S)$ denotes the cardinality of the r-ball in the Cayley graph of Γ with respect to the generating set S; the limit exists since β is submultiplicative. The *entropy* of Γ is defined to be

$$\mathrm{ent}(\Gamma) = \inf\{\log w(\Gamma, S) : S \text{ is finite and generates } \Gamma\}$$

Among other things, the group-theoretic entropy of $\mathrm{ent}(\pi_1 M)$ of a closed, Riemannian manifold M gives a lower bound for (the product of diameter times) both

---

[4] For a wonderful discussion of this kind of issue, see Barry Mazur's article [**Maz**].

the volume growth entropy of M and the topological entropy of the geodesic flow on M. See [**Harp1**] for a survey.

Eskin, Mozes and Oh [**EMO**] proved that nonsolvable, finitely-generated linear groups Γ have positive entropy. Since it is classical that the action of Mod_g on $H_1(\Sigma_g, \mathbf{Z})$ gives a surjection $\text{Mod}_g \to \text{Sp}(2g, \mathbf{Z})$, it follows immediately that $\text{ent}(\text{Mod}_g) > 0$. This method of proving positivity of entropy fails for the Torelli group $\mathcal{I}_g$ since it is in the kernel of the standard symplectic representation of Mod_g. However, one can consider the action of $\mathcal{I}_g$ on the homology of the universal abelian cover of Σ_g, considered as a (finitely generated) module over the corresponding covering group, to find a linear representation of $\mathcal{I}_g$ which is not virtually solvable. This is basically the *Magnus representation*. Again by Eskin-Mozes-Oh we conclude that $\text{ent}(\mathcal{I}_g) > 0$.

PROBLEM 3.18. *Give explicit upper and lower bounds for* $\text{ent}(\text{Mod}_g)$. *Compute the asymptotics of* $\text{ent}(\text{Mod}_g)$ *and of* $\text{ent}(\mathcal{I}_g)$ *as* $g \to \infty$. *Similarly for* $\text{ent}(\mathcal{I}_g(k))$ *as* $k \to \infty$.

4. Problems on the geometry of $\mathcal{M}_g$

It is a basic question to understand properties of complex analytic and geometric structures on $\mathcal{M}_g$, and how these structures constrain, and are constrained by, the global topology of $\mathcal{M}_g$. Such structures arise frequently in applications. For example one first tries to put a geometric structure on $\mathcal{M}_g$, such as that of a complex orbifold or of a negatively curved Riemannian manifold. Once this is done, general theory for such structures (e.g. Schwarz Lemmas or fixed point theorems) can then be applied to prove hard theorems. Arakelov-Parshin Rigidity and Nielsen Realization are two examples of this. In this section we pose a few problems about the topology and geometry of $\mathcal{M}_g$.

4.1. Isometries. Royden's Theorem [**Ro**] states that when $g \geq 3$, every isometry of Teichmüller space Teich_g, endowed with the Teichmüller metric d_{Teich}, is induced by an element of Mod_g; that is

$$\text{Isom}(\text{Teich}_g, d_{\text{Teich}}) = \text{Mod}_g$$

Note that d_{Teich} comes from a non-Riemannian Finsler metric, namely a norm on the cotangent space at each point $X \in \text{Teich}_g$. This cotangent space can be identified with the space $Q(X)$ of holomorphic quadratic differentials on X.

I believe Royden's theorem can be generalized from the Teichmüller metric to all metrics.

CONJECTURE 4.1 (Inhomogeneity of all metrics). *Let* Teich_g *denote the Teichmüller space of closed, genus* $g \geq 2$ *Riemann surfaces. Let* h *be any Riemannian metric (or any Finsler metric with some weak regularity conditions) on* Teich_g *which is invariant under the action of the mapping class group* Mod_g, *and for which this action has finite covolume. Then* $\text{Isom}(\text{Teich}_g, h)$ *is discrete; even better, it contains* Mod_g *as a subgroup of index* $C = C(g)$.

Royden's Theorem is the special case when h is the Teichmüller metric (Royden gets $C = 2$ here). A key philosophical implication of Conjecture 4.1 is that the mechanism behind the inhomogeneity of Teichmüller space is due not to fine regularity properties of the unit ball in $Q(X)$ (as Royden's proof suggests), but to the global topology of moduli space. This in turn is tightly controlled by the structure of Mod_g. As one piece of evidence for Conjecture 4.1, I would like to point out that it would follow if one could extend the main theorem of [**FW1**] from the closed to the finite volume case.

In some sense looking at Mod_g-invariant metrics seems too strong, especially since Mod_g has torsion. Perhaps, for example, the inhomogeneity of Teich_g is simply caused by the constraints of the torsion in Mod_g. Sufficiently large index subgroups of Mod_g are torsion free. Thus one really wants to strengthen Conjecture 4.1 by replacing Mod_g by any finite index subgroup H, and by replacing the constant $C = C(g)$ by a constant $C = C(g, [\text{Mod}_g : H])$. After this one can explore metrics invariant by much smaller subgroups, such as $\mathcal{I}_g$, and at least hope for discreteness of the corresponding isometry group (as long as the subgroup is sufficiently large).

If one can prove the part of Conjecture 4.1 which gives discreteness of the isometry group of any Mod_g-invariant metric on Teich_g, one can approach the stronger statement that $C_g = 1$ or $C_g = 2$ as follows. Take the quotient of Teich_g by any group Λ of isometries properly containing Mod_g. By discreteness of Λ, the quotient Teich_g / Λ is a smooth orbifold which is finitely orbifold-covered by $\mathcal{M}_g$.

CONJECTURE 4.2 ($\mathcal{M}_g$ is maximal). *For $g \geq 3$ the smooth orbifold $\mathcal{M}_g$ does not finitely orbifold-cover any other smooth orbifold.*

A much stronger statement, which may be true, would be to prove that if N is any finite cover of $\mathcal{M}_g$, then the only orbifolds which N can orbifold cover are just the covers of $\mathcal{M}_g$. Here is a related basic topology question about $\mathcal{M}_g$.

QUESTION 4.3. *Let Y be any finite cover of $\mathcal{M}_g$, and let $f : Y \to Y$ be a finite order homeomorphism. If f is homotopic to the identity, must f equal the identity?*

4.2. Curvature and Q-rank. Nonpositive sectional curvature.

There has been a long history of studying metrics and curvature on $\mathcal{M}_g$[5]; see, e.g., [**BrF**, **LSY**, **Mc2**]. A recurring theme is to find aspects of negative curvature in $\mathcal{M}_g$. While $\mathcal{M}_g$ admits no metrics of negative curvature, even in a coarse sense, if $g \geq 2$ (see [**BrF**]), the following remains a basic open problem.

CONJECTURE 4.4 (Nonpositive curvature). *For $g \geq 2$ the orbifold $\mathcal{M}_g$ admits no complete, finite volume Riemannian metric with nonpositive sectional curvatures uniformly bounded away from $-\infty$.*

One might be more ambitious in stating Conjecture 4.4, by weakening the finite volume condition, by dropping the uniformity of the curvature bound, or by extending the statement from $\mathcal{M}_g$ to any finite cover of $\mathcal{M}_g$ (or perhaps even to certain infinite covers). It would also be interesting to extend Conjecture 4.4 beyond the Riemannian realm to that of CAT(0) metrics; see, e.g., [**BrF**] for a notion of finite volume which extends to this context.

In the end, it seems that we will have to make do with various relative notions of nonpositive or negative curvature, as in [**MM1**, **MM2**], or with various weaker notions of nonpositive curvature, such as holomorphic, Ricci, or highly singular versions (see, e.g.,[**LSY**]), or isoperimetric type versions such a Kobayashi or Kahler hyperbolicity (see [**Mc2**]). Part of the difficulty with trying to fit $\mathcal{M}_g$ into the "standard models" seems to be the topological structure of the cusp of $\mathcal{M}_g$.

---

[5]As $\mathcal{M}_g$ is an orbifold, technically one studies Mod_g-invariant metrics on the Teichmüller space Teich_g.

Scalar curvature and Q-rank. Let S be a genus g surface with n punctures, and let $\mathcal{M}(S)$ denote the corresponding moduli space. We set

$$d(S) = 3g - 3 + n$$

The constant $d(S)$ is fundamental in Teichmüller theory: it is the complex dimension of the Teichmüller space Teich(S); it is also the number of curves in any pair-of-pants decomposition of S. While previous results have concentrated on sectional and holomorphic curvatures, Shmuel Weinberger and I have recently proven the following; see [**FW2**].

THEOREM 4.5 (Positive scalar curvature). *Let M be any finite cover of $\mathcal{M}(S)$. Then M admits a complete, finite-volume Riemannian metric of (uniformly bounded) positive scalar curvature if and only if $d(S) \geq 3$.*

The analogous statement was proven for locally symmetric arithmetic manifolds $\Gamma \backslash G/K$ by Block-Weinberger [**BW**], where $d(S)$ is replaced by the **Q**-rank of Γ. When $d(S) \geq 3$, the metric on M has the quasi-isometry type of a ray, so that it is not quasi-isometric to the Teichmüller metric on $\mathcal{M}_g$. It seems likely that this is not an accident.

CONJECTURE 4.6. *Let S be any surface with $d(S) \geq 1$. Then M does not admit a finite volume Riemannian metric of (uniformly bounded) positive scalar curvature in the quasi-isometry class of the Teichmüller metric.*

The analogue of Conjecture 4.6 in the $\Gamma \backslash G/K$ case was proven by S. Chang in [**Ch**]. The same method of proof as in [**Ch**] should reduce Conjecture 4.6 to the following discussion, which seems to be of independent interest, and which came out of discussions with H. Masur.

What does $\mathcal{M}_g$, endowed with the Teichmüller metric d_{Teich}, look like from far away? This can be formalized by Gromov's notion of *tangent cone at infinity*:

$$(10) \qquad \text{Cone}(\mathcal{M}_g) := \lim_{n \to \infty} (\mathcal{M}_g, \frac{1}{n} d_{\text{Teich}})$$

where the limit is taken in the sense of Gromov-Hausdorff convergence of pointed metric spaces; here we have fixed a basepoint in $\mathcal{M}_g$ once and for all. This limit is easily shown to make sense and exist in our context. To state our conjectural answer as to what $\text{Cone}(\mathcal{M}_g)$ looks like, we will need the complex of curves on Σ_g. Recall that the *complex of curves* $\mathcal{C}_g$ for $g \geq 2$ is the simplicial complex with one vertex for each nontrivial, nonperipheral isotopy class of simple closed curves on Σ_g, and with a k-simplex for every $(k+1)$-tuple of such isotopy classes for which there are mutually disjoint representatives. Note that Mod_g acts by simplicial automorphisms on $\mathcal{C}_g$.

CONJECTURE 4.7 (**Q**-rank of moduli space). *$\text{Cone}(\mathcal{M}_g)$ is homeomorphic to the (open) cone on the quotient $\mathcal{C}_g/\text{Mod}_g$[6].*

One can pose a stronger version of Conjecture 4.7 that predicts the precise bilipschitz type of the natural metric on $\text{Cone}(\mathcal{M}_g)$; an analogous statement for quotients $\Gamma \backslash G/K$ of symmetric spaces by lattices was proven by Hattori [**Hat**]. H. Masur and I have identified the right candidate for a coarse fundamental domain

---

[6]Note: This statement is a slight cheat; the actual version requires the language of orbi-complexes.

needed to prove Conjecture 4.7; its description involves certain length inequalities analogous to those on roots defining Weyl chambers. Further, the (conjectured) dimensions of the corresponding tangent cones are $\mathbf{Q}$-rank(Γ) and $d(S)$, respectively. Thus we propose the following additions to the list of analogies between arithmetic lattices and Mod_g.

arithmetic lattices	Mod_g
$\mathbf{Q}$-rank(Γ)	$d(S)$
root lattice	{simple closed curves}
simple roots	top. types of simple closed curves
Cone$(\Gamma\backslash G/K)$	Cone$(\mathcal{M}_\mathrm{g})$

As alluded to above, Conjecture 4.7 should imply, together with the methods in [**Ch**], the second statement of Conjecture 4.5.

4.3. The Kahler group problem. Complete Kahler metrics on $\mathcal{M}_g$ with finite volume and bounded curvatures have been found by Cheng-Yau, McMullen and others (see, e.g., [**LSY**] for a survey). The following conjecture, however, is still not known. I believe this is a folklore conjecture.

CONJECTURE 4.8 (Mod_g is Kahler). *For $g \geq 3$, the group Mod_g is a Kahler group, i.e. it is isomorphic to the fundamental group of a compact Kahler manifold.*

It was shown in [**Ve**] that Mod_2 is not a Kahler group. This is proven by reducing (via finite extensions) to the pure braid group case; these groups are not Kahler since they are iterated extensions of free groups by free groups.

A natural place to begin proving Conjecture 4.8 is the same strategy that Toledo uses in [**To**] for nonuniform lattices in $\mathrm{SU}(n,1), n \geq 3$. The main point is the following. One starts with a smooth open variety V and wants to prove that $\pi_1 V$ is a Kahler group. The first step is to find a compactification $\overline{V}$ of V which is projective, and for which $\overline{V} - V$ has codimension *at least* 3. This assumption guarantees that the intersection of the generic 2-plane P in projective space with $\overline{V}$ misses V. The (weak) Lefschetz Theorem then implies that the inclusion $i : \overline{V} \cap P \hookrightarrow V$ induces an isomorphism on fundamental groups, thus giving that $\pi_1 V$ is a Kahler group.

One wants to apply this idea to the Deligne-Mumford compactification $\overline{\mathcal{M}_g}$ of moduli space $\mathcal{M}_g$. This almost works, except that there is a (complex) codimension *one* singular stratum of $\overline{\mathcal{M}_g}$, so that the above does not apply. Other compactifications of $\mathcal{M}_g$ are also problematic in this regard.

What about the Torelli group $\mathcal{I}_g$? This group, at least for $g \geq 6$, is not known to violate any of the known constraints on Kahler groups. Most notably, Hain [**Ha3**] proved the deep result that for $g \geq 6$ the group $\mathcal{I}_g$ has a quadratically presented Malcev Lie algebra; this is one of the more subtle properties possessed by Kahler groups. Note Akita's theorem (Theorem 5.13 above) that the classifying space of $\mathcal{I}_g, g \geq 7$ does not have the homotopy type of a finite complex shows that these $\mathcal{I}_g$ are not fundamental groups of closed *aspherical* manifolds. There are of course Kahler groups (e.g. finite Kahler groups) with this property. In contrast to Conjecture 4.8, I have recently proven [**Fa3**] the following.

THEOREM 4.9. *$\mathcal{I}_g$ is not a Kahler group for any $g \geq 2$.*

Denoting the symplectic representation by $\pi : \mathrm{Mod}_g \to \mathrm{Sp}(2g, \mathbf{Z})$, the next question is to ask which of the groups $\pi^{-1}(\mathrm{Sp}(2k, \mathbf{Z}))$ for $1 \leq k \leq 2g$ interpolating between the two extremes $\mathcal{I}_g$ and Mod_g are Kahler groups. My only guess is that when $k = 1$ the group is not Kahler.

4.4. The period mapping. To every Riemann surface $X \in \text{Teich}_g$ one attaches its *Jacobian* $\text{Jac}(X)$, which is the quotient of the dual $(\Omega^1(X))^* \approx \mathbf{C}^g$ of the space of holomorphic 1-forms on X by the lattice $H_1(X, \mathbf{Z})$, where $\gamma \in H_1(X, \mathbf{Z})$ is the linear functional on $\Omega^1(X)$ given by $\omega \mapsto \int_\gamma \omega$. Now $\text{Jac}(X)$ is a complex torus, and Riemann's period relations show that $\text{Jac}(X)$ is also an algebraic variety, i.e. $\text{Jac}(X)$ is an *abelian variety*. The algebraic intersection number on $H_1(X, \mathbf{Z})$ induces a symplectic form on $\text{Jac}(X)$, which can be thought of as the imaginary part of a positive definite Hermitian form on $\mathbf{C}^g$. This extra bit of structure is called a *principal polarization* on the abelian variety $\text{Jac}(X)$.

The space $\mathcal{A}_g$ of all g-dimensional (over $\mathbf{C}$) principally polarized abelian varieties is parameterized by the locally symmetric orbifold $\text{Sp}(2g, \mathbf{Z}) \backslash \text{Sp}(2g, \mathbf{R})/\text{U}(g)$. The *Schottky problem*, one of the central classical problems of algebraic geometry, asks for the image of the *period mapping*

$$\Psi : \mathcal{M}_g \to \mathcal{A}_g$$

which sends a surface X to its Jacobian $\text{Jac}(X)$. In other words, the Schottky problem asks: which principally polarized abelian varieties occur as Jacobians of some Riemann surface? Torelli's Theorem states that Ψ is injective; the image $\Psi(\mathcal{M}_g)$ is called the *period locus*. The literature on this problem is vast (see, e.g., [**D**] for a survey), and goes well beyond the scope of the present paper.

Inspired by the beautiful paper [**BS**] of Buser-Sarnak, I would like to pose some questions about the period locus from a different (and probably nonstandard) point of view. Instead of looking for precise algebraic equations describing $\Psi(\mathcal{M}_g)$, what if we instead try to figure out how to tell whether or not it contains a given torus, or if we try to describe what the period locus *roughly* looks like? Let's make these questions more precise.

The data determining a principally polarized abelian variety is not combinatorial, but is a matrix of real numbers. However, one can still ask for algorithms involving such data by using the *complexity theory over* $\mathbf{R}$ developed by Blum-Shub-Smale; see, e.g., [**BCSS**]. Unlike classical complexity theory, here one assumes that real numbers can be given and computed precisely, and develops algorithms, measures of complexity, and the whole theory under this assumption. In the language of this theory we can then pose the following.

PROBLEM 4.10 (Algorithmic Schottky problem). *Give an algorithm, in the sense of complexity theory over* $\mathbf{R}$, *which takes as input a $2g \times 2g$ symplectic matrix representing a principally polarized abelian variety, and as output tells whether or not that torus lies in the period locus.*

One might also fix some $\epsilon = \epsilon(g)$, and then ask whether or not a given principally polarized abelian variety lies within ϵ (in the locally symmetric metric on $\mathcal{A}_g$) of the period locus.

It should be noted that S. Grushevsky [**Gr**] has made the KP-equations solution to the Schottky problem effective in an algebraic sense. This seems to be different than what we have just discussed, though.

We now address the question of what the Schottky locus looks like from far away. To make this precise, let $\text{Cone}(\mathcal{A}_g)$ denote the tangent cone at infinity (defined in (10) above) of the locally symmetric Riemannian orbifold $\mathcal{A}_g$. Hattori [**Hat**] proved that $\mathrm{Cone}(\mathcal{A}_g)$ is homeomorphic to the open cone on the quotient of the Tits boundary of the symmetric space $\mathrm{Sp}(2g,\mathbf{R})/\mathrm{U}(g)$; indeed it is isometric to a Weyl chamber in the symmetric space, which is just a Euclidean sector of dimension g.

PROBLEM 4.11 (Coarse Schottky problem). *Describe, as a subset of a g-dimensional Euclidean sector, the subset of $\mathrm{Cone}(\mathcal{A}_g)$ determined by the Schottky locus in $\mathcal{A}_g$.*

Points in $\mathrm{Cone}(\mathcal{A}_g)$ are recording how the relative sizes of basis vectors of the tori are changing; it is precisely the "skewing parameters" that are being thrown away. It doesn't seem unreasonable to think that much of the complexity in describing the Schottky locus is coming precisely from these skewing parameters, so that this coarsification of the Schottky problem, unlike the classical version, may have a reasonably explicit solution.

There is a well-known feeling that the Schottky locus is quite distorted in $\mathcal{A}_g$. Hain and Toledo (perhaps among others) have posed the problem of determining the second fundamental form of the Schottky locus, although they indicate that this would be a rather difficult computation. We can coarsify this question by extending the definition of distortion of subgroups given in Subsection 3.2 above to the context of subspaces of metric spaces. Here the distortion of a subset S in a metric space Y is defined by comparing the restriction of the metric d_Y to S versus the induced path metric on S.

PROBLEM 4.12 (Distortion of the Schottky locus). *Compute the distortion of the Schottky locus in $\mathcal{A}_g$.*

A naive guess might be that it is exponential.

5. The Torelli group

Problems about the Torelli group $\mathcal{I}_g$ have a special flavor of their own. As one passes from Mod_g to $\mathcal{I}_g$, significant and beautiful new phenomena occur. One reason for the richness of this theory is that the standard exact sequence

$$1 \to \mathcal{I}_g \to \mathrm{Mod}_g \to \mathrm{Sp}(2g,\mathbf{Z}) \to 1$$

gives an action $\psi : \mathrm{Sp}(2g,\mathbf{Z}) \to \mathrm{Out}(\mathcal{I}_g)$, so that any natural invariant attached to $\mathcal{I}_g$ comes equipped with an $\mathrm{Sp}(2g,\mathbf{Z})$-action. The most notable examples of this are the cohomology algebra $H^*(\mathcal{I}_g,\mathbf{Q})$ and the Malcev Lie algebra $\mathcal{L}(\mathcal{I}_g) \otimes \mathbf{Q}$, both of which become $\mathrm{Sp}(2g,\mathbf{Q})$-modules, allowing for the application of symplectic representation theory. See, e.g., [**Jo1, Ha3, Mo4**] for more detailed explanations and examples.

In this section I present a few of my favorite problems. I refer the reader to the work of Johnson, Hain and Morita for other problems about $\mathcal{I}_g$; see, e.g., [**Mo1, Mo3**].

5.1. Finite generation problems.
For some time it was not known if the group $\mathcal{K}_g$ generated by Dehn twists about bounding curves was equal to, or perhaps a finite index subgroup of $\mathcal{I}_g$, until Johnson found the *Johnson homomorphism* τ and proved exactness of:

(11) $$1 \to \mathcal{K}_g \to \mathcal{I}_g \xrightarrow{\tau} \wedge^3 H/H \to 1$$

where $H = H_1(\Sigma_g; \mathbf{Z})$ and where the inclusion $H \hookrightarrow \wedge^3 H$ is given by the map $x \mapsto x \wedge \hat{i}$, where $\hat{i}$ is the intersection from $\hat{i} \in \wedge^2 H$. Recall that a *bounding pair map* is a composition $T_a \circ T_b^{-1}$ of Dehn twists about *bounding pairs*, i.e. pairs of disjoint, nonseparating, homologous simple closed curves $\{a, b\}$. Such a homeomorphism clearly lies in $\mathcal{I}_g$. By direct computation Johnson shows that the τ-image of such a map is nonzero, while $\tau(\mathcal{K}_g) = 0$; proving that $\ker(\tau) = \mathcal{K}_g$ is much harder to prove.

Johnson proved in [**Jo2**] that $\mathcal{I}_g$ is finitely generated for all $g \geq 3$. McCullough-Miller [**McM**] proved that $\mathcal{K}_2$ is not finitely generated; Mess [**Me**] then proved that $\mathcal{K}_2$ is in fact a free group of infinite rank, generated by the set of symplectic splittings of $H_1(\Sigma_2, \mathbf{Z})$. The problem of finite generation of $\mathcal{K}_g$ for all $g \geq 3$ was recently solved by Daniel Biss and me in [**BF**].

THEOREM 5.1. *The group $\mathcal{K}_g$ is not finitely generated for any $g \geq 2$.*

The following basic problem, however, remains open (see, e.g., [**Mo3**], Problem 2.2(ii)).

QUESTION 5.2 (Morita). *Is $H_1(\mathcal{K}_g, \mathbf{Z})$ finitely generated for $g \geq 3$?*

Note that Birman-Craggs-Johnson (see, e.g., [**BC, Jo1**]) and Morita [**Mo2**] have found large abelian quotients of $\mathcal{K}_g$.

The proof in [**BF**] of Theorem 5.1, suggests an approach to answering to Question 5.2. Let me briefly describe the idea. Following the the outline in [**McM**], we first find an action of $\mathcal{K}_g$ on the first homology of a certain abelian cover Y of Σ_g; this action respects the structure of $H_1(Y, \mathbf{Z})$ as a module over the Galois group of the cover. The crucial piece is that we are able to reduce this to a representation

$$\rho : \mathcal{K}_g \to \mathrm{SL}_2(\mathbf{Z}[t, t^{-1}])$$

on the special linear group over the ring of integral laurent polynomials in one variable. This group acts on an associated Bruhat-Tits-Serre tree, and one can then analyze this action using combinatorial group theory.

One might now try to answer Question 5.2 in the negative by systematically computing more elements in the image of ρ, and then analyzing more closely the action on the tree for SL_2. One potentially useful ingredient is a theorem of Grunewald-Mennike-Vaserstein [**GMV**] which gives free quotients of arbitrarily high rank for the group $\mathrm{SL}_2(\mathbf{Z}[t])$ and the group $\mathrm{SL}_2(K[s,t])$, where K is an arbitrary finite field.

Since we know for $g \geq 3$ that $\mathcal{I}_g$ is finitely generated and $\mathcal{K}_g$ is not, it is natural to ask about the subgroups interpolating between these two. To be precise, consider the exact sequence (11). Corresponding to each subgroup $L < \wedge^3 H/H$ is its pullback $\pi^{-1}(L)$. The lattice of such subgroups L can be thought of as a kind of interpolation between $\mathcal{I}_g$ and $\mathcal{K}_g$.

PROBLEM 5.3 (Interpolations). *Let $g \geq 3$. For each subgroup $L < \wedge^3 H/H$, determine whether or not $\pi^{-1}(L)$ is finitely generated.*

As for subgroups deeper down than $\mathcal{K}_g = \mathcal{I}_g(2)$ in the Johnson filtration $\{\mathcal{I}_g(k)\}$, we would like to record the following.

THEOREM 5.4 (Johnson filtration not finitely generated). *For each $g \geq 3$ and each $k \geq 2$, the group $\mathcal{I}_g(k)$ is not finitely generated.*

Proof. We proceed by induction on k. For $k = 2$ this is just the theorem of [**BF**] that $\mathcal{K}_g = \mathcal{I}_g(2)$ is not finitely generated for any $g \geq 3$. Now assume the theorem is true for $\mathcal{I}_g(k)$. The kth *Johnson homomorphism* is a homomorphism
$$\tau_g(k) : \mathcal{I}_g(k) \to \mathfrak{h}_g(k)$$
where $\mathfrak{h}_g(k)$ is a certain finitely-generated abelian group, coming from the kth graded piece of a certain graded Lie algebra; for the precise definitions and constructions, see, e.g., §5 of [**Mo3**]. All we will need is Morita's result (again, see §5 of [**Mo3**]) that $\ker(\tau_g(k)) = \mathcal{I}_g(k+1)$. We thus have an exact sequence
$$(12) \qquad 1 \to \mathcal{I}_g(k+1) \to \mathcal{I}_g(k) \to \tau_g(k)(\mathcal{I}_g(k)) \to 1$$
Now the image $\tau_g(k)(\mathcal{I}_g(k))$ is a subgroup of the finitely generated abelian group $\mathfrak{h}_g(k)$, and so is finitely generated. If $\mathcal{I}_g(k+1)$ were finitely-generated, then by (12) we would have that $\mathcal{I}_g(k)$ is finitely generated, contradicting the induction hypothesis. Hence $\mathcal{I}_g(k+1)$ cannot be finitely generated, and we are done by induction. ⋄

The Johnson filtration $\{\mathcal{I}_g(k)\}$ and the lower central series $\{(\mathcal{I}_g)_k\}$ do not coincide; indeed Hain proved in [**Ha3**] that there are terms of the former not contained in any term of the latter. Thus the following remains open.

CONJECTURE 5.5. *For each $k \geq 1$, the group $(\mathcal{I}_g)_k$ is not finitely generated.*

Another test of our understanding of the Johnson filtration is the following.

PROBLEM 5.6. *Find $H_1(\mathcal{I}_g(k), \mathbf{Z})$ for all $k \geq 2$.*

Generating sets for $\mathcal{I}_g$. One difficulty in working with $\mathcal{I}_g$ is the complexity of its generating sets: any such set must have at least $\frac{1}{3}[4g^3 - g]$ elements since $\mathcal{I}_g$ has abelian quotients of this rank (see [**Jo5**], Corollary after Theorem 5). Compare this with Mod_g, which can always be generated by $2g + 1$ Dehn twists (Humphries), or even by 2 elements (Wajnryb [**Wa**])! How does one keep track, for example, of the (at least) 1330 generators for $\mathcal{I}_{10}$? How does one even give a usable naming scheme for working with these? Even worse, in Johnson's proof of finite generation of $\mathcal{I}_g$ (see [**Jo2**]), the given generating set has $O(2^g)$ elements. The following therefore seems fundamental; at the very least it seems that solving it will require us to understand the combinatorial topology underlying $\mathcal{I}_g$ in a deeper way than we now understand it.

PROBLEM 5.7 (Cubic genset problem). *Find a generating set for $\mathcal{I}_g$ with $O(g^d)$ many elements for some $d \geq 3$. Optimally one would like $d = 3$.*

In fact in §5 of [**Jo2**], Johnson explicitly poses a much harder problem: for $g \geq 4$ can $\mathcal{I}_g$ be generated by $\frac{1}{3}[4g^3 - g]$ elements? As noted above, this would be a sharp result. Johnson actually obtains this sharp result in genus three, by finding ([**Jo2**], Theorem 3) an explicit set of 35 generators for $\mathcal{I}_g$. His method of converting his $O(2^g)$ generators to $O(g^3)$ becomes far too unwieldy when $g > 3$.

One approach to Problem 5.7 is to follow the original plan of [**Jo2**], but using a simpler generating set for Mod_g. This was indeed the motivation for Brendle and me when we found in [**BFa1**] a generating set for Mod_g consisting of 6 involutions, i.e. 6 elements of order 2. This bound was later improved by Kassabov [**Ka**] to 4 elements of order 2, at least when $g \geq 7$. Clearly Mod_g is never generated by 2 elements of order two, for then it would be a quotient of the infinite dihedral group, and so would be virtually abelian. Since the current known bounds are so close to being sharp, it is natural to ask for the sharpest bounds.

PROBLEM 5.8 (Sharp bounds for involution generating sets). *For each $g \geq 2$, prove sharp bounds for the minimal number of involutions required to generate Mod_g. In particular, for $g \geq 7$ determine whether or not Mod_g is generated by 3 involutions.*

5.2. Higher finiteness properties and cohomology. While there has been spectacular progress in understanding $H^*(\mathrm{Mod}_g, \mathbf{Z})$ (see [**MW**]), very little is known about $H^*(\mathcal{I}_g, \mathbf{Z})$, and even less about $H^*(\mathcal{K}_g, \mathbf{Z})$. Further, we do not have answers to the basic finiteness questions one asks about groups.

Recall that the *cohomological dimension* of a group Γ, denoted $\mathrm{cd}(\Gamma)$, is defined to be
$$\mathrm{cd}(\Gamma) := \sup\{i : H^i(\Gamma, V) \neq 0 \text{ for some } \Gamma\text{-module } V\}$$
If a group Γ has a torsion-free subgroup H of finite index, then the *virtual* cohomological dimension of Γ is defined to be $\mathrm{cd}(H)$; Serre proved that this number does not depend on the choice of H. It is a theorem of Harer, with earlier estimates and a later different proof due to Ivanov, that Mod_g has virtual cohomological dimension $4g - 5$; see [**Iv1**] for a summary.

PROBLEM 5.9 (Cohomological Dimension). *Compute the cohomological dimension of $\mathcal{I}_g$ and of $\mathcal{K}_g$. More generally, compute the cohomological dimension of $\mathcal{I}_g(k)$ for all $k \geq 1$.*

Note that the cohomological dimension $\mathrm{cd}(\mathcal{I}_g)$ is bounded above by the (virtual) cohomological dimension of Mod_g, which is $4g - 5$. The following is a start on some lower bounds.

THEOREM 5.10 (Lower bounds on cd). *For all $g \geq 2$, the following inequalities hold:*
$$\mathrm{cd}(\mathcal{I}_g) \geq \begin{cases} (5g-8)/2 & \text{if } g \text{ is even} \\ (5g-9)/2 & \text{if } g \text{ is odd} \end{cases}$$

$$\mathrm{cd}(\mathcal{K}_g) \geq 2g - 3$$

$$\mathrm{cd}(\mathcal{I}_g(k)) \geq g - 1 \quad \text{for } k \geq 3$$

Proof. Since for any group Γ with $\mathrm{cd}(\Gamma) < \infty$ we have $\mathrm{cd}(\Gamma) \geq \mathrm{cd}(H)$ for any subgroup $H < \Gamma$, an easy way to obtain lower bounds for $\mathrm{cd}(\Gamma)$ is to find large free abelian subgroups of Γ. To construct such subgroups for $\mathcal{I}_g$ and for $\mathcal{K}_g$, take a maximal collection of mutually disjoint separating curves on Σ_g; by an Euler characteristic argument it is easy to see that there are $2g - 3$ of these, and it is not hard to find them. The group generated by Dehn twists on these curves is isomorphic to $\mathbf{Z}^{2g-3}$, and is contained in $\mathcal{K}_g < \mathcal{I}_g$.

For $\mathcal{I}_g$ we obtain the better bounds by giving a slight variation of Ivanov's discussion of *Mess subgroups*, given in §6.3 of [**Iv1**], adapted so that the constructed subgroups lie in $\mathcal{I}_g$. Let Mod_g^1 denote the group of homotopy classes of orientation-preserving homeomorphisms of the surface Σ_g^1 of genus g with one boundary component, fixing $\partial \Sigma_g^1$ pointwise, up to isotopies which fix $\partial \Sigma_g^1$ pointwise. We then have a well-known exact sequence (see, e.g. [**Iv1**], §6.3)

(13) $$1 \to \pi_1 T^1 \Sigma_g \to \mathrm{Mod}_g^1 \xrightarrow{\pi} \mathrm{Mod}_g \to 1$$

where $T^1 \Sigma_g$ is the unit tangent bundle of Σ_g. Now suppose $g \geq 2$. Let C_2 and C_3 be maximal abelian subgroups of $\mathcal{I}_2$ and $\mathcal{I}_3$, respectively; these have ranks 1 and 3, respectively. We now define C_g inductively, beginning with C_2 if g is even, and with C_3 if g is odd. Let C_g^1 be the pullback $\pi^{-1}(C_g)$ of C_g under the map π in (13). Note that, since the copy of $\pi_1 T^1 \Sigma_g$ in Mod_g^1 is generated by "point pushing" and the twist about $\partial \Sigma_g$, it actually lies in the corresponding Torelli group $\mathcal{I}_g^1$. The inclusion $\Sigma_g^1 \hookrightarrow \Sigma_{g+2}$ induces an injective homomorphism $i : \mathcal{I}_g^1 \hookrightarrow \mathcal{I}_{g+2}$ via "extend by the identity". The complement of Σ_g^1 in Σ_{g+2} clearly contains a pair of disjoint separating curves. Now define C_{g+2} to be the group by the Dehn twists about these curves together with $i(C_g^1)$. Thus $C_{g+2} \approx C_g^1 \times \mathbf{Z}^2$. The same exact argument as in the proof of Theorem 6.3A in [**Iv1**] gives the claimed answers for $\mathrm{cd}(C_g)$.

Finally, for the groups $\mathcal{I}_g(k)$ with $k, g \geq 3$ we make the following construction. Σ_g admits a homeomorphism f of order $g - 1$, given by rotation in the picture of a genus one subsurface V with $g - 1$ boundary components, with a torus-with-boundary attached to each component of ∂V. It is then easy to see that there is a collection of $g - 1$ mutually disjoint, f-invariant collection of simple closed curves which decomposes Σ_g into a union of $g - 1$ subsurfaces $S_1, \ldots, S_{g-1}$, each having genus one and two boundary components, with mutually disjoint interiors.

Each S_i contains a pair of separating curves α_i, β_i with $i(\alpha_i, \beta_i) \geq 2$. The group generated by the Dehn twists about α_i and β_i thus generates a free group L_i of rank 2 (see, e.g. [**FMa**]). Nonabelian free groups have elements arbitrarily far down in their lower central series. As proven in Lemma 4.3 of [**FLM**], any element in the kth level of the lower central series for any L_i gives an element γ_i lying in $\mathcal{I}_g(k)$. Since $i(\gamma_i, \gamma_j) = 0$ for each i, j, it follows that the group A generated by Dehn twists about each γ_i is isomorphic to $\mathbf{Z}^{g-1}$. As A can be chosen to lie in in any $\mathcal{I}_g(k)$ with $k \geq 3$, we are done. $\diamond$

Since $\cap_{k=1}^\infty \mathcal{I}_g(k) = 0$, we know that there exists $K > 1$ with the property that the cohomological dimension of $\mathcal{I}_g(k)$ is constant for all $k \geq K$. It would be interesting to determine the smallest such K. A number of people have different guesses about what the higher finiteness properties of $\mathcal{I}_g$ should be.

PROBLEM 5.11 (Torelli finiteness). *Determine the maximal number $f(g)$ for which there is a $K(\mathcal{I}_g, 1)$ space with finitely many cells in dimensions $\leq f(g)$.*

Here is what is currently known about Problem 5.11:
(1) $f(2) = 0$ since $\mathcal{I}_2$ is not finitely generated (McCullough-Miller [**McM**]).
(2) $f(3) \leq 3$ (Johnson-Millson, unpublished, referred to in [**Me**]).
(3) For $g \geq 3$, combining Johnson's finite generation result [**Jo2**] and a theorem of Akita (Theorem 5.13 below) gives $1 \leq f(g) \leq 6g - 5$.

One natural guess which fits with the (albeit small amount of) known data is that $f(g) = g - 2$. As a special case of Problem 5.11, we emphasize the following, which is a folklore conjecture.

CONJECTURE 5.12. *$\mathcal{I}_g$ is finitely presented for $g \geq 4$.*

One thing we do know is that, in contrast to Mod_g, neither $\mathcal{I}_g$ nor $\mathcal{K}_g$ has a classifying space which is homotopy equivalent to a finite complex; indeed Akita proved the following stronger result.

THEOREM 5.13 (Akita [**Ak**]). *For each $g \geq 7$ the vector space $H_*(\mathcal{I}_g, \mathbf{Q})$ is infinite dimensional. For each $g \geq 2$ the vector space $H_*(\mathcal{K}_g, \mathbf{Q})$ is infinite dimensional.*

Unfortunately the proof of Theorem 5.13 does not illuminate the reasons why the theorem is true, especially since the proof is far from constructive. In order to demonstrate this, and since the proof idea is simple and pretty, we sketch the proof.

Proof sketch of Theorem 5.13 for $\mathcal{I}_g$. We give the main ideas of the proof, which is based on a similar argument made for $\text{Out}(F_n)$ by Smillie-Vogtmann; see [**Ak**] for details and references.

If $\dim_\mathbf{Q}(H_*(\mathcal{I}_g, \mathbf{Q})) < \infty$, then the multiplicativity of the Euler characteristic for group extensions, applied to

$$1 \to \mathcal{I}_g \to \text{Mod}_g \to \text{Sp}(2g, \mathbf{Z}) \to 1$$

gives that

(14) $$\chi(\mathcal{I}_g) = \chi(\text{Mod}_g)/\chi(\text{Sp}(2g, \mathbf{Z}))$$

Each of the groups on the right hand side of (14) has been computed; the numerator by Harer-Zagier and the denominator by Harder. Each of these values is given as a value of the Riemann zeta function ζ. Plugging in these values into (14) gives

(15) $$\chi(\mathcal{I}_g) = \frac{1}{2 - 2g} \prod_{k=1}^{g-1} \frac{1}{\zeta(1 - 2k)}$$

It is a classical result of Hurwitz that each finite order element in Mod_g acts nontrivially on $H_1(\Sigma_g, \mathbf{Z})$; hence $\mathcal{I}_g$ is torsion-free. Thus $\chi(\mathcal{I}_g)$ is an integer. The rest of the proof of the theorem consists of using some basic properties of ζ to prove that the right hand side of (15) is not an integer. ◇

The hypothesis $g \geq 7$ in Akita's proof is used only in showing that the right hand side of (15) is not an integer. This might still hold for $g < 7$.

PROBLEM 5.14. *Extend Akita's result to $2 < g < 7$.*

Since Akita's proof produces no explicit homology classes, the following seems fundamental.

PROBLEM 5.15 (Explicit cycles). *Explicitly construct infinitely many linearly independent cycles in $H_*(\mathcal{I}_g, \mathbf{Q})$ and $H_*(\mathcal{K}_g, \mathbf{Q})$.*

So, we are still at the stage of trying to find explicit nonzero cycles. In a series of papers (see [**Jo1**] for a summary), Johnson proved the quite nontrivial result:

(16) $$H^1(\mathcal{I}_g, \mathbf{Z}) \approx \frac{\wedge^3 H}{H} \oplus B_2$$

where B_2 consists of 2-torsion. While the $\wedge^3 H/H$ piece comes from purely algebraic considerations, the B_2 piece is "deeper" in the sense that it is purely topological, and comes from the Rochlin invariant (see [**BC**] and [**Jo1**]); indeed the former appears in H_1 of the "Torelli group" in the analogous theory for $\text{Out}(F_n)$, while the latter does not.

Remark on two of Johnson's papers. While Johnson's computation of $H_1(\mathcal{I}_g, \mathbf{Z})$ and his theorem that $\ker(\tau) = \mathcal{K}_g$ are fundamental results in this area, I believe that the details of the proofs of these results are not well-understood. These results are proved in [**Jo4**] and [**Jo3**], respectively; the paper [**Jo4**] is a particularly dense and difficult read. While Johnson's work is always careful and detailed, and so the results should be accepted as true, I think it would be worthwhile to understand [**Jo3**] and [**Jo4**], to exposit them in a less dense fashion, and perhaps even to give new proofs of their main results. For [**Jo3**] this is to some extent accomplished in the thesis [**vdB**] of van den Berg, where she takes a different approach to computing $H_1(\mathcal{I}_g, \mathbf{Z})$.

Since dimension one is the only dimension $i \geq 1$ for which we actually know the ith cohomology of $\mathcal{I}_g$, and since very general computations seem out of reach at this point, the following seems like a natural next step in understanding the cohomology of $\mathcal{I}_g$.

PROBLEM 5.16. *Determine the subalgebras of $H^*(\mathcal{I}_g, K)$, for $K = \mathbf{Q}$ and $K = \mathbf{F}_2$, generated by $H^1(\mathcal{I}_g, K)$.*

Note that $H^*(\mathcal{I}_g, K)$ is a module over $\text{Sp}(2g, K)$. When $K = \mathbf{Q}$ this problem has been solved in degree 2 by Hain [**Ha3**] and degree 3 (up to one unknown piece) by Sakasai [**Sa**]. Symplectic representation theory (over $\mathbf{R}$) is used as a tool in these papers to greatly simplify computations. When $K = \mathbf{F}_2$, the seemingly basic facts one needs about representations are either false or they are beyond the current methods of modular representation theory. Thus computations become more complicated. Some progress in this case is given in [**BFa2**], where direct geometric computations, evaluating cohomology classes on abelian cycles, shows that each of the images of

$$\sigma^* : H^2(B_3, \mathbf{F}_2) \to H^2(\mathcal{I}_g, \mathbf{F}_2)$$
$$(\sigma|_{\mathcal{K}_g})^* : H^2(B_2, \mathbf{F}_2) \to H^2(\mathcal{K}_g, \mathbf{F}_2)$$

has dimension at least $O(g^4)$.

5.3. Automorphisms and commensurations of the Johnson filtration.
The following theorem indicates that all of the algebraic structure of the mapping class group Mod_g is already determined by the infinite index subgroup $\mathcal{I}_g$, and indeed the infinite index subgroup $\mathcal{K}_g$ of $\mathcal{I}_g$. Recall that the *extended mapping class group*, denoted $\text{Mod}_g^\pm$, is defined as the group of homotopy classes of *all* homeomorphisms of Σ_g, including the orientation-reversing ones; it contains Mod_g as a subgroup of index 2.

THEOREM 5.17. *Let $g \geq 4$. Then $\text{Aut}(\mathcal{I}_g) \approx \text{Mod}_g^\pm$ and $\text{Aut}(\mathcal{K}_g) \approx \text{Mod}_g$. In fact $\text{Comm}(\mathcal{I}_g) \approx \text{Mod}_g^\pm$ and $\text{Comm}(\mathcal{K}_g) \approx \text{Mod}_g$.*

The case of $\mathcal{I}_g, g \geq 5$ was proved by Farb-Ivanov [**FI**]. Brendle-Margalit [**BM**] built on [**FI**] to prove the harder results on $\mathcal{K}_g$. The cases of Aut, where one can use explicit relations, were extended to $g \geq 3$ by McCarthy-Vautaw [**MV**]. Note too that $\mathrm{Aut}(\mathrm{Mod}_g) = \mathrm{Mod}_g^{\pm}$, as shown by Ivanov (see §8 of [**Iv1**]).

QUESTION 5.18. *For which $k \geq 1$ is it true that* $\mathrm{Aut}(\mathcal{I}_g(k)) = \mathrm{Mod}_g^{\pm}$? *that* $\mathrm{Comm}(\mathcal{I}_g(k)) = \mathrm{Mod}_g^{\pm}$?

Theorem 5.17 answers the question for $k = 1, 2$. It would be remarkable if all of Mod_g could be reconstructed from subgroups deeper down in its lower central series.

5.4. Graded Lie algebras associated to $\mathcal{I}_g$. Fix a prime $p \geq 2$. For a group Γ let $P_i(\Gamma)$ be defined inductively via $P_1(\Gamma) = \Gamma$ and
$$P_{i+1}(\Gamma) := [\Gamma, P_i(\Gamma)]\Gamma^p \text{ for } i \geq 1$$
The sequence $\Gamma \supseteq P_2(\Gamma) \supseteq \cdots$ is called the *lower exponent p central series*. The quotient $\Gamma/P_2(\Gamma)$ has the universal property that any homomorphism from Γ onto an abelian p-group factors through $\Gamma/P_2(\Gamma)$; the group $\Gamma/P_{i+1}(\Gamma)$ has the analgous universal property for homomorphisms from Γ onto class i nilpotent p-groups. We can form the direct sum of vector spaces over the field $\mathbf{F}_p$:
$$\mathcal{L}_p(\Gamma) := \bigoplus_{i=1}^{\infty} \frac{P_i(\Gamma)}{P_{i+1}(\Gamma)}$$
The group commutator on Γ induces a bracket on $\mathcal{L}_p(\Gamma)$ under which it becomes a graded Lie algebra over $\mathbf{F}_p$. See, e.g. [**Se**] for the basic theory of Lie algebras over $\mathbf{F}_p$.

When $p = 0$, that is when $P_{i+1}(\Gamma) = [\Gamma, P_i(\Gamma)]$, we obtain a graded Lie algebra $\mathcal{L}(\Gamma) := \mathcal{L}_0(\Gamma)$ over $\mathbf{Z}$. The Lie algebra $\mathcal{L}(\Gamma) \otimes \mathbf{R}$ is isomorphic to the associated graded Lie algebra of the Malcev Lie algebra associated to Γ. In [**Ha3**] Hain found a presentation for the infinite-dimensional Lie algebra $\mathcal{L}(\mathcal{I}_g) \otimes \mathbf{R}$: it is (at least for $g \geq 6$) the quotient of the free Lie algebra on $H_1(\mathcal{I}_g, \mathbf{R}) = \wedge^3 H/H, H := H_1(\Sigma_g, \mathbf{R})$, modulo a finite set of *quadratic relations*, i.e. modulo an ideal generated by certain elements lying in $[P_2(\mathcal{I}_g)/P_3(\mathcal{I}_g)] \otimes \mathbf{R}$. Each of these relations can already be seen in the Malcev Lie algebra of the pure braid group.

The main ingredient in Hain's determination of $\mathcal{L}(\mathcal{I}_g) \otimes \mathbf{R}$ is to apply Deligne's *mixed Hodge theory*. This theory is a refinement and extension of the classical Hodge decomposition. For each complex algebraic variety V it produces, in a functorial way, various filtrations with special properties on $H^*(V, \mathbf{Q})$ and its complexification. This induces a remarkably rich structure on many associated invariants of V. A starting point for all of this is the fact that $\mathcal{M}_g$ is a complex algebraic variety. Since, at the end of the day, Hain's presentation of $\mathcal{L}(\mathcal{I}_g) \otimes \mathbf{R}$ is rather simple, it is natural to pose the following.

PROBLEM 5.19. *Give an elementary, purely combinatorial-topological and group-theoretic, proof of Hain's theorem.*

It seems that a solution to Problem 5.19 will likely require us to advance our understanding of $\mathcal{I}_g$ in new ways. It may also give a hint towards attacking the following problem, where mixed Hodge theory does not apply.

PROBLEM 5.20 (Hain for $\mathrm{Aut}(F_n)$). *Give an explicit finite presentation for the Malcev Lie Algebra $\mathcal{L}(\mathrm{IA}_n)$, where IA_n is the group of automorphisms of the free group F_n acting trivially on $H_1(F_n, \mathbf{Z})$.*

There is a great deal of interesting information at the prime 2 which Hain's theorem does not address, and indeed which remains largely unexplored. While Hain's theorem tells us that reduction mod 2 gives us a large subalgebra of $\mathcal{L}_2(\mathcal{I}_{g,1})$ coming from $\mathcal{L}_0(\mathcal{I}_{g,1})$, the Lie algbera $\mathcal{L}_2(\mathcal{I}_{g,1})$ over $\mathbf{F}_2$ is much bigger. This can already be seen from (16). As noted above, the 2-torsion B_2 exists for "deeper" reasons than the other piece of $H_1(\mathcal{I}_{g,1}, \mathbf{Z})$, as it comes from the Rochlin invariant as opposed to pure algebra. Indeed, for the analogous "Torelli group" IA_n for $\mathrm{Aut}(F_n)$, the corresponding "Johnson homomorphism" gives all the first cohomology. Thus the 2-torsion in $H^1(\mathcal{I}_g, \mathbf{Z})$ is truly coming from 3-manifold theory.

PROBLEM 5.21 (Malcev mod 2). *Give an explicit finite presentation for the $\mathbf{F}_2$-Lie algebra $\mathcal{L}_2(\mathcal{I}_{g,1})$.*

We can also build a Lie algebra using the Johnson filtration. Let
$$\mathfrak{h}_g := \bigoplus_{k=1}^{\infty} \frac{\mathcal{I}_g(k)}{\mathcal{I}_g(k+1)} \otimes \mathbf{R}$$
Then $\mathfrak{h}$ is a real Lie algebra. In §14 of [**Ha3**], Hain proves that the Johnson filtration is not cofinal with the lower central series of $\mathcal{I}_g$. He also relates $\mathfrak{h}_g$ to $\mathfrak{t}_g$. The following basic question remains open.

QUESTION 5.22 (Lie algebra for the Johnson filtration). *Is $\mathfrak{h}_g$ a finitely presented Lie algebra? If so, give an explicit finite presentation for it.*

5.5. Low-dimensional homology of principal congruence subgroups.
Recall that the *level L congruence subgroup* $\mathrm{Mod}_g[L]$ is defined to be the subgroup of Mod_g which acts trivially on $H_1(\Sigma_g, \mathbf{Z}/L\mathbf{Z})$. This normal subgroup has finite index; indeed the quotient of Mod_g by $\mathrm{Mod}_g[L]$ is the finite symplectic group $\mathrm{Sp}(2g, \mathbf{Z}/L\mathbf{Z})$. When $L \geq 3$ the group $\mathrm{Mod}_g[L]$ is torsion free, and so the corresponding cover of moduli space is actually a manifold. These manifolds arise in algebraic geometry as they parametrize so-called "genus g curves with level L structure"; see [**Ha2**].

PROBLEM 5.23. *Compute $H_1(\mathrm{Mod}_g[L]; \mathbf{Z})$.*

McCarthy and (independently) Hain proved that $H_1(\mathrm{Mod}_g[L], \mathbf{Z})$ is finite for $g \geq 3$; see, e.g. Proposition 5.2 of [**Ha2**][7]. As discussed in §5 of [**Ha2**], the following conjecture would imply that the (orbifold) Picard group for the moduli spaces of level L structures has rank one; this group is finitely-generated by the Hain and McCarthy result just mentioned.

CONJECTURE 5.24 (Picard number one conjecture for level L structures). *Prove that $H_2(\mathrm{Mod}_g[L]; \mathbf{Q}) = \mathbf{Q}$ when $g \geq 3$. More generally, compute $H_2(\mathrm{Mod}_g[L]; \mathbf{Z})$ for all $g \geq 3, L \geq 2$.*

Harer [**Har2**] proved this conjecture in the case $L = 1$. This generalization was stated (for Picard groups) as Question 7.12 in [**HL**]. The case $L = 2$ was claimed in [**Fo**], but there is apparently an error in the proof. At this point even the $(g, L) = (3, 2)$ case is open.

Here is a possible approach to Conjecture 5.24 for $g \geq 4$. First note that, since $\mathrm{Mod}_g[L]$ is a finite index subgroup of the finitely presented group Mod_g, it is

---
[7]Actually, Hain proves a much stronger result, computing $H^1(\mathrm{Mod}_g[L], V)$ for V any finite-dimensional symplectic representation.

finitely presented. As we have a lot of explicit information about the finite group $\mathrm{Sp}(2g, \mathbf{Z}/L\mathbf{Z})$, it seems possible in principle to answer the following, which is also a test of our understanding of $\mathrm{Mod}_g[L]$.

PROBLEM 5.25 (Presentation for level L structures). *Give an explicit finite presentation for* $\mathrm{Mod}_g[L]$.

Once one has such a presentation, it seems likely that it would fit well into the framework of Pitsch's proof [**Pi**] that $\mathrm{rank}(H_2(\mathrm{Mod}_{g,1}, \mathbf{Z})) \leq 1$ for $g \geq 4$. Note that Pitsch's proof was extended to punctured and bordered case by Korkmaz-Stipsicsz; see [**Ko**]. What Pitsch does is to begin with an explicit, finite presentation of $\mathrm{Mod}_{g,1}$, and then to apply Hopf's formula for groups Γ presented as the quotient of a free group F by the normal closure R of the relators:

$$(17) \qquad H_2(\Gamma, \mathbf{Z}) = \frac{R \cap [F, F]}{[F, R]}$$

In other words, elements of $H_2(\Gamma, \mathbf{Z})$ come precisely from commutators which are relators, except for the trivial ones. Amazingly, one needs only write the form of an arbitrary element of the numerator in (17), and a few tricks reduces the computation of (an upper bound for) space of solutions to computing the rank of an integer matrix. In our case this approach seems feasible, especially with computer computation, at least for small L. Of course one hopes to find a general pattern.

6. Linear and nonlinear representations of Mod_g

While for $g > 2$ it is not known whether or not Mod_g admits a faithful, finite-dimensional linear representation, there are a number of known linear and nonlinear representations of Mod_g which are quite useful, have a rich internal structure, and connect to other problems. In this section we pose a few problems about some of these.

6.1. Low-dimensional linear representations.
It would be interesting to classify all irreducible complex representations $\psi : \mathrm{Mod}_g \to \mathrm{GL}(m, \mathbf{C})$ for m sufficiently small compared to g. This was done for representations of the n-strand braid group for $m \leq n - 1$ by Formanek [**For**]. There are a number of special tricks using torsion in Mod_g and so, as with many questions of this type, one really wants to understand low-dimensional irreducible representations of the (typically torsion-free) finite index subgroups of Mod_g. It is proven in [**FLMi**] that no such *faithful* representations exist for $n < 2\sqrt{g-1}$.

One question is to determine if the standard representation on homology $\mathrm{Mod}_g \to \mathrm{Sp}(2g, \mathbf{Z})$ is minimal in some sense. Lubotzky has found finite index subgroups $\Gamma < \mathrm{Mod}_g$ and surjections $\Gamma \to \mathrm{Sp}(2g-2, \mathbf{Z})$. I believe that it should be possible to prove that representations of such Γ in low degrees must have traces which are algebraic integers. This problem, and various related statements providing constraints on representations, reduce via now-standard methods to the problem of understanding representations $\rho : \mathrm{Mod}_g \to \mathrm{GL}(n, K)$, where K is a discretely valued field such as the p-adic rationals. The group $\mathrm{GL}(n, K)$ can be realized as a group of isometries of the corresponding Bruhat-Tits affine building; this is a non-positively curved (in the CAT(0) sense), $(n-1)$-dimensional simplicial complex. The general problem then becomes:

PROBLEM 6.1 (Actions on buildings). *Determine all isometric actions ψ : $\mathrm{Mod}_g \to \mathrm{Isom}(X^n)$, where X^n is an n-dimensional Euclidean building, and n is sufficiently small compared to g.*

For example, one would like conditions under which ψ has a *global fixed point*, that is, a point $x \in X^n$ such that $\psi(\mathrm{Mod}_g) \cdot x = x$. One method to attack this problem is the so-called "Helly technique" introduced in [**Fa4**]. Using standard CAT(0) methods, one can show that each Dehn twist T_α in Mod_g has a nontrivial fixed set F_α under the ψ-action; F_α is necessarily convex. Considering the nerve of the collection $\{F_\alpha\}$ gives a map $\mathcal{C}_g \to X^n$ from the complex of curves to X^n. Now $\mathcal{C}_g$ has the homotopy type of a wedge of spheres (see, e.g., [**Iv1**]), while X^n is contractible. Hence the spheres in the nerve must be filled in, which gives that many more elements $\psi(T_\alpha)$ have common fixed points. The problem now is to understand in an explicit way the spheres inside $\mathcal{C}_g$.

6.2. Actions on the circle. It was essentially known to Nielsen that $\mathrm{Mod}_{g,1}$ acts faithfully by orientation-preserving homeomorphisms on the circle. Here is how this works: for $g \geq 2$ any homeomorphism $f \in \mathrm{Homeo}^+(\Sigma_g)$ lifts to a quasi-isometry $\tilde{f}$ of the hyperbolic plane $\mathbf{H}^2$. Any quasi-isometry of $\mathbf{H}^2$ takes geodesic rays to a uniformly bounded distance from geodesic rays, thus inducing a map $\partial \tilde{f} : S^1 \to S^1$ on the boundary of infinity of $\mathbf{H}^2$, which is easily checked to be a homeomorphism, indeed a quasi-symmetric homeomorphism. If $h \in \mathrm{Homeo}^+(\Sigma_g)$ is homotopic to f, then since homotopies are compact one sees directly that $\tilde{h}$ is homotopic to $\tilde{f}$, and so these maps are a bounded distance from each other in the sup norm. In particular $\partial \tilde{h} = \partial \tilde{f}$; that is, $\partial \tilde{f}$ depends only the homotopy class of f. It is classical that quasi-isometries are determined by their boundary values, hence $\partial \tilde{f} = Id$ only when f is homotopically trivial. Now there are $\pi_1 \Sigma_g$ choices for lifting any such f, so the group $\Gamma_g \subset \mathrm{Homeo}^+(S^1)$ of all lifts of all homotopy classes of $f \in \mathrm{Homeo}^+(\Sigma_g)$ gives an exact sequence

$$(18) \qquad 1 \to \pi_1 \Sigma_g \to \Gamma_g \to \mathrm{Mod}_{g,1} \to 1$$

Since each element of $\mathrm{Mod}_{g,1}$ fixes a common marked point on Σ_g, there is a canonical way to choose a lift of each f; that is, we obtain a section $\mathrm{Mod}_{g,1} \to \Gamma_g$ splitting (18). In particular we have an injection

$$(19) \qquad \mathrm{Mod}_{g,1} \hookrightarrow \mathrm{Homeo}^+(S^1)$$

This inclusion provides a so-called (left) circular ordering on $\mathrm{Mod}_{g,1}$ - see [**Cal**]. Note that no such inclusion as in (19) exists for Mod_g since any finite subgroup of $\mathrm{Homeo}^+(S^1)$ must be cyclic[8], but Mod_g has noncyclic finite subgroups.

The action given by (19) gives a dynamical description of $\mathrm{Mod}_{g,1}$ via its action on S^1. For example, each pseudo-Anosov in $\mathrm{Mod}_{g,1}$ acts on S^1 with finitely many fixed points, with alternating sources and sinks as one moves around the circle (see, e.g., Theorem 5.5 of [**CB**]). There is an intrinsic non-smoothness to this action, and indeed in [**FF**] it is proven that any homomorphism $\rho : \mathrm{Mod}_{g,1} \to \mathrm{Diff}^2(S^1)$ has trivial image; what one really wants to prove is that no finite index subgroup of Mod_g admits a faithful C^2 action on S^1. It would be quite interesting to prove that the action (19) is canonical, in the following sense.

---

[8] One can see this by averaging any Riemannian metric on S^1 by the group action.

QUESTION 6.2 (Rigidity of the $\operatorname{Mod}_{g,1}$ action on S^1). *Is any faithful action $\rho : \operatorname{Mod}_{g,1} \to \operatorname{Homeo}^+(S^1)$ conjugate in $\operatorname{Homeo}^+(S^1)$ to the standard action, given in (19)? What about the same question for finite index subgroups of $\operatorname{Mod}_{g,1}$?*

Perhaps there is a vastly stronger, topological dynamics characterization of $\operatorname{Mod}_{g,1}$ inside $\operatorname{Homeo}^+(S^1)$, in the style of the Convergence Groups Conjecture (theorem of Tukia, Casson-Jungreis and Gabai), with "asymptotically source – sink" being replaced here by "asymptotically source – sink – $\cdots$ – source – sink", or some refinement/variation of this.

Now, the group of lifts of elements of $\operatorname{Homeo}^+(S^1)$ to homeomorphisms of $\mathbf{R}$ gives a central extension

$$1 \to \mathbf{Z} \to \widetilde{\operatorname{Homeo}}(S^1) \to \operatorname{Homeo}^+(S^1) \to 1$$

which restricts via (19) to a central extension

(20) $$1 \to \mathbf{Z} \to \widetilde{\operatorname{Mod}_{g,1}} \to \operatorname{Mod}_{g,1} \to 1$$

Note that $\operatorname{Mod}_{g,1}$ has torsion. Since $\widetilde{\operatorname{Homeo}}(S^1) \subset \operatorname{Homeo}^+(\mathbf{R})$ which clearly has no torsion, it follows that (20) does not split. In particular the extension (20) gives a nonvanishing class $\xi \in H^2(\operatorname{Mod}_{g,1}, \mathbf{Z})$. Actually, it is not hard to see that ξ is simply the "euler cocycle", which assigns to any pointed map $\Sigma_h \to \mathcal{M}_g$ the euler class of the pullback bundle of the "universal circle bundle" over $\mathcal{M}_g$.

The torsion in $\operatorname{Mod}_{g,1}$ and in Mod_g preclude each from having a left-ordering, or acting faithfully on $\mathbf{R}$. As far as we know this is the only obstruction; it disappears when one passes to appropriate finite index subgroups.

QUESTION 6.3 (orderability). *Does $\operatorname{Mod}_g, g \geq 2$ have some finite index subgroup which acts faithfully by homeomorphisms on S^1? Does either Mod_g or $\operatorname{Mod}_{g,1}$ have a finite index subgroup which acts faithfully by homeomorphisms on $\mathbf{R}$?*

Note that Thurston proved that braid groups are orderable. Since $\mathcal{I}_g$ and $\mathcal{I}_{g,1}$ are residually torsion-free nilpotent, they are isomorphic to a subgroup of $\operatorname{Homeo}^+(\mathbf{R})$; in fact one can show that (20) splits when restricted to $\mathcal{I}_{g,1}$. On the other hand, Witte [**Wi**] proved that no finite index subgroup of $\operatorname{Sp}(2g, \mathbf{Z})$ acts faithfully by homoeomorphisms on S^1 or on $\mathbf{R}$.

Non-residual finiteness of the universal central extension. The Lie group $\operatorname{Sp}(2g, \mathbf{R})$ has infinite cyclic fundamental group. Its universal cover $\widetilde{\operatorname{Sp}(2g, \mathbf{R})}$ gives a central extension

$$1 \to \mathbf{Z} \to \widetilde{\operatorname{Sp}(2g, \mathbf{R})} \to \operatorname{Sp}(2g, \mathbf{R}) \to 1$$

which restricts to a central extension

(21) $$1 \to \mathbf{Z} \to \widetilde{\operatorname{Sp}(2g, \mathbf{Z})} \to \operatorname{Sp}(2g, \mathbf{Z}) \to 1$$

The cocycle $\zeta \in H^2(\operatorname{Sp}(2g, \mathbf{Z}), \mathbf{Z})$ defining the extension (21) is nontrivial and bounded; this comes from the fact that it is proportional to the Kahler class on the corresponding locally symmetric quotient (which is a $K(\pi, 1)$ space). Deligne proved in [**De**] that $\widetilde{\operatorname{Sp}(2g, \mathbf{Z})}$ is *not residually finite*. Since there is an obvious surjection of exact sequences from (19) to (21), and since both central extensions give a bounded cocycle, one begins to wonder about the following.

QUESTION 6.4 ((Non)residual finiteness). *Is the (universal) central extension* $\widetilde{\mathrm{Mod}}_{g,1}$ *of* $\mathrm{Mod}_{g,1}$ *residually finite, or not?*

Note that an old result of Grossman states that Mod_g and $\mathrm{Mod}_{g,1}$ are both residually finite. The group $\mathrm{Sp}(2g,\mathbf{Z})$ is easily seen to be residually finite; indeed the intersection of all congruence subgroups of $\mathrm{Sp}(2g,\mathbf{Z})$ is trivial.

6.3. The sections problem.
Consider the natural projection $\pi : \mathrm{Homeo}^+(\Sigma_g) \to \mathrm{Mod}_g$, and let H be a subgroup of Mod_g. We say that π *has a section over H* if there exists a homomorphism $\sigma : \mathrm{Mod}_g \to \mathrm{Homeo}^+(\Sigma_g)$ so that $\pi \circ \sigma = \mathrm{Id}$. This means precisely that H has a section precisely when it can be realized as a group of homeomorphisms, not just a group of homotopy classes of homeomorphisms. The general problem is then the following.

PROBLEM 6.5 (The sections problem). *Determine those subgroups $H \leq \mathrm{Mod}_g$ for which π has a section over H. Do this as well with $\mathrm{Homeo}^+(\Sigma_g)$ replaced by various subgroups, such as $\mathrm{Diff}^r(S)$ with $r = 1, 2, \ldots, \infty, \omega$; similarly for the group of area-preserving diffeomorphisms, quasiconformal homeomorphisms, etc..*

Answers to Problem 6.5 are known in a number of cases.

(1) When H is free then sections clearly always exist over H.
(2) Sections to π exist over free abelian H, even when restricted to $\mathrm{Diff}^\infty(\Sigma_g)$. This is not hard to prove, given the classification by Birman-Lubotzky-McCarthy [**BLM**] of abelian subgroups of Mod_g.
(3) Sections exist over any finite group $H < \mathrm{Mod}_g$, even when restricted to $\mathrm{Diff}^\omega(\Sigma_g)$. This follows from the Nielsen Realization Conjecture, proved by Kerckhoff [**Ke**], which states that any such H acts as a group of automorphisms of some genus g Riemann surface.
(4) In contrast, Morita showed (see, e.g., [**Mo5**]) that π does not have a section with image in $\mathrm{Diff}^2(\Sigma_2)$ over all of Mod_g when $g \geq 5$. The C^2 assumption is used in a crucial way since Morita uses a putative section to build a codimension 2 foliation on the universal curve over $\mathcal{M}_g$, to whose normal bundle he applies the Bott vanishing theorem, contradicting nonvanishing of a certain (nontrivial!) Miller-Morita-Mumford class. It seems like Morita's proof can be extended to finite index subgroups of Mod_g.
(5) Markovic [**Mar**] has recently proven that $H = \mathrm{Mod}_g$ does not even have a section into $\mathrm{Homeo}(\Sigma_g)$, answering a well-known question of Thurston.

As is usual when one studies representations of a discrete group Γ, one really desires a theorem about all finite index subgroups of Γ. One reason for this is that the existence of torsion and special relations in a group Γ often highly constrains its possible representations. Markovic's proof in [**Mar**] uses both torsion and the braid relations in what seems to be an essential way; these both disappear in most finite index subgroups of Mod_g. Thus it seems that a new idea is needed to answer the following.

QUESTION 6.6 (Sections over finite index subgroups). *Does the natural map $\mathrm{Homeo}^+(\Sigma_g) \to \mathrm{Mod}_g$ have a section over a finite index subgroup of Mod_g, or not?*

Of course the ideas in [**Mar**] are likely to be pertinent. Answers to Problem 6.5 even for specific subgroups (e.g. for $\mathcal{I}_g$ or more generally $\mathcal{I}_g(k)$) would be interesting. It also seems reasonable to believe that the existence of sections is affected greatly by the degree of smoothness one requires.

Instead of asking for *sections* in the above questions, one can ask more generally whether there are any actions of Mod_g on Σ_g.

QUESTION 6.7. *Does Mod_g or any of its finite index subgroups have any faithful action by homeomorphisms on Σ_g?*

7. Pseudo-Anosov theory

Many of the problems in this section come out of joint work with Chris Leininger and Dan Margalit, especially that in the paper [**FLM**].

7.1. Small dilatations. Every pseudo-Anosov mapping class $f \in \text{Mod}_g$ has a *dilatation* $\lambda(f) \in \mathbf{R}$. This number is an algebraic integer which records the exponential growth rate of lengths of curves under iteration of f, in any fixed metric on S; see [**Th**]. The number $\log(\lambda(f))$ equals the minimal topological entropy of any element in the homotopy class f; this minimum is realized by a pseudo-Anosov homeomorphism representing f (see [**FLP**, Exposé 10]). $\log(\lambda(f))$ is also the translation length of f as an isometry of the *Teichmüller space of S* equipped with the Teichmüller metric. Penner considered the set

$$\text{spec}(\text{Mod}_g) = \{\log(\lambda(f)) : f \in \text{Mod}_g \text{ is pseudo-Anosov}\} \subset \mathbf{R}$$

This set can be thought of as the *length spectrum* of $\mathcal{M}_g$. We can also consider, for various subgroups $H < \text{Mod}_g$, the subset $\text{spec}(H) \subset \text{spec}(\text{Mod}_g)$ obtained by restricting to pseudo-Anosov elements of H. Arnoux–Yoccoz [**AY**] and Ivanov [**Iv2**] proved that $\text{spec}(\text{Mod}_g)$ is discrete as a subset of $\mathbf{R}$. It follows that for any subgroup $H < \text{Mod}_g$, the set $\text{spec}(H)$ has a least element $L(H)$. Penner proved in [**Pe**] that

$$\frac{\log 2}{12g - 12} \leq L(\text{Mod}_g) \leq \frac{\log 11}{g}$$

In particular, as one increases the genus, there are pseudo-Anosovs with stretch factors arbitrarily close to one. In contrast to our understanding of the asymptotics of $L(\text{Mod}_g)$, we still do not know the answer to the following question, posed by McMullen.

QUESTION 7.1. *Does $\lim_{g \to \infty} gL(\text{Mod}_g)$ exist?*

Another basic open question is the following.

QUESTION 7.2. *Is the sequence $\{L(\text{Mod}_g)\}$ monotone decreasing? strictly so?*

Explicit values of $L(\text{Mod}_g)$ are known only when $g = 1$. In this case one is simply asking for the minimum value of the largest root of a polynomial as one varies over all integral polynomials $x^2 - bx + 1$ with $b \geq 3$. This is easily seen to occur when $b = 3$. For $g = 2$ Zhirov [**Zh**] found the smallest dilatation for pseudo-Anosovs with orientable foliation. It is not clear if this should equal $L(\text{Mod}_2)$ or not.

PROBLEM 7.3. *Compute $L(\text{Mod}_g)$ explicitly for small $g \geq 2$.*

In principle $L(\text{Mod}_g)$ can be computed for any given g. The point is that one can first bound the degree of $L(\text{Mod}_g)$, then give bounds on the smallest possible value $\lambda(\alpha)$, where α ranges over all algebraic integers of a fixed range of degrees, and $\lambda(\alpha)$ denotes the largest root of the minimal polynomial of α. One then finds all train tracks on Σ_g, and starts to list out all pseudo-Anosovs. It is possible to give bounds for when the dilatations of these become large. Now one tries to match up the two lists just created, to find the minimal dilatation pseudo-Anosov on Σ_g. Of course actually following out this procedure, even for small g, seems to be impracticable.

QUESTION 7.4. *Is there a unique (up to conjugacy) minimal dilation pseudo-Anosov in Mod_g?*

Note that this is true for $g = 1$; the unique minimum is realized by the conjugacy class of the matrix $\begin{pmatrix} 2 & 1 \\ 1 & 1 \end{pmatrix}$.

Here is a natural refinement of the problem of finding $L(\text{Mod}_g)$. Fix a genus g. Fix a possible r-tuple $(k_1, \ldots, k_r)$ of *singularity data* for Σ_g. By this we mean to consider possible foliations with r singularities with $k_1, \ldots, k_r$ prongs, respectively. For a fixed g, there are only finitely many possible tuples, as governed by the Poincare-Hopf index theorem. Masur-Smillie [**MS**] proved that, for every admissible tuple, there is some pseudo-Anosov on Σ_g with stable foliation having the given singularity data. Hence the following makes sense.

PROBLEM 7.5 (Shortest Teichmüller loop in a stratum). *For each fixed $g \geq 2$, and for each r-tuple as above, give upper and lower bounds for*

$$\lambda_g(k_1, \ldots, k_r) := \inf\{\log \lambda(f) : f \in \text{Mod}_g$$
whose stable foliation has data $(k_1, \ldots, k_r)\}$

This problem is asking for bounds for the shortest Teichmüller loop lying in a given substratum in moduli space (i.e. the projection in $\mathcal{M}_g$ of the corresponding substratum in the cotangent bundle).

$$L(\text{Mod}_g) = \min\{\lambda_g(k_1, \ldots, k_r)\}$$

where the min is taken of all possible singularity data.

7.2. Multiplicities. The set $\text{spec}(\text{Mod}_g)$ has unbounded multiplicity; that is, given any $N > 0$, there exists $r \in \text{spec}(\text{Mod}_g)$ such that there are at least n conjugacy classes $f_1, \ldots f_n$ of pseudo-Anosovs in Mod_g having $\log(\lambda(f_i)) = r$. The reason for this is that $\mathcal{M}_g$ contains isometrically embedded finite volume hyperbolic 2-manifolds X, e.g. the so-called *Veech curves*, and any such X has (hyperbolic) length spectrum of unbounded mulitplicity.

A related mechanism which produces length spectra with unbounded multiplicities is the *Thurston representation*. This gives, for a pair of curves a, b on Σ_g whose union fills Σ_g, an injective representation $\rho :< T_a, T_b > \to \text{PSL}(2, \mathbf{R})$ with the following properties: image(ρ) is discrete; each element of image(ρ) is either pseudo-Anosov or is a power of $\rho(T_a)$ or $\rho(T_b)$; and $\text{spec}(< T_a, T_b >)$ is essentially the length spectrum of the quotient of $\mathbf{H}^2$ by image(ρ). Again it follows that $\text{spec}(< T_a, T_b >)$ has unbounded multiplicity. Since one can find a, b as above, each of which is in addition separating, it follows that $\text{spec}(\mathcal{I}_g)$ and even $\text{spec}(\mathcal{I}_g(2))$ have unbounded multiplicity.

QUESTION 7.6. *Does* spec($\mathcal{I}_g(k)$) *have bounded multiplicity for* $k \geq 3$?

One way to get around unbounded multiplicities is to look at the *simple* length spectrum, which is the subset of spec(Mod_g) coming from pseudo-Anosovs represented by simple (i.e. non-self-intersecting) geodesic loops in $\mathcal{M}_g$.

QUESTION 7.7 (Simple length spectrum). *Does the simple length spectrum of* $\mathcal{M}_g$, *endowed with the Teichmüller metric, have bounded multiplicity? If so, how does the bound depend on* g?

Of course this question contains the corresponding question for (many) hyperbolic surfaces, which itself is still open. These questions also inspire the following.

PROBLEM 7.8. *Give an algorithm which tells whether or not any given pseudo-Anosov is represented by a simple closed Teichmüller geodesic, and also whether or not this geodesic lies on a Veech curve.*

Note that the analogue of Question 7.7 is not known for hyperbolic surfaces, although it is true for a generic set of surfaces in $\mathcal{M}_g$.

7.3. Special subgroups. In [**FLM**] we provide evidence for the principle that algebraic complexity implies dynamical complexity. A paradigm for this is the following theorem.

THEOREM 7.9 ([**FLM**]). *For* $g \geq 2$, *we have*
$$.197 < L(\mathcal{I}_g) < 4.127$$

The point is that $L(\mathcal{I}_g)$ is universally bounded above and below, independently of g. We extend this kind of universality to every term of the Johnson filtration, as follows.

THEOREM 7.10 ([**FLM**]). *Given* $k \geq 1$, *there exist* $M(k)$ *and* $m(k)$, *where* $m(k) \to \infty$ *as* $k \to \infty$, *so that*
$$m(k) < L(\mathcal{I}_g(k)) < M(k)$$
for every $g \geq 2$.

QUESTION 7.11. *Give upper and lower bounds for* $L(\mathcal{I}_g(k))$ *for all* $k \geq 2$ *which are of the same order of magnitude.*

In [**FLM**] bounds on $L(H)$ are given for various special classes of subgroups $H < \mathrm{Mod}_g$. It seems like there is much more to explore in this direction. One can also combine these types of questions with problems such as Problem 7.5.

7.4. Densities in the set of dilatations. Let P be a property which pseudo-Anosov homeomorphisms might or might not have. For example, P might be the property of lying in a fixed subgroup of $H < \mathrm{Mod}_g$, such as $H = \mathcal{I}_g$; or P might be the property of having dilatation which is an algebraic integer of a fixed degree. One can then ask the natural question: how commonly do the dilatations of elements with P arise in spec(Mod_g)?

To formalize this, recall that the *(upper) density* $d^*(A)$ of a subset A of the natural numbers $\mathbf{N}$ is defined as

$$d^*(A) := \limsup_{N \to \infty} \frac{\#A \cap [0,n]}{n}$$

This notion can clearly be extended from $\mathbf{N}$ to any countable ordered set $\mathcal{S}$ once an order-preserving bijection $\mathcal{S} \to \mathbf{N}$ is fixed.

Now fix $g \geq 2$, and recall that $\text{spec}(\text{Mod}_g) \subset \mathbf{R}^+$ is defined to be the set of (logs of) dilatations of pseudo-Anosov homeomorphisms of Mod_g. The set $\text{spec}(\text{Mod}_g)$ comes with a natural order $\lambda_1 < \lambda_2 < \cdots$, which determines a fixed bijection $\text{spec}(\text{Mod}_g) \to \mathbf{N}$. If we wish to keep track of all pseudo-Anosovs, and not just their dilatations, we can simply consider the (total) ordering on the set of all pseudo-Anosovs $\mathcal{P}_g$ given by $f \leq g$ if $\lambda(f) \leq \lambda(g)$.

QUESTION 7.12. *For various subgroups $H < \text{Mod}_g$, compute the density of $\text{spec}(H)$ in $\text{spec}(\text{Mod}_g)$ and the density of $H \cap \mathcal{P}_g$ in $\mathcal{P}_g$. In particular, what is the density of $\text{spec}(\text{Mod}_g[L])$ in $\text{spec}(\text{Mod}_g)$? What about $H = \mathcal{I}_g(k)$ with $k \geq 1$?*

References

[Ac] R. Accola, On the number of automorphisms of a closed Riemann surface, *Trans. Amer. Math. Soc.* 131 1968 398–408.

[Ak] T. Akita, Homological infiniteness of Torelli groups, *Topology*, Vol. 40 (2001), no. 2, 213–221.

[AY] P. Arnoux and J-C. Yoccoz, Construction de difféomorphismes pseudo-Anosov, *C. R. Acad. Sci. Paris* 292 (1981) 75–78.

[BC] J. Birman and R. Craggs, The μ-invariant of 3-manifolds and certain structural properties of the group of homeomorphisms of a closed, oriented 2-manifold, *Trans. Amer. Math. Soc.*, Vol. 237 (1978), pp. 283-309.

[Be] M. Belolipetsky, Estimates for the number of automorphisms of a Riemann surface, *Siberian Math. J.* 38 (1997), no. 5, 860–867.

[BF] D. Biss and B. Farb, K_g is not finitely generated, *Invent. Math.*, Vol. 163, pp. 213–226 (2006).

[BFa1] T. Brendle and B. Farb, Every mapping class group is generated by 6 involutions, *Jour. of Algebra*, Vol. 278 (2004), 187-198.

[BFa2] T. Brendle and B. Farb, The Birman-Craggs-Johnson homomorphism and abelian cycles in the Torelli group, preprint, Dec. 2005.

[BH] M. Burger and P. de la Harpe, Constructing irreducible representations of discrete groups, *Proc. Indian Acad. Sci. Math. Sci.* 107 (1997), no. 3, 223–235.

[Bi] J. Birman, *Braids, links, and mapping class groups*, Annals of Mathematics Studies, No. 82, 1974.

[BL] H. Bass and A. Lubotzky, Linear-central filtrations on groups, in *The mathematical legacy of Wilhelm Magnus: groups, geometry and special functions (Brooklyn, NY, 1992)*, 45–98, Contemp. Math., 169, AMS, 1994.

[BLM] J. Birman, A. Lubotzky, and J. McCarthy, Abelian and solvable subgroups of the mapping class group, *Duke Math. Jour.*, Vol. 50, No.4 (1983), pp.1107-1120.

[BM] T. Brendle and D. Margalit, Commensurations of the Johnson kernel, *Geom. and Top.*, Vol. 8, pp.1361–1384, 2004.

[Br] M. Bridson, Finiteness properties for subgroups of $\text{GL}(n, \mathbf{Z})$, *Math. Ann.* 317 (2000), no. 4, 629–633.

[BrF] J. Brock and B. Farb, Curvature and rank of Teichmüller space, *Amer. Jour. Math.*, Vol. 128 (2006), no. 1, 1–22.

[BS] P. Buser and P. Sarnak, On the period matrix of a Riemann surface of large genus, with an appendix by J. H. Conway and N. J. A. Sloane, *Invent. Math.* 117 (1994), no. 1, 27–56.

[BCSS] L. Blum, F. Cucker, M. Shub and S. Smale, *Complexity and real computation*, foreword by R.M. Karp, Springer-Verlag, 1998.

[BW] J. Block and S. Weinberger, Arithmetic manifolds of positive scalar curvature, *J. Differential Geom.* 52 (1999), no. 2, 375–406.

[Ca] J. Cannon, Almost convex groups, *Geom. Dedicata* 22 (1987), no. 2, 197–210.

[Cal] D. Calegari, Circular groups, planar groups, and the Euler class, *Geom. and Top. Monographs*, Vol. 7:Proc. of the Casson Fest, 431–491.

[CB]   A. Casson and S. Bleiler, *Automorphisms of surfaces after Nielsen and Thurston*, London Math. Soc. Student Texts, Vol. 9, 1988.
[CF]   J. Crisp and B. Farb, The prevalence of surface subgroups in mapping class groups, in preparation.
[Ch]   S. Chang, Coarse obstructions to positive scalar curvature in noncompact arithmetic manifolds, *J. Differential Geom.* 57 (2001), no. 1, 1–21.
[D]   O. Debarre, The Schottky problem: an update, in *Current topics in complex algebraic geometry* (Berkeley, CA, 1992/93), 57–64, MSRI Publ., Vol. 28, 1995.
[De]   P. Deligne, Extensions centrales non résiduellement finies de groupes arithmétiques, *C. R. Acad. Sci. Paris* Sr. A-B, 287 (1978), no. 4, A203–A208.
[DW]   G. Daskalopoulos and R. Wentworth, Harmonic maps and Teichmüller theory, preprint, April 2006.
[EMO] A. Eskin, S. Mozes and H. Oh, On uniform exponential growth for linear groups, *Invent. Math.*, Vol. 160 (2005), no. 1, 1–30.
[Fa1] B. Farb, The Extrinsic geometry of subgroups and the generalized word problem, *Proc. London Math. Soc.* (3) 68 (1994), 577-593.
[Fa2] B. Farb, Relatively hyperbolic groups, GAFA, *Geom. Funct. Anal.*, Vol. 8 (1998), pp.1-31.
[Fa3] B. Farb, The Torelli group, Kahler groups, and normal subgroups of mapping class groups, in preparation.
[Fa4] B. Farb, Group actions and Helly's theorem, in preparation.
[FaM] B. Farb and H. Masur, Superrigidity and mapping class groups, *Topology*, Vol. 37, No.6 (1998), 1169-1176.
[FF]   B. Farb and J. Franks, Groups of homeomorphisms of one-manifolds, I: Actions of non-linear groups, preprint, 2001.
[FI]   B. Farb and N. Ivanov, The Torelli geometry and its applications, *Math. Res. Lett.*, 12 (2005), no. 2-3, 293–301.
[FLM] B. Farb, C. Leininger, and D. Margalit, The lower central series and pseudo-Anosov dilatations, preprints, February 2006.
[FLMi] B. Farb, A. Lubotzky and Y. Minksy, Rank one phenomena for mapping class groups, *Duke Math. Jour.*, Vol. 106, No.3 (2001), 581–597.
[FLP] *Travaux de Thurston sur les surfaces*, volume 66 of *Astérisque*. Société Mathématique de France, Paris, 1979. Séminaire Orsay.
[FMa] B. Farb and D. Margalit, *A primer on mapping class groups*, book in preparation, currently 200 pages, available at http://www.math.utah.edu/∼margalit/ma7853/primer.html.
[FMo] B. Farb and L. Mosher, Convex cocompact subgroups of mapping class groups, *Geom. and Top.*, Vol. 6 (2002), 91-152.
[Fo]   J. Foisy, The second homology group of the level 2 mapping class group and extended Torelli group of an orientable surface, *Topology*, Vol. 38, No. 6, 1175-1207 (1999).
[For] E. Formanek, Braid group representations of low degree, *Proc. London Math. Soc.* (3) 73 (1996), no. 2, 279–322.
[FW1] B. Farb and S. Weinberger, Isometries, rigidity and universal covers, preprint, June 2005.
[FW2] B. Farb and S. Weinberger, Positive scalar curvature metrics on the moduli space of Riemann surfaces, in preparation.
[GH]   G. Gonzalez-Diez and W. Harvey, Surface subgroups inside mapping class groups, *Topology*, Vol. 38, No. 1. 57–69, 1999.
[Gr]   S. Grushevsky, Effective algebraic Schottky problem, preprint, June 2002.
[Gro] M. Gromov, Hyperbolic groups, in *Essays in group theory*, pp. 75–263, Math. Sci. Res. Inst. Publ., 8, Springer, 1987.
[GMV] F. Grunewald, J. Mennicke, and L. Vaserstein, On the groups $SL_2(Z[x])$ and $SL_2(k[x,y])$, *Israel J. Math.* 86 (1994), no. 1-3, 157–193.
[Ha1] R. Hain, Locally symmetric families of curves and Jacobians, in *Moduli of curves and abelian varieties*, 91–108, Aspects Math., E33, 1999.
[Ha2] R. Hain, Torelli groups and geometry of moduli spaces of curves, in *Current topics in complex algebraic geometry (Berkeley, CA, 1992/93)*, 97–143, MSRI Publ., 28, 1995.
[Ha3] R. Hain, Infinitesimal presentations of the Torelli groups, *Jour. Amer. Math. Soc.*, Vol. 10, No. 3 (1997), pp. 597-651.

[Ham]   U. Hamenstadt, Geometry of the mapping class group II: subsurfaces, preprint, Dec. 2005.
[Har1] J. Harer, The virtual cohomological dimension of the mapping class group of an orientable surface, *Invent. Math.* 84 (1986), no. 1, 157–176.
[Har2] J. Harer, The second homology group of the mapping class group of an orientable surface, *Invent. Math.*, Vol. 72 (1983), pp. 221-239.
[Harp1] P. de la Harpe, Uniform growth in groups of exponential growth, in Proc. Conf. on Geometric and Combinatorial Group Theory, Part II (Haifa, 2000), *Geom. Dedicata* 95 (2002), 1–17.
[Harp2] P. de la Harpe, *Topics in geometric group theory*, Chicago Lectures in Mathematics, Univ. of Chicago Press, 2000.
[Hat]   T. Hattori, Asymptotic geometry of arithmetic quotients of symmetric spaces, *Math. Zeit.* 222 (1996), no. 2, 247–277.
[He]   G. Hemion, On the classification of homeomorphisms of 2-manifolds and the classification of 3-manifolds, *Acta Math.* 142 (1979), no. 1-2, 123–155.
[HL]   R. Hain and E. Looijenga, Mapping class groups and moduli spaces of curves, in *Algebraic geometry—Santa Cruz 1995*, pp. 97–142, Proc. Sympos. Pure Math., 62, Part 2, AMS, 1997.
[Iv1]   N.V. Ivanov, Mapping class groups, in *Handbook of geometric topology*, 523–633, 2002.
[Iv2]   N. V. Ivanov, Coefficients of expansion of pseudo-Anosov homeomorphisms, *J. Soviet Math.* 52 (1990), no. 1, 2822–2824.
[Iv3]   N. Ivanov, Subgroups of Teichmüller modular groups, translated from the Russian by E. J. F. Primrose and revised by the author, Translations of Math. Monographs, 115, AMS, 1992.
[Jo1]   D. Johnson, A survey of the Torelli group, *Contemp. Math.*, Vol. 20 (1983), 165-17.
[Jo2]   D. Johnson, The structure of the Torelli group I: A finite set of generators for $\mathcal{I}$, *Annals of Math.*, Vol. 118, No. 3 (1983), pp. 423-442.
[Jo3]   D. Johnson, The structure of the Torelli group II: A characterization of the group generated by twists on bounding curves, *Topology*, Vol. 24, No. 2 (1985), pp. 113-126.
[Jo4]   D. Johnson, The structure of the Torelli group III: the abelianization of $\mathcal{I}$, *Topology*, Vol. 24, No. 2 (1985), pp. 127-144.
[Jo5]   D. Johnson, Quadratic forms and the Birman-Craggs homomorphisms, *Trans. Amer. Math. Soc.*, Vol. 261, No. 1 (1980), pp. 423-422.
[Ka]   M. Kassabov, Generating mapping class groups by involutions, *Trans. Amer. Math. Soc.*, to appear.
[Ke]   S. Kerckhoff, The Nielsen Realization Problem, *Annals of Math.* 117 (1983), pp. 235-265.
[Ko]   M. Korkmaz, Low-dimensional homology groups of mapping class groups: a survey, in Proc. 8th Gokova Geom.-Top. Conf., *Turkish Jour.Math.*, Vol. 26, no. 1 (2002), 101–114.
[Ku]   R. Kulkarni, Riemann surfaces admitting large automorphism groups, in *Extremal Riemann surfaces*, Contemp. Math., Vol. 201, 63–79.
[La]   M. Larsen, How often is $84(g-1)$ achieved?, *Israel J. Math. 126* (2001), 1–16.
[LM]   C. Leininger and D.B. McReynolds, Separable subgroups of mapping class groups, *Topology Applic.*, to appear.
[LR]   C. Leininger and A. Reid, A combination theorem for Veech subgroups of the mapping class group, *Geom. Funct. Anal.* (GAFA), to appear.
[LSY]   K. Liu, X. Sun, and S-T. Yau, Geometric aspects of the moduli space of Riemann surfaces, *Sci. China Ser. A*, Vol. 48 (2005), suppl., 97–122.
[Mah]   J. Maher, Random walks on the mapping class group, preprint, Dec. 2005.
[Ma]   G.A. Margulis, *Discrete subgroups of semisimple Lie groups*, Springer-Verlag, 1990.
[Mar]   V. Markovic, Realization of the mapping class group by homeomorphisms, preprint, 2005.
[Maz]   B. Mazur, It is a story, preprint.
[Mc1]   C. McMullen, From dynamics on surfaces to rational points on curves, *Bulletin AMS*, Vol. 37, No. 2, 119–140 (1999).
[Mc2]   C. McMullen, The moduli space of Riemann surfaces is Kahler hyperbolic, *Annals of Math.* (2), 151(1):327–357, 2000.
[McM]   D. McCullough and A. Miller, The genus 2 Torelli group is not finitely generated, *Topology Appl.* 22 (1986), no. 1, 43–49.
[Me]   G. Mess, The Torelli group for genus 2 and 3 surfaces, *Topology* 31 (1992), pp. 775-790.

[Mi] C.F. Miller III, Decision problems for groups—survey and reflections, in *Algorithms and classification in combinatorial group theory* (Berkeley, CA, 1989), 1–59, Math. Sci. Res. Inst. Publ., Vol. 23, Springer, 1992.

[Mir] M. Mirzakhani, Simple geodesics and Weil-Petersson volumes of moduli spaces of bordered Riemann surfaces, *Annals of Math.*, to appear.

[MM1] H. Masur and M. Minsky, Geometry of the complex of curves I: Hyperbolicity, *Invent. Math.* 138 (1999), no. 1, 103–149.

[MM2] H. Masur and M. Minsky, Geometry of the complex of curves II: Hierarchical structure, GAFA, *Geom. Funct. Anal.*, Vol. 10 (2000), 902–974.

[Mo1] S. Morita, Mapping class groups of surfaces and three-dimensional manifolds, *Proc. of the International Congress of Mathematicians*, Vol. I, II (Kyoto, 1990), 665–674, Math. Soc. Japan, Tokyo, 1991.

[Mo2] S. Morita, Casson's invariant for homology 3-spheres and characteristic classes of surface bundles. I. *Topology* 28 (1989), no. 3, 305–323.

[Mo3] S. Morita, Structure of the mapping class groups of surfaces: a survey and a prospect. Proceedings of the Kirbyfest (Berkeley, CA, 1998), 349–406, *Geom. Topol. Monogr.* 2, 1999.

[Mo4] S. Morita, Structure of the mapping class group and symplectic representation theory, in *Essays on geometry and related topics*, Vol. 1, 2, 577–596, Monogr. Enseign. Math., 38, 2001.

[Mo5] S. Morita, *Geometry of characteristic classes*, AMS Translations of Math. Monographs, Vol. 199, 2001.

[Mor] I. Morrison, Subvarieties of moduli spaces of curves: open problems from an algebro-geometric point of view, in *Mapping class groups and moduli spaces of Riemann surfaces (Gttingen, 1991/Seattle, WA, 1991)*, pp. 317–343, Contemp. Math., 150, AMS, 1993.

[Mos] L. Mosher, Mapping class groups are automatic, *Annals of Math.* (2) 142 (1995), no. 2, 303–384.

[MS] H. Masur and J. Smillie, Quadratic differentials with prescribed singularities and pseudo-Anosov diffeomorphisms, *Comment. Math. Helv.* 68 (1993), no. 2, 289–307.

[MV] J.D. McCarthy and W. Vautaw, Automorphisms of Torelli groups, preprint, Nov. 2003.

[MW] I. Madsen and M. Weiss, The stable moduli space of Riemann surfaces: Mumford's conjecture, *Annals of Math.*, to appear.

[Ni] J. Nielsen, *Om Regning med ikke Kommutative Faktorer og dens Anvendelse i Gruppeteorien*, Matematisk Tidsskrift B (1921), pp. 77-94. English translation: *Math. Scientist* **6**, (1981), pp. 73-85.

[Pa] L. Paris, Actions and irreducible representations of the mapping class group, *Math. Ann.* 322 (2002), no. 2, 301–315.

[Pe] R. Penner, Bounds on least dilatations, *Proc. Amer. Math. Soc.* 113 (1991) 443–450.

[Pi] W. Pitsch, Un calcul élémentaire de $H_2(\mathcal{M}_{g,1}, \mathbf{Z})$ pour $g \geq 4$, *C.R. Acad. Sci. Paris.*, t. 329, Série I, 667-670 (1999).

[Ri] I. Rivin, Counting reducible matrices, preprint, April 2006.

[Ro] H.L. Royden, Automorphisms and isometries of Teichmüller spaces, in *Advances in the theory of Riemann surfaces*, Ann. of Math. Studies 66 (1971), pp. 369-383.

[Sa] T. Sakasai, The Johnson homomorphism and the third rational cohomology group of the Torelli group, *Topology Appl.* 148 (2005), no. 1-3, 83–111.

[Sc] P. Scott, Ends of pairs of groups, *J. Pure Appl. Algebra* 11 (1977/78), no. 1-3, 179–198.

[Se] J.P. Serre, *Lie algebras and Lie groups*, 1964 lectures given at Harvard University, Lecture Notes in Math., Vol. 1500, Springer-Verlag, 1992.

[Su] M. Suzuki, The Magnus representation of the Torelli group $\mathcal{I}_{g,1}$ is not faithful for $g \geq 2$, *Proc. Amer. Math. Soc.* 130 (2002), no. 3, 909–914.

[Su2] M. Suzuki, A class function on the Torelli group, *Kodai Math. J.* 26 (2003), no. 3, 304–316.

[Th] W.P. Thurston, On the geometry and dynamics of diffeomorphisms of surfaces, *Bull. Amer. Math. Soc. (N.S.)*, 19(2):417–431, 1988.

[To] D. Toledo, Examples of fundamental groups of compact Kahler manifolds, *Bull. London Math. Soc.* 22 (1990), 339–343.

[vdB] B. van den Berg, On the abelianization of the Torelli group, thesis, Universiteit Utrecht, 2003.

[Ve] R. Veliche, Genus 2 mapping class groups are not Kahler, preprint, March 2004.
[Wa] B. Wajnryb, Mapping class group of a surface is generated by two elements, *Topology*, Vol. 35, No. 2 (1996), pp. 377-383.
[Wh] K. Whittlesey, Normal all pseudo-Anosov subgroups of mapping class groups, *Geom. and Top.*, Vol. 4 (2000), 293–307.
[Wi] D. Witte, Arithmetic groups of higher **Q**-rank cannot act on 1-manifolds, *Proc. Amer. Math. Soc.*, Vol.122, No.2, Oct. 1994, pp.333-340.
[Ye] S. Yeung, Representations of semisimple lattices in mapping class groups, *Int. Math. Res. Not.* 2003, no. 31, 1677–1686.
[Zh] A. Yu Zhirov, On the minimum dilation of pseudo-Anosov diffeomorphisms of a double torus, translation in Russian Math. Surveys 50 (1995), no. 1, 223–224.
[Zo] R. Zomorrodian, Nilpotent automorphism groups of Riemann surfaces, *Trans. Amer. Math. Soc.*, Vol. 288, No. 1 (1985), 241–255.

DEPT. OF MATHEMATICS, UNIVERSITY OF CHICAGO, 5734 UNIVERSITY AVE., CHICAGO, IL 60637

E-mail address: farb@math.uchicago.edu

Finiteness and Torelli Spaces

Richard Hain

Torelli space $\mathcal{T}_g$ ($g \geq 2$) is the quotient of Teichmüller space by the Torelli group T_g. It is the moduli space of compact, smooth genus g curves C together with a symplectic basis of $H_1(C;\mathbb{Z})$ and is a model of the classifying space of T_g. Mess, in his thesis [12], proved that $\mathcal{T}_2$ has the homotopy type of a bouquet of a countable number of circles. Johnson and Millson (cf. [12]) pointed out that a similar argument shows that $H_3(\mathcal{T}_3)$ is of infinite rank. Akita [2] used an indirect argument to prove that $\mathcal{T}_g$ does not have the homotopy type of a finite complex for (almost) all $g \geq 2$. However, the infinite topology of $\mathcal{T}_g$ is not well understood. The results of Mess and Johnson-Millson are the only ones I know of that explicitly describe some infinite topology of any Torelli space. Moreover, although Johnson [10] proved that T_g is finitely generated when $g \geq 3$, there is not one $g \geq 3$ for which it is known whether T_g is finitely presented or not.

The goal of this note is to present a suite of problems designed to probe the infinite topology of Torelli spaces in all genera. These are presented in the fourth section of the paper. The second and third sections present background material needed in the discussion of the problems.

To create a context for these problems, we first review the arguments of Mess and Johnson-Millson. As explained in Section 2, Torelli space in genus 2 is the complement of a countable number of disjoint smooth divisors D_α (i.e., codimension 1 complex subvarieties) in $\mathfrak{h}_2$, the Siegel upper half plane of rank 2. More precisely,

$$\mathcal{T}_2 = \mathfrak{h}_2 - \bigcup_{\phi \in \mathrm{Sp}_2(\mathbb{Z})} \phi(\mathfrak{h}_1 \times \mathfrak{h}_1)$$

where $\mathrm{Sp}_g(R)$ denotes the automorphisms of R^{2g} that fix the standard symplectic inner product.[1] One can now argue, as Mess did [12], that $\mathcal{T}_2$ is homotopy equivalent to a countable wedge of circles. Such an explicit description of $\mathcal{T}_g$ is unlikely in higher genus due to the increasing complexity of the image of the period mapping and the lack of an explicit description of its closure. So it should be fruitful to ponder the source of the infinite rank homology without appeal to this explicit description of $\mathcal{T}_2$. To this end, consider the exact sequence

$$\cdots \to H^4_c(\mathfrak{h}_2) \to H^4_c(D) \to H^5_c(\mathcal{T}_2) \to H^5_c(\mathfrak{h}_2) \to \cdots$$

Supported in part by grants from the National Science Foundation.

[1] Note that $\mathrm{Sp}_1 = \mathrm{SL}_2$.

©2006 American Mathematical Society

where $D = \cup D_\alpha$ is a locally finite union of smooth divisors indexed by $\mathrm{Sp}_2(\mathbb{Z})$ mod the stabilizer $S_2 \ltimes \mathrm{Sp}_1(\mathbb{Z})^2$ of $\mathfrak{h}_1 \times \mathfrak{h}_1$.[2] Since $\mathfrak{h}_2$ is a contractible complex manifold of real dimension 6,
$$H_c^k(\mathfrak{h}_2) = H_{6-k}(\mathfrak{h}_2) = 0 \text{ when } k < 6.$$
Thus
$$H_1(\mathcal{T}_2) = H_c^5(\mathcal{T}_2) \cong H_c^4(D) \cong \bigoplus_{\alpha \in \mathrm{Sp}_2(\mathbb{Z})/S_2 \ltimes \mathrm{Sp}_1(\mathbb{Z})^2} \mathbb{Z}[D_\alpha],$$
a free abelian group of countable rank.

The situation is genus 3 is similar, but slightly more complicated. In this case, the period mapping $\mathcal{T}_3 \to \mathfrak{h}_3$ is two-to-one, branched along the locus of hyperelliptic curves. The image is the complement $\mathcal{J}_3$ of a countable union Z of submanifolds Z_α of complex codimension 2:
$$\mathcal{J}_3 = \mathfrak{h}_3 - Z = \mathfrak{h}_3 - \bigcup_{\phi \in \mathrm{Sp}_3(\mathbb{Z})/\mathrm{Sp}_1(\mathbb{Z}) \times \mathrm{Sp}_2(\mathbb{Z})} \phi(\mathfrak{h}_1 \times \mathfrak{h}_2);$$
it is the set of framed jacobians of smooth genus 3 curves. By elementary topology,
$$H_\bullet(\mathcal{J}_3; \mathbb{Q}) \cong H_\bullet(\mathcal{T}_3; \mathbb{Q})^{\mathbb{Z}/2\mathbb{Z}}$$
where $\mathbb{Z}/2\mathbb{Z}$ is the group of automorphisms of the ramified covering $\mathcal{T}_3 \to \mathcal{J}_3$. If $H_k(\mathcal{J}_3; \mathbb{Q})$ is infinite dimensional, then so is $H_k(\mathcal{T}_3; \mathbb{Q})$. As in the genus 2, we have an exact sequence
$$0 = H_c^8(\mathfrak{h}_3) \to H_c^8(Z) \to H_c^9(\mathcal{J}_3) \to H_c^9(\mathfrak{h}_3) = 0$$
so that
$$H_3(\mathcal{J}_3) = H_c^9(\mathcal{J}_3) \cong H_c^8(Z) = \bigoplus_{\alpha \in \mathrm{Sp}_3(\mathbb{Z})/\mathrm{Sp}_1(\mathbb{Z}) \times \mathrm{Sp}_2(\mathbb{Z})} \mathbb{Z}[Z_\alpha].$$

Three important ingredients in these arguments are:
- $\mathcal{T}_g$ or a space very closely related to it (e.g., $\mathcal{J}_g$) is the complement $X - Z$ in a manifold X of a countable union $Z = \cup Z_\alpha$ of smooth subvarieties;
- The topology of the manifold X is very simple — in the cases above, $X = \mathfrak{h}_g$, which is contractible;
- the "topology at infinity" of X is very simple — in the cases above the boundary of $X = \mathfrak{h}_g$ is a sphere;
- the topology of X, Z and $\mathcal{T}_g$ are related by a Gysin sequence of compactly supported cohomology, and the relationship between the compactly supported cohomology and ordinary cohomology is entwined with the "topology at infinity" of X.

What is special in genus 2 and 3 is that the closure $\mathcal{J}_g^c$ of the image of the period mapping $\mathcal{T}_g \to \mathfrak{h}_g$ is very simple — it is all of $\mathfrak{h}_g$, a topological ball, which is contractible and has very simple topology at infinity. This fails to be true in genus 4, where $\mathcal{J}_g^c$ is a singular subvariety of complex codimension 1. Here, even though there is an explicit equation for the $\mathcal{J}_4^c$ in $\mathfrak{h}_4$, we do not understand its topology. The situation only gets worse in higher genus.

---

[2] The symmetric group on 2 letters, S_2, acts on $\mathfrak{h}_1 \times \mathfrak{h}_1$ by swapping the two factors.

1. Preliminaries

Suppose that $2g - 2 + n > 0$. Fix a compact oriented surfaced S of genus g and a finite subset P of n distinct points in S. The corresponding mapping class group is
$$\Gamma_{g,n} = \pi_0 \operatorname{Diff}^+(S, P).$$
By a complex curve, or a curve for short, we shall mean a Riemann surface. Denote the Teichmüller space of marked, n-pointed, compact genus g curves by $\mathcal{X}_{g,n}$. As a set $\mathcal{X}_{g,n}$ is
$$\left\{ \begin{array}{l} \text{orientation preserving diffeomorph-} \\ \text{isms } f : S \to C \text{ to a complex curve} \end{array} \right\} \Big/ \text{isotopies, constant on } P.$$
This is a complex manifold of dimension $3g - 3 + n$.

The mapping class group $\Gamma_{g,n}$ acts properly discontinuously on $\mathcal{X}_{g,n}$. The quotient $\Gamma_{g,n} \backslash \mathcal{X}_{g,n}$ is the moduli space $\mathcal{M}_{g,n}$ of n-pointed curves of genus g.

Set $H_R = H_1(S; R)$. The intersection pairing $H_R^{\otimes 2} \to R$ is skew symmetric and unimodular. Set
$$\operatorname{Sp}(H_R) = \operatorname{Aut}(H_R, \text{intersection pairing}).$$
The choice of a symplectic basis of H_R gives an isomorphism $\operatorname{Sp}(H_R) \cong \operatorname{Sp}_g(R)$ of $2g \times 2g$ symplectic matrices with entries in R. The action of $\Gamma_{g,n}$ on S induces a homomorphism
$$\rho : \Gamma_{g,n} \to \operatorname{Sp}(H_{\mathbb{Z}})$$
which is well-known to be surjective.

The *Torelli group* $T_{g,n}$ is defined to be the kernel of ρ. It is torsion free. The Torelli space $\mathcal{T}_{g,n}$ is defined by
$$\mathcal{T}_{g,n} = T_{g,n} \backslash \mathcal{X}_{g,n}.$$
It is the moduli space of n-pointed Riemann surfaces $(C; x_1, \ldots, x_n)$ of genus g together with a symplectic basis of $H_1(C; \mathbb{Z})$. Since $T_{g,n}$ is torsion free, it acts freely on Teichmüller space. Torelli space $\mathcal{T}_{g,n}$ is thus a model of the classifying space $BT_{g,n}$. Note that $T_{1,1}$ is trivial and that $\mathcal{X}_{1,1} = \mathcal{T}_{1,1}$ is just the upper half plane, $\mathfrak{h}_1$. Torelli spaces also exist in genus 0 provided $n \geq 3$. In this case, $\mathcal{T}_{0,n} = \mathcal{M}_{0,n}$.

The Siegel upper half space of rank g
$$\mathfrak{h}_g := \operatorname{Sp}_g(\mathbb{R})/U(g) \cong \{\Omega \in \mathcal{M}_g(\mathbb{C}) : \Omega = \Omega^T, \operatorname{Im}\Omega > 0\}$$
is the symmetric space of $\operatorname{Sp}_g(\mathbb{R})$. It is a complex manifold of dimension $g(g+1)/2$. It can usefully be regarded as the moduli space of g-dimensional principally polarized abelian varieties (A, θ) together with a symplectic (with respect to the polarization $\theta : H_1(A; \mathbb{Z})^{\otimes 2} \to \mathbb{Z}$) basis of $H_1(A; \mathbb{Z})$. The group $\operatorname{Sp}_g(\mathbb{Z})$ acts on the framings, and the quotient $\operatorname{Sp}_g(\mathbb{Z}) \backslash \mathfrak{h}_g$ is the moduli space $\mathcal{A}_g$ of principally polarized abelian varieties of dimension g.

The decoration n in $\Gamma_{g,n}, T_{g,n}, \mathcal{T}_{g,n}$, etc will be omitted when it is zero.

2. Geography

2.1. The period mapping. A *framed* genus g curve is a compact Riemann surface of genus g together with a symplectic basis $(a_1, \ldots, a_g, b_1, \ldots, b_g)$ of $H_1(C; \mathbb{Z})$.

For each framed curve, there is a unique basis $w_1, \ldots, w_g$ of the holomorphic differentials $H^0(C, \Omega_C^1)$ on C such that
$$\int_{a_j} w_k = \delta_{j,k}.$$
The period matrix of $(C; a_1, \ldots, a_g, b_1, \ldots, b_g)$ is the $g \times g$ matrix
$$\Omega = \left(\int_{b_j} w_k \right).$$
It is symmetric and has positive definite imaginary part.

As remarked above, $\mathcal{T}_g$ is the moduli space of framed genus g curves. The *period mapping*
$$\mathcal{T}_g \to \mathfrak{h}_g$$
takes a framed curve to its period matrix. It is holomorphic and descends to the mapping
$$\mathcal{M}_g = \mathrm{Sp}_g(\mathbb{Z}) \backslash \mathcal{T}_g \to \mathrm{Sp}_g(\mathbb{Z}) \backslash \mathfrak{h}_g = \mathcal{A}_g$$
that takes the point $[C]$ of $\mathcal{M}_g$ corresponding to a curve C to the point $[\mathrm{Jac}\, C]$ of $\mathcal{A}_g$ corresponding to its jacobian.

2.2. The jacobian locus. This is defined to be the image $\mathcal{J}_g$ of the period mapping $\mathcal{T}_g \to \mathfrak{h}_g$. It is a locally closed subvariety of $\mathfrak{h}_g$. It is important to note, however, that it is *not* closed in $\mathfrak{h}_g$.

To explain this, we need to introduce the locus of *reducible principally polarized abelian varieties*. This is
$$\mathcal{A}_g^{\mathrm{red}} = \bigcup_{h=1}^{\lfloor g/2 \rfloor} \mathrm{im}\{\mu_h : \mathcal{A}_h \times \mathcal{A}_{g-h} \to \mathcal{A}_g\}$$
where $\mu_h(A, B) = A \times B$.[3] Set
$$\mathfrak{h}_g^{\mathrm{red}} = \text{inverse image in } \mathfrak{h}_g \text{ of } \mathcal{A}_g^{\mathrm{red}} = \bigcup_{h=0}^{\lfloor g/2 \rfloor} \bigcup_{\phi \in \mathrm{Sp}_g(\mathbb{Z}) / \mathrm{Sp}_g(\mathbb{Z})_{\mathfrak{h}_h \times \mathfrak{h}_{g-h}}} \phi(\mathfrak{h}_h \times \mathfrak{h}_{g-h})$$
where $\mathfrak{h}_h \times \mathfrak{h}_{g-h}$ is imbedded into $\mathfrak{h}_g$ by taking (Ω_1, Ω_2) to the matrix $\Omega_1 \oplus \Omega_2$; the group $Sp_g(\mathbb{Z})_{\mathfrak{h}_h \times \mathfrak{h}_{g-h}}$ denotes the subgroup of $\mathrm{Sp}_g(\mathbb{Z})$ that fixes it set-wise.

For a subset N of $\mathfrak{h}_g$, set $N^{\mathrm{red}} = N \cap \mathfrak{h}_g^{\mathrm{red}}$. By [8, Prop. 6],
$$\mathcal{J}_g = \mathcal{J}_g^c - \mathcal{J}_g^{c,\mathrm{red}}$$
from which it follows that
$$\mathcal{J}_g^{c,\mathrm{red}} = \bigcup_{\phi \in \mathrm{Sp}_g(\mathbb{Z})} \bigcup_{h=1}^{\lfloor g/2 \rfloor} \phi\bigl(\mathcal{J}_h^{c,\mathrm{red}} \times \mathcal{J}_{g-h}^{c,\mathrm{red}}\bigr).$$

---

[3] Note that $\mathrm{im}\, \mu_h = \mathrm{im}\, \mu_{g-h}$. This is why we need only those h between 1 and $\lfloor g/2 \rfloor$.

2.3. Curves of compact type.

A genus g curve C of *compact type* is a connected, compact, nodal curve[4] satisfying:

(i) the dual graph of C is a tree — this guarantees that the jacobian $\operatorname{Jac} C$ of C is a principally polarized abelian variety;

(ii) the sum of the genera of the components of C is g.

These are precisely the stable curves of genus g whose generalized jacobian is compact. The generalized jacobian a genus g curve of compact type is the product of the jacobians of its components. It is an abelian variety of dimension g.

An n-pointed nodal curve of genus g is a nodal genus g curve C together with n labeled points in the smooth locus of C. An n-pointed nodal curve (C, P) of genus g is *stable* if its automorphism group is finite. This is equivalent to the condition that each connected component of

$$C - (C^{\mathrm{sing}} \cup P)$$

has negative Euler characteristic.

Using the deformation theory of stable curves, one can enlarge $\mathcal{T}_{g,n}$ to the moduli space $\mathcal{T}_{g,n}^c$ of framed stable n-pointed genus g curves of compact type. This is a complex manifold that contains $\mathcal{T}_g$ as a dense open subset and on which $\operatorname{Sp}_g(\mathbb{Z})$ acts (via its action on framings). The quotient $\operatorname{Sp}_g(\mathbb{Z})\backslash\mathcal{T}_{g,n}^c$ is the space $\mathcal{M}_{g,n}^c$ of stable n-pointed, genus g curves of compact type.

Note that when $n \geq 3$, $\mathcal{T}_{0,n}^c = \overline{\mathcal{M}}_{0,n}$, the moduli space of stable n-pointed curves of genus 0.

PROPOSITION 1. *If $2g - 2 + n > 0$, then $\mathcal{T}_{g,n}^c$ is a smooth complex analytic variety of complex dimension $3g - 3 + n$ and*

$$\mathcal{T}_{g,n} = \mathcal{T}_{g,n}^c - Z$$

where Z is a countable union of transversally intersecting smooth divisors.

The strata of Z of complex codimension k are indexed by the k-simplices of the quotient $T_{g,n}\backslash K^{\mathrm{sep}}(S, P)$ of the complex of separating curves $K^{\mathrm{sep}}(S, P)$ of the n-pointed reference surface (S, P) by the Torelli group. This correspondence will be made more explicit in the next paragraph.

Since $\mathcal{T}_{g,n} = \mathcal{T}_{g,n}^c - Z$, where Z has complex codimension 1, the mapping

$$\pi_1(\mathcal{T}_{g,n}, *) \to \pi_1(\mathcal{T}_{g,n}^c, *)$$

is surjective. Its kernel is generated by the conjugacy classes of small loops about each of the components of Z. But these are precisely the conjugacy classes of Dehn twists on separating simple closed curves (SCCs).

PROPOSITION 2. *For all (g, n) satisfying $2g - 2 + n > 0$,*

$$\pi_1(\mathcal{T}_{g,n}^c, *) \cong T_{g,n}/\{\text{subgroup generated by Dehn twists on separating SCCs}\}.$$

As in the previous section, we set $H_{\mathbb{Z}} = H_1(S; \mathbb{Z})$, where S is the genus g reference surface. Denote the image of $u \in \Lambda^3 H_{\mathbb{Z}}$ under the quotient mapping

$$\Lambda^3 H_{\mathbb{Z}} \to (\Lambda^3 H_{\mathbb{Z}})/(\theta \wedge H_{\mathbb{Z}})$$

by $\overline{u}$.

---

[4] A nodal curve is a complex analytic curve, all of whose singularities are nodes — that is, of the form $zw = t$.

The surjectivity of the "Johnson homomorphism" $\tau : T_{g,1} \to \Lambda^3 H_\mathbb{Z}$ and Johnson's result [**11**] that its kernel is generated by Dehn twists on separating SCCs implies quite directly that:

COROLLARY 3. *If $g \geq 1$ and $2g - 2 + n > 0$, then*
$$\pi_1(\mathcal{T}_{g,n}^c, *) = H_1(\mathcal{T}_{g,n}^c; \mathbb{Z}) \cong \{(u_1, \ldots, u_n) \in (\Lambda^3 H_\mathbb{Z})^n : \overline{u}_1 = \cdots = \overline{u}_n\},$$
which is a torsion free abelian group.

2.4. The complex of separating curves, $K^{\text{sep}}(S, P)$. As above S is a compact oriented surface of genus g and P is a subset of cardinality n, where $2g - 2 + n > 0$. A simple closed curve γ in $S - P$ is separating if $S - (P \cup \gamma)$ is not connected. An SCC is *cuspidal* if it bounds a once-punctured disk in $S - P$ and *trivial* if it bounds a disk.

The simplicial complex $K^{\text{sep}}(S, P)$ has vertices the isotopy classes of non-trivial separating SCCs γ in $S - P$ that are not cuspidal. The isotopy classes of the SCCs $\gamma_0, \ldots, \gamma_k$ of non-cuspidal separating SCCs span a k-simplex of $K^{\text{sep}}(S, P)$ if they are disjoint and lie in distinct isotopy classes. When P is empty, we shall abbreviate $K^{\text{sep}}(S, P)$ by $K^{\text{sep}}(S)$.

The correspondence between $T_{g,n}$ orbits of $K^{\text{sep}}(S, P)$ and the strata of $\mathcal{T}_{g,n}^{c,\text{red}}$ is given as follows. Given a k-simplex $\vec{\gamma} = (\gamma_0, \ldots, \gamma_k)$ of $K^{\text{sep}}(S, P)$, one can contract each of the SCCs γ_j. The resulting space $(S/\vec{\gamma}, P)$ is the topological model of a stable, n-pointed, genus g (complex) curve of compact type. Every topological type of stable complex curve of compact type arises in this way.

A marked n-pointed, genus g curve of compact type is a homotopy class of homeomorphisms
$$(S/\vec{\gamma}, P) \to (C, \{x_1, \ldots, x_n\})$$
to an n-pointed, genus g stable curve $(C, \{x_1, \ldots, x_n\})$. For each $\vec{\gamma} \in K^{\text{sep}}(S, P)$, one can add a connected "rational boundary component" $\mathcal{X}_\gamma$ to the Teichmüller space $\mathcal{X}_{g,n}$ of (S, P) to obtain a topological space. The mapping class group $\Gamma_{g,n}$ acts on this enlarged Teichmüller space $\mathcal{X}_{g,n}^c$ and the quotient is $\mathcal{M}_g^c$ from which it follows that $\mathcal{T}_{g,n}^c = T_{g,n} \backslash \mathcal{X}_{g,n}^c$.

The stratum of $\mathcal{T}_g^{c,\text{red}}$ that has codimension k in $\mathcal{T}_g^c$ is the locus of stable curves of compact type with precisely k double points. These correspond to the $k - 1$ simplices of $K^{\text{sep}}(S, P)$. It follows that the strata of $\mathcal{T}_{g,n}^c$ correspond to the simplices of $T_{g,n} \backslash K^{\text{sep}}(S, P)$.

REMARK 4. Farb and Ivanov [**4**] have shown that $K^{\text{sep}}(S)$ is connected whenever $g \geq 3$.

2.5. Singularities and dimension. Note that $\mathcal{J}_g$ is the quotient of $\mathcal{T}_g$ by the involution
$$\sigma : (C; a_1, \ldots, a_g, b_1, \ldots, b_g) \to (C; -a_1, \ldots, -a_g, -b_1, \ldots, -b_g)$$
Since
$$(C; a_1, \ldots, a_g, b_1, \ldots, b_g) \cong (C; -a_1, \ldots, -a_g, -b_1, \ldots, -b_g)$$
if and only if C is hyperelliptic, this mapping is ramified along the locus $\mathcal{H}_g$ of hyperelliptic curves. Since this has complex codimension $g - 2$ in $\mathcal{T}_g$, we know that $\mathcal{J}_g$ is singular along the locus of hyperelliptic jacobians when $g \geq 4$. It is the quotient of the manifold $\mathcal{T}_g$ by $\mathbb{Z}/2$, so $\mathcal{J}_g$ is always a $\mathbb{Z}[1/2]$-homology manifold.

The period mapping $\mathcal{T}_g^c \to \mathcal{J}_g^c$ has positive dimensional fibers over $\mathcal{J}_g^{c,\mathrm{red}}$ when $g \geq 3$. As a result, $\mathcal{J}_g^c$ is singular along $\mathcal{J}_g^{c,\mathrm{red}}$ when $g \geq 4$, and is not a rational homology manifold.[5]

Since $\mathcal{J}_g^c$ has dimension $3g-3$ and $\mathfrak{h}_g$ dimension $g(g+1)/2$, $\mathcal{J}_g^c$ is a proper subvariety of $\mathfrak{h}_g$ when $g \geq 4$. This and the fact that $\mathcal{J}_g^c$ is not a rational homology manifold when $g \geq 4$ help explain the difficulty of generalizing Mess' arguments to any $g \geq 4$.

On the positive side, we can say that, since $\mathcal{J}_g^c$ is a closed analytic subvariety of the Stein manifold $\mathfrak{h}_g$, it is a Stein space. Consequently, by a result of Hamm [**9**] we have:

PROPOSITION 5. *$\mathcal{J}_g^c$ has the homotopy type of a CW-complex of dimension at most $3g-3$.* □

3. Homological Tools

This section can be omitted or skimmed on a first reading.

3.1. A Gysin sequences. Since $\mathcal{T}_g$ is obtained from the manifold $\mathcal{T}_g^c$ by removing a countable union of closed subvarieties, we have to be a little more careful than usual when constructing the Gysin sequence.

Suppose that X is a PL manifold of dimension m and that Y is a closed PL subset of X. Suppose that A is any coefficient system. For each compact (PL) subset K of X, we have the long exact sequence

$$\cdots \to H^{m-k-1}(Y, Y-(Y\cap K); A) \to H^{m-k}(X, Y\cup (X-K); A)$$
$$\to H^{m-k}(X, X-K; A) \to H^{m-k}(Y, Y-(Y\cap K); A) \to \cdots$$

of the triple $(X, Y\cup(X-K), X-K)$. Taking the direct limit over all such K, we obtain the exact sequence

$$\cdots \to H_c^{m-k-1}(Y; A) \to H_c^{m-k}(X-Y; A) \to H_c^{m-k}(X; A) \to H_c^{m-k}(Y; A) \to \cdots$$

When X is oriented, we can apply Poincaré duality to obtain the following version of the Gysin sequence:

PROPOSITION 6. *If A is any coefficient system, X an oriented PL-manifold of dimension m and Y is a closed PL subset of X, then there is a long exact sequence*

$$\cdots \to H_c^{m-k-1}(Y;A) \to H_k(X-Y;A) \to H_k(X;A) \to H_c^{m-k}(Y;A) \to \cdots$$

3.2. A spectral sequence. In practice, we are also faced with the problem of computing $H_c^\bullet(Y)$ in the Gysin sequence above when Y is singular. Suppose that

$$Y = \bigcup_{\alpha \in I} Y_\alpha$$

is a locally finite union of closed PL subspaces of a PL manifold X. Set

$$Y_{(\alpha_0,\ldots,\alpha_k)} := Y_{\alpha_0} \cap Y_{\alpha_1} \cap \cdots \cap Y_{\alpha_k}$$

and

$$\mathcal{Y}_k = \coprod_{\alpha_0 < \alpha_1 < \cdots < \alpha_k} Y_{(\alpha_0,\ldots,\alpha_k)}.$$

---

[5]An explicit description of the links of the singularities of the top stratum of $\mathcal{J}_g^{c,\mathrm{red}}$ can be found in [**7**, Prop. 6.5]. There are similar descriptions in higher codimension.

The inclusions
$$d_j : Y_{(\alpha_0,\ldots,\alpha_k)} \hookrightarrow Y_{(\alpha_0,\ldots,\widehat{\alpha_j},\ldots,\alpha_k)}$$
induce face maps
$$d_j : \mathcal{Y}_k \to \mathcal{Y}_{k-1} \quad j = 0,\ldots,k$$
With these, $\mathcal{Y}_\bullet$ is a strict simplicial space.

PROPOSITION 7. *There is a spectral sequence*
$$E_1^{s,t} \cong H_c^t(\mathcal{Y}_s; A) \implies H_c^{s+t}(Y; A)$$
whose E_1 differential
$$d_1 : H_c^t(\mathcal{Y}_s; A) \to H_c^t(\mathcal{Y}_{s+1}; A)$$
is $\sum_j (-1)^j d_j^*$.

PROOF. This follows rather directly from the standard fact that the natural "augmentation"
$$\epsilon : |\mathcal{Y}_\bullet| \to Y$$
from the geometric realization[6] is a homotopy equivalence. Since Y is a locally finite union of the closed subspaces Y_α, the natural mapping
$$S_c^\bullet(Y; A) \to S_c^\bullet(\mathcal{Y}_\bullet; A)$$
is a quasi-isomorphism. The spectral sequence is that of the double complex $S_c^\bullet(\mathcal{Y}_\bullet; A)$. The quasi-isomorphism implies that the spectral sequence abuts to $H_c^\bullet(Y; A)$. $\square$

Suppose now that each Y_α is an oriented PL submanifold of dimension $2d$ and codimension 2 in X. In addition, suppose that the components of Y_α of Y intersect transversally in X, so that each component of $\mathcal{Y}_s$ has dimension $2d - 2s$. By duality, the E_1 term of the spectral sequence can we written as
$$E_1^{s,t} = H_{2d-2s-t}(\mathcal{Y}_s; A)$$
The differential
$$d_1 : H_{2d-2s-t}(\mathcal{Y}_s; A) \to H_{2d-2s-t-2}(\mathcal{Y}_{s+1}; A)$$
is the alternating sum of the Gysin mappings $d_j^* : \mathcal{Y}_{s+1} \to \mathcal{Y}_s$.

PROPOSITION 8. *There is a natural isomorphism $H_c^{2d}(Y; A) \cong H_0(\mathcal{Y}_0; A)$ and an exact sequence*
$$H_c^{2d-2}(Y; A) \to H_2(\mathcal{Y}_0; A) \to H_0(\mathcal{Y}_1; A) \to H_c^{2d-1}(Y; A) \to H_1(\mathcal{Y}_0; A) \to 0$$

---

[6]This is the quotient of $\coprod_{k \geq 0} \mathcal{Y}_k \times \Delta^k$ by the natural equivalence relation generated by $(y, \partial_j \xi) \cong (d_j y, \xi)$, where $(y, \xi) \in \mathcal{Y}_k \times \Delta^{k-1}$, where Δ^d denotes the standard d-simplex, and $\partial_j : \Delta^{k-1} \to \Delta^k$ is the inclusion of the jth face.

3.3. Cohomology at infinity.
For a topological space X and an R-module (or local coefficient system of R-modules) A, set

$$H^\bullet_\infty(X;A) = \varinjlim_{\substack{K \subseteq X \\ \text{compact}}} H^\bullet(X - K; A).$$

We shall call it the *cohomology at infinity* of X. When X is a compact manifold with boundary, then there is a natural isomorphism

$$H^\bullet_\infty(X - \partial X; A) \cong H^\bullet(\partial X; A).$$

REMARK 9. I am not sure if this definition appears in the literature, although I would be surprised if it does not. Similar ideas have long appeared in topology, such as in the paper Bestvina and Feighn [3], where they introduce the notion of a space being "r-connected at infinity."

The direct limit of the long exact sequence of the pairs $(X, X - K)$, where K is compact, is a long exact sequence

(1) $$\cdots \to H^{k-1}_\infty(X;A) \to H^k_c(X;A) \to H^k(X;A) \to H^k_\infty(X;A) \to \cdots$$

When X is an oriented manifold of dimension m, Poincaré duality gives an isomorphism

$$H_{m-k}(X;A) \to H^k_c(X;A).$$

Thus, if X has the homotopy type of a CW-complex of dimension d, then

$$H^k_c(X;A) = 0 \text{ when } k < m - d \text{ and } H^k(X;A) = 0 \text{ when } k > d.$$

Plugging these into the long exact sequence (1), we see that

$$H^k_\infty(X;A) \cong H^k(X;A) \text{ when } k < m - d - 1$$

and

$$H^k_\infty(X;A) \cong H^{k+1}_c(X;A) \text{ when } k > d.$$

Moreover, if R is a field, then

$$H^k(X;R) \to \operatorname{Hom}_R\left(H^{m-k}_c(X;R), R\right)$$

is an isomorphism, which gives an isomorphism

$$H^k_\infty(X;R) \to \operatorname{Hom}_R\left(H^{m-k-1}_\infty(X;R), R\right)$$

when $k < m - d - 1$.

LEMMA 10. *Suppose that R is a field. If $H^k_\infty(X;R)$ is a countably generated R-module when $a \leq k \leq b$, then $H^k(X;R)$ is a finite dimensional R-module when $a \leq k \leq b$.*

PROOF. Since X has the homotopy type of a countable CW-complex, each $H_k(X;R)$ is countably generated. Poincaré duality then implies that $H^k_c(X;R)$ is also countably generated. The Universal Coefficient Theorem implies that $H^k(X;R)$ is the dual of $H_k(X;R)$. Consequently, if $H^k(X;R)$ is countably generated, it is finitely generated. The result now follows from the exact sequence (1). □

4. Discussion and Problems

One natural approach to the problem of understanding the topology of $\mathcal{T}_g$ is to view it as the complement of the normal crossings divisor $\mathcal{T}_g^{c,\mathrm{red}}$ in $\mathcal{T}_g^c$. The space $\mathcal{T}_g^c$ can in turn be studied via the period mapping $\mathcal{T}_g^c \to \mathcal{J}_g^c$. There are two ways to factorize this. The first is to take the quotient $\mathcal{Q}_g^c := \mathcal{T}_g^c / \langle \sigma \rangle$ of $\mathcal{T}_g^c$ by the involution

$$\sigma : [C; a_1, \ldots, a_g, b_1, \ldots, b_g] \to [C; -a_1, \ldots, -a_g, -b_1, \ldots, -b_g].$$

The mapping $\mathcal{T}_g^c \to \mathcal{Q}_g^c$ is branched along the hyperelliptic locus $\mathcal{H}_g^c$. The second is to consider the Stein factorization (cf. [**6**, p. 213])

$$\mathcal{T}_g^c \to \mathcal{S}_g^c \to \mathcal{J}_g^c$$

of the period mapping. The important properties of this are that $\mathcal{S}_g^c$ is a complex analytic variety, the first mapping has connected fibers, while the second is finite (in the sense of analytic geometry). The two factorizations are related by the diagram

where all spaces are complex analytic varieties, all mappings are proper, holomorphic and surjective. The horizontal mappings have connected fibers and the vertical mappings are finite and two-to-one except along the hyperelliptic locus.

Perhaps the first natural problem is to understand the topology of $\mathcal{J}_g^c$.

PROBLEM 1 (*Topological Schottky Problem*). Understand the homotopy type of $\mathcal{J}_g^c$ and use it to compute $H^\bullet(\mathcal{J}_g^c)$ and $H_c^\bullet(\mathcal{J}_g^c)$.

The first interesting case is when $g = 4$, where $\mathcal{J}_g^c$ is a singular divisor in $\mathfrak{h}_4$. It would also be interesting and natural to compute the intersection homology of $\mathcal{J}_g^c$.

A knowledge of the topology of $\mathcal{J}_g$, $\mathcal{J}_g^c$ or $\mathcal{Q}_g^c$ should help with the computation of the σ-invariant homology $H_\bullet(\mathcal{T}_g)^{\langle \sigma \rangle}$.[7] When 2 is invertible in the coefficient ring R, we can write

$$H_\bullet(\mathcal{T}_g; R) = H_\bullet(\mathcal{T}_g; R)^+ \oplus H_\bullet(\mathcal{T}_g; R)^-$$

where σ acts as the identity on $H_\bullet(\mathcal{T}_g; R)^+$ and as -1 on $H_\bullet(\mathcal{T}_g; R)^-$.

PROBLEM 2. Determine whether or not $H_\bullet(\mathcal{T}_g; \mathbb{Z}[1/2])^-$ is always a finitely generated $\mathbb{Z}[1/2]$-module. Does the infinite topology of $\mathcal{T}_g$ comes from $\mathcal{J}_g$?

To get one's hands on $H_\bullet(\mathcal{T}_g)^-$, it is necessary to better understand the topology of $\mathcal{T}_g^c$ or $\mathcal{S}_g^c$. Since $\mathcal{J}_g^c$ is a Stein space, so is $\mathcal{S}_g^c$. Hamm's result [**9**] (see also [**5**, p. 152]) implies that $\mathcal{S}_g^c$ has the homotopy type of a CW-complex of dimension at most $3g - 3$. Since $\mathcal{S}_g^c$ is not a rational homology manifold when $g \geq 3$, it is probably most useful to study the topology of the manifold $\mathcal{T}_g^c$ as Poincaré duality will then be available.[8]

---

[7]When $g = 3$, $\mathcal{Q}_3^c$ is obtained from $\mathfrak{h}_3$ by first blowing up the singular locus of $\mathfrak{h}_3^{\mathrm{red}}$, which is smooth of complex codimension 3, and then blowing up the proper transforms of the components of $\mathfrak{h}_3^{\mathrm{red}}$. The strata of $\mathfrak{h}_3^{\mathrm{red}}$ are described in detail in [**8**].

[8]If one uses intersection homology instead, then duality will still be available. For this reason it is natural to try to compute the intersection homology of $\mathcal{S}_g^c$ and $\mathcal{J}_g^c$.

PROBLEM 3. Determine good bounds for the homological dimension (or the CW-dimension) of $\mathcal{T}_g^c$.

The best upper bound that I know of is obtained using Stratified Morse Theory:

PROPOSITION 11. *If $g \geq 1$ and $2g - 2 + n > 0$, then the dimension of $\mathcal{T}_{g,n}^c$ as a CW-complex satisfies*

$$2(g - 2 + n) \leq \dim_{\mathrm{CW}} \mathcal{T}_{g,n}^c \leq \left\lfloor \frac{7g - 8}{2} \right\rfloor + 2n.$$

PROOF. Suppose that $g \geq 2$. Since $\mathcal{T}_{g,n}^c \to \mathcal{T}_g^c$ is proper with fibers of complex dimension n,

$$\dim_{\mathrm{CW}} \mathcal{T}_{g,n}^c = 2n + \dim_{\mathrm{CW}} \mathcal{T}_g^c.$$

To establish the lower bound, we need to exhibit a topological $(2g - 4)$-cycle in $\mathcal{T}_g^c$ that is non-trivial in $H_\bullet(\mathcal{T}_g^c)$. Observe that

$$Y := (\mathcal{T}_{1,1}^c)^2 \times (\mathcal{T}_{1,2}^c)^{g-2}$$

is a component of the closure of the locus of chains of g elliptic curves in $\mathcal{T}_g^c$. Let E be any elliptic curve. Then $E \subset \mathcal{T}_{1,2}$ and Y therefore contains the projective subvariety E^{g-2}. The rational homology class of a compact subvariety of a Kähler manifold is always non-trivial (just integrate the appropriate power of the Kähler form over it). Since $\mathcal{T}_g^c$ covers $\mathcal{M}_g^c$, which is a Kähler orbifold, $\mathcal{T}_g^c$ is a Kähler manifold and the class of E^{g-2} is non-trivial in $H_\bullet(\mathcal{T}_g^c)$. This establishes the lower bound.

The upper bound is a direct application of Stratified Morse Theory [5] to the proper mapping $\mathcal{T}_g^c \to \mathcal{J}_g^c$. Since $\mathcal{J}_g^c$ is a Stein space, it is a closed analytic subvariety of $\mathbb{C}^N$ for some N. The constant c in [5, Thm. 1.1*, p. 152] is thus 1. Consequently, $\dim_{\mathrm{CW}} \mathcal{T}_{g,n}$ is bounded by

$$d(g, n) := \sup_{k \geq 0} \bigl(2k + f(k)\bigr)$$

where $f(k)$ is the maximal complex dimension of a subvariety of $\mathcal{T}_{g,n}^c$ over which the fiber of $\mathcal{T}_{g,n}^c \to \mathcal{J}_g^c$ is k dimensional. Observe that $d(g, n) = d(g) + 2n$, when $g \geq 2$, where $d(g) = d(g, 0)$. Since $\mathcal{T}_{1,n}^c \to \mathfrak{h}_1$ is proper with fibers of dimension $n - 1$, we have $d(1, n) = 1 + 2(n - 2) = 2n - 1$.

Since the mapping $\mathcal{T}_g \to \mathcal{J}_g$ is finite and since the components of $\mathcal{T}_g^{c,\mathrm{red}} \to \mathcal{J}_g^{c,\mathrm{red}}$ are the $\mathrm{Sp}_g(\mathbb{Z})$ orbits of the period mappings

$$\mathcal{T}_{h,1}^c \times \mathcal{T}_{g-h,1}^c \to \mathcal{J}_h^c \times \mathcal{J}_{g-h}^c$$

we have

$$d(g) = \max\bigl[3g - 3, \max\{d(h, 1) + d(g - h, 1) : 1 \leq h \leq g/2\}\bigr].$$

The formula for the upper bound $d(g)$ is now easily proved by induction on g. □

The upper bound is not sharp when $g \leq 2$; for example, $\mathcal{T}_{1,1}^c = \mathfrak{h}_1$ and $\mathcal{T}_2^c = \mathfrak{h}_2$, which are contractible. I suspect it is not sharp in higher genus as well. The first interesting case is to determine the CW-dimension of $\mathcal{T}_3^c$.

The upper bound on the homological dimension of $\mathcal{T}_g^c$ implies that

$$H_k(\mathcal{T}_g^c) \cong H_\infty^{6g-7-k}(\mathcal{T}_g^c) \text{ when } k < \left\lceil \frac{5g - 6}{2} \right\rceil$$

So the low dimensional homology of $\mathcal{T}_g^c$ is related to the topology at infinity of $\mathcal{T}_g^c$.

PROBLEM 4. Try to understand the "topology at infinity" of $\mathcal{T}_g^c$. In particular, try to compute $H_\infty^k(\mathcal{T}_g^c)$ for k in some range $k \geq d_o$. Alternatively, try to compute the homology $H_\bullet(\mathcal{T}_g^c)$ in lower degrees.

Note that $\mathcal{T}_2^c$ is a manifold with boundary S^5. The boundary of $\mathcal{J}_3^c$ is S^{11}. The first interesting case is in genus 3.

PROBLEM 5. Compute $H_\infty^\bullet(\mathcal{T}_3^c)$.

The homology of $\mathcal{T}_g^c$ is related to that of $\mathcal{T}_g$ via the Gysin sequence. In order to apply it, one needs to understand the topology of the divisor $\mathcal{T}_g^{c,\mathrm{red}}$. This is built up out of products of lower genus Torelli spaces of compact type. The combinatorics of the divisor is given by the complex $K^{\mathrm{sep}}(S)/T_g$.

PROBLEM 6. Compute the $\mathrm{Sp}_g(\mathbb{Z})$-module $H_c^k(\mathcal{T}_g^{c,\mathrm{red}})$ in some range $k \geq k_o$.

We already know that $H_c^{6g-7}(\mathcal{T}_g^{c,\mathrm{red}}) = H_0(\mathcal{B}_g^c)$ and that there is a surjection $H_c^{6g-8}(\mathcal{T}_g^{c,\mathrm{red}}) \to H_1(\mathcal{B}_g^c)$, where $\mathcal{B}_g^c$ denotes the normalization (i.e., disjoint union of the irreducible components) of $\mathcal{T}_g^{c,\mathrm{red}}$. In concrete terms:

$$\mathcal{B}_g^c = \coprod_{h=1}^{\lfloor g/2 \rfloor} \coprod_{\phi \in \mathrm{Sp}_g(\mathbb{Z})_{(\mathcal{T}_{h,1}^c \times \mathcal{T}_{g-h,1}^c)}} \phi(\mathcal{T}_{h,1}^c \times \mathcal{T}_{g-h,1}^c).$$

Lurking in the background is the folk conjecture

$$H_k(\mathcal{T}_g) \text{ is finitely generated when } k < g - 1.$$

If true, this places strong conditions on the finiteness of the topology of $\mathcal{T}_g^c$ and $\mathcal{T}_g^{c,\mathrm{red}}$. It is worthwhile to contemplate (for $g \geq 3$) the Gysin sequence:

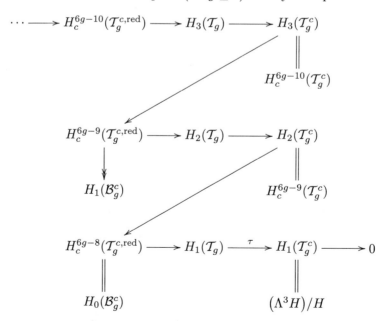

Here τ denotes the Johnson homomorphism, realized as the map on H_1 induced by the inclusion $\mathcal{T}_g \hookrightarrow \mathcal{T}_g^c$.

Finally, it is interesting to study the topology of the branching locus of $\mathcal{T}_g^c \to \mathcal{J}_g^c$. This is the locus $\mathcal{H}_g^c$ of hyperelliptic curves of compact type. Using A'Campo's result [1] that the image of the hyperelliptic mapping class group[9] Δ_g in $\mathrm{Sp}_g(\mathbb{Z})$ contains the level two subgroup

$$\mathrm{Sp}_g(\mathbb{Z})[2] = \{A \in \mathrm{Sp}_g(\mathbb{Z})[2] : A \equiv I \bmod 2\}$$

of $\mathrm{Sp}_g(\mathbb{Z})$ and the fact that the image Δ_g in $\mathrm{Sp}_g(\mathbb{F}_2)$ is S_{2g-2}, the symmetric group on the Weierstrass points, one can see that $\mathcal{H}_g^c$ has

$$|\mathrm{Sp}_g(\mathbb{F}_2)|/|S_{2g+2}| = \frac{2^{g^2} \prod_{k=1}^g (2^{2k}-1)}{(2g+2)!}$$

components. Each component of $\mathcal{H}_g^c$ is smooth and immerses in $\mathfrak{h}_g$ via the period mapping. The irreducible components of $\mathcal{H}_g$ are disjoint in $\mathcal{J}_g$. In genus 3, $\mathcal{H}_3^c$ has 36 components. Their images are cut out by the 36 even theta nulls $\vartheta_\alpha : \mathfrak{h}_3 \to \mathbb{C}$.

CONJECTURE 1. Each component of $\mathcal{H}_g^c$ is simply connected.

This is trivially true in genus 2, where there is one component which is all of $\mathfrak{h}_2$. If true in genus 3, it implies quite directly the known fact that T_3 is generated by $35 = 36 - 1$ bounding pair elements. The number 35 is the rank of $H_1(\mathcal{J}_3, \mathcal{H}_3)$. The inverse images of generators of $\pi_1(\mathcal{J}_3, \mathcal{H}_3)$, once oriented, generate T_3.[10]

The conjecture has an equivalent statement in more group theoretic terms. Define the hyperelliptic Torelli group to be the intersection of the hyperelliptic mapping class group and the Torelli group:

$$T\Delta_g := \Delta_g \cap T_g = \ker\{\Delta_g \to \mathrm{Sp}_g(\mathbb{Z})\}.$$

It is a subgroup of the Johnson subgroup $K_g := \ker\{\tau : T_g \to \Lambda^3 H/H\}$. Examples of elements in $T\Delta_g$ are Dehn twists on separating simple closed curves that are invariant under the hyperelliptic involution σ. If $\mathcal{H}_{g,\alpha}$ is a component of $\mathcal{H}_g$, then there is an isomorphism

$$\pi_1(\mathcal{H}_{g,\alpha}, *) \cong T\Delta_g.$$

The conjugacy classes of twists on a σ-invariant separating SCC correspond to loops about components of the divisor $\mathcal{H}_g^{c,\mathrm{red}}$.

Van Kampen's Theorem implies that

$$\pi_1(\mathcal{H}_{g,\alpha}^c, *) \cong \pi_1(\mathcal{H}_{g,\alpha})/\text{these conjugacy classes} \cong T\Delta_g/\text{these conjugacy classes}.$$

The conjecture is thus equivalent to the statement that $T\Delta_g$ is generated by the conjugacy classes of Dehn twists on σ-invariant separating SCCs.

It is important to understand the topology of the loci of hyperelliptic curves as it is the branch locus of the period mapping and also because it is important in its own right.

PROBLEM 7. Investigate the topology of $\mathcal{H}_g$ and $\mathcal{H}_g^c$ and their components. Specifically, compute their homology and the cohomology at infinity of $\mathcal{H}_{g,\alpha}^c$.

---

[9]The hyperelliptic mapping class is the centralizer of a hyperelliptic involution in Γ_g. It is the orbifold fundamental group of the moduli space of smooth hyperelliptic curves of genus g.

[10]As the genus increases, the number of components of $\mathcal{H}_g$ increases exponentially while the minimum number of generators of T_g increases polynomially. The failure of the genus 3 argument presented above in higher genus suggests that $\mathcal{J}_g^c$ is not, in general, simply connected.

The period mapping immerses each $\mathcal{H}_{g,\alpha}^c$ in $\mathfrak{h}_g$ as a closed subvariety. Consequently, each $\mathcal{H}_{g,\alpha}^c$ is a Stein manifold and thus has the homotopy type of a CW-complex of dimension equal to its complex dimension, which is $2g - 1$.

References

[1] N. A'Campo: *Tresses, monodromie et le groupe symplectique*, Comment. Math. Helv. 54 (1979), 318–327.
[2] T Akita: *Homological infiniteness of decorated Torelli groups and Torelli spaces*, Tohoku Math. J. (2) 53 (2001), 145–147.
[3] M. Bestvina, M. Feighn: *The topology at infinity of* $\mathrm{Out}(F_n)$, Invent. Math. 140 (2000), 651–692.
[4] B. Farb, N. Ivanov: *The Torelli geometry and its applications*, research announcement, Math. Res. Lett. 12 (2005), 293–301.
[5] M. Goresky, R. MacPherson: *Stratified Morse theory*. Ergebnisse der Mathematik und ihrer Grenzgebiete (3), 14. Springer-Verlag, Berlin, 1988.
[6] H. Grauert, R. Remmert: *Coherent analytic sheaves*. Grundlehren der Mathematischen Wissenschaften, 265, Springer-Verlag, Berlin, 1984.
[7] R. Hain: *Locally symmetric families of curves and Jacobians*, in Moduli of curves and abelian varieties, 91–108, Aspects Math., E33, Vieweg, Braunschweig, 1999.
[8] R. Hain: *The rational cohomology ring of the moduli space of abelian 3-folds*, Math. Res. Lett. 9 (2002), 473–491.
[9] H. Hamm: *Zur Homotopietyp Steinscher Räume*, J. Reine Angew. Math. 338 (1983), 121–135.
[10] D. Johnson: *The structure of the Torelli group, I: A finite set of generators for* $\mathcal{I}$, Ann. of Math. (2) 118 (1983), 423–442.
[11] D. Johnson: *The structure of the Torelli group, II: A characterization of the group generated by twists on bounding curves*, Topology 24 (1985), 113–126.
[12] G. Mess: *The Torelli groups for genus 2 and 3 surfaces*, Topology 31 (1992), 775–790.

DEPARTMENT OF MATHEMATICS, DUKE UNIVERSITY, DURHAM, NC 27708-0320
E-mail address: hain@math.duke.edu

Fifteen Problems about the Mapping Class Groups

Nikolai V. Ivanov

Let S be a compact orientable surface, possibly with boundary. We denote by Mod_S the *mapping class group* of S, i.e. the group $\pi_0(\text{Diff}(S))$ of isotopy classes of diffeomorphisms of S. This group is also known as the *Teichmüller modular group* of S, whence the notation Mod_S. Note that we include the isotopy classes of orientation-reversing diffeomorphisms in Mod_S, so our group Mod_S is what is sometimes called the *extended* mapping class group.

For any property of discrete groups one may ask if Mod_S has this property. Guidance is provided by the analogy, well-established by now, between the mapping class groups and arithmetic groups. See, for example, [I1] or [I7] for a discussion.

After the 1993 Georgia Topology Conference, R. Kirby was preparing a new version of his famous problem list in low-dimensional topology. In response to his appeal, I prepared a list of ten problems about the mapping class groups, which was informally circulated as a preprint [I5]. Some of these problems ended up in Kirby's list [Ki] in a somewhat modified form, some did not. In the present article I will indicate the current status of these problems and also will present some additional problems. Only Problems 4, 6, and 9 of the original list of ten problems are by now completely solved. (For the convenience of the readers familiar with [I5] I preserved the numbering of these ten problems.)

I tried to single out some specific questions, leaving aside such well-known problems as the existence of finite dimensional faithful linear representations or the computation of the cohomology groups.

1. The Congruence Subgroups Problem

This is Problem 2.10 in Kirby's list.

Suppose that S is closed. Recall that a subgroup Γ of a group G is called *characteristic* if Γ is invariant under all automorphisms of G. If Γ is a characteristic subgroup of $\pi_1(S)$, then there is a natural homomorphism $\text{Out}(\pi_1(S)) \to \text{Out}(\pi_1(S)/\Gamma)$, where for a group G we denote by $\text{Out}(G)$ the quotient of the group $\text{Aut}(G)$ of all automorphisms of G by the (automatically normal) subgroup of all inner automorphisms. Clearly, if Γ is of finite index in $\pi_1(S)$, then the kernel of this homomorphism is also of finite index. Note that any subgroup of finite index in a finitely generated group (in particular, in $\pi_1(S)$) contains a characteristic subgroup of finite index. Since by the Dehn–Nielsen theorem Mod_S is canonically isomorphic

---

Supported in part by the NSF Grants DMS-9401284 and DMS-0406946.

to $\mathrm{Out}(\pi_1(S))$, our construction gives rise to a family of subgroups of finite index in Mod_S. By analogy with the classical arithmetic groups we call them the *congruence subgroups*.

CONJECTURE. *Every subgroup of finite index in* Mod_S *contains a congruence subgroup*

V. Voevodsky had indicated (in a personal communication) a beautiful application of this conjecture. Namely, the conjecture implies that a smooth algebraic curve over $\mathbf{Q}$ is determined up to isomorphism by its algebraic fundamental group (which is isomorphic to the profinite completion of $\pi_1(S)$) considered together with the natural action of the absolute Galois group $\mathrm{Gal}(\overline{\mathbf{Q}}/\mathbf{Q})$ on it. This corollary was apparently first conjectured by A. Grothendieck. I am not aware of any publication where this conjecture of Grothendieck is deduced from the solution of the congruence subgroup problem for the mapping class groups. The Grothendieck conjecture itself was proved by S. Mochizuki [**Mo1**] after the initial work by A. Tamagawa [**Tam**]. See also [**Mo3**] and the expository accounts by S. Mochizuki [**Mo2**] and G. Faltings [**F**]. The proof of the Grothendieck conjecture lends some additional credibility (beyond the analogy with the classical congruence-subgroup problem) to the above congruence subgroup conjecture.

2. Normal Subgroups

This is Problem 2.12 (B) in Kirby's list.

If a subgroup Γ of $\pi_1(S)$ is characteristic, then the kernel of the natural homomorphism $\mathrm{Out}(\pi_1(S)) \to \mathrm{Out}(\pi_1(S)/\Gamma)$ from the Problem 1 is a normal subgroup of $\mathrm{Out}(\pi_1(S))$. So, this construction gives rise to a family of normal subgroups of Mod_S. In general, these subgroups have infinite index. For example, the Torelli subgroup is a subgroup of this type.

QUESTION. *Is it true that any normal subgroup is commensurable with such a subgroup?*

Recall that two subgroups Γ_1 and Γ_2 of a group G are *commensurable* if the intersection $\Gamma_1 \cap \Gamma_2$ has finite index in both Γ_1 and Γ_2.

This problem was suggested by a discussion with H. Bass.

3. Normal Subgroups and pseudo-Anosov Elements

This is Problem 2.12 (A) in Kirby's list.

QUESTION. *Is it possible that all nontrivial elements of a* normal *subgroup of* Mod_S *are pseudo-Anosov?*

To the best of my knowledge, this question was posed independently by D. D. Long, J. D. McCarthy, and R. C. Penner in the early eighties. I learned it from R. C. Penner in 1984.

Moderate progress in the direction of the solution of this problem is due to K. Whittlesey [**Whi**], who constructed examples of such subgroups in the case of spheres with 5 or more punctures and for closed surfaces of genus 2 (in the latter case the mapping class group is well-known to be intimately connected with the mapping class group of the sphere with 6 punctures). Unfortunately, there is, apparently, no hope of extending the method of proof to the other cases, especially to the closed surfaces of higher genus (which were the focus of all previous attempts at this problem, I suspect).

4. Conjecture (Mostow-Margulis superrigidity)

This is Problem 2.15 in Kirby's list.

CONJECTURE. *If Γ is an irreducible arithmetic group of rank ≥ 2, then every homomorphism $\Gamma \to \operatorname{Mod}_S$ has finite image.*

For many arithmetic groups Γ the conjecture can be proved by combining some well-known information about arithmetic groups with equally well-known properties of Mod_S. But, for example, for cocompact lattices in $SU(p,q)$ this straightforward approach seems to fail. (I owe this specific example to G. Prasad.) In any case, by now this conjecture is completely proved.

V. A. Kaimanovich and H. Masur [**KaM**] proved that so-called non-elementary subgroups (subgroups containing a pair of pseudo-Anosov elements with disjoint sets of fixed points in the Thurston boundary) of Mod_S are not isomorphic to irreducible arithmetic groups of rank ≥ 2. The proof is based on the theory of random walks on Mod_S developed by Masur and Kaimanovich–Masur and on the results of H. Furstenberg about random walks on arithmetic groups (see the references in [**KaM**]). If combined with the Margulis finiteness theorem and the techniques of [**I4**], this result can be used to prove the conjecture.

A proof close in the spirit to the one alluded to in the previous paragraph is due to B. Farb and H. Masur [**FM**].

In addition, S-K. Yeung [**Y**] completed the picture by proving that any homomorphism from a lattice Γ in either $\operatorname{Sp}(m,1)$ or F_4^{-20} (the isometry group of the Cayley plane) into a mapping class group has finite image. The lattices in the remaining simple Lie groups of rank 1 often admit non-trivial homomorphisms to **Z**, and, therefore, homomorphisms to the mapping class groups with infinite image.

5. Dehn multitwists

For a nontrivial circle α on S (i.e. a submanifold α of S diffeomorphic to the standard circle S^1) let us denote by t_α the (left) Dehn twist along α; $t_\alpha \in \operatorname{Mod}_S$. A *Dehn multi-twist* is any composition of the form $t_{\alpha_1}^{\pm 1} \circ t_{\alpha_2}^{\pm 1} \circ \cdots \circ t_{\alpha_n}^{\pm 1}$ for disjoint circles $\alpha_1, \alpha_2, \ldots, \alpha_n$.

QUESTION. Is there a constant N_S, depending only on S, such that the following holds? Let $f \in \operatorname{Mod}_S$ and let $t = t_{\alpha_1}^{\pm 1} \circ t_{\alpha_2}^{\pm 1} \circ \cdots \circ t_{\alpha_n}^{\pm 1}$ be a Dehn multi-twist. If $A = \{f^m(\alpha_i) : 1 \leq i \leq n, m \in \mathbf{Z}\}$ fills S (i.e. for any nontrivial circle γ there exists an $\alpha \in A$ such that the geometric intersection number $\mathrm{i}(\gamma, \alpha) \neq 0$), then only a finite number of elements of the form $t^j \circ f$, $j \in \mathbf{Z}$, are not pseudo-Anosov, and the corresponding j are among some N_S consecutive integers.

By a theorem of A. Fathi [**Fath**], if t is a Dehn *twist*, then this is true for $N_S = 7$. This theorem of A. Fathi was preceded by a theorem of D. Long and H. Morton [**LM**] to the effect that for a Dehn twist t, only a finite number of element $t^j \circ f$ are not pseudo-Anosov under the above assumptions, without a uniform bound on the number of exceptions.

A weaker version of this question is still interesting: under the same conditions, is it true that no more than N_S of $t^j \circ f$ are not pseudo-Anosov?

The initial motivation for this question was that a positive answer would allow to prove the Conjecture 4 in some nontrivial cases. This motivation is now (completely?) obsolete, but I still consider this question to be interesting. I believe that Fathi's paper [**Fath**] is one of the deepest and the most underappreciated works in the theory of the mapping class groups. In fact, I am aware of only one application of his results: the author's theorem [**I2**] to the effect that the mapping class groups have rank 1 in an appropriate sense (note that the Long–Morton theorem [**L**] is not sufficient here); see [**I7**], Section 9.4 for a discussion. The above question may be considered as a test of our understanding of Fathi's ideas.

6. Automorphisms of Complexes of Curves

The *complex of curves* $C(S)$ of a surface S is a simplicial complex defined as follows. The vertices of $C(S)$ are the isotopy classes of nontrivial (i.e., not bounding a disk and non deformable into the boundary) circles on S. A set of vertices forms a simplex if and only if they can be represented by disjoint circles. This notion was introduced by W. J. Harvey [**Ha**]. Clearly, Mod_S acts on $C(S)$, and this action is almost always effective (the main exception is the case of a closed surface of genus 2, where the hyperelliptic involution acts on $C(S)$ trivially). If the genus of S is at least 2, then all automorphisms of $C(S)$ come from Mod_S, according to a well-known theorem of the author [**I3**], [**I6**]. The problem was to prove the same for surfaces of genus 1 and 0.

In his 1996 MSU Ph.D. thesis M. Korkmaz proved that all automorphisms of $C(S)$ come from Mod_S for all surfaces of genus 0 and 1 with the exception of spheres with ≤ 4 holes and tori with ≤ 2 holes. See his paper [**Ko1**].

Later on, F. Luo [**Luo2**] suggested another approach to the above results about automorphisms of $C(S)$, still based on the ideas of [**I3**], and also on a multiplicative structure on the set of vertices of $C(S)$ introduced in [**Luo1**]. He also observed that $\text{Aut}(C(S))$ is not equal to Mod_S if S is a torus with 2 holes. The reason is very simple: if $S_{1,2}$ is a torus with 2 holes, and $S_{0,5}$ is a sphere with 5 holes, then $C(S_{1,2})$ is isomorphic to $C(S_{0,5})$, but $\text{Mod}_{S_{1,2}}$ is not isomorphic to $\text{Mod}_{S_{0,5}}$.

So, the problem is solved completely. The theorem about the automorphisms of $C(S)$ had stimulated a lot of further results about the automorphisms of various objects related to surfaces; the proofs are usually based on a reduction of the problem in question to the theorem about the automorphisms of $C(S)$. I will not attempt to survey these results. Instead of this, I will state a metaconjecture.

METACONJECTURE. *Every object naturally associated to a surface S and having a sufficiently rich structure has Mod_S as its groups of automorphisms. Moreover, this can be proved by a reduction to the theorem about the automorphisms of $C(S)$.*

7. The First Cohomology Group of Subgroups of Finite Index

This is Problem 2.11 (A) in Kirby's list.

QUESTION. Is it true that $H^1(\Gamma) = 0$ for any subgroup Γ of finite index in Mod_S?

It is well known that $H^1(\mathrm{Mod}_S) = 0$. In his 1999 MSU Ph.D. thesis F. Taherkhani carried out some extensive computer calculation aimed at finding a subgroup of finite index with non-zero first cohomology group. For genus 2 he found several subgroups Γ with $H^1(\Gamma) \neq 0$, but in genus 3 all examined subgroups Γ turned out to have $H^1(\Gamma) = 0$. The higher genus cases apparently were well beyond the computer resources available at the time. See [**Tah**].

J. D. McCarthy [**McC2**] proved that if S is a closed surface of genus ≥ 3 and Γ is a subgroup of finite index in Mod_S containing the Torelli subgroup, then the first cohomology group $H^1(\Gamma)$ is trivial. His methods are based on D. Johnson results about the Torelli subgroup, the solution of the congruence subgroups problem for $Sp_{2g}(\mathbf{Z})$, $g \geq 3$, and the Kazhdan property (T) of $Sp_{2g}(\mathbf{Z})$, $g \geq 3$.

8. Kazhdan Property (T)

This is Problem 2.11 (B) in Kirby's list.

QUESTION. Does Mod_S have the Kazhdan property (T)?

A positive answer would imply the positive answer to the previous question, but this problem seems to be much more difficult. Most of the known proofs of the property (T) for discrete groups are eventually based on the relations of these discrete groups with Lie groups and on the representation theory of Lie groups. Such an approach is not available for Mod_S. New approaches to the Kazhdan Property (T) (see, for example Y. Shalom [**Shal**], A. Żuk [**Z**] and the N. Bourbaki Seminar report of A. Valette [**V**]) hold a better promise for our problem, but, to the best of my knowledge, no serious work in this direction has been done.

The problem remains completely open.

9. Unipotent Elements

This is Problem 2.16 in Kirby's list.

Let $d_W(\cdot, \cdot)$ be the word metric on Mod_S with respect to some finite set of generators. Let $t \in \mathrm{Mod}_S$ be a Dehn twist.

QUESTION. What is the growth rate of $d_W(t^n, 1)$?

One would expect that either the growth is linear, or $d_W(t^n, 1) = O(\log n)$. In the arithmetic groups case, the logarithmic growth corresponds to virtually unipotent elements of arithmetic groups of rank ≥ 2, according to a theorem of A. Lubotzky, S. Moses and M.S. Raghunathan [**LMR**].

B. Farb, A. Lubotzky and Y. Minsky [**FLM**] proved that the growth is linear, so the problem is solved completely.

10. Nonorientable Surfaces

Most of the work on the mapping class groups is done for orientable surfaces only. Some exceptions are the paper of M. Scharlemann [**Scha**], and the computation of the virtual cohomological dimension for the mapping class groups of nonorientable surfaces in the author's work [**I1**]. One may look for analogues of other theorems about Mod_S for nonorientable surfaces. For example, what are the automorphisms of mapping class groups of nonorientable surfaces?

Of course, one may expect that some results will extend to the nonorientable case more or less automatically. This applies, for example, to the author's results about the subgroups of the mapping class groups reported in [**I4**]. The only reason why these results are stated only in the orientable case is the fact that expositions of Thurston's theory of surfaces exist only in the orientable case. But in other problems some new phenomena appear. See, for example, the result of M. Korkmaz [**Ko2**], [**Ko3**]. N. Wahl [**Wha**] recently reported some work in progress aimed at proving a homology stability theorem for the mapping class groups of nonorientable surfaces, where some new difficulties appear in comparison with the orientable case.

11. Free subgroups generated by pseudo-Anosov elements

One of the first results in the modern theory of the mapping class groups was the theorem to the effect that for two independent pseudo-Anosov elements (where *independence* means that their sets of fixed points in the Thurston boundary are disjoint) their sufficiently high powers generate a free group (with two generators). This observation was done independently by J.D. McCarthy and the author and for the author it was the starting point of the theory presented in [**I4**]. This result clearly extends to any finite number of independent elements and even to an infinite collection of pairwise independent elements, *provided that we do not impose a bound* on the powers to which our elements are raised. Now, what if we do impose a uniform bound?

CONJECTURE. *If f is pseudo-Anosov element of a mapping class group Mod_S with sufficiently large dilatation coefficient, then the subgroup of Mod_S normally generated by f is a free group having as generators the conjugates of f. More cautiously, one may conjecture that the above holds for a sufficiently high power $g = f^N$ of a given pseudo-Anosov element f.*

This conjecture is motivated by a theorem of M. Gromov (see [**Gr**], Theorem 5.3.E). According to this theorem, the subgroup of a hyperbolic group normally generated by a hyperbolic element (with sufficiently big translation length) is a free group having as generators the conjugates of this element. Of course, it is well-known that mapping class groups are not hyperbolic. But, as M. Gromov noticed, what is essential for his proof is not the global negative curvature (hyperbolicity), but the negative curvature around the loop representing the considered element in an Eilenberg-MacLane space of the group in question. So, the hope is that the Teichmüller spaces have enough negative curvature along the axes of pseudo-Anosov elements for a similar conclusion to hold. For a more direct and elementary approach to Gromov's result see the work of Th. Delzant [**D**].

This conjecture may be relevant to Problem 3 above, since one may expect that all nontrivial elements of such subgroups are pseudo-Anosov.

12. Deep relations

It is well-known that mapping class groups of closed surfaces are generated by the Dehn twists (this result is due to Dehn himself, see [**I7**] for a detailed discussion), and that some specific relations among Dehn twists, such as the Artin (braid) relations and the lantern relations play a crucial role in the proofs of many results about the mapping class groups. In fact, according to a theorem of S. Gervais [**Ge**] (a hint to such a theorem is contained already in J. Harer's paper [**H**]) Mod_S admits a presentation having all Dehn twists as generators and only some standard relations between them as relations. There is also a similar presentation having only Dehn twists about nonseparating curves as generators. See [**Ge**] for exact statements. (For surfaces with boundary simple additional generators are needed, but the following discussion of the relations between Dehn twists equally makes sense for surfaces with boundary.)

Almost all of these relations disappear if we replace Dehn twists by powers of Dehn twists (for example, if we are forced to do so by trying to prove something about subgroups of finite index). Apparently, only the relations

$$T_\alpha T_\beta T_\alpha^{-1} = T_{T_\alpha(\beta)},$$

survive, where T_γ for a circle γ denotes some power t_γ^L of a Dehn twist t_γ about γ. We are primarily interested in the case of powers independent of circles, since there are no other natural way to assign powers to circles. Notice that these relations include the commutation relations $T_\alpha T_\beta = T_\beta T_\alpha$ for disjoint α, β, and for the circles α, β intersecting once transversely they imply the Artin relation $t_\alpha t_\beta t_\alpha = t_\beta t_\alpha t_\beta$.

QUESTION. Are there any other relations between N-th powers $T_\gamma = t_\gamma^N$ of Dehn twists for sufficiently large N? In other words, do the above relations provide a presentation of the group generated by the N-th powers of Dehn twists? If not, what are the additional relations?

13. Burnside groups

Note that the subgroup generated by the N-th powers of Dehn twists is often of infinite index in Mod_S, as it follows from the results of L. Funar [**Fu**]. (See also G. Masbaum [**Mas**] for a simple proof of more precise results.)

QUESTION. Is the subgroup of Mod_S generated by the N-th powers of *all* elements of Mod_S of infinite index in Mod_S for sufficiently large N?

Notice that such a subgroup is obviously normal and if it is of infinite index, then the quotient group is an infinite Burnside group.

14. Homomorphisms between subgroups of finite index

In general, one would like to understand homomorphisms between various mapping class groups and their finite index subgroups. Here is one specific conjecture.

CONJECTURE. *Let S and R be closed surfaces. Let Γ be a subgroup of finite index in Mod_S. If the genus of R is less than the genus of S, then there is no homomorphism $\Gamma \to \mathrm{Mod}_R$ having as an image a subgroup of finite index in Mod_R.*

In fact, one may hope that any such homomorphism $\Gamma \to \mathrm{Mod}_R$ has a finite image. But while Problem 7 about $H^1(\Gamma)$ is unresolved, a more cautious conjecture seems to be more appropriate and more accessible, because if $H^1(\Gamma)$ is infinite, Γ admits a homomorphism onto $\mathbf{Z}$, and therefore, a lot of homomorphisms into Mod_R, in particular, with infinite image.

In the special case when $\Gamma = \mathrm{Mod}_S$, the conjecture was recently proved by W. Harvey and M. Korkmaz [**HaK**]. Their methods rely on the use of elements of finite order in $\Gamma = \mathrm{Mod}_S$, and therefore cannot be extended to general subgroups Γ of finite index.

15. Frattini subgroups

Recall that the Frattini subgroup $\Phi(G)$ of a group G is the intersection of all proper (i.e. different from G) maximal subgroups of G. According to a theorem of Platonov [**Pl**] the Frattini subgroup $\Phi(G)$ of a finitely generated linear group G is nilpotent. An analogue of this result for subgroups of mapping class groups was proved by the author [**I4**] after some initial results of D. D. Long [**L**]. Namely, for any subgroup G of Mod_S, its Frattini subgroup $\Phi(G)$ is nilpotent; see [**I4**], Chapter 10.

Despite the similarly sounding statements, the theorems of Platonov and of the author have a completely different nature. In order to explain this, let us denote by $\Phi_f(G)$ the intersection of all maximal subgroups of *finite index* in G, and by $\Phi_i(G)$ the intersection of all maximal subgroups of *infinite index* in G; clearly, $\Phi(G) = \Phi_f(G) \cap \Phi_i(G)$. It turns out that the main part of the argument in [**I4**] proves that $\Phi_i(G)$ is virtually abelian for subgroups of Mod_S (and so $\Phi(G)$ is both virtually abelian and nilpotent). At the same time, Platonov actually proves that $\Phi_f(G)$ is nilpotent for finitely generated linear groups G; subgroups of infinite index play no role in his arguments. (Actually, the very existence of maximal subgroups of infinite index in linear groups is a highly nontrivial result due to G. A. Margulis and G. A. Soifer [**MaSo**], and it was proved much later than Platonov's theorem.)

CONJECTURE. *For every finitely generated subgroups G of Mod_S, the group $\Phi_f(G)$ is nilpotent.*

For a more detailed discussion, see [**I4**], Section 10.10. I repeated here this old problem in order to stress that our understanding of the subgroups of finite index in Mod_S is rather limited.

References

[D] Th. Delzant, Sous-groupes distingus et quotients des groupes hyperboliques, Duke Math. J., V. 83, No. 3 (1996), 661–682.

[F] G. Faltings, Curves and their fundamental groups (following Grothendieck, Tamagawa and Mochizuki). Séminaire Bourbaki, Vol. 1997/98, Astérisque No. 252 (1998), Exp. No. 840, 131–150.

[FM] B. Farb and H. Masur, Superrigidity and mapping class groups, Topology, V. 37, No. 6 (1998), 1169- 1176.

[FLM] B. Farb, A. Lubotzky, and Y. Minsky, Rank-1 phenomena for mapping class groups, Duke Math. J., V. 106, No. 3 (2001), 581–597.

[Fath] A. Fathi, Dehn twists and pseudo-Anosov diffeomorphism, Inventiones Math., V. 87, No. 1 (1987), 129–151.

[Fu] L. Funar, On the TQFT representations of the mapping class groups. Pacific J. Math., V. 188, No. 2 (1999), 251–274.

[Ge] S. Gervais, Presentation and central extensions of mapping class groups, Trans. Amer. Math. Soc., V. 348, No. 8 (1996), 3097–3132.

[Gr] M. Gromov, Hyperbolic groups, in *Essays in Group Theory*, Ed. by S. Gersten, MSRI Publications V. 8 (1987), 75–265.

[H] J. L. Harer, The second homology group of the mapping class group of an orientable surface, Inventiones Math., V. 72, F. 2 (1983), 221–239.

[Ha] W. J. Harvey, Boundary structure of the modular group, in: *Riemann surfaces and related topics: Proceedings of the 1978 Stony Brook conference*, Ed. by I. Kra and B. Maskit, Annals of Mathematics Studies, No. 97 (1981), 245–251.

[HaK] W. J. Harvey and M. Korkmaz, Homomorphisms from mapping class groups, Bull. London Math. Soc. V. 37, No. 2 (2005), 275–284.

[I1] N. V. Ivanov, Complexes of curves and the Teichmüller modular group, Uspekhi Mat. Nauk, V. 42, No. 3 (1987), 49–91; English transl.: Russian Math. Surveys, V. 42, No. 3 (1987), 55–107.

[I2] N. V. Ivanov, Rank of Teichmüller modular groups, Mat. Zametki, V. 44, No. 5 (1988), 636–644; English transl.: Math. Notes, V. 44, Nos. 5-6 (1988), 829–832.

[I3] N. V. Ivanov, Automorphisms of complexes of curves and of Teichmüller spaces, Preprint IHES/M/89/60, 1989, 13 pp.; Also in: *Progress in knot theory and related topics*, Travaux en Cours, V. 56, *Hermann, Paris,* 1997, 113–120.

[I4] N. V. Ivanov, *Subgroups of Teichmüller Modular Groups*, Translations of Mathematical Monographs, Vol. 115, *AMS*, 1992.

[I5] N. V. Ivanov, Ten Problems on the Mapping Class Groups, Preprint, 1994.

[I6] N. V. Ivanov, Automorphisms of complexes of curves and of Teichmüller spaces, International Mathematics Research Notices, 1997, No. 14, 651–666.

[I7] N. V. Ivanov, Mapping class groups, *Handbook of Geometric Topology,* Ed. by R. Daverman and R. Sher, *Elsevier,* 2001, 523–633.

[KaM] V. A. Kaimanovich and H. Masur, The Poisson boundary of the mapping class group, Inventiones Math., V. 125, F. 2 (1996), 221–264.

[Ki] R. Kirby, Problems in low-dimensional topology. Ed. by Rob Kirby. AMS/IP Stud. Adv. Math., V. 2.2, Geometric topology (Athens, GA, 1993), 35–473, Amer. Math. Soc., Providence, RI, 1997.

[Ko1] M. Korkmaz, Automorphisms of complexes of curves on punctured spheres and on punctured tori. Topology Appl. V. 95, No. 2 (1999), 85–111.

[Ko2] M. Korkmaz, First homology group of mapping class group of nonorientable surfaces, Mathematical Proc. Cambridge Phil. Soc., V. 123, No. 3 (1998), 487–499.

[Ko3] M. Korkmaz, Mapping class groups of nonorientable surfaces, Geom. Dedicata V. 89 (2002), 109–133.

[L] D. D. Long, A note on the normal subgroups of mapping class groups, Math. Proc. Cambridge Philos. Soc., V. 99, No. 1 (1986), 79–87.

[LM] D. D. Long, H. R. Morton, Hyperbolic 3-manifolds and surface automorphisms. Topology V. 25, No. 4 (1986), 575–583.

[LMR] A. Lubotzky, Sh. Mozes, M. S. Raghunathan, The word and Riemannian metrics on lattices of semisimple groups. Publ. Math. IHES, No. 91 (2000), 5–53.

[Luo1] F. Luo, On nonseparating simple closed curves in a compact surface, Topology, V. 36 (1997), 381–410.

[Luo2] F. Luo, Automorphisms of the complex of curves, Topology V.39, No. 2 (2000), 283–298.

[MaSo] G. A. Margulis, G. A. Soifer, Maximal subgroups of infinite index in finitely generated linear groups, J. of Algebra, V. 69, No. 1 (1981), 1–23.

[Mas] G. Masbaum, An element of infinite order in TQFT-representations of mapping class groups, *Proceedings of the Conference on Low Dimensional Topology in Funchal, Madeira, 1998,* Contemp. Math. V. 233 (1999), 137–139.

[McC1] J. D. McCarthy, A "Tits-alternative" for subgroups of surface mapping class groups, Trans. AMS, V.291, No. 2 (1985), 583–612.

[McC2] J. D. McCarthy, On the first cohomology group of cofinite subgroups in surface mapping class groups, Topology V. 40, No. 2 (2001), 401–418.

[Mo1] Sh. Mochizuki, The profinite Grothendieck conjecture for closed hyperbolic curves over number fields, J. Math. Sci. Univ. Tokyo, V. 3 (1996), 571–627.

[Mo2] Sh. Mochizuki, The intrinsic Hodge theory of p-adic hyperbolic curves. Proceedings of the International Congress of Mathematicians, Vol. II (Berlin, 1998). Doc. Math. 1998, Extra Vol. II, 187–196.

[Mo3] Sh. Mochizuki, The local pro-p anabelian geometry of curves. Invent. Math. V. 138, F. 2 (1999), 319–423.

[Pl] V. P. Platonov, Frattini subgroups of linear groups and finitary approximability, Dokl. Akad. Nauk SSSR, V. 171 (1966), 798–801; English transl.: Soviet Math. Dokl., V. 7, No. 6 (1966), 1557–1560.

[Scha] M. Scharlemann, The complex of curves of a nonorientable surface, J. London Math. Soc., V. 25 (1982), 171–184.

[Shal] Y. Shalom, Bounded generation and Kazhdan's property (T), Publ. Math. IHES, No. 90 (1999), 145–168.

[Tah] F. Taherkhani, The Kazhdan property of the mapping class groups of closed surfaces and the first cohomology group of their cofinite subgroups, Journal of Experimental Mathematics, V. 9, No. 2 (2000,) 261–274.

[Tam] A. Tamagawa, The Grothendieck conjecture for affine curves, Compositio Mathematica, V. 109 (1997), 135–194.

[V] A. Valette, Nouvelles approches de la propriété (T) de Kazhdan, Astérisque No. 294, (2004), 97–124.

[Wha] N. Wahl, Mapping class groups of non-orientable surfaces, Talk at the Spring 2005 Midwest Topology Seminar, University of Chicago, April 15–19, 2005.

[Whi] K. Whittlesey, Normal all pseudo-Anosov subgroups of mapping class groups, Geometry and Topology, Vol. 4 (2000), Paper No. 10, 293–307.

[Y] S.-K. Yeung, Representations of semisimple lattices in mapping class groups, International Mathematical Research Notices, 2003, No. 31, 1677–1686.

[Z] A. Żuk, Property (T) and Kazhdan constants for discrete groups, Geom. Funct. Anal. V. 13, No. 3 (2003), 643–670.

MICHIGAN STATE UNIVERSITY, DEPARTMENT OF MATHEMATICS, WELLS HALL, EAST LANSING, MI 48824-1027

E-mail address: ivanov@math.msu.edu

Problems on Homomorphisms of Mapping Class Groups

Mustafa Korkmaz

1. Introduction

The purpose this note is to single out some of the problems on the algebraic structure of the mapping class group. Most of our problems are on homomorphisms from mapping class groups. We also state a couple of others problems, such as those related to the theory of Lefschetz fibrations.

Let Σ be a connected (orientable or nonorientable) surface of genus g with $p \geq 0$ punctures and with $b \geq 0$ boundary components. In the case Σ is nonorientable, we assume that $b = 0$. Here, by a puncture on Σ we mean a distinguished point, so that Σ is compact. The mapping class group of Σ may be defined in several ways. For our purpose, we define the *mapping class group* of Σ to be the group of isotopy classes of (orientation-preserving if Σ is orientable) diffeomorphisms $\Sigma \to \Sigma$. Here, diffeomorphisms are assumed to be the identity on the points of the boundary and are allowed to permute the punctures. Isotopies fix both the points of the boundary and the punctures.

2. Orientable surfaces

In this section, let S denote a connected orientable surface of genus g with p. We will denote the mapping class group of S by $\mathrm{Mod}_{g,p}^b$. For simplicity, if p or/and b is zero, then we omit it from the notation.

2.1. Homomorphic images of mapping class group.
The mapping class group $\mathrm{Mod}_{g,p}$ is residually finite by the work of Grossmann [12]. See also [21] for a geometric proof this fact. Dehn [7], and independently Lickorish [34], proved that it is also finitely generated. Hence, by a theorem of Mal'cev it is hopfian, i.e. every surjective endomorphism of $\mathrm{Mod}_{g,p}$ is necessarily an automorphism. Dually, it was proved by Ivanov and McCarthy in [23] that the mapping class group is cohopfian, i.e. every injective endomorphism of $\mathrm{Mod}_{g,p}$ must be an automorphism.

Another result concerning the endomorphisms of the mapping class group was obtained by the author in [28]. It states that if $\phi : \mathrm{Mod}_{g,p} \to \mathrm{Mod}_{g,p}$ is an endomorphism such that image of ϕ is of finite index in $\mathrm{Mod}_{g,p}$, then ϕ is indeed an

---

The author was supported in part by the Turkish Academy of Sciences under Young Scientists Award Program (MK/TÜBA-GEBİP 2003-10).

automorphism. In fact, by modifying the proof of the main theorem of [28] slightly, one can extend this result to finite index subgroups of the mapping class groups as follows.

THEOREM 1. *Let S be a compact connected orientable surface of genus g with p punctures. Suppose, in addition, that if $g = 0$ then $p \geq 5$ and if $g = 1$ then $p \geq 3$. Let Γ be a subgroup of finite index in the mapping class group $\mathrm{Mod}_{g,p}$. If $\phi : \Gamma \to \Gamma$ is an endomorphism of Γ such that $\phi(\Gamma)$ is of finite index in Γ, then ϕ is the restriction of an automorphism of $\mathrm{Mod}_{g,p}$. In particular, ϕ is an automorphism of Γ.*

This theorem can proved as follows (c.f. [28]). Since the mapping class group is residually finite and finitely generated, so is any subgroup of finite index, and hence Γ. Now, by a result of Hirshon [16], the restriction of ϕ to $\phi^n(\Gamma)$ is an isomorphism onto its image for some n. It follows from a theorem (Theorem 2 below) of Ivanov [22] for $g \geq 2$ and of the author [27] for $g \leq 1$ that the restriction of ϕ to $\phi^n(\Gamma)$ is the restriction of an automorphism ψ of the mapping class group. Then it can be shown that ϕ itself is the restriction of ψ.

Theorem 1 holds, in particular, for the mapping class group itself. However, we do not know whether the assumption that $\phi^n(\Gamma)$ is of finite index in Γ is necessary for the mapping class group (or for finite index subgroups). In fact, we are not aware of any homomorphisms from the mapping class group onto an infinite index infinite subgroup.

PROBLEM 1. *Is there an endomorphism of the mapping class group of a (closed) orientable surface onto an infinite index infinite subgroup? If there is one such endomorphism f, is it true that f restricted to the image of f^n is injective for some n?*

We conjecture that there is no such f.

There are also results on injective homomorphisms from subgroups of finite index into the extended mapping class group $\mathrm{Mod}^*_{g,p}$, the group of isotopy classes of all (including orientation-reversing) diffeomorphisms of the surface S of genus g with p punctures. Let Γ be a subgroup of finite index in $\mathrm{Mod}^*_{g,p}$. Any injective homomorphism from Γ into $\mathrm{Mod}^*_{g,p}$ is the restriction of an automorphism of $\mathrm{Mod}^*_{g,p}$. This result was proved by Irmak [18, 19] for $g = 2$, $p \geq 2$ and for $g \geq 3$, and by Bell and Margalit [1] for $g = 0$.

One more result about the homomorphisms from the mapping class group we mention is the following.

THEOREM 2. *Let S be a compact connected orientable surface of genus g with p punctures. Suppose, in addition, that if $g = 0$ then $p \neq 4$ and if $g = 1$ then $p \geq 3$. Let Γ_1 and Γ_2 be two subgroups of finite index in the mapping class group $\mathrm{Mod}_{g,p}$. If $\phi : \Gamma_1 \to \Gamma_2$ is an isomorphism, then ϕ is the restriction of an automorphism of $\mathrm{Mod}_{g,p}$. In particular, any automorphism of a finite index subgroup Γ is the restriction of an automorphism of $\mathrm{Mod}_{g,p}$.*

Theorem 2 was proved by Ivanov [22] for $g \geq 2$ and by the author [27] for the remaining nontrivial cases. It was shown in [31] that it does not hold for $(g, b) \in \{(0, 4), (1, 0), (1, 1)\}$ either; in each of these three cases there is a finite index subgroup Γ in $\mathrm{Mod}_{g,p}$ and an automorphism $\phi : \Gamma \to \Gamma$ such that ϕ is not the restriction of any automorphism of $\mathrm{Mod}_{g,p}$. Thus the only unknown case is $(g,b)=(1,2)$.

PROBLEM 2. *Let Γ be a subgroup of finite index in the mapping class group $\mathrm{Mod}_{1,2}$ and let $\phi:\Gamma\to\Gamma$ be an automorphism. Is ϕ the restriction of an automorphism of $\mathrm{Mod}_{1,2}$?* [1]

In all situations we have discussed so far, the domain and the range of the homomorphism lie in the same mapping class group. A result about homomorphisms between the mapping class groups of surfaces of different genera is due to Harvey and the author [14]: if $g > h$, then the image of any homomorphism $\mathrm{Mod}_g \to \mathrm{Mod}_h$ is trivial, with only one exception (in the case $g = 2$, the order of the image is at most 2). It is now natural to ask the following question.

PROBLEM 3. *Let $g > h$ and let Γ be a finite index subgroup of Mod_g. Assume that $g \geq 3$ and $\phi : \Gamma \to \mathrm{Mod}_h$ is a homomorphism.*
  (a) *Is the image of ϕ necessarily finite?*
  (b) *(A weaker version of (a)) Is the image of ϕ necessarily of infinite index in Mod_h.*

On the other hand, it is well known that $H^1(\mathrm{Mod}_g; \mathbb{Z}) = 0$ for all g. Another way of saying this is that any homomorphism $\mathrm{Mod}_g \to \mathbb{Z}$ is trivial. A problem of Ivanov in [25], Problem 2.11 (A), asks whether the same conclusion holds for finite index subgroups:

PROBLEM 4. *Let Γ be a finite index subgroup of Mod_g. Is it true that $H^1(\Gamma; \mathbb{Z}) = 0$?*

There are some partial answers to Problem 4. If $g \geq 3$ and if Γ contains the Torelli group, it was proved by McCarthy [37] that $H^1(\Gamma; \mathbb{Z}) = 0$. However, if $g = 2$, this is not true anymore; the first examples of finite index subgroups in Mod_2 with nontrivial first cohomology were constructed by McCarthy in [37] and by Taherkhani in [43]. Moreover, if $g = 1$ or 2, it was shown in [31] that for each positive integer $n \geq 2$ the mapping class group Mod_g contains a subgroup Γ_n of finite index which admits a homomorphism onto a free group of rank n. In particular, the rank of $H^1(\Gamma_n; \mathbb{Z})$ is at least n.

We note that a positive answer to Problem 1 (a) implies a positive answer to Problem 4. Note also that since there is a subgroup of finite index in Mod_2 mapping onto a finite index subgroup of Mod_1, the hypothesis $g \geq 3$ in Problem 1 is necessary.

A special case of Problem 4 is still interesting. Suppose that $g \geq 3$ and that Γ is s subgroup of finite index in Mod_g containing the Johnson group (or Johnson kernel) $\mathcal{K}_g$, the subgroup of Mod_g generated of by Dehn twists about separating simple closed curves. Is it true that any homomorphism $\Gamma \to \mathbb{Z}$ is trivial?

---

[1] After this paper is written up, Behrstock and Margalit announced in [2] that there exists an isomorphism between finite index subgroups of $\mathrm{Mod}_{1,2}^*$ which is not the restriction of an inner automorphism.

2.2. Problems related to Lefschetz fibrations.

By a result of Donaldson [8], every closed symplectic 4-manifold, perhaps after blowing up, admits a genus g Lefschetz fibration over the 2-sphere S^2 for some g. A Lefschetz fibration on a closed oriented smooth 4-manifold X is a smooth map $f : X \to S^2$ with certain type of singularities. Around each critical point, f is modeled as $f(z_1, z_2) = z_1^2 + z_2^2$ for some orientation-preserving complex coordinates (z_1, z_2). It turns out that the monodromy around each critical value is a right Dehn twist about some simple closed curve.

After fixing a regular fiber S, which a closed oriented surface of genus g, and a certain generating set for the fundamental group of S^2 minus the critical values, a genus g Lefschetz fibration gives a factorization of the identity into right Dehn twists in the mapping class group Mod_g of S. Conversely, such a factorization of the identity in Mod_g gives rise to a genus g Lefschetz fibration. If $g \geq 2$ then by a theorem of Gompf [11] the total space of the Lefschetz fibration admits a symplectic structure. Thus, symplectic 4-manifolds are completely determined by factorizations of the identity in the mapping class group Mod_g. For the definition and the details on Lefschetz fibrations, the reader is referred to [11].

We note that a genus g Lefschetz fibration over a closed orientable surface of genus h can also be defined similarly. In that case, the Lefschetz fibration is determined by a factorization of a product of h commutators into right Dehn twists in Mod_g.

We now discuss some problems motivated by this correspondence.

Let S be a closed connected oriented surface of genus g. For a simple closed curve a, let us denote by t_a the isotopy class of a right Dehn twist along a.

PROBLEM 5. *Suppose that $t_{a_1} t_{a_2} \cdots t_{a_n} = 1$ in Mod_g, where $n \geq 1$. Let G denote the quotient $H_1(S)/\langle [a_1], [a_2], \ldots, [a_n] \rangle$, where $[a_i]$ denotes the homology class of the simple closed curve a_i. Is it true that $b_1(G) \leq g$?*

A positive answer to this problem imply that the total space of every genus g Lefschetz fibration over the 2-sphere has the first Betti number $b_1 \leq g$, whenever the Lefschetz fibration has at least one singular fiber. To the best knowledge of the author, all known examples of Lefschetz fibrations satisfy this conclusion.

Assume that $g \geq 2$. On the closed connected oriented surface S, let us mark a point P. Let Mod_g^1 denote the mapping class group S with one puncture P. In this case, by forgetting that P is marked, we get an epimorphism $\varphi : \mathrm{Mod}_g^1 \to \mathrm{Mod}_g$, whose kernel is isomorphic to the fundamental group of S at the point P.

PROBLEM 6. *Given a factorization $t_{a_1} t_{a_2} \cdots t_{a_n} = 1$ of the identity into a product of right Dehn twists in Mod_g, is it always possible to lift this factorization to a factorization of the identity into a product of n right Dehn twists in Mod_g^1? That is, is it always possible to choose simple closed curves $b_1, b_2, \ldots, b_n$ on S disjoint from P such that $t_{b_1} t_{b_2} \cdots t_{b_n} = 1$ in Mod_g^1 and $\varphi(t_{b_i}) = t_{a_i}$, i.e. b_i is isotopic to a_i through isotopies not fixing P?*

The existence of a lifting of the factorization $t_{a_1} t_{a_2} \cdots t_{a_n} = 1$ in Mod_g to Mod_g^1 implies that the corresponding Lefschetz fibration has a section; by lifting the relation to Mod_g^1, we specify a point on each fiber. Thus, Problem 6 may be rephrased as follows: Does every Lefschetz fibration over S^2 admit a section?

PROBLEM 7. *Suppose that a mapping class $f \in \operatorname{Mod}_g^b$, $b \geq 1$, is a product of right Dehn twists. Does there exist a constant C_f, depending on f, such that whenever f can be written as a product of N right Dehn twists, then $N \leq C_f$?*

It was conjectured by Ozbagci and Stipsicz ([**38**], Conjecture 3.9) that there exists such a constant C_f. If there is no such C_f for some mapping class f, then this would imply that the set of Euler characteristics of all Stein fillings of a fixed contact three manifold is unbounded. It was pointed out to the author by Stipsicz that whether the existence of C_f for all such f implies the boundedness of Euler characteristics of Stein fillings is also an open problem.

For a group G, let $[G, G]$ denote the subgroup generated by all commutators $xyx^{-1}y^{-1}$. For an element of $x \in [G, G]$, let $c(x)$ denote the *commutator length* of x, the minimum number of factors needed to express x as a product of commutators. It is easy to verify that the sequence $c(x^n)$ is sub-additive, i.e., $c(x^{n+m}) \leq c(x^n) + c(x^m)$ for all n, m. From this, it is can be seen easily that the limit

$$\lim_{n \to \infty} \frac{c(x^n)}{n}$$

exists. This limit is called the *stable commutator length* of x, and we denote it by $||x||$.

It is well known that the mapping class group Mod_g is perfect for $g \geq 3$, i.e. every element in Mod_g can be written as a product of commutators. On the other hand, it is not uniformly perfect; there are elements in Mod_g with arbitrarily large commutator length. In fact, it was shown by the author in [**30**] and by Endo and Kotschick in [**9**] that for any nontrivial simple closed curve a on a closed connected orientable surface S, the stable commutator length of a Dehn twist t_a in Mod_g satisfies $\frac{1}{18g-6} \leq ||t_a||$ for $g \geq 3$. In particular, the sequence $c(t_a^n)$ is unbounded. An upper bound for $||t_a||$ is also given in [**30**]; $||t_a|| \leq \frac{3}{20}$ if a is nonseparating, and $||t_a|| \leq \frac{3}{4}$ if a is separating.

For a separating nontrivial simple closed curve a, let the *genus* of a, denoted by $G(a)$, be the minimum of the genera of the two subsurfaces that a bounds. For convention, we set $G(a) = 0$ if a is a nonseparating simple closed curve. Thus, for any two simple closed curves a and b, there exists a homeomorphism of S mapping a to b if and only if $G(a) = G(b)$. This is also equivalent to requiring that t_a is conjugate to t_b in Mod_g. Therefore, if $G(a) = G(b)$ then we have $||t_a|| = ||t_b||$. Now, for any two integers g and n with $0 \leq n \leq \frac{g}{2}$ ($g \geq 3$), we may define a function $\varphi(g, n) = ||t_a||$, where a is any simple closed curve on the genus g surface S with $G(a) = n$. By the previous paragraph, we have

$$\frac{1}{18g - 6} \leq \varphi(g, n) \leq M_n,$$

where $M_0 = \frac{3}{20}$, and $M_n = \frac{3}{4}$ if $n > 0$.

PROBLEM 8. *Compute $\varphi(g, n)$. Is it constant? If not, for $g < h$, compare $\varphi(g, n)$ and $\varphi(h, n)$.*

PROBLEM 9. *Let $g \geq 3$ and $b \geq 1$. Let $a_1, a_2, \ldots$ be an infinite sequence of nonseparating simple closed curves on an oriented surface S of genus g with b boundary components. In the mapping class group Mod_g^b of S, does the limit*

$$\lim_{n \to \infty} \frac{c(t_{a_1} t_{a_2} \cdots t_{a_n})}{n}$$

exist? If it does, is it always positive? More generally, is there a positive number K such that
$$c(t_{a_1} t_{a_2} \cdots t_{a_n}) \geq nK$$
for all n?

Suppose that there is a constant K such that $c(t_{a_1} t_{a_2} \cdots t_{a_n}) \geq nK$ for any sequence of nonseparating simple closed curves $a_1, a_2, \ldots$ and for all n. If $f \in \text{Mod}_g^b$ can be written as a product of right Dehn twists and if $f = t_{a_1} t_{a_2} \cdots t_{a_k}$, then $c(f) = c(t_{a_1} t_{a_2} \cdots t_{a_k}) \geq kK$. Hence, $k \leq c(f)/K$. Therefore, a positive answer to Problem 9, with K independent of sequences, implies a positive answer to Problem 7.

We would like to note that the hypothesis $b \geq 1$ in Problem 9 is necessary. Otherwise, the identity element in Mod_g can be written as a product of right Dehn twists, say $1 = t_{a_1} t_{a_2} \cdots t_{a_n}$. Then for the sequence
$$a_1, a_2, \ldots, a_n, a_1, a_2, \ldots, a_n, a_1, a_2, \ldots, a_n, \ldots$$
the limit in Problem 9 is zero. We would also like to note that when $b \geq 1$, the identity in Mod_g^b cannot be expressed as a product of right Dehn twists. By embedding the surface with boundary into a closed surface of bigger genus, the proof of this follows from a theorem of Smith [41]; on a closed surface S, if $1 = t_{a_1} t_{a_2} \cdots t_{a_n}$, then the curves $a_1, a_2, \ldots, a_n$ fill up S. (I first learned this fact from Rostislav Matveyev in June 2000.)

2.3. Other problems.
We now state a few other problems on the mapping class group of an orientable surface.

PROBLEM 10. *Let r be a positive integer and let Γ_r be the (normal) subgroup of Mod_g generated by the rth powers of all Dehn twists. Is Γ_r of infinite index?*[2]

If we let N_r to be the normal closure in the mapping class group Mod_g of the rth power of a single Dehn twist about a nonseparating simple closed curve, then by a result of Humphries [17] the index of N_r is finite in Mod_g for all g when $r = 2$ and for $(g, r) = (2, 3)$. Thus, the answer to Problem 10 is no in these cases. It is also known from [17] and [10] that N_r is of infinite index if $r \neq 2, 3, 4, 6, 8, 12$. Clearly, N_r is contained in Γ_r.

Let $\text{PMod}_{g,p}^b$ denote the subgroup of $\text{Mod}_{g,p}^b$ consisting of the isotopy classes of those diffeomorphisms which fix each puncture. If $g \geq 4$, then the homology group $H_2(\text{PMod}_{g,p}^b; \mathbb{Z})$ is isomorphic to $\mathbb{Z}^{p+1}$. (cf. [13], [39] and [33].) (We note that the roles of the subscript p and the superscript b are interchanged in [33].)

PROBLEM 11. (a) *It is known from [33] that $H_2(\text{Mod}_3; \mathbb{Z})$ and $H_2(\text{Mod}_3^1; \mathbb{Z})$ are either $\mathbb{Z}$ or $\mathbb{Z} \oplus \mathbb{Z}_2$. What are they? Also compute $H_2(\text{PMod}_{3,p}^b; \mathbb{Z})$ for all p and b.*
(b) *It is also known that $H_2(\text{Mod}_2; \mathbb{Z}) \cong H_2(\text{Mod}_2^1; \mathbb{Z}) \cong \mathbb{Z}_2$. Compute $H_2(\text{PMod}_{2,p}^b; \mathbb{Z})$ for $b \geq 2$ and for all p.*

---

[2]The author was informed by Louis Funar that the answer to this problem is yes.

3. Nonorientable surfaces

Let N be a connected nonorientable surface of genus g with p punctures. We denote the mapping class group of N by $\mathrm{Mod}(N)$. Note that, in this case, we include all diffeomorphisms $N \to N$ into the definition.

Every problem about the mapping class group of an orientable surface may be asked for that of nonorientable surfaces as well. Here, we give a few of them.

It was shown by Wajnryb [44] that, the mapping class groups Mod_g is generated by two elements. These two generators may be chosen to be torsion elements [32]. By the work of Brendle and Farb [5], and Kassabov [24], it can also be generated by four involutions. Also, the extended mapping class groups Mod_g^* is generated by two elements [32] and by three involutions [40]. These results hold true in the presence of one puncture or boundary component as well.

The problem of finding a set of generators for the mapping class group $\mathrm{Mod}(N)$ for a closed surface N was first considered by Lickorish [35]. He proved that $\mathrm{Mod}(N)$ is generated by Dehn twists together with the isotopy class of a so-called Y-homeomorphism. Chillingworth [6] obtained a finite generating set for $\mathrm{Mod}(N)$. Recently, Szepietowski [42] proved that the mapping class group $\mathrm{Mod}(N)$ of a closed nonorientable surface N is generated by three elements and by four involutions.

The first homology groups of $\mathrm{Mod}(N)$ were computed in [26]. If N is of genus four, then $H_1(\mathrm{Mod}_N; \mathbb{Z})$ is isomorphic to $\mathbb{Z}_2 \oplus \mathbb{Z}_2 \oplus \mathbb{Z}_2$. In particular, $\mathrm{Mod}(N)$ cannot be generated by two elements. So the minimal number of generators for $\mathrm{Mod}(N)$ is three when $g = 4$. It is also known that the mapping class group of a Klein bottle is isomorphic to $\mathbb{Z}_2 \oplus \mathbb{Z}_2$ [35].

PROBLEM 12. (a) *Is it possible to generate the mapping class group of a closed nonorientable surface by two elements?*

(b) *Is it possible to generate the mapping class group of a closed nonorientable surface by two torsion elements? three involutions?*

PROBLEM 13. *Compute the (outer) automorphism group of* $\mathrm{Mod}(N)$.

PROBLEM 14. *Let $g > h$, and let N and N' denote the closed nonorientable surfaces of genera g and h respectively. Is it true that any homomorphism $\phi : \mathrm{Mod}(N) \to \mathrm{Mod}(N')$ has finite image?*

PROBLEM 15. *Let $\phi : \mathrm{Mod}(N) \to \mathrm{Mod}(N)$ be a homomorphism such that the image of ϕ is of finite index. Is ϕ necessarily an automorphism? How about if we take the domain of ϕ to be a subgroup of finite index of $\mathrm{Mod}(N)$?*

PROBLEM 16. *Study homomorphisms* $\mathrm{Mod}_g \to \mathrm{Mod}(N)$ *and* $\mathrm{Mod}(N) \to \mathrm{Mod}_g$.

It is known by the work of Birman and Chillingworth [3] that the mapping class group $\mathrm{Mod}(N)$ of a closed nonoreintable surface N of genus g is isomorphic to a subgroup of the extended mapping class group Mod_{g-1}^* of the orientation double cover of N modulo an element of order two. It follows that there are finite index subgroups in $\mathrm{Mod}(N)$ which are isomorphic to a subgroup of Mod_{g-1}.

Acknowledgments. I would like to thank to Columbia University Mathematics Department, where the final writings of the paper was carried out when I was visiting in the spring of 2005. Problem 3 (a) was asked to me by Nikolai Ivanov in an e-mail correspondence about two years ago. Also, I became aware of Problem 6 after an e-mail from Andras Stipsicz. Stipsicz read one of the earlier versions of the paper and made a number of suggestions, especially on problems related to Lefschetz fibrations. Sergey Finashin made several comments on the problems. The referee read the paper very carefully and made many suggestions which improved the presentation. I thank Finashin, Ivanov, Stipsicz and the referee for their suggestions and efforts.

References

[1] B.W. Bell, D. Margalit, *Injections of Artin groups*. Preprint arxiv:math.GR/0501051.

[2] J. Behrstock, D. Margalit, *Curve complexes and finite index subgroups of mapping class groups*. Preprint.

[3] J.S. Birman, D.R.J. Chillingworth, *On the homeotopy group of a non-orientable surface*. Proc. Cambridge Philos. Soc. **71** (1972), 437–448.

[4] J.S. Birman, D.R.J. Chillingworth, *Erratum: "On the homeotopy group of a non-orientable surface"* [Proc. Cambridge Philos. Soc. **71** (1972), 437–448]. Math. Proc. Cambridge Philos. Soc. **136** (2004), 441.

[5] T. Brendle, B. Farb, *Every mapping class group is generated by 6 involutions*. J. Algebra **278** (2004), 187–198.

[6] D.R.J. Chillingworth, *A finite set of generators for the homeotopy group of a non-orientable surface*. Proc. Cambridge Philos. Soc. **65** (1969), 409–430.

[7] M. Dehn, *Die gruppe der abdildungsklassen*. Acta Math. **69** (1938), 135–206.

[8] S. Donaldson, Lefschetz fibrations in symplectic geometry. Proceedings of the International Congress of Mathematicians, Vol. II (Berlin, 1998). Doc. Math. 1998, Extra Vol. II, 309–314

[9] H. Endo, D. Kotschick, *Bounded cohomology and non-uniform perfection of mapping class groups*. Invent. Math. **144** (2001), 169–175.

[10] L. Funar, *On the TQFT representations of the mapping class groups*. Pacific J. Math. **188** (1999), 251–274.

[11] R.E. Gompf, A.I. Stipsicz, 4-manifolds and Kirby calculus. Graduate Studies in Mathematics, 20. American Mathematical Society, Providence, RI, 1999.

[12] E.K. Grossman, *On the residual finiteness of certain mapping class groups*. J. London Math. Soc. (2) **9** (1974/75), 160–164.

[13] J. Harer, *The second homology group of the mapping class group of an orientable surface*. Invent. Math. **72** (1983), 221–239.

[14] W. Harvey, M. Korkmaz, *Homomorphisms from mapping class groups*. Bull. London Math. Soc. **37** (2005), 275–284.

[15] A. Hatcher, W. Thurston, *A presentation for the mapping class group of a closed orientable surface*. Topology **19** (1980), 221–237.

[16] R. Hirshon, *Some properties of endomorphisms in residually finite groups*. J. Austral. Math. Soc. Ser. A **24** (1977), 117–120

[17] S.P. Humphries, *Normal closures of powers of Dehn twists in mapping class groups*. Glasgow Math. J. **34** (1992), 313–317.

[18] E. Irmak, *Superinjective simplicial maps of complexes of curves and injective homomorphisms of subgroups of mapping class groups*. Topology **43** (2004), 513–541.

[19] E. Irmak, *Superinjective simplicial maps of complexes of curves and injective homomorphisms of subgroups of mapping class groups II*. Topology Appl. **153** (2006), 1309–1340.

[20] N.V. Ivanov, *Automorphisms of Teichmller modular groups*. in Topology and Geometry—Rohlin Seminar, Lecture Notes in Math. 1346, Springer-Verlag, Berlin, 1988, 199–270.

[21] N.V. Ivanov, *Residual finiteness of modular Teichmller groups*. Siberian Math. J. **32** (1991), 148–150.

[22] N.V. Ivanov, *Automorphisms of complexes of curves and of Teichmuller spaces*. International Mathematics Research Notices **1997**, 651–666.

[23] N.V. Ivanov, J. D. McCarthy, *On injective homomorphisms between Teichmüller modular groups. I*. Invent. Math. **135** (1999), 425–486.

[24] M. Kassabov, *Generating mapping class groups by involutions*. Preprint arxiv:math.GT/0311455.

[25] R. Kirby, *Problems in low-dimensional topology*. in Geometric Topology (W. Kazez ed.) AMS/IP Stud. Adv. Math. vol 2.2, American Math. Society, Providence 1997.

[26] M. Korkmaz, *First homology group of mapping class groups of nonorientable surfaces.* Math. Proc. Camb. Phil. Soc. **123** (1998), 487-499.

[27] M. Korkmaz, *Automorphisms of complexes of curves on punctured spheres and on punctured tori.* Topology and its Applications **95** (2) (1999), 85–111.

[28] M. Korkmaz, *On endomorphisms of surface mapping class groups.* Topology **40** (2001), 463-467.

[29] M. Korkmaz, *Mapping class groups of nonorientable surfaces.* Geometriae Dedicata **89** (2002), 109-133.

[30] M. Korkmaz, *Stable commutator length of a Dehn twist.* Michigan Math. J. **52** (2004), 23–31.

[31] M. Korkmaz, *On cofinite subgroups of mapping class groups.* Turkish Journal of Mathematics **27** (2003), 115-123.

[32] M. Korkmaz, *Generating the surface mapping class group by two elements.* Trans. Amer. Math. Soc. **357** (2005), 3299–3310.

[33] M. Korkmaz, A. Stipsicz, *The second homology groups of mapping class groups of orientable surfaces.* Math. Proc. Camb. Phil. Soc. **134** (2003), 479-489.

[34] W. B. R. Lickorish, *A finite set of generators for the homeotopy group of a 2-manifold.* Math. Proc. Camb. Phil. Soc. **60** (1964), 769–778.

[35] W. B. R. Lickorish, *Homeomorphisms of non-orientable two-manifolds.* Proc. Camb. Phil. Soc. **59** (1962), 307-317.

[36] J. D. McCarthy, *Automorphisms of surface mapping class groups. A recent theorem of N. Ivanov.* Invent. Math. **84** (1986), 49–71.

[37] J. D. McCarthy, *On the first cohomology group of cofinite subgroups in surface mapping class groups.* Topology **40** (2001), 401–418.

[38] B. Ozbagci, A.I. Stipsicz, *Contact 3-manifolds with infinitely many Stein fillings.* Proc. Amer. Math. Soc. **132** (2004), 1549–1558.

[39] W. Pitsch *Un calcul élémentaire de $H_2(\mathcal{M}_{g,1}, Z)$ pour $g \geq 4$.* C. R. Acad. Sci. Paris Sr. I Math. **329** (1999), 667–670.

[40] M. Stukow, *The extended mapping class group is generated by 3 symmetries.* C. R. Math. Acad. Sci. Paris **338** (2004), 403–406.

[41] I. Smith, *Geometric monodromy and the hyperbolic disc.* Q. J. Math. **52** (2001), 217–228.

[42] B. Szepietowski, *The mapping class group of a nonorientable surface is generated by three elements and by four involutions.* Preprint.

[43] F. Taherkhani, *The Kazhdan property of the mapping class group of closed surfaces and the first cohomology group of its cofinite subgroups.* Experimental Mathematics **9** (2000), 261-274.

[44] B. Wajnryb, *Mapping class group of a surface is generated by two elements.* Topology **35** (1996), 377-383.

DEPARTMENT OF MATHEMATICS, MIDDLE EAST TECHNICAL UNIVERSITY, 06531 ANKARA, TURKEY

E-mail address: korkmaz@arf.math.metu.edu.tr

The Mapping Class Group and Homotopy Theory

Ib Madsen

1. The problems and questions to be discussed below all relate to the recent developments around the generalized Mumford conjecture. In a nutshell, the new homotopical results about the stable mapping class group are based upon tools from high dimensional manifold theory, algebraic K-theory and infinite loop space theory. The papers most relevant for the discussions in this note include [2], [4], [5], [6], [13], [14] and [19].

2. I will use the notation $\Gamma^s_{g,b}$ for the mapping class group of a genus g surface with b boundary circles and s marked points; the boundary and the marked points are left pointwise fixed. For $g > 3$, $\Gamma^s_{g,b}$ is a perfect group, so we may apply Quillen's plus construction to its classifying space. This turns $B\Gamma^s_{g,b}$ into a simply connected space $(B\Gamma^s_{g,b})^+$ and there is a homology equivalence

$$B\Gamma^s_{g,b} \to \left(B\Gamma^s_{g,b}\right)^+.$$

For $b > 0$ one has the standard group homomorphisms from $\Gamma^s_{g,b}$ to $\Gamma^s_{g+1,b}$ and $\Gamma^s_{g,b-1}$, respectively. The map into $\Gamma_{g+1,b}$ glues a torus with two discs removed to one of the boundary circles; the map into $\Gamma_{g,b-1}$ glues a disc to one of the boundary circles. They give maps

(1) $$B\Gamma^s_{g,b} \to B\Gamma^s_{g+1,b}, \quad B\Gamma^s_{g,b} \to B\Gamma^s_{g,b-1}$$

that induce isomorphisms on integral homology in degrees less than $(g-1)/2$. This is Harer's stability theorem [7] and Ivanov's improvement [11].

The direct limit of $\Gamma^s_{g,b}$ as $g \to \infty$ is the stable mapping class group $\Gamma^s_{\infty,b}$. Harer stability shows that $(B\Gamma^s_{\infty,b})^+$ is independent of b up to homotopy equivalence. We write $(B\Gamma^s_\infty)^+$ for the common homotopy type. There is a fibration

$$B\Gamma_{g,b+s} \to B\Gamma^s_{g,b} \to (\mathbb{C}P^\infty)^s$$

(associated to the extension $\mathbb{Z}^s \to \Gamma_{g,b+s} \to \Gamma^s_{g,b}$). We have the forgetful map $\Gamma^s_{g,b} \to \Gamma_{g,b}$ and by stability the composition

$$(B\Gamma_{g,b+s})^+ \to (B\Gamma^s_{g,b})^+ \to (B\Gamma_{g,b})^+$$

is $(g-1)/2$ connected. Thus

(2) $$(B\Gamma^s_\infty)^+ \simeq (B\Gamma_\infty)^+ \times (\mathbb{C}P^\infty)^s.$$

The generalized Mumford conjecture identifies the homotopy type of $(B\Gamma_\infty)^+$.

Let $G(d,n)$ be the Grassmann manifold of oriented d-planes in $\mathbb{R}^{d+n}$ and define
$$U_{d,n}^\perp = \{(v,V) \mid V \in G(d,n), v \perp V\}.$$
This is an n-dimensional vector bundle over $G(d,n)$ that restricts to $U_{d,n-1}^\perp \times \mathbb{R}$ over $G(d, n-1)$. The one-point compactification or Thom space of $U_{d,n}^\perp$ is denoted $\mathrm{Th}(U_{d,n}^\perp)$ and we have maps
$$\sigma_n : \mathrm{Th}(U_{d,n-1}^\perp) \wedge S^1 \to \mathrm{Th}(U_{d,n}^\perp).$$
The family $\{\mathrm{Th}(U_{d,n}^\perp), \sigma_n\}_{n \geq 0}$ defines a spectrum $\mathbb{G}_{-d}$, graded so that the Thom class sits in degree $-d$. The associated infinite loop space is the direct limit
$$(3) \qquad \Omega^\infty \mathbb{G}_{-d} = \operatorname*{colim}_n \Omega^{n+d} \mathrm{Th}(U_{d,n}^\perp).$$
For $d = 2$, we have $G(2,\infty) \simeq \mathbb{C}P^\infty$ and it is customary to use complex grading and write $\mathbb{C}P_{-1}^\infty$ instead of $\mathbb{G}_{-2}$ and $\Omega^\infty \mathbb{C}P_{-1}^\infty$ instead of $\Omega^\infty \mathbb{G}_{-2}$.

THEOREM 2.1 ([14]). *There is a homotopy equivalence*
$$\alpha : \mathbb{Z} \times B\Gamma_\infty^+ \to \Omega^\infty \mathbb{C}P_{-1}^\infty.$$

3. The action of $\Gamma_{g,1}$ on $H_1(F_{g,1}, \partial)$ induces a representation of $\Gamma_{g,1}$ into $Sp_{2g}(\mathbb{Z})$. Its kernel is the Torelli group $T_{g,1}$. For $g \to \infty$ we get a fibration
$$BT_{\infty,1} \xrightarrow{t} B\Gamma_{\infty,1} \xrightarrow{\rho} BSp(\mathbb{Z})$$
with a highly non-trivial action of $Sp(\mathbb{Z})$ on $H_*(BT_{\infty,1})$. Applying the plus construction yields
$$(4) \qquad \rho : \mathbb{Z} \times B\Gamma_\infty^+ \to \mathbb{Z} \times BSp(\mathbb{Z})^+.$$
The target is the symplectic algebraic K-theory $KSp(\mathbb{Z})$. It maps to $K(\mathbb{Z}) = \mathbb{Z} \times BGl(\mathbb{Z})^+$. I refer to [16] for a (conjectural) discussion of the homotopy type $K(\mathbb{Z})$. Presumably one can give a similar discussion of $KSp(\mathbb{Z})$. In particular the homotopy type of the p-localization of $KSp(\mathbb{Z})$ should be within reach for regular primes p, pending a proof of the Lichtenbaum-Quillen conjecture.

The cohomology of $\Omega^\infty \mathbb{C}P_{-1}^\infty$, and hence of $B\Gamma_\infty$, is well understood both rationally and with finite coefficients, [4].

PROBLEM 3.1. *Give a homotopy theoretic construction of a map*
$$\rho_h : \Omega^\infty \mathbb{C}P_{-1}^\infty \to KSp(\mathbb{Z})$$
with $\rho \simeq \rho_h \circ \alpha$, at least after localization at a regular prime. (The 2-local case is of particular interest; cf. [12]).

The rational cohomology of $\Omega^\infty \mathbb{C}P_{-1}^\infty$ is easy to list. For each connected component,
$$H^*(\Omega_k^\infty \mathbb{C}P_{-1}^\infty; \mathbb{Q}) = \mathbb{Q}[\kappa_1', \kappa_2', \ldots]$$
where $\kappa_i = \alpha^*(\kappa_i')$ are the standard Miller-Morita-Mumford classes of degree $2i$, see [5], [15], [17] for further details. The rational cohomology of $BSp(\mathbb{Z})$ was calculated by Borel:
$$H^*(BSp(\mathbb{Z}); \mathbb{Q}) \cong H^*(Sp/U; \mathbb{Q}) \cong \mathbb{Q}[c_1, c_3, \ldots].$$

where c_{2i+1} is the image of the $(2i + 1)$st Chern class under the natural map from Sp/U to BU. Moreover, $\rho^*(c_{2i+1})$ is a non-zero multiple of κ'_{2i+1} so that κ_{2i+1} restricts to zero in the rational cohomology of the Torelli group, [17].

QUESTION 3.2. Is $\kappa_{2i} = 0$ in $H^*(BT_\infty; \mathbb{Q})$?[1]

Very little is known about the cohomology of the Torelli group past dimension 1, and as far as I know it might be possible (although hard to believe) that

(5) $$BT_\infty \to hofiber(B\Gamma_\infty^+ \to BSp(\mathbb{Z})^+)$$

is identically zero on cohomology.

The calculations in [4] show that $H^*(B\Gamma_\infty; \mathbb{Z})$ contains a wealth of torsion classes and that there are torsion classes of any order. Some of these classes might be of interest in other areas of mathematics. But the difficulty is that the description of torsion classes from [4] is indirect and complicated, and very hard to communicate to non-specialists.

PROBLEM 3.3. Find a direct description of some particular simple torsion classes in $H^*(B\Gamma_\infty; \mathbb{Z})$.

There is an interesting connection between the higher Reidemeister torsion classes from [10] and the classes κ_i that require further study. Indeed both [10] and [14] use parametrized Morse theory. Could Problem 3.3 be related to "modular higher Reidemeister torsion"?

Let F_g denote the free groups on g generators and $\text{Aut}(F_g)$ its automorphism group. It obviously maps into $\text{Aut}(F_{g+1})$ and one can form $B\text{Aut}(F_\infty)$ and also its plus construction $B\text{Aut}(F_\infty)^+$. ($\text{Aut}(F_\infty)$ contains a perfect index two subgroup). A. Hatcher has shown that $\Omega^\infty S^\infty$ is a direct factor of $\mathbb{Z} \times B\text{Aut}(F_\infty)^+$ (up to homotopy), [8].

QUESTION 3.4. Is $\mathbb{Z} \times B\text{Aut}(F_\infty)^+$ homotopy equivalent to $\Omega^\infty S^\infty$?

4. The interplay between homotopy theory and the mapping class group is inspired by the study of conformal field theories, seen in [18] as representations of the surface category. The space of morphisms in Segal's surface category is the moduli space of Riemann surfaces with parametrized boundary circles. Teichmüller theory implies that this morphism space is homotopy equivalent to $\sqcup B\Gamma_{g,b}$, $g \geq 0, b \geq 0$.

Under suitable stability conditions, in our case Harer's stability theorem, there is a close relationship between the loop space of the classifying space of a category and its space of morphisms. This was used by Tillmann to show that $B\Gamma_\infty^+$ is an infinite loop space, [19], and in [13] to show that the map α of Theorem 2.1 is an infinite loop map. The situation is similar to that of algebraic K-theory, see e.g. [20].

The surface category is a special case of the cobordism category $\mathcal{C}_d$ that exists in all dimensions $d \geq 1$. The objects of $\mathcal{C}_d$ are oriented closed $(d-1)$-dimensional submanifolds of $\mathbb{R}^\infty$ and morphisms are oriented compact d-dimensional submanifolds of $[0, a] \times \mathbb{R}^\infty$, $0 < a$, that meet the walls $\{0, a\} \times \mathbb{R}^\infty$ transversely, see [6] for details. For $d = 2$, $\mathcal{C}_d$ is homotopy equivalent to Segal's surface category, i.e. they have homotopy equivalent classifying spaces, at least rationally, [13].

---

[1] This has been solved by Soren Galatius.

THEOREM 4.1. [**6**]. *For each $d \geq 1$, there is a homotopy equivalence*
$$\alpha_d : B\mathcal{C}_d \to \Omega^{\infty-1}\mathbb{G}_{-d}.$$

The target is the deloop of the space defined in Theorem 2.1 above. In particular, $B\mathcal{C}_2 \simeq \Omega^{\infty-1}\mathbb{C}P^\infty_{-1}$ and $\Omega B\mathcal{C}_2 \simeq \mathbb{Z} \times B\Gamma^+_\infty$.

For $d \neq 2$ there is no replacement of Harer stability that I know of, and therefore no obvious relationship between $\Omega B\mathcal{C}_d$ and diffeomorphism groups of d-manifolds. Nevertheless it is a natural question to ask how Theorem 4.1 is related to the traditional approach to high dimensional manifolds via surgery and pseudo-isotopy theory. These tools work simpler in the category of topological manifolds.

PROBLEM 4.2. Find an analogue of Theorem 4.1 for topological manifolds, and relate it to surgery and pseudo-isotopy theory.

Waldhausen's functor $A(X)$ is the algebraic K-theory of the "ring" $\Omega^\infty S^\infty(\Omega X_+)$. It is a homotopy theoretic construction which, quite explicitly, contains information about both $\mathrm{Diff}(X \times I, \partial X \times I \cup X \times 0)$ and $\mathrm{Top}(X \times I, \partial X \times I \cup X \times 0)$ where X is a compact d-manifold in a range of dimensions that increase with d, cf [**20**], [**21**]. Less explicitly $A(X)$ also relates to $\mathrm{Diff}(X)$ and $\mathrm{Top}(X)$, cf. [**22**].

The topological cyclic homology functor $\mathrm{TC}(X)$, introduced in [**1**], is an enriched version of Connes' cyclic homology. $A(X)$ maps to $\mathrm{TC}(X)$ via the cyclotomic trace, but $\mathrm{TC}(X)$ and the associated functor for rings (or linear categories) is of interest in its own right, cf. [**9**].

The p-adic completion $\mathrm{TC}(X)^\wedge_p$ can be expressed in terms of more standard constructions in homotopy theory involving the free loop space. For $X = pt$,

(6) $$\mathrm{TC}(pt)^\wedge_p \simeq (\Omega^\infty S^\infty \times \Omega^{\infty-1}\mathbb{C}P^\infty_{-1})^\wedge_p.$$

QUESTION 4.3. Is there a geometric map from $B\mathcal{C}_2$ into topological cyclic homology of a point?

A recent manuscript by Kevin Costello, [**3**], relates conformal field theories to (the linear) Hochschild homology, and this might well be the place to start. In the same paper Costello announces a construction of the Deligne-Mumford compactification of the moduli space of Riemann surface in terms of the open moduli space. There is also preliminary work of Søren Galatius and Yasha Eliashberg which gives a homotopical description of a partial compactification of the stable moduli space.

QUESTION 4.4. Can one generalize Theorem 2.1 to the Deligne-Mumford compactification of the moduli space of Riemann surfaces?

References

[1] M. Böksted, W.-C. Hsiang, I. Madsen, *The cyclotomic trace and algebraic K-theory of spaces*, Invent. Math. **111** (1993), 465–540.

[2] R. Cohen, I. Madsen, U. Tillmann, *The stable moduli space of Riemann surfaces in a background space*. In preparation.

[3] K. Costello, *Topological conformal field theories and Calabi-Yau categories*, arXiv:math.QA/0412149 V5.

[4] S. Galatius, *The mod p homology of the stable mapping class group*, Topology **43** (2004, 1105–1132.

[5] S. Galatius, I. Madsen, U. Tillmann, *Divisibility of the stable Miller-Morita-Mumford classes*, preprint, Aarhus University (2005), *J.AMS* (to appear).

[6] S. Galatius, I. Madsen, U. Tillmann, M. Weiss, *The homotopy type of the cobordism category*, arXiv:math.AT/0605259.

[7] J. Harer, Stability of the homology of the mapping class groups of orientable surfaces, *Annals of Math.* **121** (1985), 215–249.

[8] A. Hatcher, Homoloical stability for the automorphism groups of free groups, Comment. Math. Helv. **70** (1995), no. 1, 39–62.

[9] L. Hesselholt, I. Madsen, On the K-theory of local fields, *Annals of Math.* **158** (2003), 1–113.

[10] K. Igusa, Higher Franz-Reidemeister torsion, *Studies in Advanced Mathematics*, vol. **31**, AMS and International Press, 2000.

[11] N.V. Ivanov, Stabilization of the homology of Teichmüller modular groups, *Leningrad Math. J.* **1** (1990), 675–691.

[12] I. Madsen, C. Schlichtkrull, The circle transfer and K-theory, *AMS Contemporary Math.* **258** (2000), 307–328.

[13] I. Madsen, U. Tillmann, The stable mapping class groups and $Q(\mathbb{C}P_+^\infty)$, *Invent. Math.* **145** (2001), 509–544.

[14] I. Madsen, M. Weiss, The stable moduli space of Riemann surfaces: Mumford's conjecture, *Annals of Math.* (to appear).

[15] E. Miller, The homology of the stable mapping class group, *J. Diff. Geom.* **24** (1986), 1–14.

[16] S.A. Mitchell, On the Lichtenbaum-Quillen conjectures from a stable homotopy-theoretic viewpoint, in *Algebraic Topology and its Applications Math. Sci. Research Inst. Publications* (1994), 163–240, Springer-Verlag.

[17] S. Morita, Characteristic classes of surface bundles, *Invent. Math.* **90** (1987), 551–577.

[18] G. Segal, The definition of conformal field theory, in *Topology, Geometry and Quantum Field Theory, London Math. Soc. Lecture Notes Series* **308** (2004), 423–577.

[19] U. Tillmann, On the homotopy of the stable mapping class group, *Invent. Math.* **130** (1997), 257–275.

[20] F. Waldhausen, Algebraic K-theory of spaces, in *Algebraic and Geometric Topology*, LNM **1126** (1985), 318–419, Springer-Verlag.

[21] F. Waldhausen, Algebraic K-theory of spaces, a manifold approach, in *Current Trends in Algebraic Topology, CMS Conference Proceedings*, Part I (1982), 141–184, Amer. Math. Soc. and Can. Math. Soc.

[22] M. Weiss, B. Williams, Automorphisms of manifolds and algebraic K-theory, *K-Theory* **1** (1988), 575–626.

DEPARTMENT OF MATHEMATICAL SCIENCES, UNIVERSITY OF AARHUS

Probing Mapping Class Groups Using Arcs

R. C. Penner

ABSTRACT. The action of the mapping class group of a surface on the collection of homotopy classes of disjointly embedded curves or arcs in the surface is discussed here as a tool for understanding Riemann's moduli space and its topological and geometric invariants. Furthermore, appropriate completions, elaborations, or quotients of the set of all such homotopy classes of curves or arcs give for instance Thurston's boundary for Teichmüller space or a combinatorial description of moduli space in terms of fatgraphs. Related open problems and questions are discussed.

Introduction

One basic theme in this paper on open problems is that the action of the mapping class group on spaces of measured foliations, and in particular on weighted families of curves and arcs, is calculable and captures the dynamics of homeomorphisms of the surface both on the surface itself and on its Teichmüller space. Another basic theme is that suitable spaces of arcs can be exploited to give group-theoretic and other data about the mapping class groups as well as their subgroups and completions. The author was specifically given the task by the editor of presenting open problems on his earliest and latest works in [23] and [46]–[49], which respectively develop these two themes and are also surveyed here.

1. Definitions

Let $F = F_{g,r}^s$ denote a smooth oriented surface of genus $g \geq 0$ with $r \geq 0$ boundary components and $s \geq 0$ punctures, where $2g - 2 + r + s > 0$. The mapping class group $MC(F)$ of F is the collection of isotopy classes of all orientation-preserving homeomorphisms of F, where the isotopies and homeomorphisms necessarily setwise fix the boundary ∂F and setwise fix the collection of punctures. Let $PMC(F) < MC(F)$ denote the pure mapping class group whose homeomorphisms and isotopies pointwise fix each puncture and each boundary component.

But one aspect of Bill Thurston's seminal contributions to mathematics, [37] among other things provides a natural spherical boundary of the Teichmüller space $\mathcal{T}(F)$ of a surface F with negative Euler characteristic by an appropriate space of "projectivized measured foliations of compact support" in F. Specifically [37, 9, 32], let $\mathcal{MF}(F)$ denote the space of all isotopy classes rel ∂F of Whitehead

©2006 American Mathematical Society

equivalence classes of measured foliations in F, and let $\mathcal{MF}_0(F)$ denote the subspace comprised of those foliations with "compact support", i.e., leaves are disjoint from a neighborhood of the punctures and boundary, and no simple closed leaf is puncture- or boundary-parallel.

Define the projectivized spaces

$$\mathcal{PF}_0(F) = [\mathcal{MF}_0(F) - \{\vec{0}\}]/\mathbf{R}_{>0}$$
$$\subseteq [\mathcal{MF}(F) - \{\vec{0}\}]/\mathbf{R}_{>0} = \mathcal{PF}(F),$$

where $\vec{0}$ denotes the empty foliation and $\mathbf{R}_{>0}$ acts by homothety on transverse measures. Thus, $\mathcal{MF}(F_{g,r-1}^{s+1}) \subseteq \mathcal{MF}(F_{g,r}^s)$ and $\mathcal{MF}_0(F_{g,r}^s) \approx \mathcal{MF}_0(F_{g,0}^{r+s})$ with corresponding statements also for projectivized foliations.

A basic fact (as follows from the density of simple closed curves in $\mathcal{PF}(F)$) is that the action of $MC(F)$ or $PMC(F)$ on $\mathcal{PF}_0(F)$ has dense orbits, so the quotients are non-Hausdorff. (The action is actually minimal in the sense that every orbit is dense, and in fact, the action is ergodic for a natural measure class as independently shown by Veech [43] and Masur [44].)

We shall say that a measured foliation or its projective class *fills* F if every essential simple closed curve has positive transverse measure and that it *quasi fills* F if every essential curve with vanishing transverse measure is puncture-parallel.

Define the *pre-arc complex* $\mathcal{A}'(F)$ to be the subspace of $\mathcal{PF}(F)$ where each leaf in the underlying foliation is required to be an arc connecting punctures or boundary components, and define the open subspace $\mathcal{A}'_\#(F) \subseteq \mathcal{A}'(F)$ where the foliations are furthermore required to fill F.

In particular in the *punctured case* when $r = 0$, $s \geq 1$, and F has negative Euler characteristic the product $\mathcal{T}(F) \times \Delta^{s-1}$ of Teichmüller space with an open $(s-1)$-dimensional simplex Δ^{s-1} is canonically isomorphic to $\mathcal{A}'_\#(F)$, and in fact this descends to an isomorphism between the *filling arc complex* $\mathcal{A}_\#(F) = \mathcal{A}'_\#(F)/PMC(F)$ and the product $\mathcal{M}(F) \times \Delta^{s-1}$ of Riemann's (pure) moduli space $\mathcal{M}(F) = \mathcal{T}(F)/PMC(F)$ with the simplex [12, 15, 36, 27]. Thus, the *arc complex* $\mathcal{A}(F) = \mathcal{A}'(F)/PMC(F)$ forms a natural combinatorial compactification of $\mathcal{A}_\#(F) \approx \mathcal{M}(F) \times \Delta^{s-1}$. (In fact, this is *not* the most useful combinatorial compactification when $s > 1$, cf. [30, 46], where one chooses from among the punctures a distinguished one.)

Another special case r, s of interest here is the case of *bordered surfaces* when $r \geq 1$ and $s \geq 0$. Choose one distinguished point on each boundary component, and define the analogous complexes $Arc'(F) \subseteq \mathcal{PF}(F)$, where leaves are required to be asymptotic to the distinguished points on the boundary (and may not be asymptotic to punctures) with its quasi filling subspace $Arc'_\#(F) \subseteq Arc'(F)$ and quotients $Arc_\#(F) = Arc'_\#(F)/PMC(F) \subseteq Arc(F) = Arc'(F)/PMC(F)$. In analogy to the punctured case, $Arc(F)$ is proper homotopy equivalent to Riemann's moduli space of F (as a bordered surface with one distinguished point in each geodesic boundary component) modulo a natural action of the positive reals [31]. Furthermore [30], Arc-complexes occur as virtual links of simplices in the $\mathcal{A}$-complexes, and the local structure of the $\mathcal{A}$-complexes is thus governed by the global topology of Arc-complexes. In fact, the Arc-complexes are stratified spaces of a particular sort as explained in §5.

There are other geometrically interesting subspaces and quotients of $\mathcal{MF}(F)$ or $\mathcal{PF}(F)$, for instance the curve complex of Harvey [14] or the complex of pants decompositions of Hatcher-Thurston [13], which are surely discussed elsewhere in this volume.

As was noted before, the quotients $\mathcal{PF}_0(F)/PMC(F) \subseteq \mathcal{PF}(F)/PMC(F)$ are maximally non-Hausdorff, and yet for $r > 0$, $\mathcal{PF}(F)/PMC(F)$ contains as an open dense subset the (Hausdorff) stratified space $Arc(F) = Arc'(F)/PMC(F)$; in particular, for the surface $F = F_{0,r}^s$ with $r + s \leq 3$, $PMC(F_{0,r}^s)$ is the free abelian group generated by Dehn twists on the boundary, $Arc(F_{0,r}^s)$ is piecewise-linearly homeomorphic to a sphere of dimension $3r + 2s - 7$, and it is not difficult to understand the non-Hausdorff space $\mathcal{PF}(F)/PMC(F)$ with its natural foliation. This leads to our first problem:

PROBLEM 1. Understand either classically or as quantum geometric objects the non-Hausdorff quotients of $\mathcal{PL}_0(F)$ or $\mathcal{PL}(F)$ by $MC(F)$ or $PMC(F)$.

$\mathcal{T}(F)$ has been quantized in [5] and [17] as surveyed in [6] and [35], respectively, and $\mathcal{PL}_0(F_{1,0}^1)$ has been quantized in [6]. In interesting contrast, [21, 54] has described a program for studying real quadratic number fields as quantum tori limits of elliptic curves, thus quantizing limiting curves rather than Teichmüller space.

2. Dehn-Thurston coordinates

Fix a surface $F = F_{g,r}^s$. In this section, we introduce global Dehn-Thurston coordinates from [23] for $\mathcal{MF}_0(F)$ and $\mathcal{MF}(F)$.

Define a *pants decomposition* of F to be a collection $\mathcal{P}$ of curves disjointly embedded in F so that each component of $F - \cup \mathcal{P}$ is a *generalized pair of pants* $F_{0,r}^s$ with $r + s = 3$. One easily checks using Euler characteristics that there are $3g - 3 + 2r + s$ curves in a pants decomposition of F and $2g - 2 + r + s$ generalized pairs of pants complementary to $\mathcal{P}$.

Given a measured foliation $\mathcal{F}$ in a generalized pair of pants P, where ∂P has components ∂_i, define the triple m_i of *intersection numbers* given by the transverse measure of ∂_i of $\mathcal{F}$, for $i = 1, 2, 3$.

DEHN-THURSTON LEMMA. *Isotopy classes (not necessarily the identity on the boundary) of Whitehead equivalence classes of non-trivial measured foliations in the pair of pants P with no closed leaves are uniquely determined by the triple (m_1, m_2, m_3) of non-negative real intersection numbers, which are subject only to the constraint that $m_1 + m_2 + m_3$ is positive.*

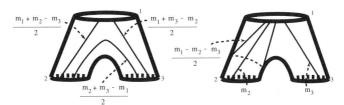

1a. Triangle inequality   **1b.** The case $m_1 > m_2 + m_3$

FIGURE 1. Constructing measured foliations in pants

PROOF. The explicit construction of a measured foliation realizing a putative triple of intersection numbers is illustrated in Figure 1 in the two representative cases that m_1, m_2, m_3 satisfy all three possible (weak) triangle inequalities in Figure 1a or perhaps $m_1 \geq m_2 + m_3$ in Figure 1b. The other cases are similar, and the projectivization of the positive orthant in (m_1, m_2, m_3)-space is illustrated in Figure 2. Elementary topological considerations show that any measured foliation of P is isotopic to a unique such foliations keeping endpoints of arcs in the boundary of P, completing the proof. □

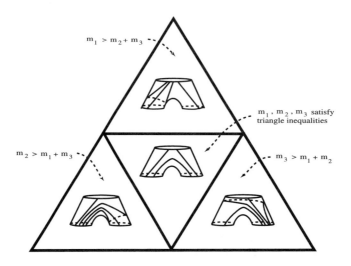

FIGURE 2. Measured foliations in pants

In order to refine the Dehn-Thurston Lemma and keep track of twisting around the boundary, we shall introduce in each component ∂_i of the boundary ∂F an arc $w_i \subseteq \partial_i$ called a *window*, for $i = 1, 2, 3$. We require that the support of the restriction to ∂P of $\mathcal{F}$ lie in the union of the windows, so-called *windowed measured foliations*. (Collapsing each window to a point gives a surface with a distinguished point in each boundary component, so a windowed measured foliation in P gives rise to an element of $Arc(P)$ in the sense of §1.) We seek the analogue of the Dehn-Thurston Lemma for windowed isotopy classes.

To this end, there are two conventions to be made:
1) when a leaf of $\mathcal{F}$ connects a boundary component to itself, i.e., when it is a loop, then it passes around a specified leg, right or left, of P as illustrated in Figure 3a-b, i.e., it contains a particular boundary component or puncture in its complementary component with one cusp;
2) when a leaf is added to the complementary component of a loop in P with two cusps, then it either follows or precedes the loop as illustrated in Figure 3c-d.

For instance in Figure 2, the conventions are: 1) around the right leg for loops; and 2) on the front of the surface. We shall call these the *standard twisting conventions*.

Upon making such choices of convention, we may associate a twisting number $t_i \in \mathbf{R}$ to $\mathcal{F}$ as follows. Choose a regular neighborhood of ∂P and consider the

Figure 3a. Around the right leg.   **Figure 3b.** Around the left leg.

Figure 3c. Arc follows loop.   **Figure 3d.** Arc precedes loop.

FIGURE 3. Twisting conventions

sub-pair of pants $P_1 \subseteq P$ complementary to this regular neighborhood. Given a weighted arc family α in P, by the Dehn-Thurston Lemma, we may perform an isotopy in P supported on a neighborhood of P_1 to arrange that $\alpha \cap P_1$ agrees with a conventional windowed arc family in P_1 (where the window in ∂P_1 arises from that in ∂P in the natural way via a framing of the normal bundle to ∂P in P).

For such a representative of $\mathcal{F}$, we finally consider its restriction to each annular neighborhood A_i of ∂_i. Choose another arc a whose endpoints are *not* in the windows (and again such an arc is essentially uniquely determined up to isotopy rel windows from a framing of the normal bundle to ∂P in P in the natural way); orient a and each component arc of $(\cup \alpha) \cap A_i$ from ∂P_1 to ∂P, and let t_i be the signed (weighted) intersection number of a with the (weighted) arc family $(\cup \alpha) \cap A_i$, for $i = 1, 2, 3$.

 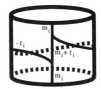

4a. Right twisting for $t_i \geq 0$ **4b.** Left twisting for $t_i \leq 0$

FIGURE 4. Windowed measured foliations in the annulus

As illustrated in Figure 4, all possible real twisting numbers $-m_i \leq t_i \leq m_i$ arise provided $m_i \neq 0$, where again in this figure, the indicated "weight" of a component arc is the width of a band of leaves parallel to the arc. By performing Dehn twists along the core of the annulus, it likewise follows that every real twisting number t_i occurs provided $m_i \neq 0$. Again, elementary topological considerations show that each windowed isotopy class of a windowed measured foliation is uniquely determined by its invariants:

LEMMA 1. *Points of $\mathcal{MF}(P)$ are uniquely determined by the triple $(m_i, t_i) \in \mathbf{R}_{\geq 0} \times \mathbf{R}$, which are subject only to the constraint that $\forall i = 1, 2, 3 (m_i = 0 \Rightarrow t_i \geq 0)$.*

One difference between the Dehn-Thurston Lemma and Lemma 1 is that closed leaves are permitted in the latter (but not in the former), where the coordinates $m_i = 0$ and $t_i = |t_i| > 0$ correspond to the class of a foliated annulus of width t_i whose leaves are parallel to ∂_i. In the topology of projective measured foliations, extensive twisting to the right or left about ∂_i approaches the curve parallel to ∂_i. One imagines identifying in the natural way the ray $\{0\} \times \mathbf{R}_{\geq 0}$ with the ray $\{0\} \times \mathbf{R}_{\leq 0}$ in the half plane $\mathbf{R}_{\geq 0} \times \mathbf{R}$ and thinks therefore of (m_i, t_i) as lying in the following quotient homeomorph of $\mathbf{R}^2$:

$$R = (\mathbf{R}_{\geq 0} \times \mathbf{R})/\text{antipodal map}.$$

We shall also require the subspace

$$Z = (\mathbf{Z}_{\geq 0} \times \mathbf{Z})/\text{antipodal map},$$

which corresponds to the collection of all disjointly embedded weighted curves and arcs in F with endpoints in the windows.

Arguing as above with an annular neighborhood of a pants decomposition, one concludes:

THEOREM 2 ([**23, 32**]). *Given an isotopy class of pants decomposition $\mathcal{P}$ of $F = F_{g,r}^s$, where each pants curve is framed, there is a homeomorphism between $\mathcal{MF}(F)$ and the space of all pairs $(m_i, t_i) \in R$ as i ranges over the elements of $\mathcal{P}$. Likewise, there is a homeomorphism between $\mathcal{MF}_0(F)$ and the space of all pairs $(m_i, t_i) \in R$ as i ranges over the elements of $\mathcal{P} - \partial F$. In particular, $\mathcal{PF}_0(F) \approx S^{3g-3+r+s}$ and $\mathcal{PF}(F) \approx S^{3g-3+2r+s}$.*

There is the following "standard problem" about which not much is known (on the torus, it devolves to greatest common divisors, cf. [**6**], and see [**11**] for genus two):

PROBLEM 2. *Given a tuple $\times_{i=1}^{N}(m_i, t_i) \in Z^N$, give a tractable expression in terms of Dehn-Thurston or other coordinates for the number of components of the corresponding weighted family of curves and arcs.*

There is an algorithm which leads to a multiply weighted curve from an integral measure on a general train track akin to that on the torus gotten by serially "splitting" the track, cf. [**6**], but we ask in Problem 2 for a more closed-form expression. See also Problem 3. A related problem which also seems challenging is to describe $\mathcal{A}'(F)$ or $Arc'(F)$ in Dehn-Thurston coordinates on $\mathcal{MF}(F)$.

This class of curve and arc component counting problems might be approachable using the quantum path ordering techniques of [**5, 6**] or with standard fermionic statistical physics [**57, 29**].

3. Mapping class action on Dehn-Thurston coordinates

As already observed by Max Dehn [**8**] in the notation of Theorem 2, a Dehn twist on the ith pants curve in a pants decomposition of $F = F_{g,r}^s$ acts linearly, leaving invariant all coordinates (m_j, t_j), for $j \neq i$, and sending

(†)
$$m_i \mapsto m_i,$$
$$t_i \mapsto t_i \pm m_i.$$

As proved by Hatcher-Thurston [**13**] using Cerf theory, the two *elementary transformations* or *moves* illustrated in Figure 5 act transitively on the set of all

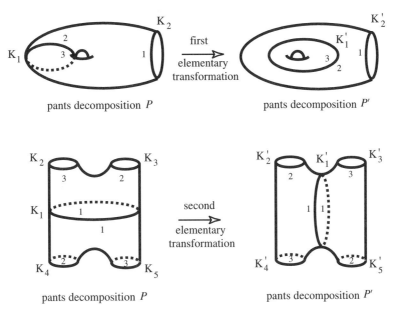

FIGURE 5. The elementary transformations

pants decompositions of any surface F. Thus, any Dehn twist acts on coordinates by the conjugate of a linear map, where the conjugating transformation is described by compositions of "elementary transformations" on Dehn-Thurston coordinates corresponding to the elementary moves. More explicitly, it is not difficult [23] to choose a finite collection of pants decompositions of F whose union contains all the curves in Lickorish's generating set [20] and calculate the several compositions of elementary moves relating them.

In any case, the calculation of the action of Dehn twist generators for $MC(F)$ thus devolves to that of the two elementary moves on Dehn-Thurston coordinates. This problem was suggested in [8], formalized in [38], and solved in [23] as follows.

Let $\vee$ and $\wedge$ respectively denote the binary infimum and supremum.

Given an arc family in the pair of pants, we introduce the notation ℓ_{ij} for the arc connecting boundary components i and j, for $i, j = 1, 2, 3$, where the windowed isotopy class of ℓ_{ii} depends upon the choice of twisting conventions. Given a weighted arc family in the pair of pants, the respective weights λ_{ij}, for $\{i, j, k\} = \{1, 2, 3\}$, on these component arcs are given in terms of the intersection numbers m_1, m_2, m_3 by the following formulas:

$$2\lambda_{ii} = (m_i - m_j - m_k) \vee 0$$
$$2\lambda_{ij} = (m_i + m_j - m_k) \vee 0, \text{ for } i \neq j,$$

and the intersection numbers are in turn given by $m_i = 2\lambda_{ii} + \lambda_{ij} + \lambda_{ik}$.

THEOREM 3 ([23, 32]). *Adopt the standard twisting conventions, the enumeration of curves indicated in Figure 5, and let (m_i, t_i) denote the Dehn-Thurston coordinates of a measured foliation with respect to the pants decomposition $\mathcal{P}$.*

FIRST ELEMENTARY TRANSFORMATION. Let λ_{ij} denote the weight of ℓ_{ij} with respect to $\mathcal{P}$ and λ'_{ij} the weight with respect to $\mathcal{P}'$, so in particular, $r = \lambda_{12} = \lambda_{13}$. Then the first elementary transformation is given by the following formulas.

$$\lambda'_{11} = (r - |t_1|) \vee 0,$$
$$\lambda'_{12} = \lambda'_{13} = L + \lambda_{11},$$
$$\lambda'_{23} = |t_1| - L,$$
$$t'_2 = t_2 + \lambda_{11} + (L \wedge t_1) \vee 0,$$
$$t'_1 = -sgn(t_1)(\lambda_{23} + L),$$

where $L = r - \lambda'_{11}$ and $sgn(x) \in \{\pm 1\}$ is the sign of $x \in \mathbf{R}$, with $sgn(0) = -1$.

(The formulas [**24**] above correct a typographical error in [**32**].)

SECOND ELEMENTARY TRANSFORMATION. Let λ_{ij} denote the weight of ℓ_{ij} in the bottom pair of pants for $\mathcal{P}$ and κ_{ij} the weight in the top pair or pants, and let λ'_{ij} denote the weight in the left pair of pants for $\mathcal{P}'$ and κ'_{ij} in the right pair or pants. The second elementary transformation is given by the following formulas.

$$\kappa'_{11} = \kappa_{22} + \lambda_{33} + (L - \kappa_{13}) \vee 0 + (-L - \lambda_{12}) \vee 0,$$
$$\kappa'_{22} = (L \wedge \lambda_{11} \wedge (\kappa_{13} - \lambda_{12} - L)) \vee 0,$$
$$\kappa'_{33} = (-L \wedge \kappa_{11} \wedge (\lambda_{12} - \kappa_{13} + L)) \vee 0,$$
$$\kappa'_{23} = (\kappa_{13} \wedge \lambda_{12} \wedge (\kappa_{13} - L) \wedge (\lambda_{12} + L)) \vee 0,$$
$$\kappa'_{12} = -2\kappa'_{22} - \kappa'_{23} + \kappa_{13} + \kappa_{23} + 2\kappa_{33},$$
$$\kappa'_{13} = -2\kappa'_{33} - \kappa'_{23} + \lambda_{12} + \lambda_{23} + 2\lambda_{22},$$
$$\lambda'_{11} = \lambda_{22} + \kappa_{33} + (K - \lambda_{13}) \vee 0 + (-K - \kappa_{12}) \vee 0,$$
$$\lambda'_{22} = (K \wedge \kappa_{11} \wedge (\lambda_{13} - \kappa_{12} - K)) \vee 0,$$
$$\lambda'_{33} = (-K \wedge \lambda_{11} \wedge (\kappa_{12} - \lambda_{13} + K)) \vee 0,$$
$$\lambda'_{23} = (\lambda_{13} \wedge \kappa_{12} \wedge (\lambda_{13} - K) \wedge (\kappa_{12} + K)) \vee 0,$$
$$\lambda'_{12} = -2\lambda'_{22} - \lambda'_{23} + \lambda_{13} + \lambda_{23} + 2\lambda_{33},$$
$$\lambda'_{13} = -2\lambda'_{33} - \lambda'_{23} + \kappa_{12} + \kappa_{23} + 2\kappa_{22},$$
$$t'_2 = t_2 + \lambda_{33} + ((\lambda_{13} - \lambda'_{23} - 2\lambda'_{22}) \wedge (K + \lambda'_{33} - \lambda'_{22})) \vee 0,$$
$$t'_3 = t_3 - \kappa'_{33} + ((L + \kappa'_{33} - \kappa'_{22}) \vee (\kappa'_{23} + 2\kappa'_{33} - \lambda_{12})) \wedge 0,$$
$$t'_4 = t_4 - \lambda'_{33} + ((K + \lambda'_{33} - \lambda'_{22}) \vee (\lambda'_{23} + 2\lambda'_{33} - \kappa_{12})) \wedge 0,$$
$$t'_5 = t_5 + \kappa_{33} + ((\kappa_{13} - \kappa'_{23} - 2\kappa'_{22}) \wedge (L + \kappa'_{33} - \kappa'_{22})) \vee 0,$$
$$t'_1 = \kappa_{22} + \lambda_{22} + \kappa_{33} + \lambda_{33} - (\lambda'_{11} + \kappa'_{11} + (t'_2 - t_2) + (t'_5 - t_5))$$
$$\quad + [(sgn(L + K + \lambda'_{33} - \lambda'_{22} + \kappa'_{33} - \kappa'_{22})](t_1 + \lambda'_{33} + \kappa'_{33}),$$

where $L = \lambda_{11} + t_1$, $K = \kappa_{11} + t_1$, and $sgn(x) \in \{\pm 1\}$ is the sign of $x \in \mathbf{R}$, with

$$sgn(0) = \begin{cases} +1, & \text{if } \lambda_{12} - 2\kappa'_{33} - \kappa'_{23} \neq 0; \\ -1, & \text{otherwise.} \end{cases}$$

These formulas are derived in [**23**], in effect, by performing explicit isotopies of arcs in certain covers of $F^0_{1,1}$ and $F^0_{0,4}$. Their computer implementation has been useful to some for analyzing specific mapping classes, e.g., fibered knot monodromies given by Dehn twists.

It is notable that the formulas are piecewise-*integral*-linear or PIL, cf. [**37**]. Furthermore, all of the "corners in the PIL structure actually occur", so the formulas are non-redundant in this sense; on the other hand, a given word in Lickorish's generators (i.e., the composition of conjugates of linear mappings (†) by specific PIL transformations given by finite compositions of the two elementary transformations) may not have "all possible corners occur". Insofar as continuous concave PIL functions are in one-to-one correspondence with tropical polynomials [**40, 41**], we are led to the following problem:

PROBLEM 3. *Give a useful (piecewise) tropical description of the two elementary transformations. One thus immediately derives a (piecewise) tropical polynomial representation of the mapping class groups. What properties does it have, for instance under iteration?*

As alternative coordinates, [**9**] describes a family of curves whose intersection numbers alone coordinatize measured foliations of compact support (but there are relations), and presumably these intersection numbers could be computed using Theorem 3.

We also wonder what are further applications or consequences of all these formulas.

4. Pseudo-Anosov maps and the length spectrum of moduli space

Thurston's original construction of pseudo-Anosov (pA) mappings [**37**] (cf. [**9**]) was generalized in [**23**] (cf. [**25, 39**]) to give the following recipe for their construction:

THEOREM 4 ([**23, 25**]). *Suppose that $\mathcal{C}$ and $\mathcal{D}$ are each families of disjointly embedded essential simple closed curves so that each component of $F - (\mathcal{C} \cup \mathcal{D})$ is either disk, a once-punctured disk, or a boundary-parallel annulus. Let w be any word consisting of Dehn twists to the right along elements of $\mathcal{C}$ and to the left along elements of $\mathcal{D}$ so that the Dehn twist on each curve of $\mathcal{C}$ or $\mathcal{D}$ occurs at least once in w. Then w represents a pseudo-Anosov mapping class.*

PROBLEM 4. *Does the recipe in Theorem 4 give virtually all pA maps? That is, given a pA map f, is there some iterate f^n, for $n \geq 1$, so that f^n arises from the recipe?*

(This question from [**23, 25**] is related [**10**] to the Ehrenpreis Conjecture, our Problem 14.)

In relation to Problem 4, let us mention that there are still other descriptions of pA maps up to iteration, for instance by Mosher [**42**] and in joint work of the author with Papadopoulos [**45**]; these descriptions are combinatorial rather than in terms of Dehn twists.

For a fixed surface F, consider the set of logarithms of dilatations of all pA maps supported on F. This characteristic "spectrum" $\Sigma(F) \subseteq \mathbf{R}_{>0}$ of F is precisely the Teichmüller geoedsic length spectrum of Riemann's moduli space $\mathcal{M}(F)$. The spectrum $\Sigma(F)$ is discrete. (In fact, dilatations occur as spectral radii of integral-linear Perron-Frobenius symplectomorphisms in a range of dimensions bounded above and below in terms of the topological type of F.)

PROBLEM 5. For a given surface F, calculate $\Sigma(F)$. More modestly, calculate the least element of $\Sigma(F)$ or the least gap among elements of $\Sigma(F)$. Characterize the number fields arising as dilatations of pA maps on F.

Problems 4 and 5 are clearly related. For instance, the recipe in Theorem 4 allows one to give estimates on least elements in Problem 5, cf. [**26, 2**]. McMullen [**50**] has also given estimates and examples.

5. Arc complexes

Refining the discussion in §1, let $F^s_{g,\vec{\delta}}$ denote a bordered surface of type $F^s_{g,r}$, with $r > 0$ and $\vec{\delta} = (\delta_1, \ldots, \delta_r)$ an r-dimensional vector of natural numbers $\delta_i \geq 1$, where there are δ_i distinguished points on the ith boundary component of $F^s_{g,\vec{\delta}}$, for $i = 1, \ldots, r$. Construct an arc complex $Arc(F)$ as before as the $PMC(F)$-orbits of isotopy classes of families of disjointly embedded essential and non-boundary parallel arcs connecting distinguished points on the boundary.

Given two bordered surfaces S_1, S_2, we consider inclusions $S_1 \subseteq S_2$, where the distinguished points and punctures of S_1 map to those of S_2, and S_1 is a complementary component to an arc family in S_2 (possibly an empty arc family if $S_1 = S_2$).

Define the *type 1 surfaces* to be the following: $F^0_{1,(1,1)}$, $F^0_{0,(1,1,1,1)}$, $F^1_{0,(1,1,1)}$, $F^2_{0,(1,1)}$.

THEOREM 5 ([**46**]). *The arc complex $Arc(F)$ of a bordered surface F is PL-homeomorphic to the sphere of dimension $6g-7+3r+2s+\delta_1+\delta_2+\cdots+\delta_r$ if and only if $M \not\subseteq F$ for any type 1 surface M. In other words, $Arc(F)$ is spherical only in the following cases: polygons ($g = s = 0$, $r = 1$), multiply punctured polygons ($g = 0$, $r = 1$), "generalized" pairs of pants ($g = 0$, $r + s = 3$), the torus-minus-a-disk ($g = r = 1$, $s = 0$), and the once-punctured torus-minus-a-disk ($g = r = s = 1$). Only the type 1 surfaces have an arc complex which is a PL-manifold other than a sphere.*

PROBLEM 6. Calculate the topological type (PL-homeomorphism, homotopy, homology... type) of the Arc-complexes.

The first non-trivial case is the calculation of the topological type of the PL-manifolds $Arc(M)$ for the four type 1 surfaces M.

Arc complexes as stratified spaces conjecturally have specific singularities and topology described recursively as follows. A PL sphere is a type zero space. A closed, connected, and simply connected manifold is a type one space provided it occurs among a list of four specific such (non-spherical) manifolds of respective dimensions 5,7,7, and 9, namely, the arc complexes of the four type one surfaces. For $n > 1$, define a type n space to be a finite polyhedron, defined up to PL-isomorphism, so that the link of each vertex in any compatible triangulation is PL isomorphic to an iterated suspension of the join of at most two spaces of type less than n.

By Theorem 5, many links of simplices are indeed of this type, and we conjecture that any arc complex is of some finite type. (As explained in [**46**], a specific collapsing argument in the "calculus of mapping cylinders" would give a proof a of this conjecture.)

The non-Hausdorff space $\mathcal{P}F(F)/PMC(F)$ thus contains the stratified space $Arc(F)$ as an open dense subset, explaining one classical (i.e., non-quantum) aspect to Problem 1. In light of the stratification of Arc-complexes in general, one might hope to apply techniques such as [**1, 34**] to address parts of Problem 6.

PROBLEM 7. Devise a matrix model (cf. [**28**]) for the calculation of the Euler characteristics of Arc-complexes.

PROBLEM 8 (Contributed by the referee). Does Theorem 5 say anything about the structure of the end of Riemann's moduli space? For instance, what is the homology near the end?

Take the one-point compactification $F^\times$ of $F = F_{g,r}^s$, where all of the $s \geq 0$ punctures of F are identified to a single point in $F^\times$, so $F^\times = F$ if and only if $s \leq 1$.

Let ∂ denote the boundary mapping of the chain complex $\{C_p(Arc) : p \geq 0\}$ of $Arc = Arc(F)$. Suppose that α is an arc family in F with corresponding cell $\sigma[\alpha] \in C_p(Arc)$. A codimension-one face of $\sigma[\alpha]$ of course corresponds to removing one arc from α, and there is a dichotomy on such faces $\sigma[\beta]$ depending upon whether the rank of the first homology of $F^\times - \cup\beta$ agrees with or differs by one from that of $F^\times - \cup\alpha$. This dichotomy decomposes ∂ into the sum of two operators $\partial = \partial^1 + \partial^2$, where ∂^2 corresponds to the latter case.

The operators ∂^1, ∂^2 are a pair of anti-commuting differentials, so there is a spectral sequence converging to $H_*(Arc)$ corresponding to the bi-grading

$$E^0_{u,v} = \{\text{chains on } \sigma[\alpha] \in C_p(Arc) : v = -\text{rank}(H_1(F_\alpha)) \text{ and } u = p - v\},$$

where $\partial_1 : E^0_{u,v} \to E^0_{u-1,v}$ and the differential of the E^0 term is $\partial_2 : E^0_{u,v} \to E^0_{u,v-1}$.

It is not quite fair to call it a problem, nor a theorem since the argument is complicated and has not been independently checked, but we believe that this spectral sequence collapses in its E^1-term to its top horizontal row except in dimension zero. Thus, the homology of Arc is the ∂_1-homology of the ∂_2-kernels in the top row, and on the other hand, it follows from [**31**] that the ∂_1-homology of the top row itself agrees with that of uncompactified Riemann's moduli space $\mathcal{M}(F)$.

As discussed in [**30, 46**], the stratified structure of the arc complexes for bordered surfaces gives a corresponding stratified structure to $\mathcal{A}(F)$ for punctured F. This may be enough to re-visit the calculations of [**19**] and [**16, 22**] with an eye towards avoiding technical difficulties with the Deligne-Mumford compactification $\bar{\mathcal{M}}(F)$.

6. Cell decompositions of $\mathcal{M}(F)$ and $\bar{\mathcal{M}}(F)$

For the next several sections unless otherwise explicitly stated, surfaces F will be taken to be once-punctured and without boundary. This is done for simplicity in order for instance that the moduli space $\mathcal{M}(F)$ itself, rather than some "decoration" of it, comes equipped with an ideal cell decomposition. Nevertheless, the theory extends, the discussion applies, and the problems and conjectures we articulate are intended in the more general setting (of multiply punctured surfaces with a distinguished puncture and no boundary).

The basic combinatorial tool for studying moduli space $\mathcal{M}(F)$ is the $MC(F)$-invariant ideal cell decomposition of the Teichmüller space $\mathcal{T}(F)$, and there are two effective constructions: from combinatorics to conformal geometry using Strebel coordinates on fatgraphs [**36, 12, 15**], and from hyperbolic geometry to combinatorics using the convex hull construction and simplicial coordinates [**27**]. (See [**31**] for further details.)

PROBLEM 9 (Bounded Distortion Conjecture). *Given a hyperbolic structure on F, associate its combinatorial invariant, namely, an ideal cell decomposition of F together with the projective simplicial coordinate assigned to each edge. Take these projective simplicial coordinates as Strebel coordinates on the dual fatgraph to build a conformal structure on F. The underlying map on Teichmüller space is of bounded distortion in the Teichmüller metric.*

As posed by Ed Witten to the author in the early 1990's, a compelling problem at that time was to find an orbifold compactification of $\mathcal{M}(F)$ which comes equipped with a cellular description in terms of suitably generalized fatgraphs. Calculations such as [**16, 19, 22**] and more might then be performed using matrix models derived from the combinatorics of this putative compactification. Perhaps the desired compactification was the Deligne-Mumford compactification or perhaps another one. The combinatorial compactification of the previous section fails to provide an orbifold but rather another stratified generalization of manifold.

Guidance from Dennis Sullivan has recently led to the following solution:

THEOREM 6. *Suppose that F has only one puncture. Then $\bar{\mathcal{M}}(F)$ is homeomorphic to the geometric realization of the partially ordered set of $MC(F)$-orbits of pairs $(\alpha, \mathcal{A})$, where α fills F, and the* screen $\mathcal{A}$ *is a collection of subsets of α so that:*

[Fulton-MacPherson nest condition] *for any two $A, B \in \mathcal{A}$ which are not disjoint, either $A \subseteq B$ or $B \subseteq A$;*

[Properness] *$\cup \mathcal{A}$ is a proper subset of α, and furthermore for any $A \in \mathcal{A}$, we likewise have $\cup \{B \in \mathcal{A} : A \neq B \subseteq A\}$ is a proper subset of A;*

[Recurrence] *for any $a \in A \in \mathcal{A}$, there is an essential simple closed curve in F, meeting $\cup \alpha$ a minimal number of times in its isotopy class, meeting only the arcs in A, and crossing a,*
where inclusion of ideal cell decompositions induces the partial ordering on the set of pairs $(\alpha, \mathcal{A})$.

In effect, the elements of the screen $\mathcal{A}$ detect how quickly hyperbolic lengths of arcs in α diverge. It is a concise new description of a combinatorial structure on $\bar{\mathcal{M}}(F)$, which bears similarities to renormalization in physics, and it would be interesting to make this precise. Let us also remark that the proof of the previous theorem depends upon working in the hyperbolic category, where the "pinch curves" in a nearly degenerate structure can be detected using the coordinates. (In fact, most of the required ideas and estimates are already described in [**30**].)

PROBLEM 10. *Though the (virtual) Euler characteristics are already known* [**28, 56**], *devise a matrix model using screens to calculate these invariants for $\bar{M}(F)$.*

7. Torelli groups

Recall [**52**],[**48**] the Torelli group $\mathcal{I}_k(F)$, defined as those mapping classes on F fixing a basepoint (taken to be a puncture) that act trivially on the kth nilpotent quotient of the fundamental group of F. Since the ideal cell decomposition of $\mathcal{T}(F)$

is invariant under $MC(F)$, it is in particular invariant under each $\mathcal{I}_k(F)$, and so the quotient "Torelli space" $T_k(F) = \mathcal{T}(F)/\mathcal{I}_k(F)$ is a manifold likewise admitting an ideal cell decomposition.

Recent work [48] with Shigeyuki Morita studies the "Torelli tower"

$$\mathcal{T}(F) \to \cdots \to T_{k+1}(F) \to T_k(F) \to \cdots \to T_1(F) \to \mathcal{M}(F)$$

of covers of Torelli spaces, each $T_k(F)$ a manifold covering the orbifold $\mathcal{M}(F)$. In particular, one essentially immediately derives infinite presentations for all of the higher Torelli groups as well as a finite presentation for instance of the "level N (classical) Torelli groups", i.e., the subgroup of $MC(F)$ which acts identically on homology with $\mathbf{Z}/N$ coefficients.

PROBLEM 11 (Level N Torelli Franchetta Problem). What is the second cohomology group of the level N Torelli group?

The cell decomposition of $\bar{\mathcal{M}}(F)$ described before is compatible with the ideal cell decompositions of Torelli spaces, i.e., the fatgraph dual to an ideal cell decomposition of F admits a "homology marking" in the sense of [48] as well as admitting the structure of screens. There are thus "DM type" boundaries of each Torelli space replete with an ideal cell decomposition. It is natural to try to understand the topology of these DM-type bordifications of Torelli spaces to approach the following class of problems:

PROBLEM 12. Calculate various group-theoretic boundaries of mapping class and Torelli groups, for instance, Tits boundaries.

Further results in [48] arise by writing an explicit one cocyle representing the first Johnson homomorphism. In effect, one checks that the putative cycle represents a crossed homomorphism by verifying the several essentially combinatorial constraints imposed by the cell decomposition, and then compares with known values of the first Johnson homomorphism. As discussed more fully in [48], one might realistically hope to find similar canonical cocycles for the higher Johnson homomorphisms as well. In particular by work of Morita [53], the Casson invariant of a homology 3-sphere is algorithmically calculable from the second Johnson homomorphism, so this is of special interest.

More explicitly, there is a combinatorial move, a "Whitehead move", that acts transitively on ideal triangulations of a fixed surface (where one removes an edge e from the triangulation to create a complementary quadrilateral and then replaces e with the other diagonal of this quadrilateral). One seeks invariants of sequences of Whitehead moves lying in an $MC(F)$-module satisfying three explicit combinatorial conditions just as for the Johnson homomorphism. There is thus a kind of "machine" here for producing cocycles with values in various modules by solving for expressions that satisfy certain explicit combinatorial constraints; several such invariant one cocycles have been produced in this way on the computer but so far without success for constructing the higher Johnson homomorphisms.

Very recent work with Shigeyuki Morita and Alex Bene solves a related problem: the Magnus representations (which are closely related to the Johnson homomorphisms [52]) in fact lift directly to the groupoid level of [48] as the Fox Jacobians of appropriately enhanced Whitehead moves. One works in the free fundamental group of the punctured surface with basepoint distinct from the puncture, as discussed in the closing remarks of [48], and the corresponding dual fatgraph comes equipped with a canonical maximal tree defined by greedily adding edges to the tree while traversing a small circle around the puncture starting from the basepoint. Perhaps these explicit calculations of the Magnus representations might be of utility for instance to address the following standard question:

PROBLEM 13. What are the kernels of the Magnus representations?

Finally in [**48**] by taking contractions of powers of our canonical one cocycle, new combinatorially explicit cycles and cocycles on $\mathcal{M}(F)$ are constructed which on the other hand generate the tautological algebra. It is natural to wonder about the extension of these classes to the screen model for $\bar{\mathcal{M}}(F)$ and to the combinatorial compactification, and to revisit [**16, 19, 22**] in this context.

8. Closed/open string theory

This section describes recent work [**49**] with Ralph Kaufmann, where it turns out that the material of Section 2 together with a further combinatorial elaboration describes a reasonable model for the phenomenology of interactions of open and closed strings.

Returning now to multiply punctured surfaces $F = F_{g,\vec{\delta}}^s$ with at least one distinguished point in each boundary component as in Section 5, let D denote the set of distinguished points in the boundary, and let S denote the set of punctures. Fix a set $\mathcal{B}$ of "brane labels", let $\mathcal{P}(\mathcal{B})$ denote its power set, and define a *brane labeling* to be a function $\beta : D \cup S \to \mathcal{P}(\mathcal{B})$ subject to the unique constraint that if the value $\emptyset \in \mathcal{P}(\mathcal{B})$ is taken at a point on the boundary of F, then this point is the only distinguished point in its boundary component. Complementary components to the distinguished points in the boundary are called "windows". Given a brane labeling β on a windowed surface F, define the set $D(\beta) = \{d \in D : \beta(d) \neq \emptyset\}$, and consider proper isotopy classes rel D of arc families with endpoints in windows, where the arcs are required to be non-boundary parallel in $F - D(\beta)$. Say that such an arc family is *exhaustive* if each window has at least one incident arc, and construct the space $\widetilde{Arc}(F, \beta)$ of positive real weights on exhaustive arc families.

There are various geometric operations on the spaces $\{\widetilde{Arc}(F, \beta)\}$ induced by gluing together measured foliations along windows provided the total weights agree on the windows to be glued (taking unions of brane labels when combining distinguished points). It is shown in [**49**] that these geometric operations descend to the level of suitable chains on the spaces $\widetilde{Arc}(F, \beta)$ and finally to the level of the integral homology groups of $\widetilde{Arc}(F, \beta)$. These algebraic operations on homology satisfy the expected "operadic" equations of open/closed string theory, and new equations can be discovered as well.

In effect, [**49**] gives the string field theory of a single point, and natural questions are already discussed in detail in [**49**] including the "passage to conformal field theory" which seems to be provided by corresponding operations not for exhaustive arc families but rather for quasi filling arc families; an obvious challenge is to perform meaningful calculations in CFT. Calculate the homology groups of exhaustive or quasi filling arc families in brane labeled windowed surfaces, and calculate the homology groups of their combinatorial or DM-type compactifications (the former problem already articulated as our Problem 6 and the latter surely also posed elsewhere in this volume). Organize and understand the many algebraic relations of open/closed strings on the level of homology. Find the BRST operator. Introduce excited strings, i.e., extend to string field theory of realistic targets. Is the Torelli structure of any physical significance?

9. Punctured solenoid

A problem well-known in the school around Dennis Sullivan is:

PROBLEM 14 (Ehrenpreis Conjecture). Given two closed Riemann surfaces, there are finite unbranched covers with homeomorphic total spaces which are arbitrarily close in the Teichmüller metric.

This section describes joint work with Dragomir Šarić on related universal constructions in Teichmüller theory [47].

As a tool for understanding dynamics and geometry, Sullivan defined the hyperbolic solenoid [33] as the inverse limit of the system of finite-sheeted unbranched pointed covers of any fixed closed oriented surface of negative Euler characteristic. Following Ahlfors-Bers theory, he developed its Teichmüller theory and studied the natural dynamics and geometry, in particular introducing two principal mapping class groups, the continuous "full mapping class group" and the countable "baseleaf preserving mapping class group". He furthermore showed that the Ehrenpreis Conjecture is equivalent to the conjecture that the latter group has dense orbits in the Teichmüller space of the solenoid. [55] contains many basic results and open problems about Sullivan's solenoid.

Following this general paradigm, one defines the punctured hyperbolic solenoid S as the inverse limit over all finite-index subgroups Γ of $PSL_2(\mathbf{Z})$ of the system of covers $\mathcal{U}/\Gamma$ over the modular curve $\mathcal{U}/PSL_2(\mathbf{Z})$, where $\mathcal{U}$ denotes the upper half-space. In effect, branching is now permitted but only at the missing punctures covering the three orbifold points of the modular curve.

Let us build a particular model space homeomorphic to the punctured solenoid. Take $\Gamma = PSL_2(\mathbf{Z})$ acting by fractional linear transformations on $\mathcal{U}$, and let $\hat{\Gamma}$ denote its profinite compleion. Thus, $\gamma \in \Gamma$ acts naturally on $\mathcal{U} \times \hat{\Gamma}$ by $\gamma \cdot (z,t) = (\gamma z, t\gamma^{-1})$, and the quotient is homeomorphic to S.

Truly the entire decorated Teichmüller theory of a punctured surface of finite type [27] extends appropriately to S: there are global coordinates, there is an ideal cell decomposition of the decorated Teichmüller space, and there is an explicit non-degenerate two-form, the latter of which are invariant under the action of the baseleaf preserving subgroup $Mod_{BLP}(S)$ of the full mapping class group $Mod(S)$; generators for $Mod_{BLP}(S)$ are provided by appropriate equivariant Whitehead moves, and a complete set of relations has recently been derived as well in further recent work with Dragomir Šarić and Sylvain Bonnot.

There are a number of standard questions (again see [55]): Is $Mod_{BLP}(S)$ finitely generated? Do the "mapping class like" elements generate $Mod_{BLP}(S)$? It seems to be a deep question which tesselations of the disk, other than the obvious so-called "TLC" ones, arise from the convex hull construction applied to the decorated solenoid in [47]. Is there any relationship between $Mod(S)$ and the absolute Galois group? (This cuts both ways since only partial information is known about either group; notice that the action of the latter on $\hat{\Gamma}$ is explicit as a subgroup of the Grothendieck-Teichmüller group [51].)

References

[1] Nils Andreas Baas, "Bordism theories with singularities", Proceedings of the Advanced Study Institute on Algebraic Topology (Aarhus Univ., Aarhus, 1970), Vol. I, pp. 1-16. Various Publ. Ser., No. 13, Mat. Inst., Aarhus Univ., Aarhus, 1970.

[2] Max Bauer, "An upper bound for the least dilatation", *Trans. Amer. Math. Soc.* **330** (1992), 361-370.

[3] Moira Chas and Dennis P. Sullivan, "String topology", to appear *Ann. Math.*, preprint math.GT/9911159,

[4] ———, "Closed string operators in topology leading to Lie bialgebras and higher string algebra", The legacy of Niels Henrik Abel, Springer, Berlin (2004), 771-784, Preprint math.GT/0212358.

[5] L. Chekhov and V. Fock, "A quantum Techmüller space", *Theor. Math. Phys.*, **120** (1999) 1245-1259; "Quantum mapping class group, pentagon relation, and geodesics" *Proc. Steklov Math. Inst.* **226** (1999) 149-163.

[6] Leonid Chekhov and R. C. Penner, "On the quantization of Teichmüller and Thurston theories", to appear: Handbook of Teichmüller theory, European Math Society, ed. A. Papadopoulos, math.AG/0403247.

[7] Ralph L. Cohen and John D.S. Jones "A homotopy theoretic realization of string topology", *Math. Ann.* **324** (2002), 773-798.

[8] Max Dehn, Lecture notes from Breslau, 1922, Archives of the University of Texas at Austin.

[9] A. Fathi, F. Laudenbach, V. Poenaru, "Travaux de Thurston sur les Surfaces", *Asterisques* **66-67** (1979), Sem. Orsay, Soc. Math. de France.

[10] Tim Gendron, "The Ehrenpreis conjecture and the moduli-rigidity gap", Complex manifolds and hyperbolic geometry (Guanajuato, 2001), *Contemp. Math.* **311**, 207-229.

[11] Andrew Haas and Perry Susskind, "The connectivity of multicurves determined by integral weight train tracks", *Trans. Amer. Math. Soc.* **329** (1992), 637-652.

[12] John L. Harer, "The virtual cohomological dimension of the mapping class group of an orientable surface", *Invent. Math.* **84** (1986), 157-176.

[13] A. Hatcher and W. Thurston, "A presentation for the mapping class group of a closed orientble surface", *Topology* **19** (1980), 221-137.

[14] William J. Harvey, "Boundary structure of the modular group", Riemann surfaces and related topics: Proceedings of the 1978 Stony Brook Conference, *Ann. of Math. Stud.* **97**, Princeton Univ. Press, Princeton, N.J., (1981), 245-251.

[15] John H. Hubbard and Howard Masur, "Quadratic differentials and foliations", *Acta Math.* **142** (1979), 221-274.

[16] Kyoshi Igusa, "Combinatorial Miller-Morita-Mumford classes and Witten cycles", *Algebr. Geom. Topol.* **4** (2004), 473-520.

[17] Rinat M. Kashaev, "Quantization of Teichmüller spaces and the quantum dilogarithm", *Lett. Math. Phys.*, **43**, (1998), 105-115; q-alg/9705021.

[18] Ralph Kaufman, Muriel Livernet, R. C. Penner, "Arc operads and arc algebras", *Geom. Topol.* **7** (2003), 511-568.

[19] Maxim Kontsevich, "Intersection theory on the moduli space of curves and the matrix Airy function", *Comm. Math. Phys.* **147** (1992), 1-23.

[20] W. B. R. Lickorish, "A finite set of generators for the homeotopy group of a two-manifold" *Proc. Camb. Phil. Soc.* **60** (1964), 769-778; Corrigendum, *Proc. Camb. Phil. Soc.* **62** (1966), 679-681.

[21] Yuri I. Manin, "Von Zahlen und Figuren", Talk at the International Conference "Géométrie au vingtième cièle: 1930–2000", Paris, Institut Henri Poincaré, Sept. 2001, 27 pp., preprint math.AG/0201005.

[22] Gabriel Mondello, "Combinatorial classes on $\overline{M}_{g,n}$ are tautological", *Int. Math. Res. Not.* (2004), 2329-2390.

[23] R. C. Penner, "The action of the mapping class group on isotopy classes of curves and arcs in surfaces", thesis, Massachusetts Institute of Technology (1982), 180 pages.

[24] ———, "The action of the mapping class group on curves in surfaces", *L'Enseignement Mathematique* **30** (1984), 39-55.

[25] ———, "A construction of pseudo-Anosov homeomorphisms", *Proceedings of the American Math Society* **104** (1988), 1-19.

[26] _____, "Bounds on least dilatations", *Trans. Amer. Math. Soc.* **113** (1991), 443-450.

[27] _____, "The decorated Teichmüller space of punctured surfaces", *Communications in Mathematical Physics* **113** (1987), 299-339.

[28] _____, "Perturbative series and the moduli space of Riemann surfaces", *Journal of Differential Geometry* **27** (1988), 35-53.

[29] _____, "An arithmetic problem in surface geometry", *The Moduli Space of Curves*, Birkhäuser (1995), eds. R. Dijgraaf, C. Faber, G. van der Geer, 427-466.

[30] _____, "The simplicial compactification of Riemann's moduli space", Proceedings of the 37th Taniguchi Symposium, World Scientific (1996), 237-252.

[31] _____, "Decorated Teichmüller space of bordered surfaces", *Communications in Analysis and Geometry* **12** (2004), 793-820, math.GT/0210326.

[32] R. C. Penner with John L. Harer *Combinatorics of Train Tracks*, Annals of Mathematical Studies **125**, Princeton Univ. Press (1992); second printing (2001).

[33] Dennis Sullivan, *Linking the universalities of Milnor-Thurston, Feigenbaum and Ahlfors-Bers*, Milnor Festschrift, Topological methods in modern mathematics (L. Goldberg and A. Phillips, eds.), Publish or Perish, 1993, 543-563. preprint.

[34] _____, "Combinatorial invariants of analytic spaces", Proc. of Liverpool Singularities–Symposium, I (1969/70) Springer, Berlin (1971), 165-168.

[35] Joerg Teschner "An analog of a modular functor from quantized Teichmüller theory", to appear Handbook of Teichmüller theory, European Math Society (ed. A. Papadopoulos), math.QA/0405332.

[36] Kurt Strebel, Quadratic Differentials, *Ergebnisse der Math. und ihrer Grenzgebiete*, Springer-Verlag, Berlin (1984).

[37] William P. Thurston, "On the geometry and dynamics of diffeomorphisms of surfaces", *Bull. Amer. Math. Soc.*, **19** (1988) 417–431.

[38] _____, "Three dimensional manifolds, Kleinian groups and hyperbolic geometry", *Bull. Amer. Math. Soc.* **6**(1986), 357-381.

[39] Albert Fathi, "Démonstration d'un théorème de Penner sur la composition des twists de Dehn", *Bull. Soc. Math. France* **120** (1992), 467-484.

[40] Anatol N. Kirillov, "Introduction to tropical combinatorics", Physics and combinatorics, 2000 (Nagoya), 82-150, World Sci. Publishing, River Edge, NJ, 2001.

[41] Jürg Richter-Gebert, Bernd Sturmfels, Thorsten Theobald, "First steps in tropical geometry", to appear in Proc. of COnf. on Idempotent Mathemaicals and Mathematical Physics, *Contemp. Math.*, math.AG/0306366.

[42] Lee Mosher, "The classification of pseudo-Anosovs" in *Low-dimensional topology and Kleinian groups* (Coventry/Durham, 1984), 13-75, London Math. Soc. Lecture Note Ser., **112**, Cambridge Univ. Press, Cambridge, 1986.

[43] William A. Veech, "Dynamics over Teichmüller space", *Bull. Amer. Math. Soc.* **14** (1986), 103-106.

[44] Howard Masur, "Ergodic actions of the mapping class group", *Proc. Amer. Math. Soc* **94** (1985), 455-459.

[45] Athanase Papadopoulos and R. C. Penner, "A construction of pseudo-Anosov homeomorphisms", *Proc. Amer. Math Soc.* **104** (1988), 1-19.

[46] R. C. Penner, "The structure and singularities of arc complexes", preprint (2004) math.GT/0410603.

[47] R. C. Penner and Dragomir Šarić, "Teichmüller theory of the punctured solenoid", preprint (2005), math.DS/0508476.

[48] S. Morita and R. C. Penner, "Torelli groups, extended Johnson homomorphisms, and new cycles on the moduli space of curves", preprint (2006), math.GT/0602461.

[49] Ralph M. Kaufmann and R. C. Penner, "Closed/open string diagrammatics" (2006), to appear in *Nucl. Phys B*, math.GT/0603485.

[50] Curt McMullen, "Billiards and Teichmüller curves on Hilbert modular surfaces", *J. Amer. Math. Soc.* **16** (2003), 857885.

[51] Leila Schneps, "The Grothendieck-Teichmüller group $\widehat{GT}$: a survey", Geometric Galois Actions (L. Schneps and P. Lochak, eds.) London Math Society Lecture Notes **242** (1997), 183-203.

[52] Shigeyuki Morita, " Abelian quotients of subgroups of the mapping class group of surfaces", *Duke Math. J.* **70** (1993), 699–726.

[53] Shigeyuki Morita, "Casson's invariant for homology 3-spheres and characteristic classes of surface bundles I" *Topology* **28** (1989), 305-323.

[54] Yuri I. Manin, "Real multiplication and noncommutative geometry", *The legacy of Niels Henrik Abel*, eds. O. A. Laudal and R. Piene, Springer Verlag, Berlin 2004, 685-727, preprint math.AG/0202109.

[55] Chris Odden, "The baseleaf preserving mapping class group of the universl hyperbolic solenoid", *Trans. A.M.S.* **357** (2004), 1829-1858.

[56] John L. Harer and Don Zagier, "The Euler characteristic of the moduli space of curves", *Invent. Math.* **85** (1986), 457-485.

[57] Claude Itzykson and Jean-Michel Drouffe, Statistical Field Theory, volume 1, Cambridge Monographs on Mathematical Physics, Cambridge Univ. Press (1989).

DEPARTMENTS OF MATHEMATICS AND PHYSICS/ASTRONOMY, UNIVERSITY OF SOUTHERN CALIFORNIA, LOS ANGELES, CA 90089

Relations in the Mapping Class Group

Bronislaw Wajnryb

1. Introduction

In this paper we pose several questions about mapping class groups of compact surfaces. These stem from well known connections with Artin groups and symplectic geometry (specifically, Lefschetz fibrations).

Let S be a compact surface, possibly with boundary. The mapping class group $M(S)$ of S is the group of the isotopy classes of diffeomorphisms of S onto itself which are pointwise fixed on the boundary of S. It is equal to the group of the isotopy classes of homeomorphisms of S onto itself. If S is orientable then we also assume that the diffeomorphisms preserve the fixed orientation of S. If S is oriented then $M(S)$ is generated by (Dehn) twists along simple closed curves. In this paper every *curve* is simple and closed unless otherwise specified. A (positive) twist along a curve α is denoted by T_α. It is different from the identity only in an annulus—a regular neighborhood of α. If a curve β intersects a curve α in one point transversely then T_α takes β onto a curve γ which moves along β towards the intersection point then turns to the right into α (it makes sense on an oriented surface) moves once along α and then continues along β (see Figure 1).

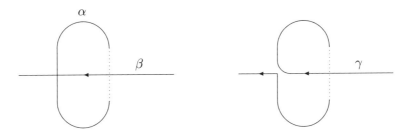

Figure 1. *Dehn Twist*

Twists along isotopic curves are isotopic, belong to the same class. Twists have the following basic properties.

PROPOSITION 1. *Suppose α and β are simple closed curves on S and $h : S \to S$ is a homeomorphism. Then*

(1) *If $\beta = h(\alpha)$ then $T_\beta = hT_\alpha h^{-1}$.*

(2) If α and β are disjoint then $T_\alpha T_\beta = T_\beta T_\alpha$.
(3) If α and β meet once then $T_\alpha T_\beta T_\alpha = T_\beta T_\alpha T_\beta$.
(4) If α and β meet more then once then there are no relations between T_α and T_β.

Parts 1 and 2 of the proposition are elementary, part 3 follows from part 1 and from a simple fact that $T_\alpha T_\beta(\alpha) = \beta$. Part 4 is a Theorem of Ishida [4].

2. Homomorphisms from Artin groups to $M(S)$.

We consider an oriented surface S and homomorphisms of Artin groups into $M(S)$. Let $M = (m_{i,j})$ be an $n \times n$ Coxeter matrix - $m_{i,j} > 0$ is an integer or ∞, $m_{i,j} = m_{j,i}$ and $m_{i,j} = 1$ if and only if $i = j$. To such a matrix corresponds an Artin group $G_M = G$ with generators $a_1, \ldots, a_n$ and relations

$$\underbrace{a_i a_j a_i \ldots}_{m_{i,j}} = \underbrace{a_j a_i a_j \ldots}_{m_{i,j}} \quad \text{for all } i, j.$$

We shall call an Artin group *small* if $m_{i,j} \leq 3$ for all i,j. For a small Artin group G we can consider a configuration of n curves on S : $\alpha_1, \ldots, \alpha_n$ such that α_i and α_j are disjoint whenever $m_{i,j} = 2$ and meet once whenever $m_{i,j} = 3$. Then a correspondence $a_i \to T_{\alpha_i}$ extends to a homomorphism of G into $M(S)$. Such a homomorphism (which takes the generators of an Artin group onto Dehn twists) is called *geometric*. We consider first a question when a geometric homomorphism is injective.

To a Coxeter matrix M corresponds a Coxeter graph Γ (also called Dynkin diagram) with n vertices numbered 1,2,...,n and with an edge between i and j whenever $m_{i,j} > 2$. If $m_{i,j} > 3$ then the edge gets a label $m_{i,j}$. We shall consider irreducible Artin groups, the groups with the connected Coxeter graph. We shall distinguish two series of Artin groups

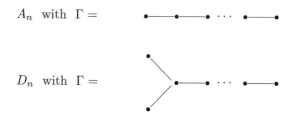

The question for geometric homomorphisms is completely solved.

THEOREM 2 (Perron-Vannier). *When G is of type A_n or D_n and S is a regular neighbourhood of the union of the curves α_i, then the geometric homomorphism is injective.*

This was proved in [10].

THEOREM 3 (Labruere-Wajnryb). *If G is an irreducible small Artin group different from A_n and D_n then any geometric homomorphism from G into $M(S)$ is not injective.*

This was proved in part by Christine Labruere in [6] and completed by myself in [13].

When I mentioned this result to Professor Brieskorn he remarked that there is a big difference between a geometric embedding and any embedding. Recently Misha Kapovich repeated this remark but he suggested that probably there are no other embeddings. This is the question I want to pose.

Question. Which Artin groups admit non-geometric embeddings into $M(S)$?

If we do not require the homomorphism to be geometric we do not need to restrict the question to small Artin groups. In view of the third part of lemma 2 it is easy to embed some Artin groups with some $m_{i,j} = \infty$ so we should restrict the question to Artin groups with finite exponents $m_{i,j}$. One example is very simple. The group B_n similar to A_n with the first edge of the Dynkin diagram having the label 4, maps into A_n if we send the first generator of B_n onto the square of the first generator of A_n. The twists $T_{\alpha_1}^2$ and T_{α_2} satisfy $T_{\alpha_1}^2 T_{\alpha_2} T_{\alpha_1}^2 T_{\alpha_2} = T_{\alpha_2} T_{\alpha_1}^2 T_{\alpha_2} T_{\alpha_1}^2$.

Any other Artin group of finite type maps into E_8 in a similar way. The generators map onto products of powers of small number of twists. We can call such homomorphisms *almost geometric*. These are the only almost geometric homomorphisms which I know.

Question. Do there exist any other examples of non commuting homeomorphisms g and h which are not both Dehn twists and satisfy a braid relation

$$\underbrace{ghg\ldots}_{m_{i,j}} = \underbrace{hgh\ldots}_{m_{i,j}} \quad \text{for some } m_{i,j} > 2?$$

If yes, then we can try to embed E_6 or a group with a triangular Dynkin diagram or find an interesting homomorphism (with a non-abelian image) of a non-small Artin group into $M(S)$. If the homeomorphisms are reducible then probably the twists along boundary curves must again satisfy braid relations and we are back to a geometric or an almost geometric embedding.

Question. Does there exist a set of at least three pseudo-Anosov homeomorpisms such that every pair satisfies a braid relation.

The referee to this paper observed that one can easily find a pair of such pseudo-Anosov homeomorphisms satisfying braid relation of length 3. We choose elements x and y of $M(S)$ of order 2 and 3 respectively and let $g = xy$ and $h = yx$. Then g and h satisfy the relation $ghg = hgh$. For suitable x and y the homeomorphisms g and h are pseudo-Anosov. If S is a torus then $M(S) = SL_2(Z)$. We may require $x^2 = -1$ and $y^3 = 1$ and almost every choice of such matrices x and y will produce Anosov matrices g and h. For a higher genus a choice of elements x and y for which g and h are pseudo-Anosov is a little more difficult.

3. Positive relations in the mapping class group

I heard this problem from Dennis Auroux. Much more about it can be found in his paper [**1**]. I'll just make few remarks. We consider again an oriented surface S. We consider positive relations in $M(S)$, products of positive powers of twists which are equal to the identity. Such considerations go back to the work of Moishezon on braid monodromy but now they gained a new importance because of the development of Symplectic Geometry. It was proven by Ivan Smith in [**11**] that on a surface with boundary we cannot have a positive relation. But we may have a product of positive twists equal to product of twists along the boundary components. Let S be an oriented surface with one boundary component. Consider products of positive twists equal to a single positive twist along the boundary.

Question (Smith). What is the maximal length of such a product? Can it be arbitrarily long?

Let us consider some examples of such products. We start with configurations of curves on a closed surface S.

Let $\alpha'_1, \alpha_1, \alpha_2, \ldots, \alpha_{2g+1}$ be a configuration of curves on a surface S of genus g corresponding to the Dynkin diagram D_{2g+2}. On a closed surface the curves α_1 and α'_1 are isotopic but a regular neighborhood of the union of all curves α_i has three boundary components, one of them separating α_1 from α'_1. If we omit α'_1 the neighborhood of the union will have two boundary components and if we also omit α_{2g+1} the regular neighborhood will have one boundary component. The corresponding Artin groups D_{2g+2}, A_{2g+1} and A_{2g} have an infinite cyclic center generated by Coxeter element (or its square) which maps to a trivial element in $M(S)$. More precisely

(1) $(T_{\alpha'_1} T_{\alpha_1} T_{\alpha_2} \ldots T_{\alpha_{2g+1}})^{2g+1} = T_{\delta_1} T_{\delta_2} T_{\delta_3}^{2g-2}$, for D_{2g+2}.

(2) $\Delta^2_{2g+1} = (T_{\alpha_1} T_{\alpha_2} \ldots T_{\alpha_{2g+1}})^{2g+2} = T_{\delta_1} T_{\delta_2}$, for A_{2g+1}.

(3) $\Delta^4_{2g} = (T_{\alpha_1} T_{\alpha_2} \ldots T_{\alpha_{2g}})^{4g+2} = T_{\delta_1}$, for A_{2g}.

Here δ_i's are boundary components of a regular neighborhood of the union of the curves and δ_3 in the first relation is the component which separates α'_1 from α_1. The curves δ_i bound disks on the closed surface S therefore on the closed surface we get positive relations. If we cap boundary components δ_2 and δ_3 with disks and leave only one boundary component we get relations considered by Smith. On such a surface we have one more relation. If we let

$$h = T_{\alpha_{2g+1}} T_{\alpha_{2g}} \ldots T_{\alpha_2} T_{\alpha_1} T_{\alpha_1} T_{\alpha_2} \ldots T_{\alpha_{2g}} T_{\alpha_{2g+1}}$$

then $\Delta^2_{2g} = h^2$. So we get another positive representation of the boundary twist

(4) $h^2 = T_{\delta_1}$.

On the closed surface S the homeomorphism h is equal to the hyperelliptic involution.

On a surface of genus 3 and 4 we have additional relations corresponding to Coxeter elements in Artin groups E_6, E_7 and E_8. They were investigated in the paper of Matsumoto [9]. Let us recall the Dynkin diagram corresponding to E_8.

$\Gamma =$

In order to get the Dynkin diagram of E_7 we omit the last vertex and in order to get the diagram of E_6 we omit the last two vertices. We call the exceptional vertex a_0 and the others are $a_1, a_2 \ldots$ from left to right. Consider configurations of curves on a surface S corresponding to these diagrams. Let N be a regular neighborhood of the union of the corresponding curves. Then N is a surface of genus 3 with one boundary component δ_1 for diagram E_6, a surface of genus 3 with two boundary components δ_1 and δ_2 for E_7 and a surface of genus 4 with one boundary component δ_1 for E_8. The Coxeter elements give the following relations:

(5) $(T_{\alpha_0}T_{\alpha_1}T_{\alpha_2}T_{\alpha_3}T_{\alpha_4}T_{\alpha_5})^{12} = T_{\delta_1}$, for E_6.
(6) $(T_{\alpha_0}T_{\alpha_1}T_{\alpha_2}T_{\alpha_3}T_{\alpha_4}T_{\alpha_5}T_{\alpha_6})^9 = T_{\delta_1}T_{\delta_2}^2$, for E_7.
(7) $(T_{\alpha_0}T_{\alpha_1}T_{\alpha_2}T_{\alpha_3}T_{\alpha_4}T_{\alpha_5}T_{\alpha_6}T_{\alpha_7})^{30} = T_{\delta_1}^k$, for E_8.

The positive power k in the last relation is most probably equal to 1 but I did not check it and I do not recall seeing it in any paper.

When we cap the components δ_2 and δ_3 with disks we can view the relations (1)–(7) as relations consider by Smith on a surface with one boundary component.

Question. Are the relations (1)–(7) the only relations which express the twist along the boundary as a product of positive twists, up to the positive equivalence in the Artin group?

The motivation for this question comes from symplectic geometry. If we are given a Symplectic Lefschetz Fibration $p : M \to D$ of a 4-manifold M over a disk D we can describe it by its monodromy product of positive Dehn twists of a fiber over the base point p on ∂D. These are twists along vanishing cycles $\alpha_1, \alpha_2, \ldots, \alpha_k$ corresponding to geometric basis of $\pi_1(D - \{x_1, x_2, \ldots, x_k\}, p)$, where x_i's are the critical values of the fibration. We consider the situation where the generic fiber has one boundary component and the monodromy along ∂D is equal to the twist along the boundary of the fiber.

I have proven in [12] that in the case of a generic Lefschetz pencil of plane curves of any fixed degree we can choose a geometric basis of $\pi_1(CP^2 - \{x_1, \ldots, x_k\})$ in such a way that the corresponding vanishing cycles α_i form a bouquet. In particular each pair of curves intersects in 0 (when tangent) or 1 point. I do not know if a similar result is true for a generic pencil of curves on an algebraic surface or even more generally for any Symplectic Lefschetz Fibration without reducible fibers but it is reasonable to ask the Smith question under these restricted conditions: we assume that the twists in our product are along curves which intersect pairwise in 0 or 1 point. In particular they lie in the image of a small Artin group by a geometric homomorphism.

Question(Smith). Let α_i, $i = 1, \ldots, n$ be a configuration of curves on a surface S of genus g with one boundary component δ such that every pair of curves intersect in 0 or 1 point and $T_{\alpha_1} \ldots T_{\alpha_n} = T_\delta$. What is the maximal length n of such a product? Can it be arbitrarily long?

In the case of genus 2 the curves are part of a D_6 configuration (with repetitions). Since the Coxeter element for D_6 gives a square of the boundary twist we probably cannot use the curve α_1' so we are left with A_5 configuration. Here we have a simple algebraic question.

Question. Consider the Artin group A_5 (the braid group on six strings) divided by the relation $(a_1a_2a_3a_4)^5 = a_5a_4a_3a_2a_1^2a_2a_3a_4a_5$. Is it true that every positive word in this group which is equal to $\Delta_4^4 = (a_1a_2a_3a_4)^{10}$ must have either 40 or 30 or 20 letters and in fact must be Hurwitz equivalent (see [1]) to Δ_4^4 or to $\Delta_5^2 = (a_1a_2a_3a_4a_5)^6$ or to $h^2 = (a_5a_4a_3a_2a_1^2a_2a_3a_4a_5)^2$?

A similar question on a closed surface has an affirmative answer. Siebert and Tian proved in their recent paper (see [1]) that any genus two Symplectic Lefschetz Fibration without reducible fibers and with transitive monodromy is holomorphic.

Therefore it corresponds to a monodromy factorization which is a product of conjugates of factorizations which are Hurwitz equivalent to Δ_5^2 or to h^2 (Δ_4^4 does not have transitive monodromy).

On a surface with boundary each such factor is equal to the twist along the boundary so there may be only one factor if the product is equal to a single twist, but the equivalence on the closed surface is weaker, has bigger equivalence classes so the answer on the surface with boundary may be different.

4. Non-orientable surfaces

Very little is known about the mapping class group of a non-orientable surface F. Lickorish found in [7] and [8] an infinite set of generators for $M(F)$. His result was improved by Chillingworth in [2] where a finite simple set of generators was established. It consists of twists along orientation preserving curves and one additional homeomorphism called Y-homeomorphism by Lickorish. Not much more has been known about $M(F)$ until very recently, when Jose Estevez established a presentation of $M(F)$ in his, not yet published Ph.D thesis. He considered an orientable surface S which covers F twice and used methods of Hatcher and Thurston applied to Morse functions on S equivariant with respect to deck transformations of the cover. Unfortunately the presentation of $M(F)$ obtained in this way is extremely complicated and even the generators appearing in the presentation are hard to understand. To find a "decent" presentation of $M(F)$ is quite challenging and important problem.

References

[1] Auroux, D. *Mapping class group factorizations and symplectic 4-manifolds: some open problems* in this volume.

[2] Chillingworth, D.R.J. *A finite set of generators for the homeotopy group of a non-orientable surface*, Proc. Camb. Phil. Soc. **65**, (1969), 409-430.

[3] Hatcher, A. and Thurston, W. *A presentation for the mapping class group of a closed orientable surface*, Topology **19**, (1980), 221-237.

[4] Ishida, A. *The structure of subgroup of mapping class group generated by two Dehn twists*, Proc. Japan Acad. **72** (1996), 240-241.

[5] Korkmaz M. *Mapping class groups of non-orientable surfaces*, Geometriae Dedicata **89**, (2002), 109-133.

[6] Labruere, C. *Generalized braid groups and mapping class groups*, Journal of Knot Theory and its Ramifications **6**, (1997) 715-726.

[7] Lickorish, W.B.R. *Homeomorphisms of non-orientable two-manifolds*, Proc. Camb. Phil. Soc. **59**, (1963), 307-317.

[8] Lickorish, W.B.R. *On the homeomorphisms of non-orientable two-manifolds*, Proc. Camb. Phil. Soc. **61**, (1965), 61-64.

[9] Matsumoto, M. *A presentation of mapping class groups in terms of Artin groups and geometric monodromy of singularities*, Math. Ann. **215** (2000) 401-418.

[10] Perron, F. and Vannier, J.P. *Groupe de monodromie géométrique de singularité simples* CRAS Paris **315** (1992), 1067-1070.

[11] Smith, I. *Geometric monodromy and the hyperbolic disc*, Quarterly J. Math. **52** (2001), 217–228. *Symplectic* Ann. of Math. **75** (2004) 183-199.

[12] Wajnryb, B.*The Lefschetz vanishing cycles on a projective nonsingular plane curve*, Math. Ann. **229** (1977), 181-197.

[13] B. Wajnryb *Artin groups and geometric monodromy*, Inventiones Mathematicae **138**, (1999), 563-571.

Part II

Connections with 3-manifolds, Symplectic Geometry and Algebraic Geometry

Mapping Class Group Factorizations and Symplectic 4-manifolds: Some Open Problems

Denis Auroux

ABSTRACT. Lefschetz fibrations and their monodromy establish a bridge between the world of symplectic 4-manifolds and that of factorizations in mapping class groups. We outline various open problems about mapping class group factorizations which translate to topological questions and conjectures about symplectic 4-manifolds.

1. Lefschetz fibrations and symplectic 4-manifolds

DEFINITION 1. *A Lefschetz fibration on a compact closed oriented smooth 4-manifold M is a smooth map $f : M \to S^2$ which is a submersion everywhere except at finitely many non-degenerate critical points $p_1, \ldots, p_r$, near which f identifies in local orientation-preserving complex coordinates with the model map $(z_1, z_2) \mapsto z_1^2 + z_2^2$.*

The fibers of a Lefschetz fibration f are compact closed oriented surfaces, smooth except for finitely many of them. The fiber through p_i presents a transverse double point, or *node*, at p_i. Without loss of generality, we can assume after perturbing f slightly that the critical values $q_i = f(p_i)$ are all distinct. Fix a reference point q_* in $S^2 \setminus \mathrm{crit}(f)$, and let $\Sigma = f^{-1}(q_*)$ be the corresponding fiber. Then we can consider the *monodromy homomorphism*

$$\psi : \pi_1(S^2 \setminus \mathrm{crit}(f), q_*) \to \mathrm{Map}(\Sigma),$$

where $\mathrm{Map}(\Sigma) = \pi_0 \mathrm{Diff}^+(\Sigma)$ is the mapping class group of Σ. The image $\psi(\gamma)$ of a loop $\gamma \subset S^2 \setminus \mathrm{crit}(f)$ is the isotopy class of the diffeomorphism of Σ induced by parallel transport (with respect to an arbitrary horizontal distribution) along the loop γ; in other terms, $\psi(\gamma)$ is the monodromy of the restriction of f to the preimage of γ.

The singular fibers of f are obtained from the nearby smooth fibers by collapsing a simple closed loop, called the *vanishing cycle*. This can be seen on the local model $(z_1, z_2) \mapsto z_1^2 + z_2^2$, whose singular fiber $\Sigma_0 = \{z_1^2 + z_2^2 = 0\}$ is obtained from the smooth fibers $\Sigma_\epsilon = \{z_1^2 + z_2^2 = \epsilon\}$ ($\epsilon > 0$) by collapsing the embedded loops $\{(x_1, x_2) \in \mathbb{R}^2, \ x_1^2 + x_2^2 = \epsilon\} = \Sigma_\epsilon \cap \mathbb{R}^2$.

The monodromy of a Lefschetz fibration around a singular fiber is the positive Dehn twist along the corresponding vanishing cycle. Choose an ordered collection

©2006 American Mathematical Society

$\eta_1, \ldots, \eta_r$ of arcs joining q_* to the various critical values of f, and thicken them to obtain closed loops $\gamma_1, \ldots, \gamma_r$ based at q_* in $S^2 \setminus \operatorname{crit}(f)$, such that each γ_i encircles exactly one of the critical values of f, and $\pi_1(S^2 \setminus \operatorname{crit}(f), q_*) = \langle \gamma_1, \ldots, \gamma_r \mid \prod \gamma_i = 1 \rangle$. Then the monodromy of f along each γ_i is a positive Dehn twist τ_i along an embedded loop $\delta_i \subset \Sigma$, obtained by parallel transport along η_i of the vanishing cycle at the critical point p_i, and in $\operatorname{Map}(\Sigma)$ we have the relation $\tau_1 \ldots \tau_r = \operatorname{Id}$.

Hence, to every Lefschetz fibration we can associate a *factorization* of the identity element as a product of positive Dehn twists in the mapping class group of the fiber (a factorization of the identity is simply an ordered tuple of Dehn twists whose product is equal to Id; we will often use the multiplicative notation, with the understanding that what is important is not the product of the factors but rather the factors themselves).

Given the collection of Dehn twists $\tau_1, \ldots, \tau_r$ (with product equal to Id), we can reconstruct the Lefschetz fibration f above a large disc Δ containing all the critical values, by starting from $\Sigma \times D^2$ and adding handles as specified by the vanishing cycles [13]. To recover the 4-manifold M we need to glue $f^{-1}(\Delta)$ and the trivial fibration $f^{-1}(S^2 \setminus \Delta) = \Sigma \times D^2$ along their common boundary, in a manner compatible with the fibration structures. In general this gluing involves the choice of an element in $\pi_1 \operatorname{Diff}^+(\Sigma)$; however the diffeomorphism group is simply connected if the genus of Σ is at least 2, and in that case the factorization $\tau_1 \ldots \tau_r = \operatorname{Id}$ determines the Lefschetz fibration $f: M \to S^2$ completely (up to isotopy).

The monodromy factorization $\tau_1 \ldots \tau_r = \operatorname{Id}$ depends not only on the topology of f, but also on the choice of an ordered collection $\gamma_1, \ldots, \gamma_r$ of generators of $\pi_1(S^2 \setminus \operatorname{crit}(f), q_*)$; the braid group B_r acts transitively on the set of all such ordered collections, by *Hurwitz moves*. The equivalence relation induced by this action on the set of mapping class group factorizations is generated by

$$(\tau_1, \ldots, \tau_i, \tau_{i+1}, \ldots, \tau_r) \sim (\tau_1, \ldots, \tau_i \tau_{i+1} \tau_i^{-1}, \tau_i, \ldots, \tau_r) \quad \forall 1 \leq i < r,$$

and is called *Hurwitz equivalence*. Additionally, in order to remove the dependence on the choice of the reference fiber Σ, we should view the Dehn twists τ_i as elements of the mapping class group Map_g of an abstract surface of genus $g = g(\Sigma)$. This requires the choice of an identification diffeomorphism, and introduces another equivalence relation on the set of mapping class group factorizations: *global conjugation*,

$$(\tau_1, \ldots, \tau_r) \sim (\gamma \tau_1 \gamma^{-1}, \ldots, \gamma \tau_r \gamma^{-1}) \quad \forall \gamma \in \operatorname{Map}_g.$$

PROPOSITION 2. *For $g \geq 2$, there is a one to one correspondence between (a) factorizations of Id as a product of positive Dehn twists in Map_g, up to Hurwitz equivalence and global conjugation, and (b) genus g Lefschetz fibrations over S^2, up to isotopy.*

The main motivation to study Lefschetz fibrations is that they seem to provide a manageable approach to the topology of symplectic 4-manifolds.

It is a classical result of Thurston that, if M is an oriented surface bundle over an oriented surface, then M is a symplectic 4-manifold, at least provided that the homology class of the fiber is nonzero in $H_2(M, \mathbb{R})$. As shown by Gompf, the argument extends to the case of Lefschetz fibrations (Theorem 10.2.18 in [12]):

THEOREM 3 (Gompf). *Let $f : M \to S^2$ be a Lefschetz fibration, and assume that the fiber represents a nonzero class in $H_2(M, \mathbb{R})$. Then M admits a symplectic structure for which the fibers of f are symplectic submanifolds; this symplectic structure is unique up to deformation.*

The assumption on the homology class of the fiber is necessary since, for example, $S^1 \times S^3$ fibers over S^2; but it only fails for non-trivial T^2-bundles without singular fibers and their blowups (see Remark 10.2.22 in [**12**]).

Conversely, we have the following result of Donaldson [**8**]:

THEOREM 4 (Donaldson). *Let (X, ω) be a compact closed symplectic 4-manifold. Then X carries a symplectic Lefschetz pencil, i.e. there exist a finite set $B \subset X$ and a map $f : X \setminus B \to \mathbb{CP}^1 = S^2$ such that f is modelled on $(z_1, z_2) \mapsto (z_1 : z_2)$ near each point of B, and f is a Lefschetz fibration with (noncompact) symplectic fibers outside of B.*

It follows immediately that the manifold $\hat{X}$ obtained from X by blowing up the points of B admits a Lefschetz fibration $\hat{f} : \hat{X} \to S^2$ with symplectic fibers, and can be described by its monodromy as discussed above.

Moreover, the fibration $\hat{f}$ has $n = |B|$ distinguished sections $e_1, \ldots, e_n$, corresponding to the exceptional divisors of the blowups. Therefore, each fiber of $\hat{f}$ comes equipped with n marked points, and the monodromy of $\hat{f}$ lifts to the mapping class group of a genus g surface with n marked points.

The fact that the normal bundles of the sections e_i have degree -1 constrains the topology in an interesting manner. For example, if $\hat{f}$ is *relatively minimal* (i.e., if there are no reducible singular fibers with spherical components), then the existence of a section of square -1 implies that $\hat{f}$ cannot be decomposed as a non-trivial fiber sum (see e.g. [**26**]). Therefore, we restrict ourselves to the preimage of a large disc Δ containing all the chosen generators of $\pi_1(S^2 \setminus \mathrm{crit}(\hat{f}))$, and fix trivializations of the normal bundles to the sections e_i over Δ. Deleting a small tubular neighborhood of each exceptional section, we can now view the monodromy of $\hat{f}$ as a morphism

$$\hat{\psi} : \pi_1(\Delta \setminus \mathrm{crit}(\hat{f})) \to \mathrm{Map}_{g,n},$$

where $\mathrm{Map}_{g,n}$ is the mapping class group of a genus g surface with n boundary components.

The product of the Dehn twists $\tau_i = \hat{\psi}(\gamma_i)$ is no longer the identity element in $\mathrm{Map}_{g,n}$. Instead, since $\prod \gamma_i$ is homotopic to the boundary of the disc Δ, and since the normal bundle to e_i has degree -1, we have $\prod \tau_i = \delta$, where $\delta \in \mathrm{Map}_{g,n}$ is the *boundary twist*, i.e. the product of the positive Dehn twists $\delta_1, \ldots, \delta_n$ along loops parallel to the boundary components.

With this understood, the previous discussion carries over, and under the assumption $2 - 2g - n < 0$ there is a one to one correspondence between factorizations of the boundary twist δ as a product of positive Dehn twists in $\mathrm{Map}_{g,n}$, up to Hurwitz equivalence and global conjugation, and genus g Lefschetz fibrations over S^2 equipped with n distinguished sections of square -1, up to isotopy.

Theorems 3 and 4 provide motivation to study the classification problem for Lefschetz fibrations, which by Proposition 2 is equivalent to the classification of mapping class group factorizations involving positive Dehn twists. Hence, various topological questions and conjectures about the classification of symplectic

4-manifolds can be reformulated as questions about mapping class group factorizations in Map_g and $\mathrm{Map}_{g,n}$. In the rest of this paper, we state and motivate a few instances of such questions for which an answer would greatly improve our understanding of symplectic 4-manifolds. Most of these questions are wide open and probably very hard.

Remarks. (**1**) The most natural invariants that one may associate to a factorization in Map_g or $\mathrm{Map}_{g,n}$ are the number r of Dehn twists in the factorization, and the normal subgroup of $\pi_1(\Sigma)$ generated by the vanishing cycles. These are both readily understood in terms of the topology of the total space M: namely, the Euler-Poincaré characteristic of M is equal to $4 - 4g + r$, and, assuming the existence of a section of the fibration, $\pi_1(M)$ is the quotient of $\pi_1(\Sigma)$ by the normal subgroup generated by the vanishing cycles (or equivalently, the quotient of $\pi_1(\Sigma)$ by the action of the subgroup $\mathrm{Im}(\psi) \subset \mathrm{Map}_g$). Similarly, from the intersection pairing between vanishing cycles in $H_1(\Sigma, \mathbb{Z})$ one can recover the intersection form on $H_2(M, \mathbb{Z})$. One invariant which might seem more promising is the number of *reducible* singular fibers, i.e. the number of vanishing cycles which are homologically trivial. However, it is of little practical value for the study of general symplectic 4-manifolds, because reducible singular fibers are a rare occurrence; in fact, the Lefschetz fibrations given by Theorem 4 can always be assumed to have no reducible fibers.

(**2**) Many of the questions mentioned below can also be formulated in terms of factorizations in the Artin braid group, or rather in the liftable subgroup of the braid group. Namely, viewing a genus g surface with boundary components as a simple branched cover of the disc, positive Dehn twists can be realized as lifts of positive half-twists in the braid group (at least as soon as the covering has degree at least 3, in the case of Dehn twists along nonseparating curves; or degree at least 4, if one allows reducible singular fibers). This corresponds to a realization of the symplectic 4-manifold X as a *branched cover* of $\mathbb{CP}^2$, from which the Lefschetz fibration can be recovered by considering the preimages of a pencil of lines in $\mathbb{CP}^2$. The reader is referred to [**3**] for a treatment of the classification of symplectic 4-manifolds from the perspective of branched covers and braid group factorizations. See also [**7**] for more background on Lefschetz fibrations and branched covers.

2. Towards a classification of Lefschetz fibrations?

In view of Proposition 2, perhaps the most important question to be asked about mapping class group factorizations is whether it is possible to classify them, at least partially. For example, it is a classical result of Moishezon and Livne [**17**] that genus 1 Lefschetz fibrations are always isotopic to holomorphic fibrations, and are classified by their number of vanishing cycles, which is always a multiple of 12 (assuming fibers to be irreducible; otherwise we also have to take into account the number of reducible fibers). In fact, all factorizations of the identity as a product of positive Dehn twists in $\mathrm{Map}_1 \simeq \mathrm{SL}(2, \mathbb{Z})$ are Hurwitz equivalent to one of the standard factorizations $(\tau_a \tau_b)^{6k} = 1$, where τ_a and τ_b are the Dehn twists along the two generators of $\pi_1(T^2) \simeq \mathbb{Z}^2$, and k is an integer.

Similarly, Siebert and Tian [**24**] have recently obtained a classification result for genus 2 Lefschetz fibrations without reducible singular fibers and with *transitive monodromy*, i.e. such that the composition of the monodromy morphism with the group homomorphism from Map_2 to the symmetric group S_6 which maps the standard generators τ_i, $1 \le i \le 5$, to the transpositions $(i, i+1)$ is surjective. Namely, these fibrations are all holomorphic, and are classified by their number of vanishing cycles, which is always a multiple of 10. In fact, all such fibrations can be obtained as fiber sums of two standard Lefschetz fibrations f_0 and f_1 with respectively 20 and 30 singular fibers, corresponding to the factorizations $(\tau_1\tau_2\tau_3\tau_4\tau_5\tau_5\tau_4\tau_3\tau_2\tau_1)^2 = 1$ and $(\tau_1\tau_2\tau_3\tau_4\tau_5)^6 = 1$ in Map_2, where $\tau_1, \ldots, \tau_5$ are the standard generating Dehn twists. (At the level of mapping class group factorizations, the fiber sum operation just amounts to concatenation: starting from two factorizations $\tau_1 \ldots \tau_r = 1$ and $\tilde\tau_1 \ldots \tilde\tau_s = 1$, we obtain the new factorization $\tau_1 \ldots \tau_r \tilde\tau_1 \ldots \tilde\tau_s = 1$.)

On the other hand, for genus ≥ 3 (or even for genus 2 if one allows reducible singular fibers) things become much more complicated, and one can build examples of Lefschetz fibrations with non-Kähler total spaces. Thus it seems hopeless for the time being to expect a complete classification of mapping class group factorizations in all generality.

A more realistic goal might be to look for criteria which can be used to determine whether two given Lefschetz fibrations, described by their monodromy factorizations, are isotopic. The main issue at stake here is the algorithmic decidability of the *Hurwitz problem*, i.e. determining whether two given factorizations in Map_g or $\mathrm{Map}_{g,n}$ are equivalent up to Hurwitz moves (or more generally, Hurwitz moves and global conjugation).

QUESTION 1. *Is the Hurwitz problem for mapping class group factorizations decidable? Are there interesting criteria which can be used to conclude that two given factorizations are equivalent, or inequivalent, up to Hurwitz moves and global conjugation?*

In broader terms, the question is whether mapping class group factorizations can be used to derive non-trivial and useful invariants of Lefschetz fibrations, or even better, of the underlying symplectic 4-manifolds.

At this point, it is worth mentioning two spectacular examples of such invariants which arise from geometric considerations (rather than purely from mapping class group theory). One is Seidel's construction of a Fukaya-type A_∞-category associated to a Lefschetz fibration [21], which seems to provide a computationally manageable approach to Lagrangian submanifolds and Fukaya categories in open 4-manifolds equipped with exact symplectic structures. The other is the enumerative invariant introduced by Donaldson and Smith, which counts embedded pseudoholomorphic curves in a symplectic 4-manifold by viewing them as sections of a "relative Hilbert scheme" associated to a Lefschetz fibration [9].

Remark. Generally speaking, it seems that the geometry of Lefschetz fibrations is very rich. An approach which has been developed extensively by Smith [27] is to choose an almost-complex structure on M which makes the fibration f pseudoholomorphic. The fibers then become Riemann surfaces (possibly nodal), and so we can view a Lefschetz fibration as a map $\phi : S^2 \to \overline{\mathcal{M}}_g$ with values in the compactified moduli space of genus g curves. The singular fibers correspond to intersections of $\phi(S^2)$ with the divisor Δ of nodal Riemann surfaces; hence, Lefschetz fibrations correspond to (isotopy classes of) smooth maps $\phi : S^2 \to \overline{\mathcal{M}}_g$ such that $\phi(S^2)$ intersects Δ *transversely* and *positively* (i.e., the local intersection number is always

+1). See [**27**] for various results arising from this description. A related question, posed by Tian, asks whether one can find special geometric representatives for the maps ϕ, e.g. as trees of conformal harmonic maps, and use these to prove that every Lefschetz fibration decomposes into holomorphic "pieces". This statement is to be taken very loosely, since it is not true that every Lefschetz fibration breaks into a fiber sum of holomorphic fibrations; on the other hand, any Lefschetz fibration over a disc is isotopic to a holomorphic fibration [**15**].

We now return to our main discussion and adopt a more combinatorial point of view. A proposed invariant of Lefschetz fibrations which, if computable, could have rich applications, comes from the notion of *matching path*, as proposed by Donaldson and Seidel [**23**]. A matching path for a Lefschetz fibration f is an embedded arc η in $S^2 \setminus \mathrm{crit}(f)$, with end points in $\mathrm{crit}(f)$, such that the parallel transports along η of the vanishing cycles at the two end points are mutually homotopic loops in the fiber of f. For example, with the notations of §1, if two of the Dehn twists $(\tau_1, \ldots, \tau_r)$ in the mapping class group factorization associated to the Lefschetz fibration are equal to each other, say $\tau_i = \tau_j$, then $\eta_i \cup \eta_j$ is a matching path. Up to the action of the braid group by Hurwitz moves, all matching paths arise in this way.

QUESTION 2 (Donaldson). *Is it possible to enumerate all matching paths in a Lefschetz fibration with given monodromy factorization?*

Geometrically, matching paths correspond to Lagrangian spheres in M (or, if considering Lefschetz fibrations with distinguished sections and their $\mathrm{Map}_{g,n}$-valued monodromy, in the blown down manifold X) [**23**].

Lefschetz fibrations often admit infinitely many matching paths, because isotopic Lagrangian spheres may be represented by different matching paths, and also because generalized Dehn twists can often be used to exhibit infinite families of non-isotopic Lagrangian spheres [**22**]. A possible solution is to ask instead which classes in $H_2(M, \mathbb{Z})$ (or $H_2(X, \mathbb{Z})$) can be represented by Lagrangian spheres arising from matching paths. Or, more combinatorially, one can look at matching paths up to the action of *automorphisms* of the Lefschetz fibration f ([**6**], §10). Namely, considering the action of the braid group B_r on tuples of Dehn twists by Hurwitz moves, an automorphism of the Lefschetz fibration is a braid $b \in B_r$ such that $b_*((\tau_1, \ldots, \tau_r)) = (\gamma \tau_1 \gamma^{-1}, \ldots, \gamma \tau_r \gamma^{-1})$ for some $\gamma \in \mathrm{Map}_g$. In other terms, the automorphism group is the *stabilizer* of the given monodromy factorization (or rather of its equivalence class up to global conjugation) with respect to the Hurwitz action of B_r. For example, if η is a matching path then it is easy to see that the half-twist supported along η is an automorphism of the fibration, which corresponds geometrically to the Dehn twist along the Lagrangian sphere associated to η.

If $b \in B_r$ is an automorphism of the fibration, then the image by b of any matching path is again a matching path; hence, automorphisms act on the set of matching paths. Thus, it may make more sense to consider the following question instead of Question 2: is it possible to find a set of generators of the automorphism group of a Lefschetz fibration with given monodromy factorization, and a collection of matching paths $\{\eta_j\}$ such that any matching path can be obtained from one of the η_j by applying an automorphism of the fibration?

Our next series of questions will be specific to factorizations in mapping class groups of surfaces with boundary, $\text{Map}_{g,n}$, with $n > 0$. In this case, the subsemigroup $\text{Map}_{g,n}^+ \subseteq \text{Map}_{g,n}$ generated by positive Dehn twists is strictly contained in the mapping class group. Geometrically, assuming $g \geq 2$ and equipping Σ with a hyperbolic metric, we can use one of the distinguished sections of the Lefschetz fibration to lift the monodromy action to the universal cover. Looking at the induced action on the boundary at infinity (i.e., on the set of geodesic rays through a given point of the hyperbolic disc), it can be observed that positive Dehn twists always rotate the boundary in the *clockwise* direction [26]. This leads e.g. to the indecomposability result mentioned in §1, but also to various questions about the finiteness or uniqueness properties of factorizations of certain elements in $\text{Map}_{g,n}$. To avoid obvious counterexamples arising from non relatively minimal fibrations, in the rest of the discussion we always make the following assumption on reducible singular fibers:

Assumption: every component of every fiber intersects at least one of the distinguished sections $e_1, \ldots, e_n$.

In other terms, we only allow Dehn twists along closed curves which either are homologically nontrivial, or separate Σ into two components each containing at least one of the n boundary components. Then we may ask:

QUESTION 3 (Smith). *Is there an a priori upper bound on the length of any factorization of the boundary twist δ as a product of positive Dehn twists in $\text{Map}_{g,n}$?*

Equivalently: is there an upper bound (in terms of the genus only) on the number of singular fibers of a Lefschetz fibration admitting a section of square -1? (In the opposite direction, various lower bounds have been established, see e.g. [28]). Unfortunately, it is hard to quantify the amount of rotation induced by a Dehn twist on the boundary of the hyperbolic disc, so it is not clear whether the approach in [26] can shed light on this question.

More generally, given an element $T \in \text{Map}_{g,n}^+$, we can try to study factorizations of T as a product of positive Dehn twists. Geometrically, such factorizations correspond to Lefschetz fibrations over the disc (with bounded fibers), such that the monodromy along the boundary of the disc is the prescribed element T. The boundary of such a Lefschetz fibration is naturally a contact 3-manifold Y equipped with a structure of *open book* [10], and the total space of the fibration is a Stein filling of Y [1, 10, 15]. Hence the classification of factorizations of T in $\text{Map}_{g,n}$ is related to (and a subset of) the classification of Stein fillings of the contact 3-manifold Y.

Some remarkable results have been obtained recently concerning the classification of symplectic fillings of lens spaces or links of singularities, using tools from symplectic geometry, and in particular pseudo-holomorphic curves (see e.g. [14, 18]); meanwhile, Lefschetz fibrations have been used to construct examples with infinitely many inequivalent fillings (see e.g. [20]). Hence we may ask:

QUESTION 4. *For which $T \in \text{Map}_{g,n}^+$ is it possible to classify factorizations of T as a product of positive Dehn twists in $\text{Map}_{g,n}$? In particular, for which T is there a unique factorization, or only finitely many factorizations, up to Hurwitz equivalence and global conjugation?*

Let us now return to factorizations of the boundary twist δ, or equivalently to Lefschetz fibrations over S^2 with distinguished sections of square -1. Whereas the classification problem seems to be beyond reach, it may be a more realistic goal to search for a minimal set of moves which can be used to relate any two Lefschetz fibrations (or mapping class group factorizations) with the same genus and the same number of singular fibers to each other. At the level of 4-manifolds, this question asks for a set of surgery operations which can be used to relate any two symplectic 4-manifolds M_1 and M_2 with the same basic topological invariants to each other.

In this context, it is not necessarily useful to require the fundamental groups of the 4-manifolds M_1 and M_2 to be the same; however, it seems natural to require the Euler-Poincaré characteristics and the signatures of M_1 and M_2 to be equal to each other (in other terms, M_1 and M_2 must have the same Chern numbers c_1^2 and c_2). Moreover, when approaching this question from the angle of Lefschetz fibrations, one should require the existence of Lefschetz fibrations with the same fiber genus and with the same number of distinguished -1-sections; if considering the fibrations given by Theorem 4, this amounts to requiring the symplectic structures on M_1 and M_2 to be integral and have the same values of $[\omega]^2$ and $c_1 \cdot [\omega]$.

The constraint on Euler-Poincaré characteristics is natural, and means that we only compare Lefschetz fibrations with identical numbers of singular fibers; additionally, for simplicity it may make sense to require all singular fibers to be irreducible (as is the case for the fibrations given by Theorem 4). The signature constraint, on the other hand, is not so easy to interprete at the level of the monodromy factorizations: determining the signature from the monodromy factorization requires a non-trivial calculation, for which an algorithm has been given by Ozbagci [**19**] (see also [**25**] for a geometric interpretation).

One way in which one can try to simplify the classification of Lefschetz fibrations of a given fiber genus and with fixed Euler-Poincaré characteristic and signature is up to *stabilization* by fiber sum operations [**4**]. However, a more intriguing and arguably more interesting question is to understand the role played by *Luttinger surgery* in the greater topological diversity of symplectic 4-manifolds compared to complex projective surfaces (see [**3**] for a discussion of this problem from the viewpoint of branched covers).

Given a Lagrangian torus T in a symplectic 4-manifold, Luttinger surgery is an operation which consists of cutting out a tubular neighborhood of T, foliated by parallel Lagrangian tori, and gluing it back via a symplectomorphism wrapping the meridian around the torus (in the direction of a given closed loop on the torus), while the longitudes are not affected [**16, 5**]. In the context of Lefschetz fibrations, an important special case is when the torus T is fibered above an embedded loop $\gamma \subset S^2 \setminus \mathrm{crit}(f)$, with fiber an embedded closed loop α in the fiber of f (invariant under the monodromy along γ). For example, this type of Luttinger surgery accounts for the difference between twisted and untwisted fiber sums of Lefschetz fibrations (i.e., concatenating the monodromy factorizations with or without first applying a global conjugation to one of them).

Consider a Lefschetz fibration $f : M \to S^2$, a system of generating loops $\gamma_1, \ldots, \gamma_r \in \pi_1(S^2 \setminus \mathrm{crit}(f))$, and integers $1 \leq k \leq l \leq r$ such that the product $\gamma_k \ldots \gamma_l$ is homotopic to a given loop $\gamma \subset S^2 \setminus \Delta$. Also consider a closed loop α in the fiber, preserved by the monodromy map $\psi(\gamma_k \ldots \gamma_l)$. Then we can build a torus $T \subset M$ by parallel transport of α along the loop γ. This torus is Lagrangian for a suitable choice of symplectic structure, and Luttinger surgery along T in the direction of α amounts to a *partial conjugation* of the monodromy of f. At the level of mapping class group factorizations this corresponds to the operation

$$(\tau_1, \ldots, \tau_r) \longrightarrow (\tau_1, \ldots, \tau_{k-1}, t_\alpha \tau_k t_\alpha^{-1}, \ldots, t_\alpha \tau_l t_\alpha^{-1}, \tau_{l+1}, \ldots, \tau_r),$$

where t_α is the Dehn twist along α (and the requirement that t_α commutes with the product $\tau_k \ldots \tau_l$ ensures that we obtain a valid factorization). So, we may ask the following question:

QUESTION 5. *Given two factorizations of the boundary twist δ as a product of positive Dehn twists along nonseparating curves in $\mathrm{Map}_{g,n}$, such that the total spaces of the corresponding Lefschetz fibrations have the same Euler characteristic and signature, is it always possible to obtain one from the other by a sequence of Hurwitz moves and partial conjugations?*

This question is the analogue for mapping class group factorizations of the question which asks whether any two compact closed integral symplectic 4-manifolds with the same $(c_1^2, c_2, [\omega]^2, c_1 \cdot [\omega])$ are related to each other via a sequence of Luttinger surgeries.

As a closing remark, let us mention that mapping class groups can shed light on the topology of symplectic manifolds not only in dimension 4, but also, with a significant amount of extra work, in dimension 6. Namely, after blowing up a finite set of points every compact closed symplectic 6-manifold can be viewed as a singular fibration over $\mathbb{CP}^2$, with smooth fibers everywhere except above a singular symplectic curve $D \subset \mathbb{CP}^2$ with cusp and node singularities [2]; the fibers above the smooth points of D are nodal. Conversely, the total space of such a singular fibration over $\mathbb{CP}^2$ can be endowed with a natural symplectic structure [11]. Therefore, while in the above discussion we have focused exclusively on mapping class group factorizations, i.e. representations of the free group $\pi_1(S^2 \setminus \{\text{points}\})$ into Map_g, it may also be worthwhile to study representations of fundamental groups of plane curve complements into mapping class groups, as a possible way to further our understanding of the topology of symplectic 6-manifolds.

Acknowledgements. The author would like to thank Paul Seidel, Ivan Smith and Gang Tian for illuminating discussions about some of the questions mentioned here. The author was partially supported by NSF grant DMS-0244844.

References

[1] S. Akbulut, B. Ozbagci, *Lefschetz fibrations on compact Stein surfaces*, Geom. Topol. **5** (2001), 319–334; Erratum: Geom. Topol. **5** (2001), 939–945.

[2] D. Auroux, *Symplectic maps to projective spaces and symplectic invariants*, Turkish J. Math. **25** (2001), 1–42 (math.GT/0007130).

[3] D. Auroux, *Some open questions about symplectic 4-manifolds, singular plane curves, and braid group factorizations*, Proc. 4th European Congress of Mathematics (Stockholm, 2004) A. Laptev Ed., European Math. Soc., 2005, 23–40 (math.GT/0410119).

[4] D. Auroux, *A stable classification of Lefschetz fibrations*, Geom. Topol. **9** (2005), 203–217.

[5] D. Auroux, S. K. Donaldson, L. Katzarkov, *Luttinger surgery along Lagrangian tori and non-isotopy for singular symplectic plane curves*, Math. Ann. **326** (2003), 185–203.

[6] D. Auroux, V. Muñoz, F. Presas, *Lagrangian submanifolds and Lefschetz pencils*, J. Symplectic Geom. **3** (2005), 171–219 (math.SG/0407126).

[7] D. Auroux, I. Smith, *Lefschetz pencils, branched covers and symplectic invariants*, to appear in Proc. CIME school "Symplectic 4-manifolds and algebraic surfaces" (Cetraro, 2003), Lecture Notes in Math., Springer (math.SG/0401021).

[8] S. K. Donaldson, *Lefschetz pencils on symplectic manifolds*, J. Differential Geom. **53** (1999), 205–236.

[9] S. Donaldson, I. Smith, *Lefschetz pencils and the canonical class for symplectic 4-manifolds*, Topology **42** (2003), 743–785.
[10] E. Giroux, *Géométrie de contact: de la dimension trois vers les dimensions supérieures*, Proc. International Congress of Mathematicians, Vol. II (Beijing, 2002), Higher Ed. Press, Beijing, 2002, pp. 405–414 (math.GT/0305129).
[11] R. E. Gompf, *A topological characterization of symplectic manifolds*, J. Symplectic Geom. **2** (2004), 177–206.
[12] R. E. Gompf, A. I. Stipsicz, *4-manifolds and Kirby calculus*, Graduate Studies in Math. **20**, Amer. Math. Soc., Providence, 1999.
[13] A. Kas, *On the handlebody decomposition associated to a Lefschetz fibration*, Pacific J. Math. **89** (1980), 89–104.
[14] P. Lisca, *On symplectic fillings of lens spaces*, to appear in Trans. Amer. Math. Soc. (math.SG/0312354).
[15] A. Loi, R. Piergallini, *Compact Stein surfaces with boundary as branched covers of B^4*, Invent. Math. **143** (2001), 325–348.
[16] K. M. Luttinger, *Lagrangian tori in $\mathbb{R}^4$*, J. Differential Geom. **42** (1995), 220–228.
[17] B. Moishezon, *Complex surfaces and connected sums of complex projective planes*, Lecture Notes in Math. **603**, Springer, Heidelberg, 1977.
[18] H. Ohta, K. Ono, *Simple singularities and symplectic fillings*, J. Differential Geom. **69** (2005), 1–42 (2002).
[19] B. Ozbagci, *Signatures of Lefschetz fibrations*, Pacific J. Math. **202** (2002), 99–118.
[20] B. Ozbagci, A. Stipsicz, *Contact 3-manifolds with infinitely many Stein fillings*, Proc. Amer. Math. Soc. **132** (2004), 1549–1558.
[21] P. Seidel, *Vanishing cycles and mutation*, Proc. 3rd European Congress of Mathematics (Barcelona, 2000), Vol. II, Progr. Math. **202**, Birkhäuser, Basel, 2001, pp. 65–85 (math.SG/0007115).
[22] P. Seidel, *Lagrangian two-spheres can be symplectically knotted*, J. Differential Geom. **52** (1999), 145–171.
[23] P. Seidel, *Fukaya categories and Picard-Lefschetz theory*, in preparation.
[24] B. Siebert, G. Tian, *On the holomorphicity of genus two Lefschetz fibrations*, Ann. Math. **161** (2005), 959–1020 (math.SG/0305343).
[25] I. Smith, *Lefschetz fibrations and the Hodge bundle*, Geom. Topol. **3** (1999), 211–233.
[26] I. Smith, *Geometric monodromy and the hyperbolic disc*, Quarterly J. Math. **52** (2001), 217–228 (math.SG/0011223).
[27] I. Smith, *Lefschetz pencils and divisors in moduli space*, Geom. Topol. **5** (2001), 579–608.
[28] A. Stipsicz, *On the number of vanishing cycles in Lefschetz fibrations*, Math. Res. Lett. **6** (1999), 449–456.

DEPARTMENT OF MATHEMATICS, M.I.T., CAMBRIDGE, MA 02139, USA
E-mail address: auroux@math.mit.edu

The Topology of 3-manifolds, Heegaard Distance and the Mapping Class Group of a 2-manifold

Joan S. Birman

We have had a long-standing interest in the way that structure in the mapping class group of a surface reflects corresponding structure in the topology of 3-manifolds, and conversely. We find this area intriguing because the mapping class group (unlike the collection of closed orientable 3-manifolds) is a group which has a rich collection of subgroups and quotients, and they might suggest new ways to approach 3-manifolds. (For example, it is infinite, non-abelian and residually finite [**14**]). In the other direction, 3-manifolds have deep geometric structure, for example the structure that is associated to intersections between 2-dimensional submanifolds, and that sort of inherently geometric structure might bring new tools to bear on open questions regarding the mapping class group. That dual theme is the focus of this article.

In §1 we set up notation and review the background, recalling some of the things that have already been done relating to the correspondence, both ways. We will also describe some important open problems, as we encounter them. We single out for further investigation a new tool which was introduced in [**17**] by Hempel as a measure of the complexity of a Heegaard splitting of a 3-manifold. His measure of complexity has its origins in the geometry of 3-manifolds. He defined it as the length of the shortest path between certain vertices in the curve complex of a Heegaard surface in the 3-manifold. In fact, the mapping class group acts on the curve complex and the action is faithful. That is, the (extended) mapping class group is isomorphic to the automorphism group of the curve complex. In §2 we will propose some number of open questions which relate to the distance, for study. In §3 we suggest approaches which could be useful in obtaining additional tools to investigate some of the problems posed in §2. Some of the 'additional tools' are in the form of additional open problems.

The symbols $\Diamond 1, \Diamond 2, \Diamond 3, \dots$ will be used to highlight known results from the literature.

Acknowledgments. We thank Jason Behrstock, Tara Brendle, Jeffrey Brock, Mustafa Korkmaz, Chris Leininger, Brendon Owens, Saul Schleimer, and Jennifer Schultens for their careful answers to our many questions, as this article was in preparation. We thank Ian Agol, John Hempel and Chan-Ho Suh for their help

---

The author is partially supported by the U.S. National Science Foundation under grant DMS-0405586.

in correcting some incomplete references in an early version of the paper and for various helpful suggestions. We also thank the referee for his/her careful reading of our first draft of this paper.

1. Background

Let S_g be a closed, connected, orientable surface, and let $\text{Diff}S_g^\pm$, (resp. $\text{Diff}S_g^+$) be the groups of diffeomorphisms (resp. orientation-preserving diffeomorphisms) of S_g. The *mapping class group* $\mathcal{M}_g$ is $\pi_0(\text{Diff}^+ S_g)$. The *extended* mapping class group $\mathcal{M}_g^\pm$ is $\pi_0(\text{Diff}^\pm S_g)$. The groups $\mathcal{M}_g$ and $\mathcal{M}_g^\pm$ are related by a split short exact sequence:

(1) $$\{1\} \longrightarrow \mathcal{M}_g \longrightarrow \mathcal{M}_g^\pm \longrightarrow \mathbb{Z}/2\mathbb{Z} \longrightarrow \{1\}.$$

We will be interested in certain subgroups of $\mathcal{M}_g$. To describe the first of these, regard S_g as the boundary of an oriented handlebody H_g. The *handlebody* subgroup $\mathcal{H}_g \subset \mathcal{M}_g$ is the (non-normal) subgroup of all mapping classes that have representatives that extend to diffeomorphisms of H_g.

We turn our attention to closed, connected orientable 3-manifolds. Let H_g be an oriented handlebody and let $H_g' = \tau(H_g)$ be a copy of H_g, with the induced orientation. We are interested in Heegaard splittings of 3-manifolds, i.e. their representations as a union of the handlebodies H_g and H_g', where H_g and H_g' are glued together along their boundaries via a diffeomorphism $\partial H_g \to \partial H_g'$. The gluing map is necessarily orientation-reversing, but if we choose a fixed orientation-reversing diffeomorphism $i : S_g \to S_g$ whose isotopy class ι realizes the splitting in the exact sequence (1), we may describe the gluing as $i \circ f$, where f is orientation-preserving. Then f determines an element $\phi \in \mathcal{M}_g$, and since the topological type of the 3-manifold which is so-obtained depends only on the mapping class ϕ of f, we use the symbol $M = H_g \cup_\phi H_g'$ to describe the *Heegaard splitting* of *genus g* of M. The surface $S_g = \partial H_g = \partial H_g'$, embedded in M, is a Heegaard surface in M. As is well-known (for example see [41] for a proof) every closed, connected, orientable 3-manifold can be obtained from a Heegaard splitting, for some $\phi \in \mathcal{M}_g$.

Since the cases $g \leq 1$ are well understood and often need special case-by-case arguments, we will assume, unless otherwise indicated, that $g \geq 2$. From now on, when we do not need to stress the genus we will omit the symbol g.

Heegaard splittings are not unique. If M admits two splittings, with defining maps ϕ_1, ϕ_2 then the splittings are *equivalent* if the splitting surfaces are isotopic, and if (assuming now that they are identical) there is a diffeomorphism $B : M \to M$ that restricts to diffeomorphisms $b_1 : H \to H$ and $b_2 : H' \to H'$. By further restricting to the common boundary of H and H' we obtain elements β_1, β_2 in the mapping class group, with

(2) $$\tau \iota \phi_2 \beta_1 = \beta_2 \tau \iota \phi_1, \quad \text{or} \quad \phi_2 = ((\tau\iota)^{-1}\beta_2(\tau\iota))(\phi_1)(\beta_1^{-1}).$$

Since β_1 and $(\tau\iota)^{-1}\beta_2(\tau\iota)$ are independent and are both in $\mathcal{H}$, it follows that the double coset $\mathcal{H}\phi_1\mathcal{H} \subset \mathcal{M}$ gives the infinitely many distinct elements in $\mathcal{M}$ which define equivalent Heegaard splittings of the same genus. Note that our 3-manifold may have other Heegaard splittings of genus g that are not in the double coset $\mathcal{H}\phi_1\mathcal{H}$, but if none such exist the splitting defined by all the gluing maps in $\mathcal{H}\phi_1\mathcal{H}$ is *unique*. For example, it was proved by Waldhausen in [53] that any two Heegaard splittings of the same genus of S^3 are equivalent.

There is another way in which Heegaard splittings are not unique. Since taking the connected sum of any 3-manifold M with S^3 preserves the topological type of M, this gives a nice way to produce, for any genus g splitting of any M infinitely many splittings of genus $g+1$, all in the same equivalence class: Choose an arbitrary Heegaard splitting $H \cup_\psi H'$ of genus 1 for S^3. Let $H \cup_\phi H'$ be any Heegaard splitting of M. Then take the connected sum $(H \cup_\phi H') \# (H \cup_\psi H')$, arranging things so that the 2-sphere that realizes the connected sum intersects the Heegaard surface in a circle, splitting it as the connect sum of surfaces of genus g and 1. Writing this as $H \cup_{\phi \# \psi} H'$, we say that this Heegaard splitting has been *stabilized*. Note that this notion immediately generalizes to include the possibility that the splitting S^2 decomposes M into the connect sum of three manifolds $M' \# M''$, both defined by Heegaard splittings, where neither M' nor M'' is S^3. In [15], Haken showed more, proving that if a 3-manifold M is the connected sum of 3-manifolds M' and M'', then any Heegaard splitting of M is equivalent to one which is a connect sum of splittings of M' and M''. Thus it also makes sense to say, when no such connect sum decomposition is possible, that a Heegaard splitting is *irreducible*. So 'irreducible' has two meanings as regards a Heegaard splitting: either the 3-manifold that it defines is not prime, or the given splitting is stabilized.

We are ready to describe some of the early work relating to the interplay between the topology of 3-manifolds and the structure of mapping class groups of surfaces. The mapping class group acts on $H_1(S_g, \mathbb{Z})$, and the induced action determines a homomorphism $\chi_g : \mathcal{M}_g \to \text{Sp}(2g, \mathbb{Z})$. The kernel of χ_g is the *Torelli subgroup* $\mathcal{I}_g \subset \mathcal{M}_g$. A good exercise for a reader who is unfamiliar with the mapping class group is to show that if M is defined by the Heegaard splitting $H_g \cup_\phi H'_g$, then every 3-manifold which has the same first homology as M has a Heegaard splitting of the form $H_g \cup_{\rho \circ \phi} H'_g$, for some $\rho \in \mathcal{I}_g$. This is the beginning of a long story, which we can only describe in the briefest way. It depends in fundamental ways on the collection of 7 deep and far-reaching papers of Johnson, written in the 1980's, about the structure of $\mathcal{I}_g$. We refer the reader to Johnson's review article [23] for an excellent guide to the results in these 7 papers.

$\diamond$1 Building on the exercise that we just assigned, Sullivan [47] used mappings classes in $\mathcal{I}_g$ and Heegaard splittings to construct 3-manifolds with the same homology as the connected sum of g copies of $S^1 \times S^2$, i.e. the manifold defined by the splitting $H_g \cup_{\text{id}} H'_g$. He then asked how the intersection ring of such a manifold differs from that of a true $\#_g(S^1 \times S^2)$? In this way Sullivan discovered a homomorphism from $\mathcal{I}_g$ to an abelian group of rank $\binom{g}{3}$. Johnson then proved that Sullivan's map lifts to a map $\tau_1 : \mathcal{I}_g \to A_1$, where A_1 is a free abelian group of rank $\binom{2g}{3}$. We are interested here in the Sullivan-Johnson homomorphism τ_1. It has a topological interpretation that is closely related to Sullivan's construction. Johnson asked whether $\tau_1(\mathcal{I}_g)$ was the abelizization of $\mathcal{I}_g$, and proved it is not. That is, A_1 is a proper quotient of $\mathcal{I}'_g$.

$\diamond$2 To say more we need to take a small detour and ask about generators of $\mathcal{I}_g$. The most obvious ones are the Dehn twists about separating curves of S_g, but they don't tell the full story. Johnson proved that for $g \geq 3$ the group $\mathcal{I}_g$ is finitely generated by certain maps which are known as 'bounding pairs'. They are determined by a pair of non-separating simple closed curves on S_g whose union divides S_g, and Johnson's generators are a pair of Dehn twists, oppositely oriented, about the two curves in a bounding pair.

This leads us, of course, to the normal subgroup $\mathcal{K}_g \subseteq \mathcal{I}_g$ that is generated by Dehn twists about all separating curves on S_g. For $g = 2$ the groups $\mathcal{I}_2$ and $\mathcal{K}_2$ coincide (because there are no bounding pairs on a surface of genus 2), but for $g \geq 3$ the group $\mathcal{K}_g$ is a proper subgroup of the Torelli group. So now we have two subgroups of $\mathcal{M}_g$ that have interest, the Torelli group $\mathcal{I}_g$ and the subgroup $\mathcal{K}_g$. To give the latter new meaning, we return to the homomorphism τ_1 that we defined in ($\Diamond 1$) above. Johnson proved that kernel(τ_1) = $\mathcal{K}_g$.

$\Diamond 3$ Since the image of τ_1 is abelian, one wonders how kernel(τ_1) is related to the commutator subgroup $[\mathcal{I}_g, \mathcal{I}_g] \subset \mathcal{I}_g$? To say more we return to the topology of 3-manifolds, and a $\mathbb{Z}/2\mathbb{Z}$-valued invariant $\mu(M^3)$ of a homology sphere M^3 that was discovered by Rohlin. Recall that, by the exercise that we assigned before the start of ($\Diamond 1$), every homology sphere may be obtained from S^3 by cutting out one of the Heegaard handlebodies and regluing it via some $\gamma \in \mathcal{I}_g$. (Here we are tacitly assuming what was proved in [**53**]: that all Heegaard splittings of any fixed genus g of S^3 are equivalent, so that it doesn't matter which splitting of S^3 you choose, initially.) We refer to the 3-manifold obtained after the regluing as M_γ. In 1978 the author and R. Craggs [**5**] showed that the function $\gamma \to \mu(M_\gamma)$ determines a finite family of homomorphisms $\delta_\gamma : \mathcal{I}_g \to \mathbb{Z}/2\mathbb{Z}$. The homomorphisms depend mildly on the choice of γ. Johnson proved that the finitely many homomorphisms $\mathcal{I}_g \to \mathbb{Z}/2\mathbb{Z}$ that were discovered in [**5**] generate Hom($\mathcal{I}_g \to \mathbb{Z}/2\mathbb{Z}$) and used them to construct another homomorphism $\tau_2 : \mathcal{I}_g \to A_2$, where A_2 is an abelian group, with the kernel of τ_2 the intersection of the kernels of the homomorphisms of [**5**]. We now know about two subgroups of $\mathcal{I}_g$ whose study was motivated by known structure in the topology of 3-manifolds, namely kernel(τ_1) and kernel(τ_2), and it turns out that kernel(τ_1) $\cap$ kernel(τ_2) = $[\mathcal{I}_g, \mathcal{I}_g]$.

$\Diamond 4$ Further work in this direction was done by Morita in [**36**] and [**37**]. Casson's invariant $\tilde{\mu}(M_\gamma)$ of a homology 3-sphere M_γ is a lift of the Rohlin-invariant $\mu(M_\gamma)$ to a $\mathbb{Z}$-valued invariant. One then has a function $\tilde{\delta}_\gamma : \mathcal{I}_g \to \mathbb{Z}$ which is defined by sending γ to $\tilde{\mu}(M_\gamma)$. Morita related this function to structure in $\mathcal{K}_g$. To explain what he did, recall that in ($\Diamond 1$) and ($\Diamond 3$) we needed the fact that the γ could be assumed to be in $\mathcal{I}_g$. In fact a sharper assertion was proved by Morita in [**36**]: we may assume that $\gamma \in \mathcal{K}_g$. Morita then went on, in [**36**] and [**37**] to prove that if one restricts to $\mathcal{K}_g$ the function $\tilde{\delta}_\gamma$ determines a homomorphism $\tilde{\tau}_2 : \mathcal{K}_g \to \mathbb{Z}$ which lifts $\tau_2|\mathcal{K}_g : \mathcal{K}_g \to \mathbb{Z}/2\mathbb{Z}$.

This brings us to our first problem, which is a vague one:

PROBLEM 1. The ideas which were just described relate to the beginning of the lower central series of $\mathcal{I}_g$. There is also the lower central series of $\mathcal{K}_g$. The correspondence between the group structure of $\mathcal{M}_g$ and 3-manifold topology, as regards the subgroups of $\mathcal{M}_g$ that have been studied, has been remarkable. It suggests strongly that there is much more to be done, with the possibility of new 3-manifold invariants as a reward.

Investigations relating to the correspondences that we just described slowed down during the period after Thurston [49] did his groundbreaking work on the topology and geometry of 3-manifolds. Recall that the 3-manifolds that we are discussing can be decomposed along embedded 2-spheres into prime summands which are unique, up to order. We learned from Thurston that there is a further canonical decomposition of prime 3-manifolds along a canonical family of incompressible tori, the 'JSJ decomposition'. In particular, Thurston conjectured that each component after the JSJ decomposition supported its own unique geometry, and the geometry was a very important aspect of the topology.[1] As a consequence, it was necessary to deal with 3-manifolds with boundary. We pause to describe the modifications that are needed to describe their Heegaard splittings.

Let X be a collection of pairwise disjoint simple closed curves on a surface S. An oriented *compression body* H_X is obtained from an oriented $S \times I$ and X by gluing 2-handles to $S \times \{1\} \subset S \times [0,1]$ along the curves in X, and then capping any 2-sphere boundary components with 3-handles. Note that if S has genus g, and if X has g components, chosen so that the closure of S split along X is a sphere with $2g$ discs removed, then g 2-handles and one 3-handle will be needed and H_X will be a handlebody with boundary S. More generally H_X will have some number of boundary components. It is customary to identify S with the 'outer boundary' of H_X, i.e. $S \times \{0\} \subset H_X$.

To construct an oriented three-manifold M with boundary we begin with $S \times I$, and a copy $\tau(S \times I)$, where τ is a homeomorphism and $\tau(S \times I)$ has the induced orientation. As in the Heegaard splitting construction, let $i : S \times \{0\} \to \tau(S \times \{0\})$ be a fixed orientation-reversing involution, and let $f : S \times \{0\} \to S \times \{0\}$ be an arbitrary orientation-preserving diffeomorphism of S. Then we may use $i \circ f$ to glue $S \times \{0\}$ to $\tau(S \times \{0\})$ along their outer boundaries. Let ϕ be the mapping class of f. This still makes sense if we attach 2-handles to $S \times I$ and $\tau(S \times I)$ along curve systems $X \subset S \times \{1\}$ and $Y \subset \tau(S \times \{1\})$ to get compression bodies H_X and H_Y. In this way we obtain an oriented 3-manifold $M = H_X \cup_\phi H_Y$ with boundary which generalizes the more familiar construction, when H_X and H_Y are handlebodies. We continue to call the more general construction a Heegaard splitting, but now it's a splitting of a 3-manifold with boundary. See [41] for a proof that every compact orientable 3-manifold with boundary arises via these generalized Heegaard splittings. Note that in particular, in this way, we obtain Heegaard splittings of the manifolds obtained after the JSJ decomposition.

An example is in order, but the most convenient way to explain the example is to pass to a slightly different way of looking at compression bodies. Dually, a compression body is obtained from $S \times [0,1]$ by attaching some number, say p, of 1-handles to $S \times \{0\} \subset S \times [0,1]$. This time we identify S with $S \times \{1\}$. An example of a Heegaard splitting of a 3-manifold with boundary is obtained when M is the complement of an open tubular neighborhood $N(K)$ of a knot K in S^3. In this case $\partial M = \partial(S^3 \setminus N(K))$ is a torus. By attaching some number, say p 1-handles to the boundary of the (in general) knotted solid torus $N(K)$, we can unknot $N(K)$, changing it to a handlebody in S^3. (Remark: the minimum such p is known as the 'tunnel number' of K.) The union of this handlebody and its complement H'

---

[1]As we write this article, over 20 years after Thurston announced his results, the main conjecture in [49], the geometrization conjecture, seems close to being proved via partial differential equations and the work of Perelman, giving new importance to the JSJ decomposition.

is a Heegaard splitting of S^3 (not of the knot space). The Heegaard surface for this splitting of S^3 will turn out to be a Heegaard surface for a related Heegaard splitting of the knot complement. To see this, let $N_0(K) \subset N(K)$ be a second neighborhood of K. Then

$$S^3 - N_0(K) = (N(K) - N_0(K)) \cup p \text{ (1-handles)} \cup H'$$
$$= (\partial N(K) \times I) \cup p \text{ (1-handles)} \cup H'.$$

The 3-manifold $H = (\partial N(K) \times I) \cup p(1-\text{handles})$ is an example of a compression body. Therefore our knot complement, which is a 3-manifold with torus boundary, has been represented as a union of a compression body H_X and a handlebody H', identified along their boundaries. By construction, $\partial H_X = \partial H'$ is a closed orientable surface of genus $p+1$. This surface is called a Heegaard surface in $S^3 \setminus N_0(K)$, and so $S^3 - N_0(K) = H_X \cup_\phi H'$, where the glueing map ϕ is an element of the mapping class group $\mathcal{M}_{p+1}$. In this way, Heegaard splittings of the components after the JSJ decomposition fit right into the existing theory.

There was also a second reason why the correspondence that is the focus of this article slowed down around the time of Thurston. In the important manuscript [11], the following new ideas (which are due to Casson and Gordon, and build on the work of Haken in [15]) were introduced in the mid-1980's. Let M be a 3-manifold which admits a Heegaard splitting $H \cup_\phi H'$. Define a *disc pair* (D, D') to be a pair of properly embedded essential discs, with $D \subset H$ and $D' \subset H'$, so that $\partial D, \partial D' \subset S = \partial H = \partial H'$. The Heegaard splitting is said to be:

- *reducible* if there exists a disc pair (D, D') with $\partial D = \partial D'$. Intuitively, either the given Heegaard splitting is stabilized, or the manifold is a nontrivial connected sum, the connected sum decomposition being consistent with the Heegaard splitting. Observe that this is identical with our earlier definition of a reducible Heegaard splitting, but with a new emphasis.
- *strongly irreducible* if and only if for every disc pair $(D, D') : \partial D \cap \partial D' \neq \emptyset$.

Also, the corresponding negations:

- *irreducible* if and only if it is not reducible. Equivalently, for every disc pair (D, D'), $\partial D \neq \partial D'$.
- *weakly reducible* if and only if it is not strongly irreducible. Equivalently, there exists a disc pair (D, D') with $\partial D \cap \partial D' = \emptyset$.

Note that any reducible splitting is also weakly reducible, and any strongly irreducible splitting is also irreducible. Here are several applications of these notions:

$\diamond 5$ In [11] Casson and Gordon proved that if a 3-manifold M has a Heegaard splitting $H \cup_\phi H'$, where H and H' are compression bodies, and if the splitting is strongly irreducible, then either the Heegaard splitting is reducible or the manifold contains an incompressible surface.

$\diamond 6$ A different application is the complete classification of the Heegaard splittings of graph manifolds. These manifolds have the property that when they are split open along the canonical tori of the JSJ decomposition, the closure of each component is a Seifert fiber space. See the paper [44], by Schultens, for a succinct and elegant presentation of the early work (much of which was her own) and the final steps in this classification. Her work uses the JSJ decompositiion, and also depends crucially on the concept of a strongly irreducible Heegaard splitting.

◇7 Casson and Gordon used the same circle of ideas to prove the existence of manifolds with irreducible Heegaard splittings of arbitrarily high genus.

For many years after [11] was published it seemed impossible to interpret the Casson-Gordon machinery in the setting of the mapping class group. As a result, the possibility of relating this very deep structure in 3-manifold topology to corresponding structure in surface mapping class groups seemed out of reach. All that changed fairly recently. because of new ideas due to John Hempel [17]. To explain his ideas, we first need to define the 'complex of curves' on a surface, a simplicial complex $\mathbf{C}(S)$ that was introduced by Harvey [16] in the late 1970's. It has proved to be of fundamental importance in the theory of Teichmüller spaces. The complex $\mathbf{C}(S)$ has as its vertices the isotopy classes of essential simple closed curves (both separating and non-separating). Distinct vertices $v_0, v_1, \ldots, v_q$ determine a q-simplex of $\mathbf{C}(S)$ if they can be represented by q pairwise disjoint simple closed curves on S. The complex $\mathbf{C}(S)$, and also its 1-skeleton, can be given the structure of a metric space by assigning length 1 to every edge and making each simplex a Euclidean simplex with edges of length 1. We have an important fact:

◇8 Aut($\mathbf{C}(S)$) was investigated by Ivanov in [20]. He proved[2] that the group Aut $\mathbf{C}(S)$ is naturally isomorphic to the extended mapping class group $\mathcal{M}^{\pm}$. This paper lead to an explosion of related results, with different complexes (see the next section for a detailed discussion). Therefore, when one talks about the complex of curves the mapping class group is necessarily nearby.

We now turn to the work of Hempel in [17]. Let X be a simplex in $\mathbf{C}(S)$. The curves that are determined by X are pairwise disjoint simple closed curves on S. One may then form a compression body H_X from $S \times [0, 1]$ by attaching 2-handles to $S \times \{1\}$ along $X \times \{1\}$ and attaching 3-handles along any 2-sphere boundary components. As before, $S \times \{0\}$ is the outer boundary of H_X. Let Y be another simplex, with associated compresion body H_Y. Then (X, Y) determine a Heegaard splitting of a 3-manifold. As before the splitting may be thought of as being determined by an element in the mapping class group $\mathcal{M}$ of S, although Hempel does not do this.

We are interested mainly in the case when H_X and H_Y are handlebodies. In this situation, using our earlier notation, X is a collection of g pairwise disjoint non-separating curves on S which decompose S into a sphere with $2g$ holes and $Y = \phi(X)$, where ϕ is the Heegaard gluing map. There is an associated *handlebody subcomplex* $\mathbf{H}_X$ of $\mathbf{C}(S)$, namely the subcomplex whose vertices are simple closed curves on S which bound discs in H_X. There is also a related subcomplex $\mathbf{H}_Y$ whose vertices are simple closed curves on S which bound discs in H_Y. Again, the latter are the image of the former under the Heegaard gluing map ϕ. Hempel's *distance* of the Heegaard splitting of a closed orientable 3-manifold M is the minimal distance in $\mathbf{C}(S)$ between vertices in $\mathbf{H}_X$ and vertices in $\mathbf{H}_Y$. He calls it $d(\mathbf{H}_X, \mathbf{H}_Y)$. In [40] the same distance is called the *handlebody distance*. It is clear that the distance is determined by the choice of the glueing map ϕ, and since our focus has been on the

---

[2]Actually, Ivanov was missing certain special cases which were later settled by Korkmaz and by Luo, however, we are only interested in the case $g \geq 2$ so this is irrelevant.

mapping class group we will use the symbol $d(\phi)$ whenever it is appropriate to do so, instead of Hempel's symbol $d(\mathbf{H}_X, \mathbf{H}_Y)$.

What does Hempel's distance have to do with the Casson-Gordon machinery?

$\diamond$9 Hempel had defined the distance to be a deliberate extension of the Casson-Gordon machinery. In particular, he observed that the triplet (S, H_X, H_Y) determines an equivalence class of Heegaard splittings of the underlying 3-manifold, and that:
> The splitting is reducible if and only if $d(\mathbf{H}_X, \mathbf{H}_Y) = 0$
> The splitting is irreducible if and only if $d(\mathbf{H}_X, \mathbf{H}_Y) \geq 1$.
> The splitting is weakly reducible if and only if $d(\mathbf{H}_X, \mathbf{H}_Y) \leq 1$.
> The splitting is strongly irreducible if and only if $d(\mathbf{H}_X, \mathbf{H}_Y)) \geq 2$.

$\diamond$10 In [**17**], Hempel shows that if M is either Seifert fibered or contains an essential torus, then every splitting of M has distance at most 2. There is also related work by Thompson [**48**], who defined a Heegaard splitting to have the *disjoint curve property* if there is a disc pair (D, D') and a simple closed curve c on the Heegaard surface such that $\partial D \cap c = \emptyset$ and $\partial D' \cap c = \emptyset$. Using this concept she then proved that if a Heegaard splitting does not have the disjoint curve property, then the manifold defined by the splitting has no embedded essential tori. Also, if the splitting is assumed to be strongly irreducible, then an essential torus forces it to have the disjoint curve property. The work in [**17**] and the work in [**48**] were done simultaneously and independently. There is some overlap in content, although Thompson was not thinking in terms of the curve complex and the results in [**48**] are more limited than those in [**17**].

$\diamond$11 There is an important consequence. From the results that we just referenced, it follows that a 3-manifold which has a splitting of distance at least 3 is irreducible, not Seifert fibered and has no embedded essential tori. Modulo the geometrization conjecture, one then concludes that M is hyperbolic if $d(\phi) \geq 3$.

$\diamond$12 In [**17**] Hempel proved that there are distance n splittings for arbitrarily large n.

As it turned out, Hempel's beautiful insight suddenly brought a whole new set of tools to 3-manifold topologists. The reason was that, at the same time that Hempel's ideas were being formulated, there were ongoing studies of the metric geometry of the curve complex that turned out to be highly relevant. The article [**31**] is a fine survey article that gives a good account of the history of the mathematics of the curve complex (which dates back to the early 1970's), continuing up to the recent contributions of Minsky, Masur, Brock, Canary and others, leading in particular to the proof of the 'Ending Lamination Conjecture'.

$\diamond$13 The complex $\mathbf{C}(S)$ can be made into a complete, geodesic metric space by making each simplex into a regular Euclidean simplex of side length 1. In [**32**] Howard Masur and Yair Minsky initiated studies of the intrinsic geometry of $\mathbf{C}(S)$. In particular, they showed that $\mathbf{C}(S)$ is a δ-hyperbolic metric space.

$\diamond$14 A subset $\mathbf{V}$ of a metric space $\mathbf{C}$ is said to be *k-quasiconvex* if for any points $p_1, p_2 \in \mathbf{V}$ the gedesic in $\mathbf{C}$ that joins them stays in a k-neighborhood of $\mathbf{V}$. The main result in [**33**] is that the handlebody subcomplex of $\mathbf{C}(S)$ is k-quasiconvex, where the constant k depends only on the genus of S.

◇15 Here is an example of how these ideas were used in 3-manifold topology: Appealing to the quasiconvexity result of [**33**] H. Namazi proved in [**40**] that if a 3-manifold which is defined by a Heegaard splitting has sufficiently large distance, then the subgroup of $\mathcal{M}$ of surface mappings that extends to both Heegaard handlebodies, H_g and H'_g, is finite. As a corollary, he proved that the mapping class group of the 3-manifold determined by the Heegaard splitting $H \cup_\phi H'$, i.e the group $\pi_0(Diff^+ M^3)$, where M^3 is the 3-manifold defined by the Heegaard splitting $H \cup_\phi H'$, is finite.

2. Some open problems

We begin with two problems that may not be either deep or interesting, although we were not sure exactly how to approach them:

PROBLEM 2. Assume, for this problem, that M is a 3-manifold with non-empty boundary. Then, on one side of the double coset $\mathcal{H}\phi\mathcal{H}$ the handlebody subgroup needs to be modified to a 'compression body subgroup'. Make this precise, by describing how to modify the double coset to take account of the handle decomposition of the compression body. What happens in the case of a knot space?

PROBLEM 3. How is the Nielsen-Thurston trichotomy related to the question of whether the distance is $0, 1, 2$ or ≥ 3?

The next 3 problems concern the very non-constructive nature of the definition of $d(\phi)$:

PROBLEM 4. Find an algorithm to compute the distance $d(\phi)$ of an arbitrary element $\phi \in \mathcal{M}$. We note that an algorithm to compute shortest paths between fixed vertices v, w in the curve complex has been presented by Shackleton in [**45**]. That problem is a small piece of the problem of computing the distance.

PROBLEM 5. Knowing that $d(\phi) \leq 1$, can we decide whether $d(\phi) = 0$? Geometrically, if a Heegaard splitting is weakly reducible, can you decide if it's reducible?

PROBLEM 6. Knowing that $d(\phi) \geq 1$, can we decide whether it is ≥ 2? Knowing that it's ≥ 2, can we decide whether it is ≥ 3?

PROBLEM 7. Schleimer has proved in [**42**] that each fixed 3-manifold M has a bound on the distances of its Heegaard splittings. Study this bound, with the goal of developing an algorithm for computing it.

Understanding the handlebody subgroup $\mathcal{H}_g$ of the mapping class group is a problem that is obviously of central importance in understanding Heegaard splittings. A finite presentation for $\mathcal{H}_g$ was given by Wajnryb in [**51**]. To the best of our knowledge, this presentation has not been simplified, except in the special case $g = 2$. Very little is known about the structure of $\mathcal{H}_g$, apart from its induced action on $H_1(2g, \mathbb{Z})$, which is a rather transparent subgroup of the symplectic group $\text{Sp}(2g, \mathbb{Z})$. We pose the problem:

PROBLEM 8. Study the handlebody subgroup of $\mathcal{M}_g$. A simplified presentation which would reveal new things about its structure, and/or anything new about its coset representatives in $\mathcal{M}_g$ would be of great interest.

PROBLEM 9. Recall that we noted, earlier, that every genus g Heegaard splitting of every homology 3-sphere is obtained by allowing φ to range over $\mathcal{I}_g$. We also noted that Morita proved in [**36**] that every genus g Heegaard splitting of every homology 3-sphere is obtained by allowing φ to range over $\mathcal{K}_g$. For these reasons it might be very useful to find generators for $\mathcal{H}_g \cap \mathcal{I}_g$ and/or $\mathcal{H}_g \cap \mathcal{K}_g$.

Our next problem is in a different direction. It concerns the classification of Heegaard splittings of graph manifolds:

PROBLEM 10. Uncover the structure in the mapping class group that relates to the classification theorem for the Heegaard splittings of graph manifolds in [**44**].

Several other complexes of curves have played a role in work on the mapping class group after 2002. We pause to describe some of them, and the role they played in recent work on the mapping class group.

◇16 The *complex of non-separating curves* **NC**(S) is the subcomplex of **C**(S) whose vertices are all non-separating simple closed curves on S. It was proved by Irmak in [**18**] that for closed surfaces of genus $g \geq 3$ its automorphism group is also isomorphic to $\mathcal{M}^{\pm}$, whereas if $g = 2$ it is isomorphic to $\mathcal{M}^{\pm}$ mod its center.

◇17 The *pants complex* **P**(S) is next. Its vertices represent pants decompositions of S, with edges connecting vertices whose associated pants decompositions differ by an elementary move and its 2-cells representing certain relations between elementary moves. In [**30**] D. Margalit proved a theorem which was much like the theorem proved by Ivanov in [**20**], namely that $\mathcal{M}^{\pm}$ is naturally isomorphic to $\text{Aut}(\mathbf{P}(S))$.

◇18 Next, there is the *Hatcher-Thurston complex* **HT**(S). Its vertices are are collections of g pairwise disjoint non-separating curves on S. Vertices are joined by an edge when they differ by a single 'elementary move'. Its 2-cells represent certain relations between elementary moves. The complex **HT**(S) was constructed by Hatcher and Thurston in order to find a finite presentation for the mapping class group, and used by Wajnryb [**52**] to find the very simple presentation that we will need later in this article. It was proved by Irmak and Korkmaz in [**19**] that $\mathcal{M}^{\pm}$ is also naturally isomorphic to $\text{Aut}(\mathbf{HT}(S))$.

◇19 The *Torelli Complex* $\mathbf{T}(S_g)$ is a simplicial complex whose vertices are either the isotopy class of a single separating curve on S_g or the isotopy class of a 'bounding pair', i.e. a pair of non-separating curves whose union separates. A collection of $k \geq 2$ vertices forms a $k-1$-simplex if these vertices have representatives which are mutually non-isotopic and disjoint. It was first proved by Farb and Ivanov, in [**13**], that the automorphism group of the Torelli subgroup $\mathcal{I}_g$ of $\mathcal{M}_g$ is naturally isomorphic to $\text{Aut}(\mathbf{T}(S))$. As it happens, $\mathcal{M}_g \cong \text{Aut}(\mathbf{T}(S_g))$, so that $\text{Aut}(\mathcal{I}_g) \cong \mathcal{M}_g^{\pm}$. Their proof used additional structure on the vertices in the form of markings, but a subsequent proof of the same result by Brendle and Margalit in [**10**] did not need the markings.

◇20 The *separating curve complex* $\mathbf{SC}(S_g)$ was used by Brendle and Margalit in [**10**] in their study of $\mathcal{K}_g$. It's a subcomplex of $\mathbf{C}(S_g)$, with vertices in one-to-one correspondence with separating simple closed curves on S_g. Brendle and Margalit used it to prove that $\text{Aut}(\mathbf{SC}(S)) \cong \text{Aut}\mathcal{K}_g \cong \mathcal{M}_g^{\pm}$

when $g \geq 4$. This result was recently extended to the case $g = 3$ by McCarthy and Vautau [**34**].

Aside: having defined all these complexes, we have a question which has little to do with the main focus of this article, but has to be asked:

PROBLEM 11. Given a normal subgroup $\mathcal{G}_g$ of $\mathcal{M}_g$, what basic properties are needed in a complex $\mathbf{G}(S_g)$ of curves on S_g so that $\mathcal{M}_g$ will turn out to be naturally isomorphic to $\mathrm{Aut}(\mathbf{G}(S))$?

We return to the central theme of this article:

PROBLEM 12. Hempel's distance function was chosen so that it would capture the geometry, and indeed it does that very well, yet in some ways it feels unnatural. The Hatcher-Thurston complex $\mathbf{HT}(S)$ seems much more natural to us, since and pairs of vertices in the latter determine a Heegaard diagram, and one gets every genus g Heegaard diagram this way. One wonders whether it is possible to redefine Heegaard distance, using $\mathbf{HT}(S)$, or perhaps even $\mathbf{P}(S)$ or one of the other complexes that has proved to be so useful in studying subgroups of $\mathcal{M}$, and whether new things will be learned that way?

We have focussed our discussion, up to now, on the 3-manifold that is determined by a choice of an element ϕ in the group $\mathcal{M}$ via the Heegaard splitting construction. A very different construction which also starts with the choice of an element in the mapping class group, say $\alpha \in \mathcal{M}_g$, produces the mapping torus of α, i.e. the surface bundle $(S \times [0,1])/\alpha$, defined by setting $(p,0) = (\alpha(p),1)$. Surface bundle structures on 3-manifolds, when they exist, are also not unique. Two surface bundles $(S \times I)/\alpha, (S \times I)/\alpha'$ are *equivalent* if and only if α, α' are in the same conjugacy class in $\mathcal{M}_g$.

In [**1**] an interesting description is given of a natural way to produce, for each $(S \times I)/\alpha$, a related Heegaard splitting $H \cup_\beta H'$. Choose a fiber S of $(S \times I)/\alpha$, say $S \times \{0\}$ and choose points $p, q \in S$, $p \neq q$, $p \neq \alpha(q)$. Let P and Q be disjoint closures of regular neighborhoods of $p \times [0, 1/2]$ and $q \times [1/2, 1]$ respectively. Set

$$H = \overline{(S \times ([0,1/2] - Q) \cup P}, \qquad H' = \overline{(S \times [1/2,1] - P) \cup Q}.$$

Note that H and H' are homeomorphic handlebodies of genus $2g+1$ which are embedded in $(S \times I)/\alpha$ and identified along their boundaries, so they give a Heegaard decomposition of $(S \times I)/\alpha$. We call it the *bundle-related* Heegaard splitting of $(S \times I)/\alpha$. It is $H \cup_\beta H'$ for some $\beta \in \mathcal{M}_{2g+1}$. We have several problems that relate to this construction:

PROBLEM 13. This one is a warm-up. Given $\alpha \in \mathcal{M}_g$, say as a product of Dehn twists, express $\beta \in \mathcal{M}_{2g+1}$ as a related product of Dehn twists. With that in hand, observe that if α, α' are equivalent in the mapping class group $\mathcal{M}_g$ then the Heegaard splittings associated to β, β' appear to be equivalent. What about the converse? And how can we tell whether an arbitrary Heegaard splitting of a 3-manifold is the bundle-related splitting of a fibered 3-manifold? What restrictions must we place on β in order to be able to reverse the construction, and produce a surface bundle from a Heegaard splitting?

PROBLEM 14. In [**43**] it is proved that in the case of the trivial genus g surface bundle, i.e. $S_g \times S^1$ the bundle-related splitting is unique, up to equivalence. Are there other cases when it is unique?

PROBLEM 15. A 3-manifold is fibered if it admits a surface bundle structure. It is *virtually fibered* if it has a finite-sheeted cover that admits a surface bundle structure. In [**49**] Thurston asked whether every finite-volume hyperbolic 3-manifold is virtually fibered. This question has turned out to be one of the outstanding open problems of the post-Thurston period in 3-manifold topology. We ask a vague question: does the distance and the very special nature of the Heegaard splitting that's associated to a 3-manifold which has a surface bundle structure give any hint about the possibility of a 3-manifold which is not fibered being virtually fibered?

With regard to Problem 15, we remark that the first examples of hyperbolic knots which are virtually fibered but not fibered were discovered 20 years after the question was posed, by Leininger [**28**] even though it seems to us that fibered knots should have been one of the easiest cases to understand. As we write this, in February 2005, there seems to be lots to learn about virtually fibered 3-manifolds.

3. Potential new tools, via the representations of mapping class groups

In this section we use the notation $\mathcal{M}_{g,b,n}$ for the mapping class group of a surface with b boundary components and n punctures, simplifying to $\mathcal{M}_g$ when we are thinking of $\mathcal{M}_{g,0,0}$.

Knowing accessible quotients of $\mathcal{M}_g$ is important, because accessible quotients have the potential to be new tools for studying aspects of $\mathcal{M}_g$. For this reason, we begin with a problem that seems very likely to tell us something new, even though it has the danger that it could be time-consuming and the new results might not even be very interesting. We note that by the main result in [**14**], $\mathcal{M}_g$ is residually finite, that is for every $\phi \in \mathcal{M}_g$ there is a homomorphism τ from $\mathcal{M}_g$ with finite image such that $\tau(\phi) \neq$ the identity. Therefore there is no shortage of finite quotients. Yet we are hard-pressed to describe any explicitly except for the finite quotients of $\mathrm{Sp}(2g,\mathbb{Z})$ which arise by passing from $\mathrm{Sp}(2g,\mathbb{Z})$ to $\mathrm{Sp}(2g,\mathbb{Z}/p\mathbb{Z})$. We are asking for data that will give substance to our knowledge that $\mathcal{M}_g$ is residually finite:

PROBLEM 16. Study, systematically and with the help of computers, the finite quotients of $\mathcal{M}_g$ which do not factor through $\mathrm{Sp}(2g,\mathbb{Z})$.

We remark that Problem 16 would simply have been impossible in the days before high-speed computers, but it is within reach now. A fairly simple set of defining relations for $\mathcal{M}_{g,0,0}$ can be found in [**52**]. As for checking whether any homomophism so-obtained factors through $\mathrm{Sp}(2g,\mathbb{Z})$, there is are two additional relations to check, namely the Dehn twist on a genus 1 separating curve for $g \geq 2$, and the Dehn twist on a genus 1 bounding pair (see [**23**]) for $g \geq 3$. One method of organization is to systematically study homomorphisms of $\mathcal{M}_g$ (maybe starting with $g = 3$) into the symmetric group Σ_n, beginning with low values of n and gradually increasing n. One must check all possible images of the generators of $\mathcal{M}_g$ in Σ_n, asking (for each choice) whether the defining relations in $\mathcal{M}_g$ and $\mathrm{Sp}(2g,\mathbb{Z})$ are satisfied. Note that if one uses Dehn twists on non-separating curves as generators, then they must all be conjugate, which places a big restriction. There are additional restrictions that arise from the orders of various generating sets, for example in [**7**] it is proved that $\mathcal{M}_g$ is generated by 6 involutions. Of course, as one proceeds with such an investigation, tools will present themselves and the calculation will organize itself, willy-nilly.

We do not mean to suggest that non-finite quotients are without interest, so for completeness we pose a related problem:

PROBLEM 17. Construct any representations of $\mathcal{M}_g$, finite or infinite, which do not factor through $Sp(2g, \mathbb{Z})$.

In a very different direction, every mathematician would do well to have in his or her pile of future projects, in addition to the usual mix, a problem to dream about. In this category I put:

PROBLEM 18. Is there a faithful finite dimensional matrix representation of $\mathcal{M}_{g,b,n}$ for any value of the triplet (g, b, n) other than $(1, 0, 0), (1, 1, 0), (1, 0, 1)$, $(0, 1, n), (0, 0, n)$ or $(2, 0, 0)$?

We have mentioned Problem 18 because we believe it has relevance for Problems 16 and 17, for reasons that relate to the existing literature. To the best of our knowledge there isn't even a known candidate for a faithful representation of $\mathcal{M}_{g,0,0}$ for $g \geq 3$, even though many experts feel that $\mathcal{M}_{g,0,0}$ is linear. This leads us to ask a question:

PROBLEM 19. Find a candidate for a faithful finite-dimensional matrix representation of $\mathcal{M}_g$ or $\mathcal{M}_{g,1,0}$.

◇21 The cases $(g, b, n) = (1, 0, 0)$ and $(1,1,0)$ are classical results which are closely related to the fact that the Burau representation of $\mathbf{B}_3$ is faithful [29]. Problem 18 received new impetus when Bigelow [2] and Krammer ([26] and [27]) discovered, in a related series of papers, that the braid groups $\mathbf{B}_n$ are all linear. Of course the braid groups are mapping class groups, namely $\mathbf{B}_n$ is the mapping class group $\mathcal{M}_{0,1,n}$, where admissible isotopies are required to fix the boundary of the surface $S_{0,1,n}$ pointwise. Passing from $\mathbf{B}_n$ to $\mathbf{B}_n/\text{center}$, and thence to the mapping class group of the sphere $\mathcal{M}_{0,0,n}$. Korkmaz [25] and also Bigelow and Budney [3] proved that $\mathcal{M}_{0,0,n}$ is linear. Using a classical result of the author and Hilden [6], which relates $\mathcal{M}_{0,0,n}$ to the so-called hyperelliptic mapping class groups, Korkmaz, Bigelow and Budney all then went on to prove that $\mathcal{M}_{2,0,0}$ is also linear. More generally the centralizers of all elements of finite order in $M_{g,0,0}$ are linear. So essentially all of the known cases are closely related to the linearity of the braid groups $\mathbf{B}_n$.

◇22 A few words are in order about the dimensions of the known faithful representations. The mapping class group $\mathcal{M}_{1,0,0}$ and also $\mathcal{M}_{1,1,0}$ have faithful matrix representations of dimension 2. The faithful representation of $\mathcal{M}_{2,0,0}$ that was discovered by Bigelow and Budney has dimension 64, which suggests that if we hope to find a faithful representation of $\mathcal{M}_{g,0,0}$ or $\mathcal{M}_{g,1,0}$ for $g > 2$ it might turn out to have very large dimension.

◇23 A 5-dimensional non-faithful representation of $\mathcal{M}_{2,0,0}$ over the ring of Laurent polynomials in a single variable with integer coefficients, was constructed in [24]. It arises from braid group representations and does not generalize to genus $g > 2$, It is not faithful, but its kernel has not been identified.

◇24 We review what we know about infinite quotients of $\mathcal{M}_g$. The mapping class group acts naturally on $H_1(S_{g,b,n})$, giving rise to the symplectic representations from $\mathcal{M}_{g,1,0}$ and $\mathcal{M}_{g,0,0}$ to $Sp(2g, \mathbb{Z})$. In [46] Sipe (and independently Trapp [**50**]), studied an extension of the symplectic representation. Trapp interpreted the new information explicitly as detecting the action of $\mathcal{M}_{g,1,0}$ on winding numbers of curves on surfaces. Much more generally, Morita [**39**] studied an infinite family of representations $\rho_k : \mathcal{M}_{g,1,0} \to G_k$ onto a group G_k, where G_k is an extension of $Sp(2g,\mathbb{Z})$. Here $k \geq 2$, and $G_2 = Sp(2g,\mathbb{Z})$ is our old friend $Sp(2g,\mathbb{Z})$. He gives a description of G_3 as a semi-direct product of $Sp(2g,\mathbb{Z})$ with a group that is closely related to Johnson's representations τ_1, τ_2 of $\mathcal{K}_g$, discussed earlier. He calls the infinite sequence of groups $G_k, k = 2, 3, 4, \ldots$ a sequence of 'approximations' to $\mathcal{M}_{g,1,0}$. Morita also has related results for the case $\mathcal{M}_{g,0,0}$.

In Problem 16 we suggested a crude way to look for interesting new quotients of the group $\mathcal{M}_g$ that don't factor through $Sp(2g,\mathbb{Z})$. In closing we note that there might be a different approach which would be more natural and geometric (but could be impossible for reasons that are unknown to us at this time). We ask:

PROBLEM 20. Is there a natural quotient complex of any one of the complexes discussed in §1 which might be useful for the construction of non-faithful representations of $\mathcal{M}_g$?

Let's suppose that we have some answers to either Problem 16 or 17 or 20. At that moment, our instincts would lead us right back to a line of investigation that was successful many years ago when, in [**4**], we used the symplectic representation and found an invariant which distinguished inequivalent minimal Heegaard splittings. In the intervening years we suggested that our students try to do something similar with other representations, but that project failed. We propose it anew. Recall that a 3-manifold M may have one or more distinct equivalence classes of Heegaard splittings. It is known that any two become equivalent after some number of stabilizations. There are many interesting unanswered questions about the collection of all equivalence classes of Heegaard splittings of a 3-manifold, of every genus. Recall that the equivalence class of the Heegaard splitting $H \cup_\phi H'$ is the double coset $\mathcal{H}\phi\mathcal{H}$ in $\mathcal{M}$.

PROBLEM 21. Study the double coset $\mathcal{H}\phi\mathcal{H}$ in $\mathcal{M}$, using new finite or infinite quotients of $\mathcal{M}$. In this regard we stress finite, because a principle difficulty when this project was attempted earlier was in recognizing the image of $\mathcal{H}$ in infinite quotients of $\mathcal{M}$, however if the quotient is finite and not too big, it suffices to know generators of $\mathcal{H} \subset \mathcal{M}$. Since a presentation for $\mathcal{H}$ was found by Waynryb in [**51**], we can compute the associated subgroup. Some of the open questions which might be revealed in a new light are:

(1) How many times must one stabilize before two inequivalent Heegaard splittings become equivalent?
(2) How can we tell whether a Heegaard splitting is not of minimal genus?
(3) How can we tell whether a Heegaard splitting is stabilized?
(4) Are any of the representations that we noted earlier useful in answering (1), (2) or (3) above?

While we have stressed the search for good working quotients of $\mathcal{M}_g$, we should not forget that in the case of homology spheres, we have already pointed out that any homology sphere may be defined by a Heegaard splitting with the Heegaard glueing map (now redefined with a new 'base point') ranging over $\mathcal{I}_g$. Even more, as was proved earlier, Morita has shown in [36] that it suffices to let the glueing map range over $\mathcal{K}_g$. This leads us to ask:

PROBLEM 22. Are there quotients of $\mathcal{I}_g$ or $\mathcal{K}_g$ in which the intersection of either $\mathcal{I}_g$ or $\mathcal{K}_g$ with the handlebody group $\mathcal{H}_g$ is sufficiently tractable to allow one to study the double cosets:

$$(\mathcal{I}_g \cap \mathcal{H}_g)(\phi)(\mathcal{I}_g \cap \mathcal{H}_g), \quad \text{where} \quad \phi \in \mathcal{I}_g,$$

or

$$(\mathcal{K}_g \cap \mathcal{H}_g)(\phi)(\mathcal{K}_g \cap \mathcal{H}_g), \quad \text{where} \quad \phi \in \mathcal{K}_g?$$

In regard to Problem 22 we note that in [36] Morita was seeking to understand how topological invariants of 3-manifolds might lead him to a better understanding of the representations of $\mathcal{I}_g$ and $\mathcal{K}_g$, but he did not ask about the potential invariants of Heegaard splittings that might, at the same time, be lurking there.

References

[1] D. Bachman and S.Schleimer, *Surface bundles versus Heegaard splittings*, arXiv:math.GT/0212104.

[2] S. Bigelow, *Braid groups are linear*, J. Amer. Math. Soc. **14** No. 2, (2001), 471-486.

[3] S. Bigelow and R. Budney, *The mapping class group of a genus 2 surface is linear*, Algebraic and Geometric Topology, **1** (2001), 699-708.

[4] J. Birman, "On the equivalence of Heegaard splittings of closed, orientable 3-manifolds", *Annals of Math Studies* **84**, Ed. L.P. Neuwirth, Princeton Univ. Press (1975), pp.137-164.

[5] J. Birman and R. Craggs, *The μ-invariant of 3-manifolds and certain structural properties of the group of homeomorphisms of a closed, oriented 2-manifold*, Trans.AMS **237** (1978), pp. 283-309.

[6] J. Birman and H. Hilden, *Isotopies of homeomorphisms of Riemann surfaces*, Annals of Math. **97**, No. 3 (1973), pp. 424-439.

[7] T. Brendle and B. Farb, *Every mapping class group is generated by 6 involutions*, Journal of Algebra **278** (2004), 187-198.

[8] ———, The Birman-Craggs-Johnson homomorphism and abelian cycles in the Torelli group, preprint, arXiv:math.GT/0601163.

[9] T. Brendle and H. Hamidi-Tehrani, *On the linearity problem for mapping class groups*, Algebraic and Geometric Topology **1** (2001), 445-468.

[10] T. Brendle and D. Margalit, *Commensurations of the Johnson kernel*, Geometry and Topology **8** (2004), 1361-1384.

[11] A. Casson and C. Gordon, *Reducing Heegaard splittings*, Topology and its Applications **27**, No. 3 (1987), 631-657.

[12] ———, unpublished notes from talks given at various conferences, constructing explicit examples of manifolds with irreducible Heegaard splittings of arbitrarily high genus.

[13] B. Farb and N. Ivanov, *The Torelli geometry and its applications*, International Math Research Letters **12**, no. 2–3 (2005), 293–301.

[14] E. Grossman, *On the residual finiteness of certain mapping class groups*, J. London Math Soc. **9**, No. 1 (1974), 160-164.

[15] W. Haken, *Some results on surfaces in 3-manifolds*, from "Studies in Modern Topology", MAA (1968), 39-98.

[16] W. Harvey,*Boundary structure of the modular group*, Ann. of Math. Studies **97** (1981), Riemann surfaces and related topics, Princeton Univ. Press, Princeton, N.J.) , pp. 245–251.

[17] J. Hempel, *3-manifolds as viewed from the curve complex*, Topology **40**, No. 3 (2001), 631-657.

[18] E. Irmak, *Complexes of non-separating curves and mapping class groups*, preprint (2004), arXiv:math.GT/0407285.

[19] E. Irmak and M. Korkmaz, *Automorphisms of the Hatcher-Thurston Complex*, preprint (2004), arXiv:math.GT/0409033.
[20] N. Ivanov, *Automorphisms of complexes of curves and of Teichmüller spaces*, Int. Math. Res. Notes **14** (1997), 651-666.
[21] N. Ivanov, *Mapping Class Groups*, Handbook of Geometric Topology (editors Daverman and Sher), Elsevier 2002, 523-634.
[22] D. Johnson, *An abelian quotient of the mapping class group $\mathcal{I}_g$*, Math. Ann. **249**, No. 3 (1980), 225- 240.
[23] _____, *A survey of the Torelli group*, Contemporary Math. **20** (1983), 165-178.
[24] V. Jones, *Hecke algebra representations of braid groups and link polynomials*, Annals of Math. **126** (1987), 335-388.
[25] M. Korkmaz, *On the linearity of certain mapping class groups*, Turkish J. Math. **24**, No. 4 (2000), 367-371.
[26] D. Krammer, *The braid group B_4 is linear*, Invent. Math. **142** No. 3, (2000), 451-486.
[27] _____, *Braid groups are linear*, Ann. of Math. (2) 155 No. 1, (2002), 131-156.
[28] C. Leininger, *Surgeries on one component of the Whitehead link are virtually fibered*, Topology **41** (2002), 307-320.
[29] W. Magnus and A. Pelluso, *On a theorem of V.I.Arnold*, Com. in Pure and Applied Mathematics **XXII** (1969), 683-692.
[30] D. Margalit, *The automorphism group of the pants complex*, Duke Mathematical Journal **121**(2004), 457-479, arXiv:math.GT/0201391
[31] Y. Minsky, *Combinbatorial and geometrical aspects of hyperbolic 3-manifolds*, in "Kleinian Groups and Hyperbolic 3-Manifolds", Lond. Math. Soc. Lec. Notes **299**, 3-40.
[32] H. Masur and Y. Minsky, *Geometry of the complex of curves I-Hyperbolicity*, Invent. Math. **138** No 1 (1999), 103-149.
[33] _____, *Quasiconvexity in the curve complex*, preprint, arXiv:math.GT/0307083.
[34] J. McCarthy and W. Vautau, work described in a talk given by McCarthy at a conference at Columbia University March 15-20 2005.
[35] Y. Moriah and J. Schultens, *Irreducible Heegaard splittings of Seifert fibered spaces are either vertical or horizontal*, Topology **37**, No. 5 (1998), 1089-1112.
[36] S. Morita, *Casson's invariant for homology 3-spheres and characteristic classes of surface bundles*, Topology **28** (1989), 305-323.
[37] _____, *On the structure of the Torelli group and the Casson invariant*, Topology **30** (1991), 603-621.
[38] _____, *The extension of Johnson's homomorphism from the Torelli group to the mapping class group*, Invent. Math. **111**, No. 1 (1993), 197-224.
[39] _____, *Abelian quotients of subgroups of the mapping class group of surfaces*, Duke Math. J. **70**, No. 3 (1993), 699-726.
[40] H. Namazi, *Big handlebody distance implies finite mapping class group*, preprint(2004), arXiv:math.GT/0406551.
[41] M. Scharlemann, *Heegaard splittings of compact 3-manifolds*, Handbook of Gometric Topology, North Holland (2002) and arXiv:math.GT/0007144.
[42] S. Schleimer, *The disjoint curve property*, Geometry and Topology **8** (2004), 77-113.
[43] J. Schultens, *The classification of Heegaard splittings of (compact orientable surface)$\times S^1$*, London Math Soc. **67** (1993), 425-448.
[44] _____, *Heegaard splittings of graph manifolds*, Geometry and Topology **8** (2004), 831-876)
[45] K. Shackleton, *Tightness and computing distances in the curve complex*, preprint(2004), arXiv:math.GT/0412078.
[46] P. Sipe, *Some finite quotients of the mapping class group of a surface*, Proc. AMS **97**, No. 3 (1986), 515-524.
[47] D. Sullivan, *On the intersection ring of compact 3-manifolds*, Topology **14**, No. 3 (1975), 275-277.
[48] A. Thompson, *The disjoint curve property and genus 2 manifolds*, Topology and its Applications **97**, No. 3 (1999), 273-279.
[49] W. Thurston, *Three-dmensional manifolds, Kleinian groups and hyperbolic geometry*, Bull. AMS **6** (1982), 357-381.

[50] R. Trapp, *A linear representation of the mapping class group* $\mathcal{M}$ *and the theory of winding numbers*, Topology and its Applications **43**, No. 1 (1992), 47-64.
[51] B. Wajnryb, *Mapping class group of a handlebody*, Fund. Math.**158**, No. 3 (1998), 195-228.
[52] _____, *An elementary approach to the mapping class group of a surface*, Geometry and Topology **3** (1999), 405-466. arXiv math.GT/9912248.
[53] F. Waldhausen, *Heegaard Zerlegungen der 3-sphäre*, Topology **7** (1968), 105-203.

Lefschetz Pencils and Mapping Class Groups

S. K. Donaldson

1. Introduction

Holomorphic maps between complex manifolds have many properties which distinguish them among general smooth maps. Consider, for example, the case of a map between Riemann surfaces. A holomorphic map is represented locally, in suitable co-ordinates, by one of the models $z \mapsto z^k$ for $k \geq 0$. These models are very different from the models of generic smooth maps between surfaces, which are, in addition to the points where the map is a local diffeomorphism, *folds* and *cusps*. It is interesting to see what happens if we perturb the holomorphic map $z \mapsto z^2$ by a small non-holomorphic term. So for $\epsilon > 0$ we define $f^{(\epsilon)} : \mathbf{C} \to \mathbf{C}$ by

$$f^{(\epsilon)}(z) = z^2 + 2\epsilon \overline{z}.$$

Thus $\frac{\partial f^{(\epsilon)}}{\partial z} = 2z$ and $\frac{\partial f^{(\epsilon)}}{\partial \overline{z}} = 2\epsilon$. The real derivative of $f^{(\epsilon)}$, has rank 2 at points where $|\frac{\partial f^{(\epsilon)}}{\partial z}| \neq |\frac{\partial f^{(\epsilon)}}{\partial \overline{z}}|$, that is to say where $|z| \neq \epsilon$. The point $z = \epsilon e^{i\theta}$ maps to the point $\gamma(\theta) = \epsilon^2(e^{2i\theta} + 2e^{-i\theta})$ and $\gamma'(\theta) = 2i\epsilon^2 e^{-i\theta}(e^{3i\theta} - 1)$. Thus γ' vanishes at the three points cube roots of unity. These three points are cusps of the map $f^{(\epsilon)}$ and the remaining points on the circle $|z| = \epsilon$ are fold points: the map $f^{(\epsilon)}$ maps this circle onto a curvilinear triangle with the three cusps as vertices. The reader is invited to visualise this map.

This example illustrates that holomorphic maps can be much simpler than typical smooth maps. We can abstract the local character of holomorphic maps and consider smooth maps which are locally modelled on holomorphic ones. This gives a way to study and exploit the topological aspects of holomorphic maps, independent of the finer details of the complex geometry and in a wider setting. For example we can consider branched covers of 2-manifolds as a class of maps, independent of the existence of Riemann surface structures. A natural setting for this theory turns out to be *symplectic topology*, and the *mapping class groups of surfaces*, and certain generalisations, enter in an essential way. In this article we will outline parts of this theory and discuss some open problems. A discussion in a somewhat similar spirit will be found in [5].

2. Holomorphic Morse theory and Dehn twists

Let $f : X \to \mathbf{C}$ be a proper, nonconstant, holomorphic map from a connected complex manifold X of complex dimension n to the complex numbers. We suppose that f is a "holomorphic Morse function", so that at each point where the derivative ∂f vanishes the Hessian $\partial^2 f$ is nondegenerate. These critical points form a discrete set in X. For simplicity we also suppose that the images of these points (the critical values) are distinct, so for each critical value there is just one corresponding critical point. Let Δ denote the set of critical values in $\mathbf{C}$. Just as in ordinary Morse Theory, the essential thing in understanding the topology of the map f is to understand the fibres $X_t = f^{-1}(t)$ where $t \in \mathbf{C}$ is either a critical value or close to a critical value. The difference from the real case is that for $t \notin \Delta$ the fibres are all diffeomorphic. Indeed if we put $X' = f^{-1}(\mathbf{C} \setminus \Delta)$ then the restriction $f : X' \to \mathbf{C} \setminus \Delta$ is a C^∞ fibration and the base $\mathbf{C} \setminus \Delta$ is connected (in contrast to the analogous situation in real geometry). Now recall the general notion of *monodromy*. Suppose $\phi : E \to B$ is a C^∞ fibration over a connected base, with base point $b_0 \in B$ and with fibre $F = \phi^{-1}(b_0)$. Let Γ_F denote the mapping class group of the fibre: the isotopy classes of self-diffeomorphisms of F. Then we have a *monodromy homomorphism*

$$\rho_\phi : \pi_1(B, b_0) \to \Gamma_F.$$

This can be defined by choosing a Riemannian metric on the total space E which gives a family of horizontal subspaces: the orthogonal complements of the tangent spaces to the fibres. (In other language we can regard this as a choice of *connection* on the bundle E regarded as a bundle with structure group $\text{Diff}(F)$.) For any smooth based loop $\gamma : [0,1] \to B$ and any point y in $\phi^{-1}(b_0)$ there is a horizontal lift $\tilde{\gamma}$ of γ starting at y. We define $R : F \to F$ by $R(y) = \tilde{\gamma}(1)$. Then R is a diffeomorphism of the fibre which, up to isotopy, is independent of the choice of metric and homotopy class of the loop γ.

Applied in our situation we get a monodromy homomorphism

$$\rho_f : \pi_1(\mathbf{C} \setminus \Delta) \to \Gamma_F,$$

where F is the fibre over some fixed base point in $\mathbf{C} \setminus \Delta$. Of course the fundamental group of $\mathbf{C} \setminus \Delta$ is a free group, with standard generators γ_i, say, winding once around a single critical value. This notion has two roots in classical complex analysis. On the one hand we can consider the case when $n = 1$, so X is a Riemann surface presented as a branched cover of $\mathbf{C}$. The fibre F is just a set of d points, where d is the degree of the map, and the mapping class group is the permutation group $\mathcal{S}_d$ on d objects. The monodromy is just the data discussed in standard Riemann surface texts, which specifies how to glue together the sheets of the branched covering. In this case $\rho(\gamma_i)$ is a transposition in $\mathcal{S}_d$. On the other hand we can consider the case where $n = 2$ so the fibres are Riemann surfaces. Thus $f : X \to \mathbf{C}$ can be regarded as a family of Riemann surfaces which degenerate over Δ. The classical topic here is not so much the monodromy in the isotopy group but its composite with the natural action of the diffeomorphisms on the homology of the fibre, which yields a homological monodromy

$$\rho_f^{H_1} : \pi_1(\mathbf{C} \setminus \Delta) \to GL(H_1(F)).$$

For example, let $z_1, \ldots, z_n$ be fixed distinct points in $\mathbf{C}$ and let X_0 be the subset of $\mathbf{C}^3$

$$X_0 = \{(z, w, \lambda) : w^2 = (z - \lambda)(z - z_1) \ldots (z - z_n)\}.$$

Let f_0 be the restriction of the projection $(z, w, \lambda) \mapsto \lambda$. In a standard way, we can compactify the fibres of f_0 to obtain a complex manifold X, containing X_0 as a dense open set, and with a extension of f_0 to a proper holomorphic map

$f : X \to \mathbf{C}$. Then the set Δ is just $\{z_1, \ldots, z_n\}$ and for $\lambda \notin \Delta$ the fibre $f^{-1}(\lambda)$ is the hyperelliptic Riemann surface defined by the equation
$$w = \sqrt{(z-\lambda)(z-z_1)\ldots(z-z_n)}.$$

The expression
$$\frac{dz}{\sqrt{(z-\lambda)\ldots(z-z_n)}}$$
defines a smoothly varying family of holomorphic 1-forms over the smooth fibres. Locally in the base we can fix a basis for the homology of the fibre and hence define the corresponding periods, by integrating the holomorphic form. Explicitly, this amounts to choosing a suitable collection of paths σ_α in $\mathbf{C}$ with end points in $\{z_1, \ldots, z_n, \lambda\}$ and the periods are then written as
$$\int_{\sigma_\alpha} \frac{dz}{\sqrt{(z-\lambda)\ldots(z-z_n)}},$$
for a choice of branch of the square root. The issue addressed by the knowledge of the monodromy homomorphism, in this special case, is how these contour integrals change when λ traces out a path encircling one of the fixed points z_i.

We can now discuss the central issue: what is the monodromy of a holomorphic Morse function around a loop about a single critical value? This is the analogue of the description of the change in the level set of a real Morse function as one crosses a critical value. As in that case, a crucial observation is that the problem can be reduced to a standard local model. Indeed if we take a very small loop γ_i about a critical value then we can choose the horizontal subspaces so that the monodromy is the identity map outside the intersection $F \cap B$ of the fibre F with a suitable small ball B in X centred on the critical value. Then we can regard the monodromy as a compactly supported diffeomorphism of $F \cap B$, defined up to compactly supported isotopy. To see what is going on take the standard local model to be the map
$$g(z_1, \ldots, z_n) = z_1^2 + \cdots + z_n^2,$$
from $\mathbf{C}^n$ to $\mathbf{C}$, and take the base point b_0 to be $1 \in \mathbf{C}$. We consider the subset of the fibre $g^{-1}(b_0)$ given by the real points $g^{-1}(b_0) \cap \mathbf{R}^n$ which we denote by V. This is just the standard unit sphere in $\mathbf{R}^n$. For any $t \in \mathbf{R}$ we can consider similarly $g^{-1}(t) \cap \mathbf{R}^n$. This is the sphere of radius $\sqrt{t}$ for $t \geq 0$ and the empty set if $t < 0$. We can choose a family of horizontal subspaces which preserves the real points so if we "parallel transport" the fibre from 1 towards 0 along the positive real axis the parallel transports of V shrink down to the critical point.

Now consider the total space TS^n of the tangent bundle of the n-sphere. The standard "generalised Dehn twist" is a compactly supported diffeomorphism of TS^n, canonical up to compactly supported isotopy. We can define it as follows. The points in TS^n can be identified with pairs (v, w) of orthogonal vectors v, w in $\mathbf{R}^n$, where $|v| = 1$. If $w \neq 0$ then for any angle θ we define the usual rotation R_θ in the (oriented) plane spanned by v, w. Now choose a function Θ on $[0, \infty)$ such that $\Theta(s) = 0$ if s is small and $\Theta(s) = \pi$ if s is large. Define a map $D : TS^n \to TS^n$ by
$$D(v, w) = R_{\Theta(|w|)}(-v, -w),$$

with the obvious interpretation if $w = 0$: *i.e.* $D(v, 0) = (-v, 0)$. Then $D(v, w) = (v, w)$ if $|w|$ is large and $D(v, w) = (-v, -w)$ if $|w|$ is small. Also, D is a diffeomorphism, with inverse
$$D^{-1}(v', w') = R_{-\Theta(|w'|)}(-v', -w').$$
Thus this model Dehn twist is a compactly supported diffeomorphism of TS^n, equal to the antipodal map on the zero section.

Now return to the sphere $V \subset g^{-1}(b_0)$ in the standard model above. This sphere is a "totally real" submanifold of the fibre; that is, multiplication of tangent vectors by I yields an identification between the tangent bundle TV and the normal bundle of V in the fibre $F = g^{-1}(b_0)$. In other words, a tubular neighbourhood N of V in F can be identified with a neighbourhood N' of the zero section in the tangent bundle of V. Now we can obviously suppose that our model map D is supported in N' and so, via this identification, we can regard D as a diffeomorphism of F, supported in the neighbourhood N'. The *fundamental fact* is that this is the monodromy around the critical value.

In general then we arrive at the following description of the topology of a map $f : X \to \mathbf{C}$ as considered above.

- In the model fibre $F = f^{-1}(b_0)$ there are "vanishing cycles" V_i associated to the critical values z_i (and the loops γ_i around the z_i). Each V_i is an embedded $(n-1)$-sphere uniquely defined up to isotopy, and we have an identification (fixed up to homotopy) of the normal bundle of V_i in F with the tangent bundle.
- The monodromy around the loop γ_i is the Dehn twist D_{V_i} about V_i, defined using an identification of a tubular neighboourhood as above (which is independent of choices, up to isotopy).

We can relate this discussion to the classical problems considered above. First, if $n = 1$ the "vanishing cycle" is just a copy of S^0, *i.e.* a pair of points, and the "Dehn twist" is just a transposition (the antipodal map on S^0). If $n = 2$ we get the familiar Dehn twists in the mapping class groups of Riemann surfaces. It is easy to see that the Dehn twist about an embedded circle V in a Riemann surface Σ acts on $H_1(\Sigma)$ by

(1) $$\alpha \mapsto \alpha + \langle V, \alpha \rangle V,$$

where $\langle \, , \, \rangle$ is the intersection form. So we arrive at the classical Picard-Lefschetz formula for the homological monodromy ρ^{H_1}.

Some words about signs may be in order here. Given an embedded circle V in a 2-manifold Σ, the Dehn twist about V is completely specified by a choice of orientation of Σ, it does not require an orientation of V. Thus in the Picard-Lefschetz formula (1) we have momentarily fixed an orientation of V, to define the homology class, but obviously the formula is unchanged if we change orientation. On the other hand, with a fixed orientation of Σ, the inverse of a Dehn twist is not a Dehn twist as we have defined things: it is a Dehn twist of the manifold Σ with the opposite choice of orientation. The same holds in higher dimensions. Note however that the usual orientation of the neighbourhood of V induced by the identification with TV differs from the standard complex orientation by a sign $(-1)^{n-1}$. Thus when n is odd the self-intersection of V, with respect to the complex orientation is -2. The Picard-Lefschetz formula (1) for the action on the middle-dimensional homology is the same in all dimensions but the significance is somewhat different depending whether n is or even or odd. The fibre F has real dimension $2(n-1)$ and the intersection form is antisymmetric if n is even and symmetric if n is odd. In the second case the Picard-Lefschetz transformation is of order 2; the reflection defined by vector $[V]$ with $[V].[V] = -2$.

In the case when $n = 1$ and X is a 2-manifold we can reverse the constructions above. That is we have the classical

PROPOSITION 1. *There is a one-to-one correspondence between equivalence classes of data:*
- *Riemann surfaces X with a proper map $f : X \to \mathbf{C}$ of degree $d \geq 1$ having only simple branch points, mapping to distinct points in $\mathbf{C}$.*
- *Discrete sets $\Delta \subset \mathbf{C}$ and homomorphisms $\rho : \pi_1(\mathbf{C} \setminus \Delta) \to \mathcal{S}_d$ mapping each standard generator to a transposition.*

(We leave it to the reader to spell out the exact equivalence relations to impose on these two kinds of data.)

The analogue of this in higher dimensions does not hold: we cannot create a pair (X, f) to realise arbitrary data (Δ, ρ). But we can extract the topology from the situation by enlarging our class of spaces.

DEFINITION 1. A Topological Lefschetz Fibration consists of a smooth oriented $2n$-dimensional manifold X and a proper map $f : X \to \mathbf{C}$ with the following properties.
- For each point x_0 of X either df_{x_0} is surjective or, when x_0 is a critical point of f, there is an oriented chart $\psi : U \to \tilde{U}$ where U is a neighbourhood of x_0 in X and $\tilde{U}$ is a neighbourhood of the origin in $\mathbf{C}^n$ such that the composite $\tilde{f} = f \circ \psi^{-1}$ is
$$\tilde{f}(z_1, \ldots, z_n) = f(x_0) + z_1^2 + \cdots + z_n^2.$$
- If x_0 and x_1 are two different critical points of f then $f(x_0) \neq f(x_1)$.

Clearly such a map has a well-defined smooth fibre F, an oriented $2(n-1)$-manifold.

We say that two Topological Lefschetz Fibrations $f_1 : X_1 \to \mathbf{C}$, $f_2 : X_2 \to \mathbf{C}$ are are equivalent "over a fixed base" if there is a diffeomorphism $\alpha : X_1 \to X_2$ with $f_1 = f_2 \circ \alpha$. This implies that the two fibrations have the same set of critical values in $\mathbf{C}$. Now fix a discrete subset $\Delta \subset \mathbf{C}$ and a set of standard generators γ_i for $\pi_1(\mathbf{C} \setminus \Delta)$. Then we have

PROPOSITION 2. *There is a one-to one correspondence between:*
- *Topological Lefschetz fibrations $f : X \to \mathbf{C}$ with fibre F and critical set Δ modulo equivalence over a fixed base.*
- *Collections of isotopy classes of embedded $(n-1)$-spheres $V_i \subset F$ with homotopy classes of isomorphisms between the normal and tangent bundles of the V_i, modulo equivalence induced by the action of a single element of Γ_F.*

The proof of the Proposition is straightforward. Suppose for example that we are given data of the second kind. Let Ω be the complement in $\mathbf{C}$ of small disjoint discs about the points of Δ. Then Ω is homotopy equivalent to a wedge of circles so by standard theory fibrations with fibre F over Ω are determined by their monodromy. So we can construct a fibration $X_\Omega \to \Omega$ with the Dehn twists in the V_i as monodromy around γ_i. On the other hand, given a single $V \subset F$, we can construct a standard model $\pi : Y_V \to \mathbf{C}$ whose monodromy around a large circle is the Dehn twist about V. Now we construct X by gluing these standard models to X_Ω to fill in the missing discs.

The following example shows that one does need to take some care in formulating this correspondence. Suppose $n = 2$ and $f : X \to \mathbf{C}$ is, say, a holomorphic Lefschetz fibration and x_0 is a point in X with $f(x_0) \notin \Delta$. Now let $\hat{X}$ be the "blow-up" of X at the point x_0 and $\hat{f}$ be the composite of the f and the canonical map $\hat{X} \to X$. The critical set $\hat{\Delta}$ is $\Delta \cup \{f(x_0)\}$. The smooth fibres of $\hat{f}$ are the same as those of f. The fibre of $\hat{f}$ over $f(x_0)$ is the union of the smooth fibre F and the exceptional sphere $E \cong S^2$, meeting at one point. What happens is that the identification between the fibres of f and $\hat{f}$ realises the diffeomorphism beween the connected sum $F \sharp S^2$ and F. The vanishing cycle associated to the extra critical value is a trivial circle in F (i.e it bounds a disc) and the resulting monodromy is trivial, up to isotopy. So $f : X \to \mathbf{C}$ and $\hat{f} : \hat{X} \to \mathbf{C}$ have in sense the "same" monodromy, even though the manifolds $X, \hat{X}$ are different.

There is another natural notion of equivalence between Topological Lefschetz fibrations. We say that $f_1 : X_1 \to \mathbf{C}, f_2 : X_2 \to \mathbf{C}$ are *equivalent* if there is a diffeomorphism $\alpha : X_1 \to X_2$ and a diffeomorphism $\beta : \mathbf{C} \to \mathbf{C}$ such that $\beta \circ f_1 = f_2 \circ \alpha$. This means that β maps the set of critical values of f_1 to that of f_2. Clearly the equivalence classes of Topological Lefschetz fibrations correspond to orbits of an a action of the group of diffeomorphisms of the plane with a marked set Δ on the set of equivalence classes over a fixed base.

3. Lefschetz pencils and symplectic four-manifolds

Holomorphic Lefschetz fibrations typically arise in the following way. For simplicity we will restrict the discussion to complex dimension 2. Suppose that $Y \subset \mathbf{CP}^N$ is a complex projective surface. We choose a generic $(N-2)$ dimensional subspace $\mathbf{CP}^{N-2} \subset \mathbf{CP}^N$, meeting Y transversely in a finite set of points $A \subset Y$, and we consider the "pencil" of hyperplanes through $\mathbf{CP}^{N-2}$. These cut out a 1-parameter family of hyperplane sections of Y. Suppose Z_i are homogeneous co-ordinates on $\mathbf{CP}^N$ and $\mathbf{CP}^{N-2}$ is defined by the equations $Z_0 = Z_1 = 0$. Then $g = Z_1/Z_0$ is a meromorphic function on Y and the hyperplane sections are the fibres $g^{-1}(\lambda)$, for different $\lambda \in \mathbf{CP}^1$. The meromorphic function g is not a well-defined map on Y but if we blow up the points of A we get a well-defined map $\tilde{g} : X \to \mathbf{CP}^1$ where X is the blow up of Y. For generic choices of the axis $\mathbf{CP}^{N-2}$ the restriction of $\tilde{g}$ to $\tilde{g}^{-1}(\mathbf{C})$ will be a holomorphic Lefschetz fibration of the kind considered before. Thus our previous discussion needs to be extended in two ways

- In place of the base $\mathbf{C}$ we have the Riemann sphere $\mathbf{CP}^1$. This makes little difference. We can arrange things so that the fibre over ∞ is smooth, then we have the additional constraint that the product of the monodromies around all the γ_i should be trivial regarded as an element of the mapping class group of the fibre (since it represents the monodromy around a large circle in $\mathbf{C}$).
- We may wish to remember the fact that the total space X arose as the blow up of Y. The exceptional spheres created by the blowing up appear

as sections of the fibration. Thus we can take our fibre to be a Riemann surface Σ with a collection of marked points $P = \{p_\alpha\}$, and our vanishing cycles to be circles in $\Sigma \setminus P$. We introduce the mapping class group $\Gamma_{\Sigma,P}$ of compactly supported diffeomorphisms of $\Sigma \setminus P$ modulo compactly supported isotopy. (Equivalently, diffeomorphims equal to the identity on fixed small discs about the p_α.)Then the condition that the sections have self-intersection -1 goes over to the condition that

(2) $$D_{V_1} \circ \ldots D_{V_\nu} = T,$$

in $\Gamma_{\Sigma,P}$ where T is the product of Dehn twists about small circles around the p_α. (Thus T is trivial in the unrestricted mapping class group, but not usually in $\Gamma_{\Sigma,P}$.)

Again we can define a topological analogue of this picture.

DEFINITION 2. A Topological Lefschetz pencil (TLP) on a compact smooth oriented 4-manifold X consists of the following data.
- Finite, disjoint subsets $A, B \subset X$.
- A smooth map $f : X \setminus A \to S^2$ which is a submersion outside $A \cup B$; such that $f(b) \neq f(b')$ for distinct $b, b' \in B$ and which is given in suitable oriented charts by the local models $(z_1, z_2) \mapsto z_2/z_1$ (in a punctured neighboourhood of a point in A) and $(z_1, z_2) \mapsto z_1^2 + z_2^2 +$ Constant (in a neighbourhood of a point in B).

The notions of "equivalence over a fixed base" and "equivalence" go over in an obvious way.

Given a Topological Lefschetz pencil we define the "hyperplane class" $h \in H^2(X; \mathbf{Z})$ to be the Poincaré dual of the homology class of a fibre of f. (More precisely, the closure in X of a fibre of f in $X \setminus A$.) We define another class $K(X, f) \in H^2(X; \mathbf{Z})$ as follows. Over $X \setminus (A \cup B)$ we have an oriented 2-plane bundle V given by the tangent space to the fibre of f. We claim that this can be extended to X and the extension is unique up to isomorphism. For if N is a small ball around a point of A or B we can extend V over N if we have a trivialisation of V over ∂N. Since $\partial N \cong S^3$ and $H^2(S^3) = 0$ such a trivialisation exists. Morover since $H^1(S^3) = 0$ any two trivialisations are homotopic and this implies that any two extensions are isomorphic. Now we set

$$K(X, f) = -(c_1(V) + 2f^*([S^2])),$$

where $c_1(V)$ is the first Chern class of V, regarded as a complex line bundle.

By straightforward algebraic topology we have:

PROPOSITION 3.
- The genus of a smooth fibre of f is $\frac{1}{2}(h.h + K(X, f).h + 2)$.
- The number of points in A is $h.h$.
- The number of points in B is $\chi(X) + h.h + 2K(X, f).h$.

So far we have been considering maps from either complex algebraic manifolds (the classical case) or general smooth manifolds. The interest of these ideas is highlighted by the connection with the intermediate class of *symplectic manifolds*. We say that a TLP on a 4-manifold X is compatible with a symplectic form ω on X if we can choose the local co-ordinates (z_1, z_2) appearing in the definition such that ω is a Kahler form in these co-ordinates (i.e. has type (1,1)) and if the fibres of f in $X \setminus (A \cup B)$ are oriented symplectic submanifolds (i.e ω is strictly positive on the fibre, with respect to the induced orientation). In general, for any symplectic manifold (X, ω) we can define a "canonical class" $K(\omega)$ to be minus the first Chern class of the tangent bundle, for any compatible almost complex structure. It is easy to see that if a pencil is compatible with a symplectic structure then $K(\omega) = K(X, f)$.

THEOREM 1. *Let X be a smooth oriented 4-manifold and let $h \in H^2(X; \mathbf{Z})$ be a class with $h^2 > 0$. Then if X has a TLP with hyperplane class h it admits a compatible symplectic form ω with $[\omega] = h$. Conversely if X admits a symplectic form with $[\omega] = h$ then for sufficiently large integers k, X admits a compatible TLP with hyperplane class kh.*

This is a composite of results of Gompf [7] and the author [4]. The interest of the result is that the question of the existence of symplectic forms is, on the face of it, a question on the borderline of differential geometry and differential topology, while the question of the existence of TLP's is, on the face of it, pure differential topology. Thus the result is a topological criterion for the existence of symplectic structures on 4-manifolds.

We will not go into the proofs of this Theorem. In one direction, Gompf's construction of a symplectic form on a manifold admitting a Lefschetz pencil is an extension of Thurston's construction of symplectic forms on fibre bundles. The proof in the other direction involves some analysis. Roughly, one considers for large k embeddings $\iota : X \to \mathbf{CP}^N$ with $\iota^*([H]) = k[h]$ where H is the standard generator of $H^2(\mathbf{CP}^N)$. One shows that, for any fixed compatible almost-complex structure on X, the embedding can be chosen to be "approximately holomorphic". This can be seen as an extension of the Kodaira embedding theorem from complex geometry to the almost-complex case. Then one constructs the pencil by choosing a sufficiently generic $\mathbf{CP}^{N-2}$ and following the procedure described above.

4. "Explicit" description of symplectic four-manifolds

The discussion of the *uniqueness* of the TLP corresponding to a sympletic form is a little more complex, partly because there is a lacuna of a rather technical nature in the foundational results proved up to now. The problem is that in the analytical theory one is lead to the notion of an "asymptotic sequence" of TLP's f_k, defined for large k. As the theory stands, the precise definition of this notion would be rather complicated: it would involve the choice of an almost complex structure and various real parameters measuring roughly "deviation from holomorphicity" and "transversality". However we can formulate our results without going into these details, making the existence of the notion part of the statement.

THEOREM 2. *Let (X, ω) be a compact symplectic 4-manifold with $[\omega]$ integral and let h be an integer lift of $[\omega]$. There is a preferred non-empty class $\mathcal{A}$ of sequences (f_k) of TLP's on X, defined for large k and where f_k has hyperplane class kh, such that if (f'_k) is another sequence in the class $\mathcal{A}$ then f_k is equivalent to f'_k for large k.*

The issue that is left open here is that given some TLP g it is hard to decide if g is a member of an asymptotic sequence (although it seems quite likely that it will be except perhaps in some special circumstances).

Next, we describe the "stabilisation operation" due to Auroux and Katzarkov [1]. Suppose that f is a TLP on X. Recall that this is determined by a smooth fibre Σ with marked points p_α and vanishing cycles $V_i \subset \Sigma \setminus \{p_\alpha\}$ such that the product of the Dehn twists about the V_i is equal to the element T (the product of Dehn twists in circles about the p_α) in the marked mapping class group $\Gamma_{\Sigma,P}$. We can think of Σ as having fixed charts around the p_α. Using these charts we form a new surface $\hat{\Sigma}$ by taking a second copy of Σ and performing the connected sum operation at each of the points p_α. Thus the genus of $\hat{\Sigma}$ is twice the genus of Σ plus the number of points p_α minus 2. We fix a standard embedding j of Σ minus small discs about the p_α into $\hat{\Sigma}$. We take four standard points on each of the cylinders making up the connected sum, giving a set $\hat{P}$ of $4|P| = (2h)^2$ marked points in $\hat{\Sigma}$. Now we work in the mapping class group $\Gamma_{\hat{\Sigma},\hat{P}}$. Here we have the element $\hat{T}$ defined by the product of the Dehn twists in small loops about the points of $\hat{P}$ and we also have the element T defined by the twists in the original loops about the p_α, regarded as loops in $\hat{\Sigma}$ via the embedding j. Auroux and Katzarkov write down two explicit and standard collections of loops $\{U_i\}, \{W_i\}$ in $\hat{\Sigma}$ such that

$$\hat{T} = ATB$$

in $\Gamma_{\hat{\Sigma},\hat{P}}$ where A is the product of Dehn twists in the U_i and B is the product of Dehn twists in the W_i. Now suppose that $V_1, \ldots, V_\nu$ are loops representing the monodromy of the pencil f. We can consider them as loops in $\hat{\Sigma}$, via the embedding j, and we have

$$\hat{T} = AD_{V_1} \ldots D_{V_\nu} B = D_{U_1} \ldots D_{U_p} D_{V_1} \ldots D_{V_\nu} D_{W_1} \ldots D_{W_q}.$$

So the U_i, V_i, W_i are data defining a TLP with fibre $\hat{\Sigma}$. We call this TLP $Sq(f)$.

PROPOSITION 4. *The 4-manifold associated to $Sq(f)$ is diffeomorphic to X.*

THEOREM 3 ([1]). *Let f_k be a sequence in $\mathcal{A}$ associated to a symplectic 4-manifold (X, ω). Then for large enough k, the TLP f_{2k} is equivalent to $Sq(f_k)$.*

The upshot of the results of the previous section is that in principle questions about the classification of symplectic 4-manifolds can be translated into questions about Dehn twists in the mapping class group of surfaces. Let us spell this out in more detail. Define a *primitive* compact symplectic 4-manifold to be a triple (X, h, ω) where X is compact 4-manifold, h is a primitive integral cohomology class and ω is a symplectic form on X with $[\omega] = h$. We give a description of the equivalence classes (in the obvious sense) of such primitive symplectic 4-manifolds.

We have explained that a TLP is specified by data consisting of loops V_i in a marked surface (Σ, P) satisfying the condition $D_{V_1} \ldots D_{V_\nu} = T^{-1}$. Now we ask when two such sets of data yield equivalent TLP's. If we just consider equivalence over a fixed base we have to consider the loops V_i up to isotopy and modulo the action of conjugation of the D_{V_i} by a single arbitrary element of $\Gamma_{\Sigma,P}$, i.e. changing each of the V_i to $g(V_i)$ for some $g \in \Gamma_{\Sigma,P}$. If we consider general equivalence we have to bring in the action of the diffeomorphisms of the 2-sphere with ν marked points, the spherical braid group. It is convenient to regard the point at infinity as fixed, so we work with the group of compactly supported diffeomorphisms of the plane with ν marked points. This is just the Braid group B_ν and has standard generators σ_i, $i = 1, \ldots \nu - 1$. The action of σ_i on the Dehn twists takes a sequence $D_{V_1}, \ldots, D_{V_\nu}$ to a sequence

$$D_{V_1}, \ldots, D_{V_{i+1}}, D^*_{V_i}, \ldots, D_{V_\nu}$$

where $D_{V_i}^*$ is the conjugate of D_{V_i}:
$$D_{V_{i+1}}^{-1} D_{V_i} D_{V_{i+1}}.$$

Note that this action preserves the product, as it should. In other words we can change a sequence of embedded circles
$$V_1 \ldots V_i, V_{i+1}, \ldots, V_\nu$$
to a new sequence
$$V_1 \ldots V_{i+1} V_i^* \ldots V_\nu,$$
where V_i^* is obtained by applying the inverse Dehn twist in V_{i+1} to V_i. Now for each g, ν, p we write $\mathcal{D}_{g,\nu,p}$ for the class of finite sequences of isotopy classes of circles $V_1, \ldots, V_\nu$, in a standard surface of genus g with p marked points, which satisfy the relation (2). So for each such sequence we can construct a 4-manifold with a TLP. Let $\mathcal{D}_{g,\nu,p}^{(0)}$ be the subclass consisting of such data which define a TLP where the class F is 2^r times a primitive class, for some r. Now let $\mathcal{C}_{g,\nu,p}$ be the quotient of $\mathcal{D}_{g,\nu,p}^{(0)}$ under the action of the braid group and the mapping class group of (Σ, p_α). We have the Auroux-Katzarkov map
$$Sq : \mathcal{C}_{g,\nu,p} \to \mathcal{C}_{\hat{g},\hat{\nu},\hat{p}},$$
where $\hat{g} = 2g + \nu - 1, \hat{\nu} = 4(g + \nu - 1), \hat{p} = 4p$. Now if we write $\chi = \nu - 2(2g - 2), \theta = \frac{2g-2-p}{\sqrt{p}}$ then we have, in an obvious notation, $\hat{\chi} = \chi, \hat{\theta} = \theta$. For fixed χ, θ we let $\mathcal{C}_{\chi,\theta}$ be the direct limit of the maps of sets $Sq : \mathcal{C}_{g,\nu,p} \to \mathcal{C}_{\hat{g},\hat{p},\hat{\nu}}$ with $\chi = \nu - 2(2g - 2), \theta = \frac{2g-2-p}{\sqrt{p}}$.

Now let $\mathcal{X}_{\chi,\theta}$ be the set of equivalence classes of data of the form (X, ω, h) where X is a smooth compact 4-manifold of Euler characteristic χ, h is a *primitive class* in $H^2(X; \mathbf{Z})$, ω is a symplectic form on X with $[\omega]$ the reduction of h and
$$\frac{K(\omega).h}{\sqrt{h.h}} = \theta.$$
Then the results of the previous section amount to

PROPOSITION 5. *There is a canonical inclusion of $\mathcal{X}_{\chi,\theta}$ into $\mathcal{C}_{\chi,\theta}$.*

It seems most likely that this inclusion is in fact a bijection. This issue is the technical lacuna in the theory referred to above, i.e. the question whether any TLP, perhaps after stabilisation by the Auroux-Katzarkov construction, arises as an element of an asymptotic sequence. Assuming this is so we see that the problem of classifying compact symplectic 4-manifolds (with integral symplectic form and a chosen lift h) is equivalent to the problem of describing the set $\mathcal{C}_{\chi,\theta}$ which is formulated entirely in terms of the mapping class group.

5. Problems

The significance of the translation of symplectic 4-manifold theory into problems about the mapping class group should not be overrated. In practice, the problem of classifying the appropriate sequences of embedded circles modulo equivalence seems very intractable and it may be unlikely that much can be done without some new idea. The situation is quite similar to that arising from the description of 3-manifolds via Heegard decompositions where again the apparent simplicity of the translated problem is largely illusory because of the complexity of the mapping class group. So far, no progress has been made in, for example, distinguishing symplectic 4-manifolds via this "combinatorial" approach. (Another way of seeing that this approach is unlikely to be immediately useful is that the classification problem, without restriction, contains the problem of classifying finitely presented groups, since any group arises as the fundamental group of a symplectic 4-manifold [7].) Nevertheless, we can state as an ambitious problem:

PROBLEM 1. Develop techniques to describe the sets $\mathcal{C}_{\chi,\theta}$.

As we have said, the real difficulty comes from the action of the braid group. One can think of this in the following way. Make a directed graph with one vertex for each element of the mapping class group $\Gamma_{\Sigma,P}$ and one oriented edge joining g and $g \circ D_V$ for each isotopy class of embedded circle V in $\Sigma \setminus P$. Thus a sequence of $V_1 \ldots V_\nu$ of the kind we want to consider is the same thing as an oriented path in the graph from the identity to the element T. Now for each vertex g and pair V_1, V_2 we have four edges

$$(g, gD_{V_1}), (gD_{V_1}, gD_{V_1}D_{V_2}), (g, gD_{V_2}), (gD_{V_2}, gD_{V_1}D_{V_2}),$$

since $D_{V_1}D_{V_2} = D_{V_2}D_{V_1}^*$ where $D_{V_1}^*$ is the conjugate as before. Now form a topological space Z by attaching a square to each such collection of four edges. Thus our braid relation is the relation on edge-paths in Z under which we are allowed to push a path across a square in the obvious way. This is somewhat similar to the usual combinatorial description of the homotopy classes of paths in Z from 1 to T^{-1}—i.e. essentially of $\pi_1(Z)$—but with the crucial difference that we are only allowed to consider "positive" paths. It seems likely that $\pi_1(Z)$ is just the integers: this would correspond to the fact that the description of TLP's becomes much simpler if we are allowed to use both positive and negative Dehn twists and cancel positive and negative pairs. Of course in formulating things this way we have not really done anything beyond restating the problem, but the point of view might be worth investigating. One could also try to fit the recent work of Auroux, Munoz and Presas [2], describing symplectomorphisms in terms of pencils, into the same mould, perhaps giving a model for the classifying space of the group of symplectomorphisms of a symplectic 4-manifold. Another related point of view is to think of a Lefschetz pencil as defining a map from the two sphere into the compactified moduli space $\overline{\mathcal{M}}_{g,P}$, [12]. Again the problem is related to a standard one, of describing $\pi_2(\overline{\mathcal{M}}_{g,P})$, but the real difficulty comes from the fact that we have to consider an equivalence defined by homotopies through maps which meet the compactification divisor with positive local intersection numbers.

Notice that the essence of the definition of $\mathcal{C}_{g,p,\nu}$ has an entirely algebraic character. We can make the same definitions given any group, an element of the centre and a preferred conjugacy class in (in our case the conjugacy class of Dehn twists). So we can ask the same kinds of questions for other groups. In the case of finite permutation groups, with the conjugacy class of transpositions, the problem essentially amounts to the classification of branched covers and was solved by Hurwitz. The definitions are natural with respect to group homomorphisms so one approach to studying the question in the mapping class group might be to consider representations, for example the permutation representation on spin structures. Or one might consider linear representations, the obvious ones being the action on homology which takes one back to the classical homological mondromy. There are also more exotic representations connected with conformal field theory and Jones invrainats, as considered by Smith [11].

Alternatively one can consider "high technology" approaches using Floer homology. This area is developing in a very exciting way through work of Seidel [10] and others, but so far has not yielded any definite results about the classification of symplectic 4-manifolds.

Taking a different direction, there are some fundamental issues left open in the basic theory described above. The main one involves understanding better the nature of asymptotic sequences which we can now formulate as

PROBLEM 2. Show that the inclusion of Proposition 5 is a bijection. This will probably require more thought about the analytical and geometric constructions which underpin the theory.

Another question is suggested by the Auroux-Katzarkov doubling formula. Suppose we know one element f_{k_0} of an asymptotic sequence. Can we describe f_l for other values of l apart from $l = 2^r k_0$? Or perhaps better

PROBLEM 3. Given a topological description of f_{k_0}, f_{k_1} describe $f_{k_0+k_1}$.

A good understanding of this would enable one to drop the rather artificial introduction of the subclass $\mathcal{D}^{(0)}_{g,\nu,p}$ in the discussion above.

More generally still, if we have two pencils on a symplectic manifold X with fibre classes F_1, F_2 one can ask for a description of a TLP with fibre $F_1 + F_2$ (if such exists). This might give information about the problem of describing the classes represented by symplectic forms on a fixed 4-manifold.

Rather than trying to use the TLP description to reduce questions to combinatorics one can attempt to use it as a tool to prove general properties of symplectic 4-manifolds. So far, this has been more fruitful, giving a new approach to Taubes' results independent of the Seiberg-Witten theory [6], [13], [9]. There is also a generalisation of the TLP description to other 4-manifolds [3] and there are many things one could try here; for example to prove that any 4-manifold has "simple type". One thing that should be important to understand is the role of the canonical class $K(\omega)$. Using the Seiberg-Witten theory and pseudo-holomorphic curve techniques, a complete classification is known of symplectic 4-manifolds with $\omega.K(\omega) < 0$, i.e. with $\theta < 0$ in our notation above. The only examples are the standard ones given by rational and ruled complex surfaces [8]. It would be interesting to derive this by the Lefschtez pencil method:

PROBLEM 4. Reproduce the classification of manifolds with $\omega.K(\omega) < 0$ by studying the sets $\mathcal{C}_{\chi,\theta}$ for $\theta < 0$.

There is a network of interesting questions dealing with the borderline case when $K(\omega).\omega = 0$ or, stronger still, $K(\omega) = 0$. In the latter case the only known examples are the the standard complex tori, certain other torus bundles over tori and K3 surfaces. So we have:

PROBLEM 5. Analyse the monodromy of Lefschetz fibrations on manifolds with $K(\omega) = 0$.

Related to this is the general question of understanding the place of complex algebraic surfaces among general symplectic 4-manifolds. One can ask:

PROBLEM 6. Find special features of the monodromy of algebraic surfaces.

There is some good motivation for this coming from at least three directions

- The problem includes (in principle) the well-known problem of describing possible fundamental groups of algebraic surfaces.
- One famous constraint is the "hard Lefschetz" property, which has a well-known translation into the action of the monodromy on homology.
- From the Seiberg-Witten theory we know that there are strong restrictions on the basic classes of algebraic surfaces, and these can be translated into the TLP point of view along the lines of [6], [13].

References

[1] D. Auroux and L. Katzarkov, *A degree-doubling formula for braid monodromies and Lefschetz pencils* Preprint.

[2] D. Auroux, V. Munoz and F. Presas, *Lagrangian submanifolds and Lefschetz pencils*, SG/0407126.

[3] D. Auroux, S. Donaldson and L. Katzarkov, *Singular Lefschetz pencils*, DG/0410332.

[4] S. Donaldson, *Lefschetz pencils on symplectic manifolds* J. Differential Geometry **44** (1999), 205-36.

[5] S. Donaldson, *Polynomials, vanishing cycles and Floer homology*, In: Mathematics: Frontiers and Perspectives (Arnold, Atiyah, Lax, Mazur Eds.) Amer. Math. Soc., 2000, 55-64.

[6] S. Donaldson and I. Smith, *Lefschetz pencils and the canonical class for symplectic four-manifolds*, Topology **42** (2003), 743-85.

[7] R. Gompf and A. Stipsicz, *Four-manifolds and Kirby calculus*, Graduate Studies in Mathematics, Amer. Math. Soc., 1999.

[8] A-K. Liu, *Some new applications of general wall-crossing formula, Gompf's conjecture and applications*, Math. Res. Letters **3** (1996), 569-85.

[9] P. Ozsvath and Z. Szabo, *Holomorphic triangle invariants and the topology of symplectic four-manifolds*, Duke Math. J. **121** (2004), 1-34, SG/0201049.

[10] P. Seidel, *A long exact sequence for symplectic Floer cohomology*, Topology **42** (2003), 1003-63, SG/0105186.

[11] I. Smith, *Symplectic four-manifolds and conformal blocks*, SG/0302088.

[12] I. Smith, *Lefschetz pencils and divisors in moduli space*, Geometry and Topology **5** (2001), 579-608.

[13] I. Smith, *Serre-Taubes duality for pseudo-holomorphic curves*, Topology **42** (2003), 937-79.

Open Problems in Grothendieck-Teichmüller Theory

Pierre Lochak and Leila Schneps

§0. Introduction, definitions, notation

The present note is not intended in any way as an introduction to Grothendieck-Teichmüller theory. It is essentially a concentrated list of questions in and around this theory, most of which are open, although we have included some questions which are natural to ask but easy to answer, and a few others which were open but are now settled. In order for the reader to appreciate the relative depth, difficulty, and interest of these problems, and their position within the theory, some previous knowledge is required. We do give some important facts and definitions, but they are intended to remind the reader of relatively well-known elements of the theory, to give something of the flavor of the objects concerned and to make statements unambiguous. They are not sufficient to provide a deep understanding of the theory.

To describe the main idea of the theory in a few words (see e.g. [L2] for more and references), one takes a category $\mathcal{C}$ of geometric objects (of finite type) defined over a field k (of characteristic zero for simplicity); these can be k-varieties, k-schemes, or k-algebraic stacks, and a collection of k-morphisms between them, for instance all k-morphisms of k-varieties. Let $\pi_1^{geom}(X)$ denote the geometric fundamental group of X, that is the algebraic fundamental group of $X \otimes \overline{k}$, where $\overline{k}$ denotes the algebraic (or separable) closure of k; it is a finitely generated profinite group. One can view π_1^{geom} as a functor from $\mathcal{C}$ to the category of finitely generated profinite groups with continous morphisms up to inner automorphisms. One then considers the (outer) automorphism group of this functor, say $\mathrm{Out}(\pi_1^{geom}(\mathcal{C}))$. Concretely speaking its elements consists of collections $(\phi_X)_X$ with $\phi_X \in \mathrm{Out}(\pi_1^{geom}(X))$, indexed by objects $X \in \mathcal{C}$, and compatible with morphisms. One usually has additional requirements, namely that the ϕ_X satisfy some Galois-style properties, like the preservation of conjugacy classes of inertia groups. Since there is a canonical outer action of the absolute Galois group $G_k = \mathrm{Gal}(\overline{k}/k)$ on $\pi_1^{geom}(X)$ for each X, and it is compatible with morphisms, one gets a natural homomorphism $G_k \to \mathrm{Out}(\pi_1^{geom}(\mathcal{C}))$, which is injective in all the interesting cases. If in fact it is an isomorphism, one thus in principle gets a geometric description of the arithmetic group G_k. In the case where $\mathcal{C}$ is the category of regular quasiprojective $\mathbb{Q}$-varieties with all $\mathbb{Q}$-morphisms between them, F. Pop has shown that $\mathrm{Out}(\pi_1^{geom}(\mathcal{C}))$ is indeed equal to $G_\mathbb{Q}$ (2002, unpublished; the result is actually more general and stronger).

©2006 American Mathematical Society

The specificity of *Grothendieck-Teichmüller theory* is that Grothendieck suggested (in [G1] and [G2]) studying the category $\underline{M}$ of moduli spaces of curves with marked points, all of which are viewed as algebraic stacks defined over $\mathbb{Q}$. One does not *a priori* consider all possible $\mathbb{Q}$-morphisms between them, but only a certain family of morphisms coming from topological operations on the topological curves themselves, such as erasing points, cutting out subsurfaces by simple closed loops, or quotienting by finite-order diffeomorphisms. All these operations on topological curves yield natural morphisms between the associated moduli spaces (which include the classical Knudsen morphisms); these in turn yield homomorphisms between their geometric fundamental groups, which are nothing other than the profinite completions of the mapping class groups studied in this volume.

Perhaps the most insightful remark of Grothendieck on this topic is that the (outer) automorphism group $\mathrm{Out}(\pi_1^{geom}(\underline{M}))$ of this category can actually be described explicitly, essentially as elements of the free profinite group on two generators satisfying a small finite number of equations, the reason for this being that in fact only the moduli spaces of dimensions 1 and 2 are important, the automorphism group remaining unchanged when the higher dimensional ones are added to the category. It is not known whether Grothendieck actually wrote down the defining equations of this group, which in essence is the Grothendieck-Teichmüller group.

However, in the seminal paper [Dr], V.Drinfel'd gave the definition of a profinite group $\widehat{GT}$ which is (essentially) equal to $\mathrm{Out}(\pi_1^{geom}(\underline{M}_0))$ where $\underline{M}_0$ is the category of moduli spaces of genus zero curves with marked points, whose geometric fundamental groups are essentially profinite braid groups. The argument above shows that there is a homomorphism $G_{\mathbb{Q}} \to \widehat{GT}$, which is injective by Belyi's celebrated theorem, and one of the essential goals of Grothendieck-Teichmüller theory is to compare these two groups. Another, somewhat alternative goal is to refine the definition of $\widehat{GT}$ to discover the automorphism group of the category of moduli spaces of all genus equipped with "as many $\mathbb{Q}$-morphisms as possible". This has been realized when the morphisms are point-erasing and cutting along simple closed loops, partially realized when quotients by finite-order diffeomorphisms are added, and in other, somewhat more general situations (see §2). But it is always possible to display other $\mathbb{Q}$-morphisms respected by $G_{\mathbb{Q}}$ and ask if any version of $\widehat{GT}$ also respects them.

Considering weaker profinite versions (pro-ℓ, pronilpotent), as well as proalgebraic (pro-unipotent, Lie algebra) versions of Grothendieck-Teichmüller theory has yielded new results, new conjectures and most interestingly, new links with aspects of number theory not visible in the full profinite situation. The later sections of this article are devoted to these.

Let us mention a handful of references which will provide the newcomer with entry points into the subject. For inspiration, we recommend reading parts 2 and 3 of Grothendieck's *Esquisse d'un Programme* ([G1]). The papers [Dr] and [I1] (as well as [De], although in a different vein) are certainly foundational for the subject. They still make very interesting, perhaps indispensable reading. Introductions to most of the main themes of the *Esquisse* are contained in the articles of [GGA]. In particular, introductions to the Grothendieck-Teichmüller group can be found in the article [S2] of [GGA] (see also [LS1], [L2]). The original article [Dr] of Drinfel'd introducing $\widehat{GT}$ is filled with impressive insights, but the point of view of moduli spaces is hardly touched upon, whereas the geometry of these spaces (in all genera) became central in [HLS] and [NS]. They can help make the bridge with the subject matter of the present volume.

In the rest of this section we will list some of the main definitions and terms of notation. Some of the objects are not defined from scratch, so that the exposition is not completely self-contained, however they are meant to make the subsequent statements understandable and unambiguous.

We start with a short list of the main geometric objects, which are also the main objects of study in the present volume:

– $\mathcal{M}_{g,n}$ (resp. $\mathcal{M}_{g,[n]}$) denotes the moduli space of smooth curves of genus g with n ordered (resp. unordered) marked points. These spaces can be considered as analytic orbifolds or as algebraic stacks over $\mathbb{Z}$, a fortiori over $\mathbb{Q}$ or any field of characteristic 0. We will make it clear what version we have in mind according to the context.

– $\Gamma_{g,n} = \pi_1^{orb}(\mathcal{M}_{g,n})$ (resp. $\Gamma_{g,[n]} = \pi_1^{orb}(\mathcal{M}_{g,[n]})$ denotes the orbifold fundamental group of the above space, as a complex orbifold. These are nothing but the mapping class groups of topologists, also called (Teichmüller) modular groups in the algebro-geometric context.

– $\widehat{\Gamma}_{g,n}$ (resp. $\widehat{\Gamma}_{g,[n]}$) are the profinite completions of the above groups. They are the geometric fundamental groups of $\mathcal{M}_{g,n}$ and $\mathcal{M}_{g,[n]}$ respectively, i.e. the fundamental groups of these spaces as $\overline{\mathbb{Q}}$- or $\mathbb{C}$-stacks.

We now pass to the Grothendieck-Teichmüller group in some of its most important versions. Others will appear in the course of the text. We start with the full profinite version $\widehat{GT}$, already mentioned above. Note that the profinite completion contains the maximum amount of information compared to the other completions and versions considered here.

First note that $\mathcal{M}_{0,4} \simeq \mathbb{P}^1 \setminus \{0, 1, \infty\}$. Then identify the topological fundamental group of the latter space with the free group F_2 on two generators x and y. This (non-canonical) identification amounts to picking two loops around 0 and 1 which generate the fundamental group of $\mathbb{C} \setminus \{0, 1\}$. With this identification we also identify $\pi_1^{geom}(\mathcal{M}_{0,4})$ with the profinite completion $\widehat{F}_2$, and we get a monomorphism:

$$\mathrm{Out}(\pi_1^{geom}(\underline{\mathcal{M}})) \hookrightarrow \mathrm{Out}(\widehat{F}_2).$$

In order to get $\widehat{GT}$, as originally defined in [Dr], replace $\underline{\mathcal{M}}$ with $\underline{\mathcal{M}}_0$, that is, use only genus 0 moduli spaces and pick a (tangential) basepoint in order to replace outer by bona fide actions. Finally require that the action preserve conjugacy classes of inertia groups, as the Galois action does. This produces again a monomorphism:

$$\widehat{GT} \hookrightarrow \mathrm{Aut}^*(\widehat{F}_2),$$

where the upper star refers to this inertia preservation condition. Concretely speaking, an element of $\widehat{GT}$ is given as a pair $F = (\lambda, f)$ with $\lambda \in \widehat{\mathbb{Z}}^*$ (invertible elements of $\widehat{\mathbb{Z}}$) and $f \in \widehat{F}_2'$ (topological derived subgroup of $\widehat{F}_2$). The action on $\widehat{F}_2$ is defined by:

$$F(x) = x^\lambda, \qquad F(y) = f^{-1} y^\lambda f.$$

One requires that these formulas define an *automorphism*, that is an invertible morphism, and there is no effective way to ensure this. Finally and most importantly the pair (λ, f) has to satisfy the following three relations (for the geometric origin of these relations, we refer to the introductions quoted above):

(I) $f(x,y)f(y,x) = 1$;
(II) $f(x,y)x^m f(z,x)z^m f(y,z)y^m = 1$ where $xyz = 1$ and $m = (\lambda - 1)/2$;
(III) $f(x_{12}, x_{23})f(x_{34}, x_{45})f(x_{51}, x_{12})f(x_{23}, x_{34})f(x_{45,51}) = 1$,

where in (III) (the pentagonal relation), the $x_{i,i+1}$ are the standard generators of the group $\widehat{\Gamma}_{0,5}$. We should also explain how substitution of variables is intended; for any homomorphism of profinite groups $\phi : \widehat{F}_2 \to G$ mapping $x \mapsto a$ and $y \mapsto b$, we write $\phi(f) = f(a,b)$ for $f \in \widehat{F}_2$ (f itself is equal to $f(x,y)$).

Thus $\widehat{GT}$ is the subgroup of $\mathrm{Aut}^*(\widehat{F}_2)$ whose elements are pairs $F = (\lambda, f)$ acting as above and satisfying (I), (II) and (III). Note that these are usually refered to as "relations" although "equations" would be more correct: indeed, $\widehat{GT}$ is a subgroup, not a quotient of $\mathrm{Aut}^*(\widehat{F}_2)$. We also mention that (I) is actually a consequence of (III), as was noted by H. Furusho, but we keep (I) in the definition nevertheless because of its geometric meaning.

There is a natural map $\widehat{GT} \to \widehat{\mathbb{Z}}^*$ defined by $F = (\lambda, f) \mapsto \lambda$. It is surjective and the kernel is denoted $\widehat{GT}^1$, which is an important subgroup of $\widehat{GT}$, the analog of which shows up in the various versions of the Grothendieck-Teichmüller group considered below.

The group $\widehat{GT}$ is profinite, as it is a closed sugroup of $\mathrm{Aut}(\widehat{F}_2)$ and any automorphism group of a finitely generated profinite group G has itself a natural structure of profinite group. Indeed, characteristic open subgroups are cofinal in G and $\mathrm{Aut}(G)$ can be written as the inverse limit:

$$\mathrm{Aut}(G) = \varprojlim_N \mathrm{Im}(\mathrm{Aut}(G) \to \mathrm{Aut}(G/N)),$$

where N runs through the open characteristic subgroups of G. Here we did not mention topology, because by a recent and fundamental result ([NiSe]), *any* automorphism of G is actually continuous. Applying this result to $\widehat{F}_2$, it makes the definition of $\widehat{GT}$ purely algebraic.

Starting from the profinite group $\widehat{GT}$, one can define interesting quotients in standard ways. In particular, one can define GT^{nil}, the maximal pronilpotent quotient of $\widehat{GT}$; it is in fact the direct product of the pro-ℓ quotients $GT^{(\ell)}$, when ℓ runs through the prime integers. However, we have very little control over these quotients. More accessible are the groups $GT_{(\ell)}$ which are defined exactly like $\widehat{GT}$ except that we take $(\lambda, f) \in (\mathbb{Z}_\ell^*, F_2^{(\ell)})$ (where $F_2^{(\ell)}$ is the pro-ℓ completion of F_2). Comparison of $GT^{(\ell)}$ with $GT_{(\ell)}$ is an open question which we will record explicitly below.

We now pass to the proalgebraic setting, which will be useful especially in the later sections. All the algebraic groups $\underline{G}$ that we encounter, including those which make their appearance in the last sections only, in connection with motives and multiple zeta values, will be of the following type. The group $\underline{G}$ is linear proalgebraic over a field of characteristic 0, usually $\mathbb{Q}$ (sometimes $\underline{G}$ can actually be regarded as a progroup-scheme over $\mathbb{Z}$). It is an extension of the multiplicative group $\mathbb{G}_m$ by its prounipotent radical $\underline{G}^1$; the usual equivalence between unipotent algebraic groups and their Lie algebras extends to the proalgebraic setting, so $\underline{G}^1$

is isomorphic to its Lie algebra $\text{Lie}(G^1)$. Moreover the latter is equipped with an action of $\mathbb{G}_m$ coming from the definition of G as an extension, and this action provides a natural grading, so that we can also consider the graded version of that Lie algebra, which is more amenable to concrete computations.

In the case of the Grothendieck-Teichmüller group, we encounter the same phenomenon as with the profinite versions mentioned above. We could consider the pro-ℓ quotient $GT^{(\ell)}$ and construct from it a prounipotent completion (over $\mathbb{Q}_\ell$). But again this is not easily accessible. So following [Dr], one first defines the prounipotent (or Malčev) completion $\underline{F}_2$. Then one defines $\underline{GT}$ by describing its k-points for k a field of characteristic 0. These are given again by pairs (λ, f) satisfying the relations as above, but now with $(\lambda, f) \in k^* \times \underline{F}_2(k)$.

The prounipotent radical $\underline{GT}^1$ is then defined as above. The associated Lie algebra is denoted $\mathfrak{gt}$ and its graded version $\mathfrak{grt}$. The latter is an especially important object, allowing for quite explicit computations. It was first defined and studied by V. Drinfeld and Y. Ihara. It is naturally defined over $\mathbb{Q}$, although Ihara showed it can in fact be defined over $\mathbb{Z}$, and this integral structure leads to very interesting arithmetic problems which we do not address in this note (see [I4], [McCS]).

Let us give here the explicit definition of $\mathfrak{grt}$, obtained by linearizing and truncating the defining relations of the group. Namely, the graded Lie algebra $\mathfrak{grt}$ is generated as a $\mathbb{Q}$-vector space by the set of homogeneous Lie polynomials $f(x, y)$ in two variables satisfying:

(i) $f(x, y) + f(y, x) = 0$;
(ii) $f(x, y) + f(z, x) + f(y, z) = 0$ with $x + y + z = 0$;
(iii) $f(x_{12}, x_{23}) + f(x_{23}, x_{34}) + f(x_{34}, x_{45}) + f(x_{45}, x_{51}) + f(x_{51}, x_{12}) = 0$,

where the x_{ij} generate the Lie algebra of the pure sphere 5-strand braid group.

This finishes our survey of the main definitions. Other objects will occur in the text, especially in the later sections. We remark that we refrained from explicitly using in this note the variant $\underline{GRT}$ of $\underline{GT}$, although it is conceptually quite significant. We refer to [Dr] and especially to [F1,F2] for more information on this point. Finally we note that we will sometimes use the bare letters GT as an abbreviation for "Grothendieck-Teichmüller" or as a "generic" version of the Grothendieck-Teichmüller group, so that this is not to be considered as a piece of mathematical notation.

Acknowledgments. Many of the questions and problems listed below arise naturally and were raised recurrently and independently by various people. We warmly (albeit anonymously) thank them all for sharing their preoccupations with us through the years. It is a pleasure to thank B. Enriquez and I. Marin for their interest and for suggesting interesting questions. We included some of these below (see §1 and §7) although in a simplified and incomplete version in order to minimize the necessary background. We are also delighted to thank D. Harbater, H. Nakamura and the referee for many useful corrections and remarks.

§1. Group theoretical questions on $\widehat{GT}$

The fundamental result concerning the group $\widehat{GT}$ is that there is an injective homomorphism
$$G_\mathbb{Q} \hookrightarrow \widehat{GT}.$$

In some sense this is built into the definitions, via Belyi's result (cf. [G1] p.4; "[...] à vrai dire elle [l'action] est fidèle déjà sur le premier 'étage' [...]"). Drinfel'd indicated this fact in his original article [Dr], and Ihara gave the first complete proof. Let us recall a basic minimum. In order to associate an element $F_\sigma = (\lambda_\sigma, f_\sigma) \in \widehat{GT}$ to $\sigma \in G_{\mathbb{Q}}$, recall that there is a canonical outer $G_{\mathbb{Q}}$-action on the geometric fundamental group of $\mathbb{P}^1 \setminus \{0, 1, \infty\}$ which is inertia preserving. Proceedings as in the introduction, we get an element $F_\sigma \in \text{Aut}^*(\widehat{F}_2)$ acting on the generators as:

$$F_\sigma(x) = x_\sigma^\lambda, \qquad F_\sigma(y) = f_\sigma^{-1} y_\sigma^\lambda f_\sigma.$$

Considering the abelianization of $\widehat{F}_2$ (i.e. the effect on homology) shows that $\lambda_\sigma = \chi(\sigma)$ where $\chi : G_{\mathbb{Q}} \to \widehat{\mathbb{Z}}^*$ is the cyclotomic character. As for f_σ, it becomes uniquely determined if one requires it to lie in the derived subgroup of $\widehat{F}_2$ (this is also the reason behind this requirement in the definition of $\widehat{GT}$). Ihara then went on, using geometric arguments, to prove that every such F_σ satisfies relations (I), (II) and (III), thus defining a homomorphism $G_{\mathbb{Q}} \to \widehat{GT}$. Injectivity is an easy consequence of Belyi's theorem.

Comparison between $\widehat{GT}$ and $G_{\mathbb{Q}}$ is a main goal of Grothendieck-Teichmüller theory. This comparison can be examined from various topological, geometrical and arithmetic points of view, the most straightforward of which may be direct group theory – at least in terms of questions to ask, if not to answer. For any group-theoretical property satisfied by $G_{\mathbb{Q}}$, it is natural to ask if $\widehat{GT}$ possesses the same property. Ihara began asking such questions in the early 1990's; we give a brief list here:

1.1. Let $(\lambda, f) \in \widehat{\mathbb{Z}}^* \times \widehat{F}_2'$. Does $x \mapsto x^\lambda$, $y \mapsto f^{-1} y^\lambda f$ extend to an automorphism, or can it actually determine a non-invertible endomorphism? One can ask the same question when (λ, f) satisfies (I), (II), (III).

Note that this question pertains to the full profinite setting only. Invertibility is immediately detected in the pronilpotent or proalgebraic situation.

1.2. Is $\widehat{GT}^1$ the topological derived subgroup of $\widehat{GT}$? In other words, is the abelianization of $\widehat{GT}$ obtained, like that of $G_{\mathbb{Q}}$ (by the Kronecker-Weber theorem), by the map $(\lambda, f) \mapsto \lambda$ corresponding to taking the cyclotomic character ($\lambda_\sigma = \chi(\sigma)$ for $\sigma \in G_{\mathbb{Q}}$)?

1.3. Does a version of the Shafarevitch conjecture hold for $\widehat{GT}$: is $\widehat{GT}^1$ a free profinite group on a countable number of generators?

1.4. $\widehat{GT}$ contains an element $c = (-1, 1)$ which acts on $\widehat{F}_2 \simeq \pi_1^{geom}(\mathbb{P}^1 \setminus \{0, 1, \infty\})$ as complex conjugation i.e. via $c(x) = x^{-1}$, $c(y) = y^{-1}$. Is the normalizer of c in $\widehat{GT}$ generated by c itself, as it is in $G_{\mathbb{Q}}$?

This question is natural but not open. It was resolved in the affirmative in [LS2], using methods of Serre and a profinite Kurosh theorem to compute the non-commutative cohomology group $H^1(\widehat{F}_2, \langle c \rangle)$. However, a natural analogy with $G_{\mathbb{Q}}$ leads to the further question:

Are the conjugates of $(-1, 1)$ the only elements of finite order in $\widehat{GT}$?

1.5. Compare $GT_{(\ell)}$, as defined in [Dr] and in the introduction with $GT^{(\ell)}$, the maximal pro-ℓ quotient of $\widehat{GT}$. Similar questions arise in the proalgebraic setting.

1.6. Can anything be said about the finite quotients of $\widehat{GT}$? Obviously all abelian groups arise as quotients, since $\widehat{\mathbb{Z}}^*$ is a quotient of $\widehat{GT}$. But what non-abelian groups arise?

1.7. One of the difficulties of inverse Galois theory is that it is easier to prove that a given finite group G is a quotient of $G_{\mathbb{Q}^{ab}}$ than of $G_{\mathbb{Q}}$, and that given the first result, it is not at all obvious how to deduce the second. Part of the problem is due to the difficulty of studying the outer action of $\widehat{\mathbb{Z}}^*$ on $G_{\mathbb{Q}^{ab}}$ explicitly. This outer action is given explicitly, however, by the expression for the outer action of $\widehat{\mathbb{Z}}^*$ on $\widehat{GT}^1$, which contains $G_{\mathbb{Q}^{ab}}$. Can this fact contribute to descending Galois groups over $\mathbb{Q}^{ab}(T)$ to Galois groups over $\mathbb{Q}(T)$? And in general, what can be said about the finite quotients of $\widehat{GT}$?

We close this section with a few words on the linear representations of $\widehat{GT}$. The theme would in principle require a section by itself, but since unfortunately practically nothing is known on this topic, we can remain brief. In particular, no irreducible non-abelian linear representation of any version of GT has been constructed to date. In [I2] Ihara constructed the $\widehat{GT}$ analog of the Soulé characters, so in particular of the Kummer characters. Other versions of at least some extensions of the Kummer and Soulé characters appear in [NS], [M] and a few other places. Hence the first question:

1.8. Are the various definitions of these characters equivalent (inasmuch as they overlap)? Investigate multiplicativity properties which come for free in the Galois case and are far from obvious in the $\widehat{GT}$ extensions (see [I2], §1.10).

I. Marin, in the article [M], constructs representations of $\underline{GT}^1(\mathbb{Q}_\ell)$ into $PGL_N(\mathbb{Q}_\ell((h)))$ (formal Laurent series). We do not recall the construction here, as it is quite complicated, noting only that one starts from an ℓ-adic representation of an infinitesimal braid group together with a given associator and then uses the action of $\widehat{GT}^1$ on the associators in order to produce a representation, provided a certain rigidity condition is fulfilled. The basic construction actually works in more general cases than ℓ-adic representations and is quite natural. The most important question in this context is:

1.9. Can one obtain non-abelian (projective) linear representations of $\underline{GT}^1(\mathbb{Q}_\ell)$ and GT_ℓ in this way?

Recall that there is a natural morphism $G_{\mathbb{Q}} \to \underline{GT}(\mathbb{Q}_\ell)$ (whose image is conjectured to be Zariski dense; see §8 below). This method thus also produces Galois representations. As the author explains, these representations are "often" abelian, hence the question above. Yet, in the abelian case and restricting to the Galois image, one produces characters of the Galois group. This leads naturally to the following problem.

1.10. Analyse the characters produced in [M] in terms of the Soulé characters.

Finally, given the map above, one can always pose the following (too) general question:

1.11. To what extent can one extend "classical" ℓ-adic representations (Tate modules, more generally étale cohomology etc.) into $\widehat{GT}(\mathbb{Q}_\ell)$-representations?

§2. Other versions of $\widehat{GT}$

Several "refined" versions of the Grothendieck-Teichmüller group have been defined, each with relations added in order to satisfy some geometric property that $G_\mathbb{Q}$ is already known to satisfy. We do not give the definitions here (referring instead to the original articles), but simply the properties satisfied by each of four of the most interesting of these "refined" groups. Each of the groups discussed here contains $G_\mathbb{Q}$ as a subgroup.

– The group $\Lambda \subset \widehat{GT}$ is obtained by adding a single new relation (R) to the defining relations of $\widehat{GT}$ coming from $\mathcal{M}_{1,2}$, and has the property that it acts on the mapping class groups $\widehat{\Gamma}_{g,n}$ in all genera and respects the basic point erasing and subsurface inclusion morphisms of the moduli spaces, that is basically the classical Knudsen morphisms. We refer to [HLS] for an early version with extra hypotheses, and to [NS] for the general case.

– The group $\mathbb{F} \subset \Lambda \subset \widehat{GT}$ is obtained by adding two new relations (IV) and (III') to $\widehat{GT}$, and not only has the same property as Λ (because (III) together with (IV) implies (R)), but also respects the exceptional morphism $\mathcal{M}_{0,4} \to \mathcal{M}_{1,1}$, as well as the usual degree 6 quotient morphism already respected by $\widehat{GT}$ (cf. [NS]).

– The group $\widehat{GS}$ is defined by adding two new relations to the definition of $\widehat{GT}$ corresponding to respecting morphisms mapping $\mathcal{M}_{0,4}$ to special loci of $\mathcal{M}_{0,5}$ and $\mathcal{M}_{0,6}$ corresponding to points in those moduli spaces having non-trivial automorphisms. These two relations imply (R), (IV) and (III'), so that $\widehat{GS} \subset \mathbb{F} \subset \Lambda \subset \widehat{GT}$. We refer to [S4] for the initial version with extra hypotheses, and to [T] for the general case.

– The group $\widehat{GTK}$, defined by Ihara in [I3], gives a definition obtained by adding an infinite series of relations requiring that the maps $\mathbb{P}^1 \setminus \{0, \mu_n, \infty\} \to \mathbb{P}^1 \setminus \{0, 1, \infty\}$ be respected, both by the quotient map $z \mapsto z^n$ and by the inclusion map, be respected by the $\widehat{GT}$-action on the π_1's of these curves, just as they are respected by the $G_\mathbb{Q}$-action (since all these curves and maps are defined over $\mathbb{Q}$). It is the only version of $\widehat{GT}$ which requires maps to be respected which are not maps between moduli spaces. Another version $\widehat{GTA}$ is defined in [I2] and elements of comparison between these two versions are discussed in [I3].

Faced with this dangerous explosion of versions of the original object, the most natural and pressing question is surely:

2.1. Are these groups actually different from each other? Or are some of the new relations already implied by previous ones, in particular by the original relations defining $\widehat{GT}$? And in particular – are all or any of them actually isomorphic to $G_\mathbb{Q}$?

The two-level or locality principle. One of the fundamental geometric properties of the all-genera Grothendieck-Teichmüller group $\mathbb{\Gamma}$ (resp. of $\widehat{GT}$) is that it is defined by relations coming from requiring the $\mathbb{\Gamma}$- (resp. $\widehat{GT}$-) action to respect certain morphisms between the moduli spaces (resp. the genus zero moduli spaces) of dimensions 1 and 2. These relations *imply* that the analogous morphisms between higher dimensional (resp. genus zero) moduli spaces are automatically respected. Indeed, relations (I) and (II) reflect the requirement that there exists a $\widehat{GT}$-action on $\widehat{\Gamma}_{0,4}$ and $\widehat{\Gamma}_{0,[4]}$, respecting the homomorphisms coming from the moduli space morphism $\mathcal{M}_{0,4} \to \mathcal{M}_{0,[4]}$, while relation (III) comes from the requirement that the $\widehat{GT}$-action on $\widehat{\Gamma}_{0,4}$ extends to $\widehat{\Gamma}_{0,5}$ respecting the inclusion map $\Gamma_{0,4} \hookrightarrow \Gamma_{0,5}$ corresponding to erasing the fifth point. Finally, relation (R) comes from requiring $\mathbb{\Gamma}$ to act on $\widehat{\Gamma}_{1,2}$ in such a way that the morphism $\mathcal{M}_{0,5} \to \mathcal{M}_{1,2}$ is respected.

2.2. Does the group $\widehat{GS}$ satisfy a two-level principle? In other words, does the assumption that it respects the special homomorphisms between moduli spaces in the first two levels imply that it automatically respects higher dimensional special morphisms (either in genus zero or in general)?

This is a highly mysterious property. Indeed the automorphisms of the curves in the first two levels involve essentially the platonic primes 2, 3 and 5, and so does the definition of $\widehat{GT}$ itself, whereas any finite group can be realised as the automorphism group of a suitable smooth hyperbolic curve. Is this reflected in the first two levels? In the same vein, one might ask:

2.3. Can the groups $\widehat{GTK}$ or $\widehat{GTA}$ be defined by a finite number of relations? Do they also satisfy some two-level principle?

§3. Moduli spaces of curves, mapping class groups and GT

We include in this section some questions concerning important algebraic and geometric aspects of the moduli spaces of curves which, even though not directly connected with the Grothendieck-Teichmüller group, are nonetheless closely related to the general themes of Grothendieck-Teichmüller theory. We also remark that the connection between that theory and anabelian geometry is far from clear at the moment.

3.1. A first very ambitious question is: Are the moduli spaces *anabelian*?

This is intended in the original sense introduced by Grothendieck in his seminal letter to G.Faltings (reproduced in [GGA]). Namely one is asking whether $\mathcal{M}_{g,n}$, viewed as a $\mathbb{Q}$-stack is the only $K(\pi,1)$ space (up to isomorphism) with augmented arithmetic fundamental group isomorphic to $\pi_1(\mathcal{M}_{g,n}) \to G_\mathbb{Q}$. We refer to the contribution of F.Pop in [GGA] ([Pop]) for a detailed and categorical formulation for general schemes. If indeed one prefers to work with schemes, the same question can be asked about any finite Galois étale covering of $\mathcal{M}_{g,n}$ which is a scheme.

3.2. Anabelian varieties (or schemes or stacks) should be rigid (see [N1], [Sx] for detailed discussions). This leads to the following test for anabelianity ([IN]): is it true that

$$\mathrm{Out}_{G_\mathbb{Q}}(\widehat{\Gamma}_{g,n}) = \mathrm{Aut}(\mathcal{M}_{g,n,\overline{\mathbb{Q}}})(= \mathcal{S}_n)?$$

($\mathcal{S}_n$ is the permutation group on n objects.) This statement has been proved for $g = 0$ ([N2] and references therein) and indeed in greater generality, replacing $\mathbb{Q}$ with any field finitely generated over $\mathbb{Q}$ (see also [IN], §4). This is actually one of the only known results in higher dimensional anabelian geometry.

We refer to [IN] for the formulation of other similar tests for abelianity; they are all equivalent *if and when* $\widehat{\Gamma}_{g,n}$ is centerfree. Note also that giving the outer action of $G_{\mathbb{Q}}$ on $\widehat{\Gamma}_{g,n}$ is equivalent to giving the augmented arithmetic fundamental group as in **3.1** above, if and only if again $\widehat{\Gamma}_{g,n}$ is centerfree.

3.3. There also arises the question of comparing the group $\text{Out}^*(\widehat{\Gamma}_{g,n})$ of outer automorphisms preserving conjugacy classes of Dehn twists and the subgroup $\text{Out}_{\mathcal{S}_n}(\widehat{\Gamma}_{g,n})$ of all outer automorphisms commuting with the permutation group. Are they by any chance equal? A weaker version of this question, restricted to the $G_{\mathbb{Q}}$-equivariant exterior automorphisms, is: Do $\text{Out}_{G_{\mathbb{Q}}}(\widehat{\Gamma}_{g,n})$ and $\text{Out}^*_{G_{\mathbb{Q}}}(\widehat{\Gamma}_{g,n})$ coincide?

So another ambitious and perhaps optimistic question is: does *any* (outer) automorphism of $\widehat{\Gamma}_{g,n}$ preserve inertia at infinity (conjugacy classes of Dehn twists)? This is important as the first group is more amenable to study than the second a priori larger group. This leads to asking whether one can give a group theoretic characterization of Dehn twists inside the profinite completion $\widehat{\Gamma}_{g,n}$ analogous to the one obtained by N.Ivanov in the discrete case (see [Iv] §7.5 and references therein). Note that this is strongly reminiscent of the so-called 'local correspondence' in birational anabelian geometry ([Sz]).

3.4. Is $\Gamma_{g,n}$ a good group in the sense of Serre?

Recall ([Se] §2.6) that a discrete and residually finite group G is good if the injection into its profinite completion induces an isomorphism in cohomology with finite (equivalently, torsion) coefficients. The question for $\Gamma_{g,n}$ is classical (see the contribution of T.Oda in [GGA]) and the answer is affirmative with an easy proof for $g \leq 2$. It is also easy to show that $\Gamma_{g,n}$ is good if $\Gamma_g = \Gamma_{g,0}$ is. So the problem actually arises only for Γ_g with $g \geq 3$. In [B], goodness is announced for H^k, $k \leq 4$, with a very interesting application of the first non-trivial case, namely $k = 2$ (cf. [Sx]).

The connection between goodness and anabelianity stems from the fact that good groups have many open subgroups, in the sense that any cohomology class can be made to vanish by restriction to a suitable open subgroup. Geometrically speaking, a $K(\pi, 1)$ scheme (stack) whose geometric fundamental group is good has many étale covers which should give rise to an interesting Galois action. So the idea is that a $K(\pi, 1)$ quasiprojective scheme (stack) whose geometric fundamental group is universally centerfree and good is a "good" candidate for being anabelian. A prominent "anti-example" is the moduli stack $\mathcal{A}_g$ of principally polarized abelian varieties, whose fundamental group is *not* good because of the congruence property for the symplectic group $Sp_{2g}(\mathbb{Z})$ ($g > 1$)). This stack has few étale covers in the sense that for example they are all defined over $\mathbb{Q}^{ab}$ (cf. [IN,§3]).

3.5. How does the Grothendieck-Teichmüller action (of the group $\mathbb{T}$, for example) on $\widehat{\Gamma}_{g,[n]}$ behave with respect to the finite-order elements?

Here we mean particularly those elements which come from the discrete group $\Gamma_{g,[n]}$, in other words are realisable as automorphisms of algebraic curves. It is not known whether all finite-order elements of the profinite completion come from the discrete group; one can ask:

3.6. Is every finite order element in $\widehat{\Gamma}_{g,[n]}$ conjugate to one in $\Gamma_{g,[n]}$?

This is essentially the torsion counterpart of the question about Dehn twists raised in **3.3**. Proving that $\Gamma_{g,n}$ is good would be a big step towards answering this question (see [LS2]).

Return to the $G_{\mathbb{Q}}$ and $\widehat{GT}$ actions on $\widehat{\Gamma}_{g,[n]}$; in a more detailed fashion, **3.5** actually asks whether the action is "cyclotomic" on the torsion elements (arising from the discrete group), in the sense that we can assert that for $F = (\lambda, f) \in \widehat{GT}$ and a finite-order element $\gamma \in \widehat{\Gamma}_{g,[n]}$, we have $F(\gamma) \sim \gamma^\lambda$; here $\sim$ denotes conjugation in $\widehat{\Gamma}_{g,[n]}$. The answer to this question is known to be affirmative only for $g = 0$ and for a few low dimensional spaces such as $\mathcal{M}_{1,1}$, $\mathcal{M}_{1,2}$, $\mathcal{M}_2$ (cf. [S4]). In the other cases, it is conjectured but not known, even for $G_{\mathbb{Q}}$.

3.7. It is shown in [HS] that for each $n \geq 5$, we have:

$$\widehat{GT} \simeq \mathrm{Out}^*_{\mathcal{S}_n}(\widehat{\Gamma}_{0,n}).$$

The geometric significance of $\mathcal{S}_n$ here is as in **3.2** for $g = 0$. Does the analogous all-genera isomorphism:

$$\mathbb{\Gamma} \simeq \mathrm{Out}^*_{\mathrm{Aut}(\mathcal{M}_{g,n})}(\widehat{\Gamma}_{g,n})$$

also hold? If not, is the right-hand group at least defined by a finite number of relations, possibly coming only from dimensions 1 and 2?

§4. Dessins d'enfants

The theory of dessins d'enfants and their various descriptions, topological, algebraic, combinatorial and others has been described in several articles (e.g. [S1]). Here, we use the definition of a dessin d'enfant as being equivalent to a *Belyi cover*, i.e. a finite cover:

$$\beta : X \to \mathbb{P}^1$$

of Riemann surfaces unramified outside 0, 1 and ∞. Belyi's famous theorem states that an algebraic curve over $\mathbb{C}$ is defined over $\overline{\mathbb{Q}}$ if and only if it can be realized as a Belyi cover (the 'only if' direction being the really new one). There is a natural Galois action on the set of dessins, as these are defined over $\overline{\mathbb{Q}}$. Since $\widehat{GT}$ by definition acts on $\pi_1^{geom}(\mathbb{P}^1 \setminus \{0, 1, \infty\}) \simeq \widehat{F}_2$, it also acts on the set of dessins, which are in one-to-one correspondence with conjugacy classes of finite index subgroups of $\widehat{F}_2$, and this $\widehat{GT}$-action extends the $G_{\mathbb{Q}}$-action. The first question is the basic and original one about dessins d'enfants:

4.1. Can one give a complete list of Galois invariants of dessins, i.e. enough combinatorial invariants of the Galois action on dessins to determine the Galois orbits?

The standard Galois invariants are such things as valencies (i.e. ramification indices over $0, 1, \infty$), order of the Galois group of the Galois closure of the Belyi cover, the Galois group itself, in fact, various extensions of this group, etc. None of these invariants seems to be enough to actually distinguish Galois orbits of dessins, although putting them together astutely yields more than using them singly (cf. [W]). By a combinatorial invariant, we mean one which is computable combinatorially from the two permutations defining the dessins (see [S1]). Very few have been found to date (see however [Z]) and one is not even sure whether it is possible to distinguish Galois orbits via combinatorial invariants only.

A slightly weaker question would be to give a combinatorial method for determining the number field of moduli of the dessin (or even its degree). On this subject, R. Parker expressed a remarkable conjecture. To phrase this conjecture, we consider a dessin to be given by the equivalent data of a finite group G on two generators, say of order n, together with an explicit choice a and b for the two generators. The dessin can easily be reconstructed from this by injecting the group into S_n via its action on itself by right multiplication, which gives a and b as permutations whose cycle lengths describe the valency lists and whose cycles themselves give the cyclic order of edges around these valencies.

4.2. *Richard Parker's suggestion:* Let G be a finite group generated by two elements a and b, and consider the element P in the group ring $\mathbb{Q}[G \times G]$ given by $P = \sum_{g \in G}(gag^{-1}, gbg^{-1})$. Choosing the basis of pairs (g, h) for the vector space $\mathbb{Q}[G \times G]$, right multiplication by P gives an automorphism of the vector space which can be written as a matrix M_P. Could the Galois closure of the field of moduli K of the dessin associated to G, a, b be generated over $\mathbb{Q}$ by the eigenvalues of M_P?

This can be proven without difficulty for dessins with abelian or dihedral Galois groups. It would be a good exercise to complete the genus zero case, by dealing with the remaining cases, namely the automorphism groups of the five Platonic solids. This is easy to do for the smaller ones, and confirms the conjecture. The larger ones should be easy too, except that the computations soon become gigantic...

Another natural question is whether the $\widehat{GT}$-action on dessins transmutes into an action on curves, as follows.

4.3. Let X be an algebraic curve, and suppose we have two dessins on X, i.e. two different Belyi functions $\beta_1 : X \to \mathbb{P}^1$ and $\beta_2 : X \to \mathbb{P}^1$. Then the images of these two dessins under an element $\sigma \in G_{\mathbb{Q}}$ are both dessins on the curve $\sigma(X)$. As above there is also a $\widehat{GT}$-action on dessins. If $F = (\lambda, f) \in \widehat{GT}$, do the two new dessins $F(\beta_1)$ and $F(\beta_2)$ also lie on the same Riemann surface?

Let us define an action of $\widehat{GT}$ on $\overline{\mathbb{Q}}$ in the following, rather artificial manner. First choose a fundamental domain D for the natural action of $\mathcal{S}_3$ on $\mathbb{P}^1\mathbb{C} \setminus \{0, 1, \infty\}$ and represent any elliptic curve E by its Legendre form: $y^2 = x(x-1)(x-\lambda)$ with $\lambda \in D$. Then take the dessin β on E given by the Belyi polynomial produced by starting from the function x on E and applying Belyi's original algorithm. For each $F \in \widehat{GT}$ and each $j \in \overline{\mathbb{Q}}$, define $F(j)$ to be the j-invariant of the elliptic curve underlying the dessin $F(\beta)$.

Several questions arise from this construction. Apart from the obvious "Is this an automorphism of $\overline{\mathbb{Q}}$?" and the questions of whether the value of $F(j)$ depends on the choice of λ, the choice of dessin on E, and whether there is not some more canonical way of defining this action, here are some that appear more approachable.

4.4. Does this definition of the action of $F \in \widehat{GT}$ on $\overline{\mathbb{Q}}$ fix $\mathbb{Q}$?

The answer is certainly yes and probably not hard to show, but needs to be written down.

4.5. What can one compute in the case where λ is of low degree, for instance a square root of a rational number?

4.6. Can one show that the action is at least additive?

Dessins became very popular as topological, indeed combinatorial objects on which the profinite group $G_\mathbb{Q}$ acts faithfully. But it is possible to consider more general such objects; in particular, instead of looking at covers of $\mathbb{P}^1 - \{0, 1, \infty\}$, the next natural step would be to consider covers of $\mathbb{P}^1 - 4$ points, or of elliptic curves, so that the variation in the complex structure of the base enriches the arithmetic and geometric structures of the covers.

Using elliptic curves as a base yields the "origamis" (square tiled surfaces) studied in [L1]; they naturally sit inside the moduli spaces of curves, and thus are a priori directly connected with GT. We refer to [L1], [Mö], [Sc] as well as papers by G. Schmithüsen and F. Herrlich for this material.

One can ask the same questions about origamis that one asks about dessins: determine invariants, Galois orbits etc. Let us rather mention two specific problems which seem to be within reach. Since origamis are really higher dimensional versions of dessins, as they are topological surfaces and degenerate to dessins when approaching the boundary of the appropriate moduli space, the first question or problem is naturally:

4.7. Study, geometrically and arithmetically, the degeneration of origamis into dessins.

The second problem is related and even more specific. Very few non-trivial combinatorial Galois invariants have been constructed for dessins; one of these is Zapponi's invariant constructed in [Z]. For differentials there is also one and only one invariant, which is indeed a Galois invariant, namely the parity of the spin structure (see [KZ]). Both invariants are signs (they are $\mathbb{Z}/2\mathbb{Z}$-valued). The question is:

4.8. Express the parity of the spin structure for origamis in a combinatorial way. Using **4.7**, can one relate this invariant with Zapponi's invariant?

§5. Number theory and $\widehat{GT}$: A direct approach

The previous section already shows how the study of $\widehat{GT}$ can lead naturally to number theoretic questions, although number theory is not immediately visible in the definition of $\widehat{GT}$, even viewed as an automorphism group of fundamental groups of moduli spaces. The following Galois-theoretic problem was formulated by Y. Ihara. It is a basic question about the Galois action in the pro-ℓ setting, and indeed represents the analog of Belyi's result in that context.

5.1. Let M^* denote the fixed field of the kernel of the homomorphism:

$$G_\mathbb{Q} \to \text{Out}^*(F_2^{(\ell)}).$$

Does M^* coincide with $M^{(\ell)}$, the maximal pro-ℓ extension of $\mathbb{Q}$ unramified outside ℓ?

The point is that Grothendieck's theory of the specialization of the fundamental group implies that $M^* \subset M^{(\ell)}$ (see [I1]); note that M^* is just the field of definition of the proscheme defined by the tower of ℓ-covers of $\mathbb{P}^1 \setminus \{0,1,\infty\}$.

One of the most frequently asked questions in the early days of $\widehat{GT}$ was the following: How can one see the primes, or the decomposition groups, or the Frobenius elements in $\widehat{GT}$? In other words, can one define local versions of $\widehat{GT}$, i.e. subgroups which would correspond in a natural way to the p-adic decomposition subgroups (defined up to conjugacy) $G_{\mathbb{Q}_p} = \mathrm{Gal}(\overline{\mathbb{Q}}_p/\mathbb{Q}_p)$ in $G_{\mathbb{Q}}$? We will denote such a subgroup GT_p (not to be confused with $GT_{(\ell)}$ in §§0,1). For the moment, we have two ways of defining such a GT_p at our disposal.

The first definition uses the action of $\widehat{GT}$ on $\overline{\mathbb{Q}}$ defined before **4.4** and so is rather *ad hoc*. Elements σ in a subgroup $G_{\mathbb{Q}_p}$ are characterized inside $G_{\mathbb{Q}}$ by the following property:

$$\sigma \in G_{\mathbb{Q}_p} \text{ if and only if } |j|_p \leq 1 \Rightarrow |\sigma(j)|_p \leq 1 \text{ for all } j \in \overline{\mathbb{Q}}.$$

So by analogy, we take GT_p to be the subgroup of $\widehat{GT}$ satisfying the same property.

For the second definition, we recall that Y. André defined (in [A1]) the temperate fundamental group $\pi_1^{temp}(X)$ of a p-adic manifold X, which injects naturally into the algebraic profinite fundamental group $\pi_1^{alg}(X)$. Considering $X = \mathbb{P}^1 \setminus \{0,1,\infty\}$ over $\mathbb{Q}_p$, we can define (as in [A1]) GT_p to be the subgroup of $\widehat{GT}$ which preserves the subgroup $\pi_1^{temp}(\mathbb{P}^1 \setminus \{0,1,\infty\})$. André has shown that considering $G_{\mathbb{Q}}$ as a subgroup of $\widehat{GT}$, one has:

$$GT_p \cap G_{\mathbb{Q}} = \mathrm{Gal}(\overline{\mathbb{Q}}_p/\mathbb{Q}_p).$$

5.2. Do the two definitions above coincide?

5.3. Can one describe inertia subgroups of the subgroups GT_p corresponding to those of $G_{\mathbb{Q}_p}$? Can one characterize Frobenius elements? Note that it was already observed in [I1] that even in the Galois setting, it could be hard to recognize the Frobenius elements from their geometric action.

§6. GT and Mixed Tate Motives

If we restrict to the unipotent setting, the objects considered above are motivic, and moreover of mixed Tate type. In this relatively restricted context, the paradise of motives is a reality; in particular there exists a Tannakian category $MT(\mathbb{Z})$ of Mixed Tate motives over $\mathbb{Z}$. It is equipped with a canonical (Beilinson-de Rham) fiber functor and the associated fundamental group G^{mot} is an extension of the multiplicative group $\mathbb{G}_m$ by its prounipotent radical U^{mot}. The latter is graded by the $\mathbb{G}_m$ action, and its graded Lie algebra L^{mot} is free on one generator in each odd degree $2k+1$, $k \geq 1$. This landscape is described in particular in the work of A.B.Goncharov ([Go1,2] as well as many other papers by the same author), in [DG] and in [A2]. The motivic point of view suggests many more or less "standard" conjectures, of which we will state only a few in the last three sections, without entering into all the necessary technicalities (even in the statements). We will also give some slightly less standard, more directly GT-oriented problems.

One can define the motivic fundamental group $\pi = \pi_1^{mot}(\mathbb{P}^1 \setminus \{0,1,\infty\})$ of $\mathbb{P}^1 \setminus \{0,1,\infty\}$ and its coordinate ring is an ind-object in $MT(\mathbb{Z})$ (see [DG] for details). Then one can consider the Tannakian subcategory $\langle \pi \rangle \subset MT(\mathbb{Z})$ generated by π, i.e. containing all the motives obtained from π by taking tensor products of π and its dual, direct sums and subquotients. A very strong conjecture asks:

6.1. Is the inclusion $\langle \pi \rangle \hookrightarrow MT(\mathbb{Z})$ an isomorphism? If G_π denotes the fundamental group of the Tannakian category $\langle \pi \rangle$, then there is a natural epimorphism $G^{\mathrm{mot}} \to G_\pi$, and this question is equivalent to asking whether it is an isomorphism.

In order to get closer to GT, it is natural to introduce the genus 0 moduli spaces $\mathcal{M}_{0,n}$ into the motivic landscape, together with their stable compactifications $\overline{\mathcal{M}}_{0,n}$. The motivic cohomology of these spaces belongs to $MT(\mathbb{Z})$, whereas that of higher genera moduli spaces does not. The motivic fundamental groups $\pi_1^{\mathrm{mot}}(\mathcal{M}_{0,n})$ are constructed in [DG] and also belong to $MT(\mathbb{Z})$. One can then propose the following task:

6.2. Consider the Tannakian subcategory of $MT(\mathbb{Z})$ generated by the fundamental groups of the $\mathcal{M}_{0,n}$ ($n \geq 4$); how does its fundamental group compare with G_π and $\underline{GT}$?

There exists in fact a monomorphism $G_\pi \hookrightarrow \underline{GT}$ (see e.g. [A2]) and the point is:

6.3. Is the natural monomorphism $G_\pi \hookrightarrow \underline{GT}$ an isomorphism?

Putting **6.1** and **6.3** together underlines the fact that one can reasonably ask whether the three groups G^{mot}, G_π and $\underline{GT}$ coincide. In the next two sections, we present weaker and more specific versions of this type of expectation. We believe that having the higher dimensional genus 0 moduli spaces come into play should help to approach them, in conformity with the original spirit of Grothendieck-Teichmüller theory.

§7. The Hodge side: GT and multiple zeta values

Multiple Zeta Values (hereafter simply multizeta values) are real numbers defined using either infinite series or integrals, the latter representation being much more recent. They satisfy two very different looking families of relations, namely the quadratic or (regularized) double shuffle relations and the associator (or modular, or GT) relations. We write Z_w for the $\mathbb{Q}$-vector space spanned by the multizeta values of weight $w \geq 0$ (where we formally set $\zeta(0) = 1$ and $\zeta(1) = 0$), and $Z_\bullet$ for the $\mathbb{Q}$-algebra of the multizeta values, filtered by the weight (see any paper on the subject, including [Go1,2], [F1,F2], [A2] etc.). Thus, we have $Z_0 = \langle 1 \rangle$ and $Z_1 = 0$, $Z_2 = \langle \zeta(2) = \pi^2/6 \rangle$, $Z_3 = \langle \zeta(3) \rangle$, and then the dimensions grow quickly as multizeta values appear (see **7.3** below).

In order to capture some of the main algebraic and combinatorial properties of these real numbers without confronting intractable transcendence problems, three other filtered $\mathbb{Q}$-algebras have been introduced: $Z_\bullet^{DS}$, $Z_\bullet^{GT}$ and $Z_\bullet^{\mathrm{mot}}$. The first (resp. second) of these consists in taking formal multizeta symbols which satisfy only the double shuffle (resp. associator) relations, which are known to be satisfied by the genuine multizeta values and conjectured to form a complete set of algebraic relations between them. These two algebras are graded by the weight. Note that in papers written in French or by French speaking authors (e.g. [A2], [E], [R]), 'DS' reads 'DM' or 'DMR' ('Double mélange régularisé'). The algebra of motivic multizeta values $Z_\bullet^{\mathrm{mot}}$ constructed by Goncharov is more complicated to define: Goncharov has proved in [Go2] that it is a quotient of $Z_\bullet^{DS}$, but perhaps not strictly, and no further relations are explicitly known. It is known, however, that this algebra is also graded by the weight. By contrast we have:

7.1. Conjecture: The weight induces a grading on $Z_\bullet$.

In other words, it is not even known whether there are any linear relations between real multizeta values of different weights. This conjecture immediately implies the transcendence of all multizeta values (since a minimal polynomial would yield such a linear relation), in particular of the values $\zeta(2n+1)$ at odd positive integers of Riemann's zeta function, so it is expected to be extremely difficult.

On the subject of $Z_\bullet^{\mathrm{mot}}$, Goncharov has shown that it is naturally realizable as a subalgebra of the universal enveloping algebra $\mathcal{U}(L^{\mathrm{mot}})$ of the Lie algebra L^{mot} of the unipotent part U^{mot} of the motivic Galois group G^{mot} of $MT(\mathbb{Z})$ (cf. §6 above). Goncharov conjectures that in fact

$$Z_\bullet^{\mathrm{mot}} \xrightarrow{\sim} \mathcal{U}(L^{\mathrm{mot}})^\vee.$$

This leads to the following well-known conjecture:

7.2. The algebras $Z_\bullet/\pi^2 Z_\bullet$, $Z_\bullet^{DS}$, $Z_\bullet^{GT}$, $Z_\bullet^{\mathrm{mot}}$ and $\mathcal{U}(L^{\mathrm{mot}})^\vee$ are canonically isomorphic.

Equivalently one conjectures the isomorphism of the five algebras $Z_\bullet$, $\mathbb{Q}[\pi^2] \otimes_\mathbb{Q} Z_\bullet^{\mathrm{mot}}$, etc. Moreover, it is easily seen ([Go1]) that:

$$\mathbb{Q}[\pi^2] \otimes_\mathbb{Q} \mathcal{U}(L^{\mathrm{mot}})^\vee \simeq \mathcal{U}(L[s_2, s_3])^\vee,$$

where $L[s_2, s_3]$ is the free Lie algebra on two generators in weights 2 and 3; the dimensions of the graded parts of this algebra are given by the coefficients of the generating series $1/(1 - t^2 - t^3)$, which leads to the following dimension conjecture for all five algebras:

7.3. Dimension conjecture (D. Zagier): Letting d_w denote the dimension of Z_w (resp. DS, GT, mot) as a $\mathbb{Q}$-vector space, one has $d_w = d_{w-2} + d_{w-3}$ (with $d_0 = d_2 = 1$, $d_1 = 0$).

The upper bound for $Z_\bullet$, i.e. the fact that the actual dimension of Z_w is less than or equal to the conjectured one, was proved independently by A. Goncharov and T. Terasoma (see [A2] for references and a sketch of proof following Goncharov). The statement involves no transcendence, and the proof is motivic.

Leaving aside $Z_\bullet$ itself, one can explore the possible isomorphisms between the other algebras. For instance:

7.4. Can one find an explicit isomorphism between $Z_\bullet^{DS}$ and $Z_\bullet^{GT}$, i.e. an explicit way to obtain double shuffle relations from associator relations and vice versa?

This is in principle an algebraic or even combinatorial problem, which has proved difficult and enticing.

The original multizeta values $Z_\bullet$ can be seen as periods of motives of the category $\langle \pi \rangle \subset MT(\mathbb{Z})$ (cf. §6), where Goncharov has shown how to attach a 'framing' to a mixed Tate motive, which yields a complex-valued period (modulo periods of motives of lower weights). A weaker version of **6.1** is thus:

7.5. Do the multizeta values give all the periods of (framed) mixed Tate motives over $\mathbb{Z}$?

Recall that the genus 0 moduli spaces $\mathcal{M}_{0,n}$ and their stable completions $\overline{\mathcal{M}}_{0,n}$ provide important objects of $MT(\mathbb{Z})$. In fact, Goncharov and Manin ([GoM]) have shown that there are canonical framed mixed Tate motives associated to these spaces, whose periods are the multizeta values. This also provides an alternative but equivalent definition of $Z_\bullet^{\mathrm{mot}}$.

The above question can be rephrased purely analytically. There is a map:
$$p : \overline{\mathcal{M}}_{0,n+3} \to (\mathbb{P}^1)^n,$$
obtained by successive blowups (so it is a birational isomorphism). Consider the standard real n-simplex $\Delta = \{0 < t_1 < t_2 < \ldots < t_n < 1\} \subset (\mathbb{P}^1)^n$; the topological closure of the preimage $p^{-1}(\Delta)$ is the standard associahedron $K \subset \overline{\mathcal{M}}_{0,n+3}$. Relative periods are of the form $\int_K \omega$, where ω is a top dimensional logarithmic form; these can be explicitly determined. The multizeta values correspond to very particular such ω's, those having only factors of t_i or $1 - t_i$ in the denominators. The analytic form of **7.5** now reads:

7.6. Are all such integrals given by $\mathbb{Q}$-linear combinations of multizeta values?

Of course not all such integrals converge, and one can dream up several different versions of the above, the simplest of which is to make the statement only for the convergent integrals (there is a nice geometric criterion for convergence).

The multizeta values satisfy the double shuffle relations and the associator relations. The associator relations in particular come directly from the geometry of the $\mathcal{M}_{0,n}$. Thus, from the perspective of Grothendieck-Teichmüller theory, it is very natural to ask:

7.7. Can one give generalized double shuffle and/or associator relations valid for all the relative periods of the $\overline{\mathcal{M}}_{0,n}$ and coming from the geometry of these spaces?

In other words the task consists in exploring the combinatorics of the relative periods of these spaces. This could be useful for attacking, but is logically independent of question **7.6**.

Similar or equivalent questions to the above are posed by several authors in the proalgebraic context. The algebras $Z_\bullet^{DS}$, $Z_\bullet^{GT}$ and $Z_\bullet^{\mathrm{mot}}$ are all commutative Hopf algebras which are universal enveloping algebras of Lie coalgebras; these results are due, in chronological order, to V. Drinfel'd and Y. Ihara for $Z_\bullet^{GT}$, A. Goncharov for $Z_\bullet^{\mathrm{mot}}$ and G. Racinet [R] for $Z_\bullet^{DS}$. Their spectra are three unipotent affine group schemes:
$$\underline{G}^{DS}, \quad \underline{GT}^1 \quad \text{and} \quad \underline{G}^{MZ}.$$
Thus, the isomorphism questions in **7.2**, apart from the original $Z_\bullet$, can be rephrased as:

7.8. Are the affine unipotent group schemes $\underline{G}^{DS}$, $\underline{GT}^1$, $\underline{G}^{MZ}$ and U^{mot} all isomorphic?

The two points of view explained here concern duals of universal enveloping algebras of Lie algebras and their spectra, affine unipotent group schemes. Let us now rephrase some of these ideas from the point of view of the Lie algebras/coalgebras themselves. This can be quite enlightening and leads to new results and connections. For instance, Goncharov has computed the coproduct on $Z_\bullet^{\mathrm{mot}}$ explicitly, and deduced the expression of the Lie cobracket on the vector space $Z_{>2}^{\mathrm{mot}}/(Z_{>0}^{\mathrm{mot}})^2$, making it into a Lie coalgebra. Generalizing this to a question about $Z_\bullet$ itself yields an equivalent but more striking and precise version of **7.5**:

7.9. Let $\mathfrak{nz}$ (for 'new zeta') be the $\mathbb{Q}$ vector space obtained by quotienting $Z_\bullet$ by the ideal generated by Z_0, Z_2 and $(Z_{>0})^2$ (or equivalently, quotienting $Z_{>2}$ by $(Z_{>0})^2$). Is there a surjection (isomorphism?) $(L^{\mathrm{mot}})^\vee \to \mathfrak{nz}$, thus defining a structure on $\mathfrak{nz}$ of a Lie coalgebra, whose dual would thus be (freely?) generated by one element in each odd rank?

The algebra Z^{GT} is related to the Lie algebra $\mathfrak{grt}$ (cf. §0) by the fact that Z^{GT} is the dual of the universal enveloping algebra of $\mathfrak{grt}$; equivalently, $Z_{>2}^{GT}/(Z_{>0}^{GT})^2$ is a Lie coalgebra dual to $\mathfrak{grt}$. Similarly, the double shuffle Lie algebra $\mathfrak{ds}$ is given as a vector space by $\left(Z_{>2}^{DS}/(Z_{>0}^{DS})^2\right)^\vee$; it is however also quite simple to define directly, cf. [R]. What Racinet actually proved is that this vector space is closed under the Poisson (alias Ihara) bracket, from which one deduces that the enveloping algebra and its dual are Hopf algebras, so that the spectrum $\underline{G}^{DS}$ of the dual is an affine group scheme.

In analogy with **7.9**, we have the following conjectures, the first of which was made by Ihara much before the second:

7.10. Conjecture: There are surjections, or better isomorphisms, from the free Lie algebra L^{mot} on generators $s_3, s_5, s_7, \ldots$ to the Grothendieck-Teichmüller and double shuffle Lie algebras:

$$L^{\mathrm{mot}} \twoheadrightarrow \mathfrak{grt} \quad \text{and} \quad L^{\mathrm{mot}} \twoheadrightarrow \mathfrak{ds}.$$

In other words, $\mathfrak{grt}$ and $\mathfrak{ds}$ would themselves be (freely) generated by one generator in each odd rank ≥ 3. Note, however, that the obvious depth filtration (by the descending central series) on L^{mot} would not map to the natural depth filtration on elements of $\mathfrak{grt}$ and $\mathfrak{ds}$ under such an isomorphism. There should be a more subtle depth filtration on L^{mot}.

Of course, **7.4** can now be rephrased in the Lie algebra context as:

7.11. Are $\mathfrak{grt}$ and $\mathfrak{ds}$ isomorphic?

Computations have confirmed that are they isomorphic in low ranks; the candidate isomorphism would simply be given by $f(x,y) \mapsto f(x,-y)$ (cf. [R]).

Most of the objects defined above are associated with $\mathbb{P}^1 \setminus \{0,1,\infty\}$. A natural generalization, much studied by A. Goncharov, is to consider $\mathbb{P}^1 \setminus \{0,\mu_N,\infty\}$ for any positive integer N. Then the Lie algebras $\mathfrak{grt}$ and $\mathfrak{ds}$ (alias $\mathfrak{dmr}$), along with the other attending objects can be generalized, as was done in [E] and [R] respectively, in which the authors define analogs for any N, denoted $\mathfrak{grtmd}(N)$ and $\mathfrak{dmrd}(N)$ respectively. These generalizations prompt one to ask the following questions (suggested by B. Enriquez):

7.12. Do the elements exhibited in [E] generate $\mathfrak{grtmd}(N)$ and are there relations?

About the first question, recall from above that for $N = 1$ it is not known whether $\mathfrak{grt}$ is generated by one generator in each odd degree ≥ 3 (cf. **7.10**). About the second question and contrary to the case $N = 1$, freeness is not expected.

Finally it is natural to try and compare $\mathfrak{grtm}\mathfrak{d}(N)$ and $\mathfrak{dmr}\mathfrak{d}(N)$, just as we compare $\mathfrak{grt}$ and $\mathfrak{ds}$, say in the following relatively weak form:

7.13. Does the inclusion $\mathfrak{grtm}\mathfrak{d}(N) \subset \mathfrak{dmr}\mathfrak{d}(N)$ hold true (at least for an odd prime N)?

§8. The Galois side; GT and $G_\mathbb{Q}$ once again

Recall that there is a canonical injection $G_\mathbb{Q} \hookrightarrow \widehat{GT}$. It is conceivable that the profinite group $\widehat{GT}$ (or its refinements) may be different from $G_\mathbb{Q}$, but some simpler quotient may be equal to the corresponding quotient of $G_\mathbb{Q}$. In particular, let GT^{nil} be the nilpotent quotient of $\widehat{GT}$, i.e. the inverse limit over the finite nilpotent quotients of $\widehat{GT}$.

8.1. Is GT^{nil} isomorphic to the nilpotent completion of $G_\mathbb{Q}$?

Equivalently, one can ask whether the maximal pro-ℓ quotient $GT^{(\ell)}$ of $\widehat{GT}$ is isomorphic to the maximal pro-ℓ quotient of $G_\mathbb{Q}$ for each prime ℓ?

For any prime ℓ, Deligne and Ihara constructed (independently) a graded $\mathbb{Q}_\ell$-Lie algebra from the action of $G_\mathbb{Q}$ on $\pi^{(\ell)} \simeq F_2^{(\ell)}$, the pro-$\ell$ fundamental group of $\mathbb{P}^1 \setminus \{0, 1, \infty\}$. Let $\pi^{(\ell)}[m]$ denote the descending central series of $\pi^{(\ell)}$: $\pi_1^{(\ell)}[0] = \pi^{(\ell)}$, $\pi^{(\ell)}[m+1] = [\pi^{(\ell)}, \pi^{(\ell)}[m]]$, i.e. the subgroup topologically generated by the commutators. There is a filtration of $G_\mathbb{Q}$ defined by:

$$I_\ell^k G_\mathbb{Q} = \text{Ker}\big(G_\mathbb{Q} \to \text{Out}(\pi^{(\ell)}/(\pi^{(\ell)})[k]\big).$$

It is easy to see that $I_\ell^0 G_\mathbb{Q} = G_\mathbb{Q}$ and $I_\ell^1 G_\mathbb{Q} = G_{\mathbb{Q}^{\text{ab}}}$; it is not too hard to see that also $I_\ell^2 G_\mathbb{Q} = G_{\mathbb{Q}^{\text{ab}}}$ and in fact $I_\ell^3 G_\mathbb{Q} = G_{\mathbb{Q}^{\text{ab}}}$. Set:

$$DI^{(\ell)} = \big(\text{Gr}^\bullet_{I_\ell} G_\mathbb{Q}\big) \otimes \mathbb{Q}_\ell = \bigoplus_{k \geq 0} (I_\ell^k G_\mathbb{Q}/I_\ell^{k+1} G_\mathbb{Q}) \otimes \mathbb{Q}_\ell.$$

This graded vector space is naturally equipped with a Lie bracket coming from the commutator map $(\sigma, \tau) \mapsto \sigma\tau\sigma^{-1}\tau^{-1}$ on the group $G_\mathbb{Q}$.

8.2. Conjecture (Y. Ihara). The $DI^{(\ell)}$ have a common $\mathbb{Q}$-structure, i.e. there exists a $\mathbb{Q}$-Lie algebra DI such that $DI^{(\ell)} = DI \otimes_\mathbb{Q} \mathbb{Q}_\ell$.

8.3. Furthermore, $DI \simeq \mathfrak{grt}$, i.e. $DI^{(\ell)} \simeq \mathfrak{grt} \otimes \mathbb{Q}_\ell$ for each ℓ.

These two Lie algebras can be computed explicitly in low degree. They are equal at least up to degree 13. Conjecture **8.3** is stronger than **8.2**, but we have stated **8.2** separately because it might be more accessible. However, a third very natural question which crops up here, the analog of the same question given previously concerning $\mathfrak{n}_3$ in **7.9** and $\mathfrak{grt}$ and $\mathfrak{ds}$ in **7.10**, is a theorem in the present case, thanks to a result of R.Hain and M.Matsumoto. Recall that L^{mot} is the motivic fundamental Lie algebra of $MT(\mathbb{Z})$, freely generated by one generator in each odd rank ≥ 3.

8.4. Theorem ([HM1,2]). There is a surjection $L^{\text{mot}} \otimes \mathbb{Q}_\ell \to DI^{(\ell)}$ for each prime ℓ.

Let us return briefly to the proalgebraic setting. From the ℓ-adic realization we get, for any prime ℓ, a morphism $G_\mathbb{Q} \to G^{\text{mot}}(\mathbb{Q}_\ell)$ whose image is Zariski dense ([Go1], [HM1,2]; see also [A2]). Composing with the natural surjection we find a morphism $G_\mathbb{Q} \to G_\pi(\mathbb{Q}_\ell)$, again with dense image. We may now compose with the monomorphism into $\underline{GT}$ of **6.3**, restrict to the prounipotent part of the image as this is what is really at stake, and arrive at a question already posed in [Dr].

8.5. Is the image of the map $G_{\mathbb{Q}^{ab}} \to \underline{GT}^1(\mathbb{Q}_\ell)$ Zariski dense?

Here one can in fact replace $G_{\mathbb{Q}^{ab}}$ with the Galois group of $\mathbb{Q}_\ell(\mu_{\ell^\infty})$. In [F2], Furusho also asks the proalgebraic analog of **8.2**, **8.3** (his Conjectures B and C) and proceeds to show that they are actually equivalent to their pro-ℓ versions (see his Proposition 4.3.3).

References

[A1] Y. André, On a geometric description of $\mathrm{Gal}(\overline{\mathbb{Q}}_p/\mathbb{Q}_p)$ and a p-adic avatar of $\widehat{GT}$, Duke Math. J. **119** (2003), 1-39.

[A2] Y. André, *Une introduction aux motifs*, Panoramas et Synthèses **17**, Société Mathématique de France, 2004.

[B] M. Boggi, Profinite Teichmüller theory, Profinite Teichmüller theory, Math. Nachr. **279** (2005), 1-35.

[De] P. Deligne, Le groupe fondamental de la droite projective moins trois points, in *Galois Groups over* $\mathbb{Q}$, Y. Ihara et al. eds. MSRI Publication **16**, 79-297, Springer, 1989.

[DE] *The Grothendieck theory of Dessins d'Enfants*, L.Schneps ed., London Math. Soc. Lect. Note Ser. **200**, Cambridge University Press, 1994.

[DG] P. Deligne and A.B.Goncharov, Groupes fondamentaux motiviques de Tate mixtes, Ann. Sci. École Norm. Sup. **38** (2005), 1–56.

[Dr] V.G. Drinfel'd, On quasitriangular quasi-Hopf algebras and a group closely connected with $\mathrm{Gal}(\overline{\mathbb{Q}}/\mathbb{Q})$, Leningrad Math. J. Vol. 2 (1991), No. 4, 829-860.

[E] B. Enriquez, Quasi-reflection algebras, multiple polylogarithms at roots of 1, and analogues of the group $\widehat{GT}$, preprint, 2004.

[F1] H. Furusho, The multiple zeta value algebra and the stable derivation algebra, Publ. Res. Inst. Math. Sci. **39** (2003), no. 4, 695-720.

[F2] H. Furusho, Multiple Zeta values and Grothendieck-Teichmüller groups, Preprint RIMS 1357, 2002.

[GGA] *Geometric Galois Actions*, L.Schneps and P.Lochak eds., London Math. Soc. Lect. Note Ser. **242**, Cambridge University Press, 1997.

[G1] A. Grothendieck, *Esquisse d'un Programme*, 1984, in [GGA], 5-47.

[G2] A. Grothendieck, *La longue marche à travers la théorie de Galois*, unpublished, 1981. Sections 1 to 37 are available from Jean Malgoire (malgoire@math.univ-montp2.fr).

[Go1] A.B. Goncharov, Multiple polylogarithms and mixed Tate motives, preprint, 2002.

[Go2] A.B. Goncharov, Periods and mixed motives, preprint, 2002.

[GoM] A.B. Goncharov, Yu.I.Manin, Multiple zeta motives and moduli spaces $\overline{\mathcal{M}}_{0,n}$, Compos. Math. **140** (2004), 1–14.

[HLS] A. Hatcher, P. Lochak, L. Schneps, On the Teichmüller tower of mapping class groups, J. reine und angew. Math. **521** (2000), 1-24.

[HM1] R. Hain, M. Matsumoto, Weighted completions of Galois groups and Galois actions on the fundamental group of $\mathbb{P}^1 \setminus \{0,1,\infty\}$, Compos. Math. **139** (2003), 119-167.

[HM2] R. Hain, M. Matsumoto, Tannakian fundamental groups associated to Galois groups, in *Galois groups and fundamental groups*, edited by L. Schneps, MSRI Publication **41**, 183-216, Cambridge University Press, 2003.

[HS] D. Harbater, L. Schneps, Fundamental groups of moduli and the Grothendieck-Teichmller group, Trans. Amer. Math. Soc. **352** (2000), 3117-3148.

[I1] Y. Ihara, Profinite braid groups, Galois representations and complex multiplications, Ann. of Math. **123** (1986), 43-106.

[I2] Y. Ihara, On beta and gamma functions associated with the Grothendieck-Teichmüller group, in *Aspects of Galois Theory*, London Math. Soc. Lecture Notes **256**, 144-179, Cambridge University Press, 1999.

[I3] Y. Ihara, On beta and gamma functions associated with the Grothendieck-Teichmüller group II, J. reine und angew. Math. **527** (2000), 1-11.

[I4]   Y. Ihara, Some arithmetic aspects of Galois actions in the pro-p fundamental group of $\mathbb{P}^1 \setminus \{0,1,\infty\}$, *Proc. Sympos. Pure Math.* **70**, Arithmetic fundamental groups and noncommutative algebra, Proc. Sympos. Pure Math., **70**, 247-273, AMS Publ., 2002.

[IN]   Y. Ihara, H. Nakamura, Some illustrative examples for anabelian geometry in high dimensions, in [GGA], 127-138.

[Iv]   N.V. Ivanov, Mapping class groupsr, in *Handbook of geometric topology*, 523–633, North-Holland, 2002.

[KZ]   M. Kontsevich, A. Zorich, Connected components of the moduli spaces of abelian differentials with prescribed singularities, Invent. Math. **153** (2003), 631–678.

[L1]   P. Lochak, On arithmetic curves in the moduli spaces of curves, Journal Inst. Math. de Jussieu, 2005.

[L2]   P. Lochak, Fragments of nonlinear Grothendieck-Teichmüller theory, *Woods Hole Mathematics*, 225-262, Series on Knots and Everything **34**, World Scientific, 2004.

[LS1]   P. Lochak, L. Schneps, The Grothendieck-Teichmüller group and automorphisms of braid groups, in [DE], 323-358.

[LS2]   P. Lochak, L. Schneps, A cohomological interpretation of the Grothendieck-Teichmüller group, Invent. Math. **127** (1997), 571-600.

[M]   I. Marin, Caractères de rigidité du groupe de Grothendieck-Teichmüller, preprint, 2004.

[Mö]   M. Möller, Teichmüller curves, Galois actions and $\widehat{GT}$-relations, Math. Nachr., 2005.

[McCS]   W. McCallum, R. Sharifi, A cup product in the Galois cohomology of number fields, Duke Math. J. **120** (2003), 269–310.

[MNT]   S. Mochizuki, H. Nakamura, A. Tamagawa, Grothendieck's conjectures concerning fundamental groups of algebraic curves, Sugaku **50** (1998), 113-129.

[N1]   H. Nakamura, Galois rigidity of algebraic mappings into some hyperbolic varieties, Internat. J. Math. **4** (1993), 421-438.

[N2]   H. Nakamura, Galois rigidity of profinite fundamental groups, Sugaku **47** (1995), 1-17; translated in Sugaku Expositions **10** (1997), 195-215.

[NS]   H. Nakamura, L. Schneps, On a subgroup of the Grothendieck-Teichmüller group acting on the tower of profinite Teichmüller modular groups, Invent. math. **141** (2000), 503-560.

[NiSe]   N. Nikolov, D. Segal, Finite index subgroups in profinite groups, C.R. Acad. Sci. Paris **337** (2003), 303-308.

[Pop]   F. Pop, Glimpses of Grothendieck's anabelian geometry, in [GGA], 113-126 .

[PS]   *Espaces de modules des courbes, groupes modulaires et théorie des champs*, Panoramas et Synthèses **7**, SMF Publ., 1999.

[R]   G. Racinet, Doubles mélanges des polylogarithmes multiples aux aux racines de l'unité, Publ. Math. Inst. Hautes tudes Sci. **95** (2002), 185–231.

[S1]   L. Schneps, Dessins d'enfants on the Riemann sphere, in [DE], 47–77.

[S2]   L. Schneps, The Grothendieck–Teichmller group $\widehat{GT}$: a survey, in [GGA], 183–203.

[S3]   L. Schneps, Special loci in moduli spaces of curves, in *Galois groups and fundamental groups*, edited by L.Schneps, MSRI Publication **41**, 217-275, Cambridge University Press, 2003.

[S4]   L. Schneps, Automorphisms of curves and their role in Grothendieck-Teichmüller theory, Math. Nachr., 2005.

[Sc]   G. Schmithüsen, An algorithm for finding the Veech group of an origami, Experiment. Math. **13** (2004), 459–472.

[Se]   J-P. Serre, *Cohomologie Galoisienne* (Cinquième Édition), LNM **5**, Springer Verlag, 1994.

[Sx]   J. Stix, Maps to anabelian varieties and extending curves, preprint, 2004.

[Sz]   T. Szamuely, Groupes de Galois de corps de type fini (d'après Pop), Astérisque **294** (2004), 403-431.

[T]   H. Tsunogai, Some new-type equations in the Grothendieck-Teichmüller group arising from the geometry of $\mathcal{M}_{0,5}$, preprint, 2004.

[W]   M. Wood, Belyi-extending maps and the Galois action on dessins d'enfants, preprint 2003, math.NT/0304489.

[Z]   L. Zapponi, Fleurs, arbres et cellules: un invariant galoisien pour une famille d'arbres, Compos. Math. **122** (2000), 113-133.

CNRS AND UNIVERSITÉ P. ET M.CURIE, INSTITUT DE MATHÉMATIQUES DE JUSSIEU, 175 RUE DU CHEVALERET, F-75013 PARIS
E-mail address: `lochak@math.jussieu.fr`

CNRS AND UNIVERSITÉ P. ET M.CURIE, INSTITUT DE MATHÉMATIQUES DE JUSSIEU, 175 RUE DU CHEVALERET, F-75013 PARIS
E-mail address: `leila@math.jussieu.fr`

Part III

Geometric and Dynamical Aspects

Mapping Class Group Dynamics on Surface Group Representations

William M. Goldman

ABSTRACT. Deformation spaces $\mathsf{Hom}(\pi, G)/G$ of representations of the fundamental group π of a surface Σ in a Lie group G admit natural actions of the mapping class group Mod_Σ, preserving a Poisson structure. When G is compact, the actions are ergodic. In contrast if G is noncompact semisimple, the associated deformation space contains open subsets containing the Fricke-Teichmüller space upon which Mod_Σ acts properly. Properness of the Mod_Σ-action relates to (possibly singular) locally homogeneous geometric structures on Σ. We summarize known results and state open questions about these actions.

Contents

Introduction
Acknowledgments
1. Generalities
1.1. The Symplectic Structure
1.2. The Complex Case
1.3. Singularities of the deformation space
1.4. Surfaces with boundary
1.5. Examples of relative $\mathsf{SL}(2,\mathbb{C})$-character varieties
2. Compact Groups
2.1. Ergodicity
2.2. The unitary representation
2.3. Holomorphic objects
2.4. Automorphisms of free groups
2.5. Topological dynamics
2.6. Individual elements
3. Noncompact Groups and Uniformizations
3.1. Fricke-Teichmüller space

2000 *Mathematics Subject Classification.* Primary 57M50; Secondary 58E20, 53C24.

Key words and phrases. Mapping class group, Riemann surface, fundamental group, representation variety, harmonic map, Teichmüller space, quasi-Fuchsian group, real projective structure, moduli space of vector bundles, hyperbolic manifold.

Goldman supported in part by NSF grants DMS-0103889 and DMS-0405605.

©2006 American Mathematical Society

3.2. Other components and the Euler class
3.3. The one-holed torus
3.4. Hyperbolic 3-manifolds
3.5. Convex projective structures and Hitchin representations
3.6. The energy of harmonic maps
3.7. Singular uniformizations and complex projective structures
3.8. Complex projective structures
References

Introduction

A natural object associated to a topological surface Σ is the *deformation space* of representations of its fundamental group $\pi = \pi_1(\Sigma)$ in a Lie group G. These spaces admit natural actions of the mapping class group Mod_Σ of Σ, and therefore determine linear representations of Mod_Σ.

The purpose of this paper is to survey recent results on the dynamics of these actions, and speculate on future directions in this subject.

The prototypes of this theory are two of the most basic spaces in Riemann surface theory: the Jacobian and the Fricke-Teichmüller space. The *Jacobian* $\mathsf{Jac}(M)$ of a Riemann surface M homeomorphic to Σ identifies with the deformation space $\mathsf{Hom}(\pi, G)/G$ when G is the circle $\mathsf{U}(1)$. The Jacobian parametrizes topologically trivial holomorphic complex line bundles over M, but its topological type (and symplectic structure) are invariants of the underlying topological surface Σ. The action of Mod_Σ is the action of the integral symplectic group $\mathsf{Sp}(2g, \mathbb{Z})$ on the torus $\mathbb{R}^{2g}/\mathbb{Z}^{2g}$, which is a measure-preserving chaotic (ergodic) action.

In contrast, the *Teichmüller space* $\mathfrak{T}_\Sigma$ (*Fricke space* if $\partial \Sigma \neq \emptyset$) is comprised of equivalence classes of *marked conformal structures* on Σ. A marked conformal structure is a pair (M, f) where f is a homeomorphism and M is a Riemann surface. Marked conformal structures (f_1, M_1) and (f_2, M_2) are *equivalent* if there is a biholomorphism $M_1 \xrightarrow{h} M_2$ such that $h \circ f_1$ is homotopic to f_2. Denote the equivalence class of a marked conformal structure (f, M) by

$$\langle f, M \rangle \in \mathfrak{T}_\Sigma.$$

A marking f determines a representation of the fundamental group:

$$\pi = \pi_1(\Sigma) \xrightarrow{f_*} \pi_1(M) \subset \mathsf{Aut}(\tilde{M}).$$

By the uniformization theorem (at least when $\chi(\Sigma) < 0$), these identify with *marked hyperbolic structures* on Σ, which in turn identify with conjugacy classes of *discrete embeddings* of the fundamental group π in the group $G = \mathsf{PGL}(2, \mathbb{R})$ of isometries of the hyperbolic plane. These classes form a connected component of $\mathsf{Hom}(\pi, G)/G$, which is homeomorphic to a cell of dimension $-3\chi(\Sigma)$ [**35**]. The mapping class group Mod_Σ acts properly on $\mathfrak{T}_\Sigma$. The quotient orbifold

$$\mathfrak{M}_\Sigma := \mathfrak{T}_\Sigma / \mathsf{Mod}_\Sigma$$

is the *Riemann moduli space*, consisting of biholomorphism classes of (unmarked) conformal structures on Σ.

Summarizing:
- When G is compact, $\mathsf{Hom}(\pi, G)/G$ has nontrivial homotopy type, and the action of the mapping class group exhibits nontrivial dynamics;
- When $G = \mathsf{PGL}(2, \mathbb{R})$ (or more generally a noncompact semisimple Lie group), $\mathsf{Hom}(\pi, G)/G$ contains open sets (like Teichmüller space) which are contractible and admit a proper Mod_Σ-action. Often these open sets correspond to locally homogeneous geometric structures uniformizing Σ.

Thus dynamically complicated mapping class group actions accompany nontrivial homotopy type of the deformation space. In general the dynamics exhibits properties of these two extreme cases, as will be described in this paper.

Acknowledgments. This paper is an expanded version of a lecture presented at the Special Session "Dynamics of Mapping Class Group Actions" at the Annual Meeting of the American Mathematical Society, January 6-11, 2005, in Atlanta, Georgia. I am grateful to Richard Brown for organizing this workshop, and the opportunity to lecture on this subject. I am also grateful to Benson Farb for encouraging me to write this paper, and to Jørgen Andersen, David Dumas, Lisa Jeffrey, Misha Kapovich, François Labourie, Dan Margalit, Howard Masur, Walter Neumann, Juan Souto, Pete Storm, Ser Tan, Richard Wentworth, Anna Wienhard and Eugene Xia for several suggestions and helpful comments. I wish to thank the referee for a careful reading of the paper and many useful suggestions.

1. Generalities

Let π be a finitely generated group and G a real algebraic Lie group. The set $\mathsf{Hom}(\pi, G)$ of homomorphisms $\pi \longrightarrow G$ has the natural structure of an affine algebraic set. The group

$$\mathsf{Aut}(\pi) \times \mathsf{Aut}(G)$$

acts on $\mathsf{Hom}(\pi, G)$ by left- and right- composition, preserving the algebraic structure: if $\alpha \in \mathsf{Aut}(\pi)$ and $h \in \mathsf{Aut}(G)$ are automorphisms, then the action of (α, h) on $\rho \in \mathsf{Hom}(\pi, G)$ is the composition $h \circ \rho \circ \alpha^{-1}$:

$$\pi \xrightarrow{\alpha^{-1}} \pi \xrightarrow{\rho} G \xrightarrow{h} G$$

The *deformation space* is the quotient space of $\mathsf{Hom}(\pi, G)$ (with the classical topology) by the subgroup $\mathsf{Inn}(G)$ of *inner automorphisms* of G, and is denoted $\mathsf{Hom}(\pi, G)/G$. The action of the inner automorphism ι_γ determined by an element $\gamma \in \pi$ equals $\iota_{\rho(\gamma^{-1})}(\rho)$. Therefore $\mathsf{Inn}(\pi)$ acts trivially on $\mathsf{Hom}(\pi, G)/G$ and the induced action of $\mathsf{Aut}(\pi)$ on $\mathsf{Hom}(\pi, G)/G$ factors through the quotient

$$\mathsf{Out}(\pi) := \mathsf{Aut}(\pi)/\mathsf{Inn}(\pi).$$

When Σ is a closed orientable surface with $\chi(\Sigma) < 0$, then the natural homomorphism

$$\pi_0(\mathsf{Diff}(\Sigma)) \longrightarrow \mathsf{Out}(\pi)$$

is an isomorphism. The *mapping class group* Mod_Σ is the subgroup of $\mathsf{Out}(\pi)$ corresponding to *orientation-preserving* diffeomorphisms of Σ.

When Σ has nonempty boundary with components $\partial_i \Sigma$, this deformation space admits a *boundary restriction map*

(1.1) $$\mathsf{Hom}(\pi_1(\Sigma), G)/G \longrightarrow \prod_{i \in \pi_0(\partial \Sigma)} \mathsf{Hom}\big(\pi_1(\partial_i \Sigma)), G\big)/G.$$

The fibers of the boundary restriction map are the *relative character varieties*. This action of Mod_Σ preserves this map.

1.1. The Symplectic Structure.
These spaces possess *algebraic symplectic structures*, invariant under Mod_Σ. For the moment we focus on the *smooth part* of $\mathsf{Hom}(\pi, G)$, which we define as follows. When G is reductive, the subset $\mathsf{Hom}(\pi, G)^{--}$ consisting of representations whose image does not lie in a parabolic subgroup of G is a smooth submanifold upon which $\mathsf{Inn}(G)$ acts properly and freely. The quotient $\mathsf{Hom}(\pi, G)^{--}/G$ is then a smooth manifold, with a Mod_Σ-invariant symplectic structure.

The symplectic structure depends on a choice of a nondegenerate Ad-invariant symmetric bilinear form $\mathbb{B}$ on the Lie algebra $\mathfrak{g}$ of G and an orientation on Σ. The composition

$$\pi \xrightarrow{\rho} G \xrightarrow{\mathsf{Ad}} \mathsf{Aut}(\mathfrak{g})$$

defines a π-module $\mathfrak{g}_{\mathsf{Ad}\rho}$. The Zariski tangent space to $\mathsf{Hom}(\pi, G)$ at a representation ρ is the space $Z^1(\pi, \mathfrak{g}_{\mathsf{Ad}\rho})$ of *1-cocycles*. The tangent space to the orbit $G\rho$ equals the subspace $B^1(\pi, \mathfrak{g}_{\mathsf{Ad}\rho})$ of *1-coboundaries*. These facts are due to Weil [**102**], see also Raghunathan [**88**]. If G acts properly and freely on a neighborhood of ρ in $\mathsf{Hom}(\pi, G)$, then $\mathsf{Hom}(\pi, G)/G$ is a manifold near $[\rho]$ with tangent space $H^1(\pi, \mathfrak{g}_{\mathsf{Ad}\rho})$. In that case a nondegenerate symmetric $\mathsf{Ad}(G)$-invariant bilinear form

$$\mathfrak{g} \times \mathfrak{g} \xrightarrow{\mathbb{B}} \mathbb{R}$$

defines a pairing of π-modules

$$\mathfrak{g}_{\mathsf{Ad}\rho} \times \mathfrak{g}_{\mathsf{Ad}\rho} \xrightarrow{\mathbb{B}} \mathbb{R}.$$

Cup product using $\mathbb{B}$ as coefficient pairing defines a nondegenerate skew-symmetric pairing

$$H^1(\pi, \mathfrak{g}_{\mathsf{Ad}\rho}) \times H^1(\pi, \mathfrak{g}_{\mathsf{Ad}\rho}) \xrightarrow{\mathbb{B}_*(\cup)} H^2(\pi, \mathbb{R}) \cong \mathbb{R}$$

on each tangent space

$$T_{[\rho]} \mathsf{Hom}(\pi, G)/G \cong H^1(\pi, \mathfrak{g}_{\mathsf{Ad}\rho}).$$

Here the isomorphism $H^2(\pi, \mathbb{R}) \cong \mathbb{R}$ arises from the orientation on Σ. The resulting exterior 2-form $\omega_\mathbb{B}$ is closed [**36**], and defines a symplectic structure on the smooth part $\mathsf{Hom}(\pi, G)^{--}/G$ of $\mathsf{Hom}(\pi, G)/G$. This topological definition makes it apparent that $\omega_\mathbb{B}$ is Mod_Σ-invariant. In particular the action preserves the measure μ defined by $\omega_\mathbb{B}$. When G is compact, the total measure is finite (Jeffrey-Weitsman [**61**, **62**], Huebschmann [**60**]).

1.2. The Complex Case.
When G is a complex Lie group, $\mathsf{Hom}(\pi, G)$ has a complex algebraic structure preserved by the $\mathsf{Aut}(\pi) \times \mathsf{Aut}(G)$-action. When G is a complex semisimple Lie group, the above construction, applied to a nondegenerate Ad-invariant *complex-bilinear form*

$$\mathfrak{g} \times \mathfrak{g} \xrightarrow{\mathbb{B}} \mathbb{C},$$

determines a *complex-symplectic structure* on $\mathsf{Hom}(\pi, G)^{--}/G$, that is, a closed nondegenerate holomorphic $(2,0)$-form. This complex-symplectic structure is evidently Mod_Σ-invariant. For a discussion of this structure when $G = \mathsf{SL}(2, \mathbb{C})$, see [**45**].

The choice of a marked conformal structure on Σ determines a *hyper-Kähler structure* on $\mathsf{Hom}(\pi,G)/G$ subordinate to this complex-symplectic structure.

A complex-symplectic structure on a $4m$-dimensional real manifold V is given by an integrable almost complex structure J and a closed nondegenerate skew-symmetric bilinear form
$$TM \times TM \xrightarrow{\Omega} \mathbb{C}$$
which is complex-bilinear with respect to J. Alternatively, it is defined by a reduction of the structure group of the tangent bundle TV from $\mathsf{GL}(4m, \mathbb{R})$ to the subgroup
$$\mathsf{Sp}(2m, \mathbb{C}) \subset \mathsf{GL}(4m, \mathbb{R}).$$
A hyper-Kähler structure further reduces the structure group of the tangent bundle from $\mathsf{Sp}(2m, \mathbb{C})$ to its maximal compact subgroup $\mathsf{Sp}(2m) \subset \mathsf{Sp}(2m, \mathbb{C})$. All of these structures are required to satisfy certain integrability conditions. A hyper-Kähler structure subordinate to a complex-symplectic structure (Ω, J) is defined by a Riemannian metric g and integrable almost complex structures I, K such that:

- g is Kählerian with respect to each of I, J, K,
- the complex structures I, J, K satisfy the quaternion identities,
- $\Omega(X, Y) = -g(IX, Y) + i\,g(KX)$ for $X, Y \in TM$.

Goldman-Xia [**54**], §5 describes this structure in detail when $G = \mathsf{GL}(1, \mathbb{C})$.

From this we can associate to every point in Teichmüller space $\mathfrak{T}_\Sigma$ a compatible hyper-Kähler structure on the complex-symplectic space $\mathsf{Hom}(\pi, G)^{--}/G$. However the hyper-Kähler structures are not Mod_Σ-invariant.

1.3. Singularities of the deformation space. In general, the spaces $\mathsf{Hom}(\pi, G)$ and $\mathsf{Hom}(\pi, G)/G$ are not manifolds, but their local structure admits a very explicit cohomological description. For convenience assume that G is reductive algebraic and that ρ is a *reductive representation,* that is, its image $\rho(\pi)$ is Zariski dense in a reductive subgroup of G. For $\rho \in \mathsf{Hom}(\pi, G)$, denote the centralizer of $\rho(\pi)$ by $\mathfrak{Z}(\rho)$ and the center of G by $\mathfrak{Z}$.

A representation $\rho \in \mathsf{Hom}(\pi, G)$ is a *singular point* of $\mathsf{Hom}(\pi, G)$ if and only if
$$\dim(\mathfrak{Z}(\rho)/\mathfrak{Z}) > 0.$$
Equivalently, the isotropy group of $\mathsf{Inn}(G)$ at ρ is not discrete, that is, the action of $\mathsf{Inn}(G)$ at ρ is not *locally free.*

The Zariski tangent space $T_\rho\mathsf{Hom}(\pi, G)$ equals the space $Z^1(\pi; \mathfrak{g}_{\mathsf{Ad}\rho})$ of $\mathfrak{g}_{\mathsf{Ad}\rho}$-valued 1-cocycles on π. The tangent space to the orbit $G \cdot \rho$ equals the subspace $B^1(\pi; \mathfrak{g}_{\mathsf{Ad}\rho})$ of coboundaries. Thus the *Zariski normal space* at ρ to the orbit $G \cdot \rho$ in $\mathsf{Hom}(\pi, G)$ equals the cohomology group $H^1(\pi; \mathfrak{g}_{\mathsf{Ad}\rho})$.

Here is a heuristic interpretation. Consider an analytic path $\rho_t \in \mathsf{Hom}(\pi, G)$ with $\rho_0 = \rho$. Expand it as a power series in t:

(1.2) $$\rho_t(x) = \exp\left(u_0(x)t + u_2(x)t^2 + u_3(x)t^3 + \dots\right)\rho(x)$$

where
$$\pi \xrightarrow{u_n} \mathfrak{g}$$
for $n \geq 0$. The condition

(1.3) $$\rho_t(xy) = \rho_t(x)\rho_t(y)$$

implies that the tangent vector $u = u_0$ satisfies the *cocycle condition*

(1.4) $$u(xy) = u(x) + \mathsf{Ad}\rho(x)u(y),$$

(the *linearization* of (1.3). The vector space of solutions of (1.4) is the space $Z^1(\pi;\mathfrak{g}_{\mathsf{Ad}\rho})$ of $\mathfrak{g}_{\mathsf{Ad}\rho}$-valued *1-cocycles* of π.

The Zariski tangent space to the orbit $G \cdot \rho$ equals the subspace $B^1(\pi,\mathfrak{g}_{\mathsf{Ad}\rho}) \subset Z^1(\pi,\mathfrak{g}_{\mathsf{Ad}\rho})$ consisting of *1-coboundaries*. Suppose that a path ρ_t in $\mathsf{Hom}(\pi, G)$ is induced by a conjugation by a path g_t

$$\rho_t(x) = g_t \rho(x) g_t^{-1},$$

where g_t admits a power series expansion

$$g_t = \exp(v_1 t + v_2 t^2 + \dots),$$

where $v_1, v_2, \dots \in \mathfrak{g}$. Thus the tangent vector to ρ_t is tangent to the orbit $G \cdot \rho$. Expanding the power series, this tangent vector equals

$$u(x) = v_1 - \mathsf{Ad}\rho(x)v_1,$$

that is, $u = \delta v_1 \in B^1(\pi;\mathfrak{g}_{\mathsf{Ad}\rho})$ is a coboundary.

Let $u \in T_\rho \mathsf{Hom}(\pi, G) = Z^1(\pi;\mathfrak{g}_{\mathsf{Ad}\rho})$ be a tangent vector to $\mathsf{Hom}(\pi, G)$ at ρ. We give necessary and sufficient conditions that u be tangent to an analytic path of representations.

Solving the equation (1.3) to second order gives:

(1.5) $$u_2(x) - u_2(xy) + \mathsf{Ad}\rho(x)u_2(y) = \frac{1}{2}[u(x), \mathsf{Ad}\rho(x)u(y)].$$

Namely, the function,

(1.6) $$\begin{aligned} \pi \times \pi &\longrightarrow \mathfrak{g} \\ (x, y) &\longmapsto \frac{1}{2}[u(x), \mathsf{Ad}\rho(x)u(y)] \end{aligned}$$

is a $\mathfrak{g}_{\mathsf{Ad}\rho}$-valued 2-cochain on π, This 2-cochain is the coboundary δu_2 of the 1-cochain $\pi \xrightarrow{u_2} \mathfrak{g}_{\mathsf{Ad}\rho}$. Similarly there are conditions on the coboundary of u_n in terms of the lower terms in the power series expansion (1.2).

The operation (1.6) has a cohomological interpretation as follows. π acts on $\mathfrak{g}$ by Lie algebra automorphisms, so that Lie bracket defines a pairing of π-modules

$$\mathfrak{g}_{\mathsf{Ad}\rho} \times \mathfrak{g}_{\mathsf{Ad}\rho} \xrightarrow{[,]} \mathfrak{g}_{\mathsf{Ad}\rho}.$$

The Lie algebra of $\mathfrak{z}(\rho)$ equals $H^0(\pi;\mathfrak{g}_{\mathsf{Ad}\rho})$. The linearization of the action of $\mathfrak{z}(\rho)$ is given by the cup product on $H^1(\pi;\mathfrak{g}_{\mathsf{Ad}\rho})$ with $[,]$ as coefficient pairing:

$$H^0(\pi;\mathfrak{g}_{\mathsf{Ad}\rho}) \times H^1(\pi;\mathfrak{g}_{\mathsf{Ad}\rho}) \xrightarrow{[,]_*(\cup)} H^1(\pi;\mathfrak{g}_{\mathsf{Ad}\rho}).$$

Now consider the cup product of 1-dimensional classes. The bilinear form

$$H^1(\pi;\mathfrak{g}_{\mathsf{Ad}\rho}) \times H^1(\pi;\mathfrak{g}_{\mathsf{Ad}\rho}) \xrightarrow{[,]_*(\cup)} H^2(\pi;\mathfrak{g}_{\mathsf{Ad}\rho}).$$

is symmetric; let Q_ρ be the associated quadratic form.

Suppose u is tangent to an analytic path. Solving (1.2) to second order (as in (1.5) and (1.6)) implies that

$$[,]_*(\cup)([u], [u]) = \delta u_2,$$

that is,

(1.7) $$Q_\rho([u]) = 0.$$

Under the above hypotheses, the necessary condition (1.7) is also sufficient. In fact, by Goldman-Millson [**50**], ρ has a neighborhood N in $\mathsf{Hom}(\pi, G)$ analytically equivalent to a neighborhood of 0 of the cone C_ρ in $Z^1(\pi; \mathfrak{g}_{\mathsf{Ad}\rho})$ defined by the homogeneous quadratic function

$$Z^1(\pi; \mathfrak{g}_{\mathsf{Ad}\rho}) \longrightarrow H^2(\pi; \mathfrak{g}_{\mathsf{Ad}\rho})$$
$$u \longmapsto Q_\rho([u]).$$

Then the germ of $\mathsf{Hom}(\pi, G)/G$ at $[\rho]$ is the quotient of this cone by the isotropy group $\mathfrak{Z}(\rho)$. (These spaces are special cases of *symplectic stratified spaces* of Sjamaar-Lerman [**79**].)

An explicit *exponential mapping*

$$N \xrightarrow{\mathsf{Exp}_\rho} \mathsf{Hom}(\pi, G)$$

was constructed by Goldman-Millson [**49**] using the Green's operator of a Riemann surface M homeomorphic to Σ.

The subtlety of these constructions is underscored by the following false argument, which seemingly proves that the Torelli subgroup of Mod_Σ acts identically on the whole component of $\mathsf{Hom}(\pi, G)/G$ containing the trivial representation. This is easily seen to be false, for G semisimple.

Here is the fallacious argument. The trivial representation ρ_0 is fixed by all of Mod_Σ. Thus Mod_Σ acts on the analytic germ of $\mathsf{Hom}(\pi, G)/G$ at ρ_0. At ρ_0, the coefficient module $\mathfrak{g}_{\mathsf{Ad}\rho}$ is trivial, and the tangent space corresponds to ordinary (untwisted) cohomology:

$$T_{\rho_0}\mathsf{Hom}(\pi, G) = Z^1(\pi; \mathfrak{g}) = Z^1(\pi) \otimes \mathfrak{g}.$$

The quadratic form is just the usual cup-product pairing, so any homologically trivial automorphism ϕ fixes the quadratic cone N pointwise. By Goldman-Millson [**50**], the analytic germ of $\mathsf{Hom}(\pi, G)$ at ρ_0 is equivalent to the quadratic cone N. Therefore $[\phi]$ *acts trivially on an open neighborhood of ρ in* $\mathsf{Hom}(\pi, G)$. By analytic continuation, $[\phi]$ acts trivially on the whole component of $\mathsf{Hom}(\pi, G)$ containing ρ.

The fallacy arises because the identification Exp_ρ of a neighborhood N in the quadratic cone with the germ of $\mathsf{Hom}(\pi, G)$ at ρ depends on a choice of Riemann surface M. Each point $\langle f, M \rangle \in \mathfrak{T}_\Sigma$ determines an exponential map $\mathsf{Exp}_{\rho, \langle f, M \rangle}$ from the germ of the quadratic cone to $\mathsf{Hom}(\pi, G)$, and these are *not* invariant under Mod_Σ. In particular, no family of isomorphisms of the analytic germ of $\mathsf{Hom}(\pi, G)$ at ρ_0 with the quadratic cone N is Mod_Σ-invariant.

PROBLEM 1.1. Investigate the dependence of $\mathsf{Exp}_{\rho, \langle f, M \rangle}$ on the marked Riemann surface $\langle f, M \rangle$.

1.4. Surfaces with boundary. When Σ has nonempty boundary, an Ad-invariant inner product $\mathbb{B}$ on $\mathfrak{g}$ and an orientation on Σ determines a *Poisson structure* (Fock-Rosly [**32**], Guruprasad-Huebschmann-Jeffrey-Weinstein [**55**]). The *symplectic leaves* of this Poisson structure are the level sets of the boundary restriction map (1.1).

For each component $\partial_i \Sigma$ of $\partial \Sigma$, fix a conjugacy class $C_i \subset G$. The subspace

(1.8) $$\mathsf{Hom}(\pi, G)/G_{(C_1, \ldots, C_b)} \subset \mathsf{Hom}(\pi, G)/G$$

consisting of $[\rho]$ such that

(1.9) $$\rho(\partial_i \Sigma) \subset C_i$$

has a symplectic structure. (To simplify the discussion we assume that it is a smooth submanifold.) De Rham cohomology with twisted coefficients in $\mathfrak{g}_{\mathrm{Ad}\rho}$ is naturally isomorphic with group cohomology of π. In terms of De Rham cohomology, the tangent space at $[\rho]$ to $\mathsf{Hom}(\pi, G)/G_{(C_1,\ldots,C_b)}$ identifies with

$$\mathsf{Ker}\left(H^1(\Sigma; \mathfrak{g}_{\mathrm{Ad}\rho}) \to H^1(\partial\Sigma; \mathfrak{g}_{\mathrm{Ad}\rho}) \right)$$
$$\cong \mathsf{Image}\left(H^1(\Sigma, \partial\Sigma; \mathfrak{g}_{\mathrm{Ad}\rho}) \to H^1(\Sigma; \mathfrak{g}_{\mathrm{Ad}\rho}) \right).$$

The cup product pairing

$$H^1(\Sigma; \mathfrak{g}_{\mathrm{Ad}\rho}) \times H^1(\Sigma, \partial\Sigma; \mathfrak{g}_{\mathrm{Ad}\rho}) \xrightarrow{\mathbb{B}_*(\cup)} H^2(\Sigma, \partial\Sigma; \mathbb{R})$$

induces a symplectic structure on $\mathsf{Hom}(\pi, G)/G_{(C_1,\ldots,C_b)}$.

Given a (possibly singular) foliation $\mathfrak{F}$ of a manifold X by symplectic manifolds, the Poisson structure is defined as follows. For functions $f, g \in C^\infty(X)$, their Poisson bracket is a function $\{f, g\}$ on X defined as follows. Let $x \in X$ and let L_x be the leaf of $\mathfrak{F}$ containing x. Define the value of $\{f, g\}$ at x as the Poisson bracket

$$\{f|_{L_x}, g|_{L_x}\}_{L_x},$$

where $\{,\}_{L_x}$ denotes the Poisson bracket operation on the symplectic manifold L_x, and $f|_{L_x}, g|_{L_x} \in C^\infty(L_x)$, are the restrictions of f, g to L_x.

The examples below exhibit *exterior bivector fields* ξ representing the Poisson structure. If $f, g \in C^\infty(X)$, their Poisson bracket $\{f, g\}$ is expressed as an interior product of ξ with the exterior derivatives of f, g:

$$\{f, g\} = \xi \cdot (df \otimes dg).$$

In local coordinates $(x^1, \ldots, x^n)$, write

$$\xi = \sum_{i,j} \xi^{i,j} \frac{\partial}{\partial x_i} \wedge \frac{\partial}{\partial x_j}$$

with $\xi^{i,j} = -\xi^{j,i}$. Then

$$\{f, g\} = \sum_{i,j} \left(\xi^{i,j} \frac{\partial}{\partial x_i} \wedge \frac{\partial}{\partial x_j} \right) \cdot \left(\frac{\partial f}{\partial x_i} dx^i \otimes \frac{\partial g}{\partial x_j} dx^j \right)$$
$$= \sum_{i,j} \xi^{i,j} \left(\frac{\partial f}{\partial x_i} \frac{\partial g}{\partial x_j} - \frac{\partial f}{\partial x_j} \frac{\partial g}{\partial x_i} \right).$$

1.5. Examples of relative $\mathsf{SL}(2,\mathbb{C})$-character varieties. We give a few explicit examples, when $G = \mathsf{SL}(2,\mathbb{C})$, and Σ is a three-holed or four-holed sphere, or a one-holed or two-holed torus. Since generic conjugacy classes in $\mathsf{SL}(2,\mathbb{C})$ are determined by the trace function

$$\mathsf{SL}(2,\mathbb{C}) \xrightarrow{\mathrm{tr}} \mathbb{C}$$

the relative character varieties are level sets of the mapping

$$\mathsf{Hom}(\pi, G)/G \longrightarrow \mathbb{C}^b$$
$$[\rho] \longmapsto \left[\mathsf{tr}\big(\rho(\partial_i(\Sigma))\big)\right]_{i=1,\ldots,b}.$$

1.5.1. *The three-holed sphere.* When Σ is a three-holed sphere, its fundamental group admits a presentation

$$\pi = \langle A, B, C \mid ABC = 1 \rangle$$

where A, B, C correspond to the three components of $\partial \Sigma$. Here is the fundamental result for $\mathsf{SL}(2, \mathbb{C})$-character varieties of a rank two free group:

THEOREM (Vogt [**101**],Fricke-Klein [**33**]). *The map*

$$\mathsf{Hom}(\pi, G)/G \longrightarrow \mathbb{C}^3$$
$$[\rho] \longmapsto \begin{bmatrix} \mathsf{tr}\big(\rho(A)\big) \\ \mathsf{tr}\big(\rho(B)\big) \\ \mathsf{tr}\big(\rho(C)\big) \end{bmatrix}$$

is an isomorphism of affine varieties.

In particular, the symplectic leaves are just points. See [**46**] for an elementary proof..

1.5.2. *The one-holed torus.* When Σ is a one-holed torus, its fundamental group admits a presentation

$$\pi = \langle X, Y, Z, K \mid XYZ = 1, K = XYX^{-1}Y^{-1} \rangle$$

where X, Y are simple loops intersecting once, and $K = XYX^{-1}Y^{-1}$ corresponds to the boundary. Presenting the interior of Σ as the quotient

$$\mathsf{int}(\Sigma) = (\mathbb{R}^2 - \mathbb{Z}^2)/\mathbb{Z}^2$$

the curves X, Y correspond to the $(1,0)$ and $(0,1)$-curves respectively. Once again the Vogt-Fricke theorem implies that $\mathsf{Hom}(\pi, G)/G \cong \mathbb{C}^3$, with coordinates

$$x = \mathsf{tr}\big(\rho(X)\big)$$
$$y = \mathsf{tr}\big(\rho(Y)\big)$$
$$z = \mathsf{tr}\big(\rho(Z)\big).$$

The boundary trace $\mathsf{tr}\big(\rho(K)\big)$ is:

$$\kappa(x, y, z) = x^2 + y^2 + z^2 - xyz - 2$$

so the relative character varieties are the level sets $\kappa^{-1}(t)$. This mapping class group Mod_Σ acts by polynomial transformations of $\mathbb{C}^3$, preserving the function κ (compare [**44**]). The Poisson structure is given by the bivector field

$$d\kappa \cdot \big(\partial_x \wedge \partial_y \wedge \partial_z\big) = (2x - yz)\partial_y \wedge \partial_z$$
$$+ (2y - zx)\partial_z \wedge \partial_x$$
$$+ (2z - xy)\partial_x \wedge \partial_y$$

(where ∂_x denotes $\frac{\partial}{\partial x}$, etc.).

1.5.3. The four-holed sphere.
When Σ is a four-holed sphere, the relative character varieties admit a similar description. Present the fundamental group as

$$\pi = \langle A, B, C, D \mid ABCD = 1 \rangle$$

where the generators A, B, C, D correspond to the components of $\partial\Sigma$. The elements

$$X = AB, Y = BC, Z = CA$$

correspond to simple closed curves on Σ. Denoting the trace functions $\mathsf{Hom}(\pi, G)/G \to \mathbb{C}$ corresponding to elements $A, B, C, D, X, Y, Z \in \pi$ by lower-case, the relative character varieties are defined by:

$$x^2 + y^2 + z^2 + xyz = (ab + cd)x + (bc + ad)y \\ + (ac + bd)z + (4 - a^2 - b^2 - c^2 - d^2 - abcd)$$

with Poisson structure

$$\xi = (ab + cd - 2x - yz)\partial_y \wedge \partial_z \\ + (bc + da - 2y - zx)\partial_z \wedge \partial_x \\ + (ca + bd - 2z - xy)\partial_x \wedge \partial_y.$$

1.5.4. The two-holed torus.
Presenting the fundamental group of a two-holed torus as

$$\pi = \langle A, B, X, Y \mid AXY = YXB \rangle,$$

where $A, B \in \pi$ correspond to the two components of $\partial\Sigma$, the elements

$$Z := Y^{-1}X^{-1}, \\ U := AXY = BYX, \\ V := BY, \\ W := AX$$

are represented by simple closed curves. Using the same notation for trace coordinates as above, the relative character varieties are defined by the equations:

$$a + b = xw + yv + uz - xyu \\ ab = x^2 + y^2 + z^2 + u^2 + v^2 + w^2 \\ + vwz - xyz - xuv - yuw - 4$$

and the Poisson structure is

$$(2z - xy)\partial_x \wedge \partial_y + (2x - yz)\partial_y \wedge \partial_z + (2y - yx)\partial_z \wedge \partial_x + \\ (2u - vx)\partial_v \wedge \partial_x + (2v - xu)\partial_x \wedge \partial_u + (2x - uv)\partial_u \wedge \partial_v + \\ (2u - wy)\partial_w \wedge \partial_y + (2w - yu)\partial_y \wedge \partial_u + (2y - uw)\partial_u \wedge \partial_w + \\ (2(xy - z) - vw)\partial_v \wedge \partial_w + (2(xu - v) - wz)\partial_w \wedge \partial_z \\ + (2(yu - w) - zv)\partial_z \wedge \partial_v.$$

These formulas are derived by applying the formulas for the Poisson bracket of trace functions developed in Goldman [38] in combination with the trace identities in $\mathsf{SL}(2,\mathbb{C})$ (see [46]).

2. Compact Groups

The simplest case occurs when $G = \mathsf{U}(1)$. Then

$$\mathsf{Hom}(\pi, G)/G = \mathsf{Hom}(\pi, G) \cong \mathsf{U}(1)^{2g} \cong H^1(\Sigma; \mathbb{R}/\mathbb{Z})$$

is a $2g$-dimensional torus. If M is a closed Riemann surface diffeomorphic to Σ, then $\mathsf{Hom}(\pi, G)/G$ identifies with the *Jacobi variety* of M, parametrizing topologically trivial holomorphic line bundles over M. Although the complex structures on $\mathsf{Hom}(\pi, G)/G$ vary with the complex structures on Σ, the symplectic structure is independent of M.

2.1. Ergodicity. The mapping class group action in this case factors through the symplectic representation

$$\mathsf{Mod}_\Sigma \longrightarrow \mathsf{Sp}(2g, \mathbb{Z})$$

(since the representation variety is just the ordinary cohomology group with values in $\mathbb{R}/\mathbb{Z}$), which is easily seen to be ergodic. This generalizes to arbitrary compact groups:

THEOREM 2.1. *Let G be a compact group. The $\mathsf{Out}(\pi)$-action on $\mathsf{Hom}(\pi, G)/G$ is ergodic.*

When the simple factors of G are locally isomorphic to $\mathsf{SU}(2)$ and Σ is orientable, this was proved in Goldman [43]. For general G, this theorem is due to Pickrell-Xia [82] when Σ is closed and orientable, and Pickrell-Xia [83] for compact orientable surfaces of positive genus.

The following conjecture generalizes the above ergodicity phenomenon:

CONJECTURE 2.2. *Let $\Omega^*(\mathsf{Hom}(\pi, G)/G)$ be the de Rham algebra consisting of all measurable differential forms on $\mathsf{Hom}(\pi, G)/G$. Then the symplectic structures $\omega_\mathbb{B}$ generate the subalgebra of $\Omega^*(\mathsf{Hom}(\pi, G)/G)$ consisting of Mod_Σ-invariant forms.*

Since the μ-measure of $\mathsf{Hom}(\pi, G)/G$ is finite, the representation of Mod_Σ on

$$\mathfrak{H} := L^2(\mathsf{Hom}(\pi, G)/G, \mu))$$

is unitary. Andersen has informed me that he has proved vanishing of the first cohomology group $H^1(\mathsf{Mod}_\Sigma, \mathfrak{H})$, and has raised the following conjecture generalizing Conjecture 2.2::

CONJECTURE 2.3. *Suppose*

$$C^\infty(\mathsf{Hom}(\pi, G)/G) \xrightarrow{D} C^\infty(\mathsf{Hom}(\pi, G)/G)$$

is a differential operator which commutes with the Mod_Σ-action on $\mathsf{Hom}(\pi, G)/G$. Then D is a scalar multiple of the identity operator.

2.2. The unitary representation. Ergodicity means that the only trivial subrepresentation of $\mathfrak{H}$ is the subspace $\mathbb{C}$ consisting of constants. Furthermore the action is *weak mixing,* by which we mean that $\mathbb{C}$ is the only *finite-dimensional* invariant subspace [43]. On the other hand the orthogonal complement $\mathfrak{H}_0$ to $\mathbb{C}$ in $\mathfrak{H}$ contains invariant subspaces. For example the closure of the span of trace functions of *nonseparating simple closed curves on Σ* is an invariant subspace [48].

PROBLEM 2.4. Decompose the representation on $\mathfrak{H}_0$ into irreducible representations of Mod_Σ.

When $G = \mathsf{U}(1)$, and Σ is the 2-torus, $\mathsf{Hom}(\pi, G)/G$ naturally identifies with T^2, by the functions α, β corresponding to a basis of $\pi_1(\Sigma)$. The functions
$$\phi_{m,n} := \alpha^m \beta^n,$$
forms a Hilbert basis of $\mathfrak{H}$, indexed by $(m,n) \in \mathbb{Z}^2$. The Mod_Σ-representation on $\mathfrak{H}$ arises from the linear $\mathsf{GL}(2,\mathbb{Z})$-action on its basis $\mathbb{Z}^2$. The $\mathsf{GL}(2,\mathbb{Z})$-orbits on $\mathbb{Z}^2$ are indexed by integers $d \geq 0$. The orbit of $(d, 0)$ consists of all $(m, n) \in \mathbb{Z}^2$ with $\gcd(m,n) = d$. These are Hilbert bases for irreducible constituents C_d of $\mathfrak{H}$.

The irreducible constituents C_d admit an alternate description, as follows. The d-fold covering homomorphism
$$G \xrightarrow{\Phi_d} G$$
induces a covering space
$$\mathsf{Hom}(\pi, G)/G \longrightarrow \mathsf{Hom}(\pi, G)/G.$$
Let L_d denote the closure of the image of the induced map $\mathfrak{H} \longrightarrow \mathfrak{H}$. Then
$$L_d = \widehat{\bigoplus_{d'|d} C_{d'}}$$
so C_d consists of the orthocomplement in L_d of the sum of all $L_{d'}$ for $d'|d$ but $d' \neq d$.

PROBLEM 2.5. Find a similar geometric interpretation for the irreducible constituents for compact nonabelian groups G.

2.3. Holomorphic objects.
By Narasimhan and Seshadri [78], and Ramanathan [89], a marked conformal structure (f, M) on Σ interprets $\mathsf{Hom}(\pi, G)/G$ as a moduli space of *holomorphic objects* on M. To simplify the exposition we only consider the case $G = \mathsf{U}(n)$, for which $\mathsf{Hom}(\pi, G)/G$ identifies with the moduli space $\mathfrak{U}_n(M)$ of *semistable holomorphic* $\mathbb{C}^n$-*bundles* over M of zero degree [78]. The union of all $\mathfrak{U}_n(M)$ over $\langle f, M \rangle$ in $\mathfrak{T}_\Sigma$ forms a holomorphic fiber bundle
$$\mathfrak{U}_n \longrightarrow \mathfrak{T}_\Sigma$$
with an action of Mod_Σ. The quotient $\mathfrak{U}_n/\mathsf{Mod}_\Sigma$ fibers holomorphicly over $\mathfrak{M}_\Sigma$. The Narasimhan-Seshadri theorem gives a (non-holomorphic) map
$$\mathfrak{U}_n \xrightarrow{\mathsf{hol}} \mathsf{Hom}(\pi, G)/G$$
which on the fiber $\mathfrak{U}_n(M)$ is the bijection associating to an equivalence class of semistable bundles the equivalence class of the holonomy representation of the corresponding flat unitary structure. $\mathfrak{U}_n/\mathsf{Mod}_\Sigma$ inherits a foliation $\mathfrak{F}_\mathfrak{U}$ from the the foliation of $\mathfrak{U}_n$ by level sets of hol. The dynamics of this foliation are equivalent to the dynamics of the Mod_Σ-action on $\mathsf{Hom}(\pi, G)$.

Go one step further and replace $\mathfrak{T}_\Sigma$ by its unit sphere bundle $U\mathfrak{T}_\Sigma$ and $\mathfrak{M}_\Sigma$ by its (orbifold) unit sphere bundle
$$U\mathfrak{M}_\Sigma = (U\mathfrak{T}_\Sigma)/\mathsf{Mod}_\Sigma.$$
Pull back the fibration $\mathfrak{U}^k(\Sigma)$ to $U\mathfrak{M}_\Sigma$, to obtain a flat $\mathsf{Hom}(\pi, G)/G$-bundle $U\mathfrak{U}_n$ over $U\mathfrak{M}_\Sigma$,

The *Teichmüller geodesic flow* is a vector field on $U\mathfrak{M}_\Sigma$ generating the geodesics for the Teichmüller metric on $\mathfrak{T}_\Sigma$. (Masur [74]) Its horizontal lift with respect to the flat connection is an vector field on the total space whose dynamics mirrors the dynamics of the Mod_Σ-action on $\mathsf{Hom}(\pi,G)/G$.

As the Mod_Σ-action on $\mathsf{Hom}(\pi,G)/G$ is weak-mixing, the unitary representation on $L^2\bigl(\mathsf{Hom}(\pi,G)/G,\mu\bigr)$ provides no nontrivial finite-dimensional representations. Thus these representations markedly differ from the representations obtained by Hitchin [58] and Axelrod, Della-Pietra, and Witten [4] obtained from projectively flat connections on $\mathsf{Hom}(\pi,G)/G$. Recently Andersen [2] has proved that these finite-dimensional projective representations of Mod_Σ are *asymptotically faithful*.

2.4. Automorphisms of free groups. Analogous questions arise for the outer automorphism group of a free group π of rank r. Let G be a compact connected Lie group. Then Haar measure on G defines an $\mathsf{Out}(\pi)$-invariant probability measure on $\mathsf{Hom}(\pi,G)$.

CONJECTURE 2.6. *If $r \geq 3$, the action of $\mathsf{Out}(\pi)$ on $\mathsf{Hom}(\pi,G)$ is ergodic.*

Using calculations in [43], this conjecture has been proved [47] when all of the simple factors of G are locally isomorphic to $\mathsf{SU}(2)$.

2.5. Topological dynamics. The topological theory is more subtle, since no longer may we ignore invariant subsets of measure zero. For example, if $F \subset G$ is a finite subgroup, then $\mathsf{Hom}(\pi,F)$ is finite and its image in $\mathsf{Hom}(\pi,G)/G$ is an invariant closed subset.

One might expect that if a representation $\rho \in \mathsf{Hom}(\pi,G)$ has dense image in $\mathsf{SU}(2)$, that the Mod_Σ-orbit of $[\rho]$ is dense in $\mathsf{Hom}(\pi,G)/G$. This is true if Σ is a one-holed torus (Previte-Xia [84]) and if the genus of Σ is positive (Previte-Xia [85]). In genus 0, representations ρ exist with dense image but $\mathsf{Mod}_\Sigma \cdot [\rho]$ consists of only two points.

Similar examples exist when Σ is a four-holed sphere. In [5] Benedetto and I showed that when $-2 < a,b,c,d < 2$, the set of $\mathbb{R}$-points of the relative character variety has one compact component. This component is diffeomorphic to S^2. Depending on the boundary traces (a,b,c,d), this component corresponds to either $\mathsf{SL}(2,\mathbb{R})$-representations or $\mathsf{SU}(2)$-representations. Previte and Xia [86] found representations ρ in the components corresponding to $\mathsf{SL}(2,\mathbb{R})$-representations having dense image, but whose orbit $\bigl(\mathsf{Mod}_\Sigma \cdot [\rho]\bigr)$ has two points. On the other hand, in both cases, Previte and Xia [87] showed the action is *minimal* (every orbit is dense) for a dense set of boundary traces in $[-2,2]^4$.

PROBLEM 2.7. *Determine necessary and sufficient conditions on a general representation ρ for its orbit $\mathsf{Mod}_\Sigma \cdot [\rho]$ to be dense.*

The case when $G = \mathsf{SU}(2)$ and Σ an n-holed sphere for $n > 4$ remains open.

2.6. Individual elements. For a closed surface of genus one, an individual element is ergodic on the $\mathsf{SU}(2)$-character variety if and only if it is *hyperbolic*. In his doctoral thesis [13, 14], Brown used KAM-theory to show this no longer holds for actions on relative $\mathsf{SU}(2)$-character varieties over the one-holed torus. Combining Brown's examples with a branched-cover construction suggests:

PROBLEM 2.8. *Construct an example of a pseudo-Anosov mapping class for a closed surface which is not ergodic on the $\mathsf{SU}(2)$-character variety.*

3. Noncompact Groups and Uniformizations

For noncompact G, one expects less chaotic dynamics. Trivial dynamics – in the form of *proper* Mod_Σ-actions — occur for many invariant open subsets corresponding to *locally homogeneous geometric structures,* (in the sense of Ehresmann [**27**]) or *uniformizations.* Such structures are defined by local coordinate charts into a homogeneous space G/H with coordinate changes which are restrictions of transformations from G. Such an atlas globalizes to a *developing map,* an immersion $\tilde{\Sigma} \longrightarrow G/H$ of the universal covering space $\tilde{\Sigma} \longrightarrow \Sigma$ which is equivariant with respect to a homomorphism $\pi \xrightarrow{\rho} G$.

To obtain a deformation space of such structures with an action of the mapping class group, one introduces markings for a fixed topological surface Σ, just as in the definition of Teichmüller space. The *deformation space* $\mathsf{Def}_{(G,G/H)}(\Sigma)$ consists of equivalence classes of marked $(G, G/H)$-structures with a *holonomy map*

$$\mathsf{Def}_{(G,G/H)}(\Sigma) \xrightarrow{\mathsf{hol}} \mathsf{Hom}(\pi, G)/G$$

which is Mod_Σ-equivariant. The *Ehresmann-Thurston theorem* asserts that, with respect to an appropriate topology on $\mathsf{Def}_{(G,G/H)}(\Sigma)$, the mapping hol is a local homeomorphism. (This theorem is implicit in Ehresmann [**28**] and first explicitly stated by Thurston [**98**]. More detailed proofs were given by Lok [**70**], Canary-Epstein-Green [**19**], and Goldman [**41**]. Bergeron and Gelander [**6**] give a detailed modern proof with applications to discrete subgroups.)

If $G = \mathsf{PGL}(2, \mathbb{R})$ and $G/H = \mathsf{H}^2$ is the hyperbolic plane, then $\mathsf{Def}_{(G,G/H)}(\Sigma) = \mathfrak{T}_\Sigma$.

Examples of uniformizations with proper Mod_Σ-actions include:

- $G = \mathsf{PSL}(2, \mathbb{R})$: The Teichmüller space $\mathfrak{T}_\Sigma$, regarded as the component of discrete embeddings in $\mathsf{Hom}(\pi, G)/G$;
- $G = \mathsf{PSL}(2, \mathbb{C})$: Quasi-fuchsian space $\mathcal{Q}_\Sigma$ is an open subset of $\mathsf{Hom}(\pi, G)/G$ which is equivariantly biholomorphic to $\mathfrak{T}_\Sigma \times \overline{\mathfrak{T}_\Sigma}$;
- $G = \mathsf{SL}(3, \mathbb{R})$. The deformation space $\mathfrak{C}_\Sigma$ of convex $\mathbb{R}\mathsf{P}^2$-structures is a connected component of $\mathsf{Hom}(\pi, G)/G$ (Choi-Goldman [**20**]) and the Mod_Σ-action is proper. More generally if G is a split $\mathbb{R}$-form of a semisimple group, Labourie [**69**] has shown that Mod_Σ acts properly on the contractible component of $\mathsf{Hom}(\pi, G)/G$ discovered by Hitchin [**59**].

3.1. Fricke-Teichmüller space.

A *Fuchsian representation* of π into $G = \mathsf{PSL}(2, \mathbb{R})$ is an isomorphism ρ of $\pi = \pi_1(\Sigma)$ onto a discrete subgroup of G. Since π is torsionfree and ρ is injective, $\rho(\pi)$ is torsionfree. Hence it acts freely on H^2 and the quotient $\mathsf{H}^2/\rho(\pi)$ is a complete hyperbolic surface. The representation ρ defines a homotopy equivalence

$$\Sigma \longrightarrow \mathsf{H}^2/\rho(\pi)$$

which is homotopic to a homeomorphism. Thus ρ is the holonomy homomorphism of a hyperbolic structure on Σ. The collection of $\mathsf{PGL}(2, \mathbb{R})$-conjugacy classes of such homomorphisms identifies (via the Uniformization Theorem) with the *Teichmüller space* $\mathfrak{T}_\Sigma$ of Σ. When $\partial\Sigma \neq \emptyset$, then the *Fricke space* is defined as the deformation space of complete hyperbolic structures on $\mathsf{Int}(\Sigma)$ such that each end is either a cusp or a complete collar on a simple closed geodesic (a *funnel*). These representations map each component of $\partial \Sigma$ to either a parabolic or a hyperbolic element of $\mathsf{PSL}(2,\mathbb{R})$ respectively. For details on Fricke spaces see Bers-Gardiner [8].

The Mod_Σ-action on $\mathfrak{T}_\Sigma$ is proper. This fact seems to have first been noted by Fricke [33] (see Bers-Gardiner [8] or Farb-Margalit [29]). It follows from two facts:
- Mod_Σ preserve a metric on $\mathfrak{T}_\Sigma$;
- The *simple marked length spectrum*

(3.1) $$\left\{\text{simple closed curves on } \Sigma\right\} \Big/ \mathsf{Diff}^0(\Sigma) \longrightarrow \mathbb{R}_+$$

is a proper map.

See Abikoff [1], Bers-Gardiner [8], Farb-Margalit [29] or Harvey [56] for a proof. Another proof follows from Earle-Eels [24], and the closely related fact (proved by Palais and Ebin [25]) that the full diffeomorphism group of a compact smooth manifold acts properly on the space of Riemannian metrics. Compare [53].

3.2. Other components and the Euler class. Consider the case $G = \mathsf{PSL}(2,\mathbb{R})$. Then the components of $\mathsf{Hom}(\pi,G)/G$ are indexed by the *Euler class*

$$\mathsf{Hom}(\pi,G)/G \xrightarrow{e} H^2(\Sigma;\mathbb{Z}) \cong \mathbb{Z}$$

whose image equals

$$\mathbb{Z} \cap [2-2g-b, 2g-2+b]$$

where Σ has genus g and b boundary components. Thus $\mathsf{Hom}(\pi,G)/G$ has $4g+2b-3$ connected components ([40] and Hitchin [57] when $b=0$). The main result of [35] is that the two extreme components $e^{-1}\big(\pm(2-2g-b)\big)$ consist of discrete embeddings. These two components differ by the choice of orientation, each one corresponding to $\mathfrak{T}_\Sigma$, upon which Mod_Σ acts properly. In contrast,

CONJECTURE 3.1. *Suppose that $b = 0$ (Σ is closed). For each integer $1 \leq k \leq 2g+b-2$, the Mod_Σ-action on the component $e^{-1}(2-2g+b+k)$ of $\mathsf{Hom}(\pi,G)$ is ergodic.*

When $b = 0$, the component

$$e^{-1}(3-2g) \approx \Sigma \times \mathbb{R}^{6g-8}$$

represents a $6g-6$-dimensional *thickening* of Σ, upon which Mod_Σ acts. However, Morita [81] showed that Mod_Σ cannot act smoothly on Σ itself inducing the homomorphism $\mathsf{Diff}(\Sigma) \longrightarrow \mathsf{Mod}_\Sigma$. (Recently Markovic [80] has announced that if Σ is a closed surface of genus > 5, then Mod_Σ cannot even act on Σ by homeomorphisms inducing $\mathsf{Homeo}(\Sigma) \longrightarrow \mathsf{Mod}_\Sigma$.)

PROBLEM 3.2. *Determine the smallest dimensional manifold homotopy-equivalent to Σ upon which Mod_Σ acts compatibly with the outer action of Mod_Σ on $\pi_1(\Sigma)$.*

3.3. The one-holed torus. For surfaces with nonempty boundary, the dynamics appears more complicated. When Σ is a one-holed torus ($g = b = 1$) and $G = \mathsf{PSL}(2,\mathbb{R})$ or $\mathsf{SU}(2)$, this was completely analyzed in [44].

As in §1.5.2, the $\mathsf{SL}(2,\mathbb{C})$-character variety identifies with $\mathbb{C}^3$, where the three coordinates (x,y,z) are traces of three generators of π corresponding to the generators X, Y, XY. In these coordinates, the trace of the element $K = XYX^{-1}Y$ of

π corresponding to $\partial\Sigma$ equals
$$\kappa(x,y,z) := x^2 + y^2 + z^2 - xyz - 2.$$
The *relative* $\mathsf{SL}(2,\mathbb{C})$-*character variety* of Σ is then the family of level sets $\kappa^{-1}(t)$ of $\mathbb{C}^3 \xrightarrow{\kappa} \mathbb{C}$.

The set $\kappa^{-1}(t) \cap \mathbb{R}^3$ of $\mathbb{R}$-points of $\kappa^{-1}(t)$, for boundary trace $t \in \mathbb{R}$, are of two types:

- The $\mathsf{SU}(2)$-characters, with $x,y,z \in [-2,2]$ and $t < 2$;
- The $\mathsf{SL}(2,\mathbb{R})$-characters, with either:
  - at least one of x,y,z lies in $(-\infty,-2] \cup [2,\infty)$, or
  - each x,y,z lies in $[-2,2]$ and $t \geq 2$.

If $|t| > 2$, no $\mathsf{SU}(2)$-characters lie in $\kappa^{-1}(t) \cap \mathbb{R}^3$. If $t \neq 2$, these two subsets of $\kappa^{-1}(t) \cap \mathbb{R}^3$ are disjoint. If $t = 2$, these two subsets intersect on the subset
$$[-2,2]^3 \cap \kappa^{-1}(2)$$
corresponding to $\mathsf{SO}(2)$-representations. The space of $\mathsf{SO}(2)$-characters is 2-sphere with 4 branch points of cone angle π (a tetrahedron with smoothed edges).

The Mod_Σ-action determines a dynamical system on each level set. By Keen [**65**], the Fricke space of Σ is the subset
$$\{(x,y,z) \in \mathbb{R}^3 \mid \kappa(x,y,z) \leq -2\}$$
with a proper Mod_Σ-action. Each level set $\mathbb{R}^3 \cap \kappa^{-1}(t)$, for $t < -2$, is homeomorphic to a disjoint union of four discs; the four components are distinguished by different lifts of the representation from $\mathsf{PSL}(2,\mathbb{R})$ to $\mathsf{SL}(2,\mathbb{R})$.

The level set $\mathbb{R}^3 \cap \kappa^{-1}(-2)$ has one notable feature. It has five components, four of which correspond to the Teichmüller space of Σ, and the other component $\{(0,0,0)\}$ consists of just the origin. The Teichmüller space (corresponding to the deformation space of complete *finite area* hyperbolic structures) corresponds to representations taking the boundary element of π to a parabolic transformation of trace -2. On the other hand, $\{(0,0,0)\}$ corresponds the *quaternion representation* in $\mathsf{SU}(2)$:
$$X \longmapsto \begin{bmatrix} i & 0 \\ 0 & -i \end{bmatrix},$$
$$Y \longmapsto \begin{bmatrix} 0 & -1 \\ 1 & 0 \end{bmatrix}.$$

The peripheral element $K \in \pi$ maps to the nontrivial *central* element $-I \in \mathsf{SU}(2)$.

Here we see — for the first time — the coexistence of two extremes of dynamical behavior:

- The proper action on the $\mathsf{SL}(2,\mathbb{R})$-characters;
- The entire mapping class group Mod_Σ fixes a point, in a sense, the "most chaotic" action.

This dichotomy persists for $-2 < t < 2$. The origin deforms to a compact component, consisting of characters of $\mathsf{SU}(2)$-representations with an ergodic Mod_Σ-action. Four contractible components correspond to holonomy representations of hyperbolic structures on a torus with a cone point. The cone angle θ relates to the boundary trace by
$$t = -2\cos(\theta/2).$$

The Mod_Σ-action on these components is proper.

Although Mod_Σ acts properly, none of the corresponding representations are discrete embeddings. The key property seems to be that nonseparating simple loops are mapped to hyperbolic elements, so the simple marked length spectrum (3.1) is a proper map.

PROBLEM 3.3. Find general conditions which ensure that (3.1) is proper.

The level set $\mathbb{R}^3 \cap \kappa^{-1}(2)$ consists of characters of abelian representations, and Mod_Σ is ergodic on each of the four connected components of the smooth part of $\mathbb{R}^3 \cap \kappa^{-1}(2)$. When $2 < t \leq 18$, the Mod_Σ-action on $\mathbb{R}^3 \cap \kappa^{-1}(t)$ is ergodic.

For $t > 18$, the level sets $\mathbb{R}^3 \cap \kappa^{-1}(t)$ display both proper dynamics and chaotic dynamics. The region $(-\infty, -2]^3$ consists of characters of discrete embeddings ρ where the quotient hyperbolic surface $\mathsf{H}^2/\rho(\pi)$ is homeomorphic to a three-holed sphere. Every homotopy equivalence $\Sigma \longrightarrow P$, where P is a hyperbolic surface homeomorphic to a three-holed sphere, determines such a character. Furthermore these determine closed triangular regions which are freely permuted by Mod_Σ. On the complement of these wandering domains the action is ergodic.

When $G = \mathsf{PGL}(2,\mathbb{R})$, the group of (possibly orientation-reversing) isometries of H^2, a similar analysis was begun by Stantchev [**94, 52**]. One obtains similar dynamical systems, where Mod_Σ acts now on the space of representations into the group

$$G_\pm = \mathsf{SL}(2,\mathbb{C}) \cap \left(\mathsf{GL}(2,\mathbb{R}) \cup i\,\mathsf{GL}(2,\mathbb{R})\right)$$

which doubly covers the two-component group $\mathsf{PGL}(2,\mathbb{R})$. These $G_\pm$-representations are again parametrized by traces. They comprise four components, one of which is the subset of $\mathbb{R}^3$ parametrizing $\mathsf{SL}(2,\mathbb{R})$-representations discussed above. The other three components are

$$\mathbb{R} \times i\mathbb{R} \times i\mathbb{R}, \quad i\mathbb{R} \times \mathbb{R} \times i\mathbb{R}, \quad i\mathbb{R} \times i\mathbb{R} \times \mathbb{R}$$

respectively. Consider $i\mathbb{R} \times \mathbb{R} \times i\mathbb{R}$. For $-14 \leq t < 2$, the Mod_Σ-action is ergodic, but when $t < -14$, wandering domains appear. The wandering domains correspond to homotopy-equivalences $\Sigma \longrightarrow P$, where P is a hyperbolic surface homeomorphic to a two-holed projective plane. The action is ergodic on the complement of the wandering domains.

PROBLEM 3.4. Determine the ergodic behavior of the Mod_Σ-action on the level sets

$$\left(i\mathbb{R} \times \mathbb{R} \times i\mathbb{R}\right) \cap \kappa^{-1}(t)$$

where $t > 2$. The level sets for $t > 6$ contains wandering domains corresponding to Fricke spaces of a one-holed Klein bottle.

3.4. Hyperbolic 3-manifolds. When $G = \mathsf{PSL}(2,\mathbb{C})$, the subset $\mathcal{Q}_\Sigma$ of $\mathsf{Hom}(\pi, G)/G$ corresponding to embeddings of π onto quasi-Fuchsian subgroups of G is open and Mod_Σ-invariant. Furthermore the Bers isomorphism [**7**] provides a Mod_Σ-invariant biholomorphism

$$\mathcal{Q}_\Sigma \longrightarrow \mathfrak{T}_\Sigma \times \overline{\mathfrak{T}_\Sigma}.$$

Properness of the action of Mod_Σ on $\mathfrak{T}_\Sigma$ implies properness on $\mathcal{Q}_\Sigma$.

Points on the boundary of $\mathcal{Q}_\Sigma$ also correspond to discrete embeddings, but the action is much more complicated. Recently Souto and Storm [**93**] have proved that $\partial \mathcal{Q}_\Sigma$ contains a Mod_Σ-invariant closed nowhere dense topologically perfect set upon which the action is topologically transitive. From this they deduce that every continuous Mod_Σ-invariant function on $\partial \mathcal{Q}_\Sigma$ is constant.

While for representations into $G = \mathsf{PSL}(2,\mathbb{R})$, the Mod_Σ-orbits of discrete embeddings are themselves *discrete,* the situation becomes considerably more complicated for larger G. For $G = \mathsf{PSL}(2,\mathbb{C})$, representations corresponding to the fiber of a hyperbolic mapping torus furnish points with infinite stabilizer. This is one of the easiest ways to see that Mod_Σ does not act properly on characters of discrete embeddings. Namely, if M^3 is a hyperbolic 3-manifold which admits a fibration $M^3 \xrightarrow{f} S^1$, then the class of the restriction ρ of the holonomy representation

$$\pi_1(M^3) \longrightarrow \mathsf{PSL}(2,\mathbb{C})$$

to the surface group

$$\pi := \pi_1(f^{-1}(s_0)) \cong \mathsf{Ker}\big(\pi_1(M) \xrightarrow{f_*} \mathbb{Z}\big)$$

is invariant under the monodromy automorphism $h \in \mathsf{Aut}(\pi)$ of M^3. That is, there exists $g \in \mathsf{PSL}(2,\mathbb{C})$ such that

$$\rho\big(h(\gamma)\big) = g\rho(\gamma)g^{-1}$$

for all $\gamma \in \pi$. Furthermore $[\rho]$ is a smooth point of $\mathsf{Hom}(\pi,G)/G$. Kapovich [64] proved McMullen's conjecture [75] that the derivative of the mapping class $[h]$ at $[\rho]$ is hyperbolic, that is, no eigenvalue has norm 1. This contrasts the case of *abelian representations,* since homologically trivial pseudo-Anosov mapping classes act trivially on $\mathsf{Hom}(\pi,G)$.

Thus Mod_Σ does not act properly on the set of characters of discrete embeddings. Let $[\rho]$ (as above) be the character of a discrete embedding of π as the fiber of a hyperbolic mapping torus. The stabilizer of $[\rho]$ contains the infinite cyclic group generated by the mapping class corresponding to $[h]$. (In fact $\langle [h] \rangle$ has finite index in the stabilizer of $[\rho]$.) Since stabilizers of proper actions of discrete groups are finite, Mod_Σ does not act properly.

The Souto-Storm theorem shows that this chaotic dynamical behavior pervades the entire boundary of quasi-Fuchsian space $\mathcal{Q}_\Sigma$.

In another direction, using ideas generalizing those of Bowditch [10], Tan, Wong and Zhang [97] have shown that the action of Mod_Σ on the representations satisfying the analogue of *Bowditch's Q-conditions* is proper. This also generalizes the properness of the action on the space of quasi-Fuchsian representations.

At present little is known about the dynamics of Mod_Σ acting on the $\mathsf{SL}(2,\mathbb{C})$-character variety. Conversations with Dumas led to the following problem:

PROBLEM 3.5. Find a point $\rho \in \mathsf{Hom}(\pi, \mathsf{SL}(2,\mathbb{C}))$ such that the closure of its orbit $\overline{\mathsf{Mod}_\Sigma \cdot [\rho]}$ meets both the image of the unitary characters $\mathsf{Hom}(\pi, \mathsf{SU}(2))$ and the closure $\overline{\mathcal{Q}_\Sigma}$ of the quasi-Fuchsian characters.

3.4.1. *Homological actions.* The action of Mod_Σ on the homology of $\mathsf{Hom}(\pi,G)/G$ furnishes another source of possibly interesting linear representations of Mod_Σ. With Neumann [51], we proved that for the relative $\mathsf{SL}(2,\mathbb{C})$-character varieties of the one-holed torus and four-holed sphere, the action of Mod_Σ factors through a finite group.

Atiyah-Bott [3] use infinite-dimensional Morse theory to analyze the algebraic topology of $\mathsf{Hom}(\pi,G)/G$, when G is compact. For the nonsingular components their techniques imply that the Mod_Σ-action on the rational cohomology of $\mathsf{Hom}(\pi,G)/G$ factors through the symplectic representation of Mod_Σ on $H^*(\Sigma)$. In particular Biswas [9] proved that the Torelli group acts trivially on nonsingular components. In contrast, Cappell-Lee-Miller [17, 18] proved the surprising result that that the Torelli group acts *nontrivially* on the homology of the SU(2)-character variety when Σ is closed.

3.5. Convex projective structures and Hitchin representations. When $G = \mathsf{SL}(3,\mathbb{R})$, the mapping class group Mod_Σ acts properly on the component $\mathfrak{C}_\Sigma$ of $\mathsf{Hom}(\pi,G)/G$ corresponding to convex $\mathbb{R}P^2$-structures (Goldman [42], Choi-Goldman [20, 21]). This component is homeomorphic to a cell of dimension $-8\chi(\Sigma)$, and, for a marked Riemann surface M homeomorphic to Σ, admits the natural structure of a holomorphic vector bundle over $\mathfrak{T}_\Sigma$. The work of Labourie [68] and Loftin [71, 72, 73] gives a more intrinsic holomorphic structure on $\mathfrak{C}_\Sigma$.

The existence of this contractible component is a special case of a general phenomenon discovered by Hitchin [59]. Hitchin finds, for *any* split real form of a semisimple group G, a contractible component in $\mathsf{Hom}(\pi,G)/G$. For $G = \mathsf{SL}(n,\mathbb{R})$ this component is characterized as the component containing the composition of discrete embeddings

$$\pi \longrightarrow \mathsf{SL}(2,\mathbb{R})$$

with the irreducible representation

$$\mathsf{SL}(2,\mathbb{R}) \longrightarrow \mathsf{SL}(n,\mathbb{R}).$$

Recently, Labourie has found a dynamical description [69] of representations in Hitchin's component, and has proved they are discrete embeddings. Furthermore he has shown that Mod_Σ acts properly on this component. (These closely relate to the *higher Teichmüller spaces* of Fock-Goncharov [30, 31].)

When G is the automorphism group of a Hermitian symmetric space of noncompact type, Bradlow, Garcia-Prada, and Gothen have found other components of $\mathsf{Hom}(\pi,G)/G$, for which the *Toledo invariant*, is maximal [11, 12]. Their techniques involve Morse theory along the lines of Hitchin [57]. Recently, Burger, Iozzi, and Wienhard have shown [15] that the representations of maximal Toledo invariant consist of discrete embeddings. Using results from [16], Wienhard has informed me that Mod_Σ acts properly on these components.

3.6. The energy of harmonic maps. An interesting invariant of surface group representations arises from the theory of *twisted harmonic maps* of Riemann surfaces, developed in detail in collaboration with Wentworth [53]. Namely to each *reductive representation* $\pi \xrightarrow{\rho} G$, one associates an *energy function*

$$\mathfrak{T}_\Sigma \xrightarrow{E_\rho} \mathbb{R}$$

whose qualitative properties reflect the Mod_Σ-action. Assuming that the Zariski closure of $\rho(\pi)$ in G is reductive, for every marked Riemann surface $\Sigma \longrightarrow M$, there is a ρ-equivariant harmonic map

$$\tilde{M} \longrightarrow G/K$$

where $K \subset G$ is a maximal compact subgroup (Corlette [22], Donaldson [23], Labourie [67], and Jost-Yau [63], following earlier work by Eels-Sampson [26]). Its energy density determines an exterior 2-form on Σ, whose integral is defined as $E_\rho(\langle f,M\rangle)$.

When $\rho(\pi)$ lies in a compact subgroup of G, then the twisted harmonic maps are constant, and the energy function is constantly zero. At the other extreme is the following result, proved in [**53**]:

THEOREM 3.6. *Suppose that ρ is an embedding of π onto a convex cocompact discrete subgroup of G. Then the energy function E_ρ is a proper function on $\mathfrak{T}_\Sigma$.*

Here a discrete subgroup $\Gamma \subset G$ is *convex cocompact* if there exists a geodesically convex subset $N \subset G/K$ such that $\Gamma\backslash N$ is compact. For $\mathsf{PSL}(2,\mathbb{C})$, these are just the quasi-Fuchsian representations. This result was first proved by Tromba [**99**] for Fuchsian representations in $\mathsf{PSL}(2,\mathbb{R})$, and the ideas go back to Sacks-Uhlenbeck [**90**] and Schoen-Yau [**91**].

It is easy to prove (see [**53**]) that if $\Omega \subset \mathsf{Hom}(\pi, G)/G$ is a Mod_Σ-invariant open set for which each function E_ρ is proper, for $[\rho] \in \Omega$, then the action of Mod_Σ on Ω is proper. This gives a general analytic condition implying properness.

Unfortunately, convex cocompactness is extremely restrictive; Kleiner and Leeb have proved [**66**] that in rank > 1 such groups are *never* Zariski dense. However, we know many examples (the deformation space $\mathfrak{C}_\Sigma$ of convex $\mathbb{R}\mathsf{P}^2$-structures, the Hitchin representations by Labourie [**69**], other components of maximal representations [**11, 12, 16**]) where we expect the Mod_Σ-action to be proper. The only use of geodesic convexity in the above result is that the images of harmonic maps are constrained to lie in the set N.

PROBLEM 3.7. Find a substitute for convex cocompactness in higher rank which includes the above examples of proper Mod_Σ-actions, and for which E_ρ is proper.

The work of Bonahon-Thurston on geometric tameness, and its recent extensions, implies that the energy function of a discrete embedding $\pi \longrightarrow \mathsf{PSL}(2,\mathbb{C})$ is proper if and only if it is quasi-Fuchsian [**53**].

3.7. Singular uniformizations and complex projective structures.

When $G = \mathsf{PSL}(2,\mathbb{R})$, the other components of $\mathsf{Hom}(\pi, G)$ may be studied in terms of hyperbolic structures with singularities as follows. Instead of requiring all of the coordinate charts to be local homeomorphisms, one allows charts which at isolated points look like the map

$$\mathbb{C} \longrightarrow \mathbb{C}$$
$$z \mapsto z^k$$

that is, the geometric structure has an isolated singularity of *cone angle* $\theta = 2k\pi$. Such a singular hyperbolic structure may be alternatively described as a singular Riemannian metric g whose curvature equals -1 plus Dirac distributions weighted by $2\pi - \theta_i$ at each singular point p_i of cone angle θ_i. The structure is nonsingular on the complement $\Sigma - \{p_1, \ldots, p_k\}$, and that hyperbolic structure has holonomy representation

$$\pi_1\big(\Sigma - \{p_1, \ldots, p_k\}\big) \xrightarrow{\hat\rho} \mathsf{PSL}(2,\mathbb{R})$$

such that the holonomy of a loop γ_i encircling p_i is elliptic with rotation angle θ_i.

In particular if each $\theta_i \in 2\pi\mathbb{Z}$, then $\hat{\rho}(\gamma_i) = 1$. The representation $\hat{\rho}$ extends to a representation ρ of $\pi_1(\Sigma)$:

$$\begin{array}{c} \pi_1(\Sigma - \{p_1, \ldots, p_k\}) \\ \downarrow \quad \searrow^{\hat{\rho}} \\ \pi_1(\Sigma) \dashrightarrow_{\rho} \mathsf{PSL}(2, \mathbb{R}). \end{array}$$

Applying Gauss-Bonnet to g implies that ρ has Euler class

$$e(\rho) = \chi(M) + \frac{1}{2\pi} \sum_{i=1}^{k} (\theta_i - 2\pi).$$

It is convenient to assume that each $\theta_i = 4\pi$ and the points p_i are not necessarily distinct — a cone point of cone angle 4π with multiplicity m is then a cone point with cone angle $2(m+1)\pi$. The uniformization theorem of McOwen [**77**], Troyanov [**100**], and Hitchin [**57**] asserts: given a Riemann surface $M \approx \Sigma$, there exists a unique singular hyperbolic structure in the conformal class of M with cone angle θ_i at x_i for $i = 1, \ldots, k$ as long as

$$\chi(\Sigma) + \frac{1}{2\pi} \sum_{i=1}^{k} (\theta_i - 2\pi) < 0.$$

(Hitchin only considers the case when θ_i are multiples of 2π, while McOwen and Troyanov deal with arbitrary positive angles.) The resulting *uniformization map* assigns to the collection of points $\{p_1, \ldots, p_k\}$ (where $0 \leq k \leq |\chi(\Sigma)|$) the singular hyperbolic structure with cone angles 4π (counted with multiplicity) at the p_i. The equivalence class of the holonomy representation in the component

$$e^{-1}(\chi(\Sigma) + k) \subset \mathsf{Hom}(\pi, G)/G$$

defines a map from the symmetric product $\mathsf{Sym}^k(M)$ to $e^{-1}(\chi(M) + k)$. The following result follows from Hitchin [35]:

THEOREM 3.8. *Let M be a closed Riemann surface. The above map*

$$\mathsf{Sym}^k(M) \longrightarrow e^{-1}(\chi(M) + k)$$

is a homotopy equivalence.

The union of the symmetric powers $\mathsf{Sym}^k(M)$, one for each marked Riemann surface M, over $\langle f, M \rangle \in \mathfrak{T}_\Sigma$, can be given the structure of a holomorphic fiber bundle $\mathfrak{S}^k(\Sigma)$ over $\mathfrak{T}_\Sigma$, to which the action of Mod_Σ on $\mathfrak{T}_\Sigma$ lifts. The above maps define a homotopy equivalence

$$\mathfrak{S}^k(\Sigma) \xrightarrow{\mathbb{U}} e^{-1}(\chi(\Sigma) + k),$$

which is evidently Mod_Σ-equivariant. However, since Mod_Σ acts properly on $\mathfrak{T}_\Sigma$, it also acts properly on the $(6g - 6 + 2k)$-dimensional space $\mathfrak{S}^k(\Sigma)$. The quotient $\mathfrak{S}^k(\Sigma)/\mathsf{Mod}_\Sigma$ is the total space of an (orbifold) $\mathsf{Sym}^k(\Sigma)$-bundle over the Riemann moduli space

$$\mathfrak{M}_\Sigma := \mathfrak{T}_\Sigma/\mathsf{Mod}_\Sigma.$$

The fibers of $\mathbb{U}$ define a (non-holomorphic) foliation of $\mathfrak{S}^k(\Sigma)/\mathsf{Mod}_\Sigma$, a *flat* $\mathsf{Sym}^k(\Sigma)$-*bundle,* whose dynamics mirrors the dynamics of the Mod_Σ-action on the component $e^{-1}(\chi(\Sigma) + k)$.

In general, U is *not* onto: if $\Sigma \xrightarrow{f} \Sigma'$ is a degree one map to a closed surface Σ' of smaller genus, and ρ' is a Fuchsian representation of $\pi_1(\Sigma')$, then $\rho := \rho' \circ f_*$ lies outside $\mathsf{Image}(\mathsf{U})$. The following conjecture arose in discussions with Neumann:

CONJECTURE 3.9. *If $k = 1$, then U is onto. In general a $\mathsf{PSL}(2,\mathbb{R})$-representation with dense image lies in $\mathsf{Image}(\mathsf{U})$.*

3.8. Complex projective structures. A similar construction occurs with the deformation space $\mathbb{CP}^1(\Sigma)$ of *marked $\mathbb{CP}^1$-structures* on Σ. A $\mathbb{CP}^1$-manifold is a manifold with a coordinate atlas modeled on $\mathbb{CP}^1$, with coordinate changes in $G = \mathsf{PSL}(2,\mathbb{C})$. The space $\mathbb{CP}^1(\Sigma)$ consists of equivalence classes of marked $\mathbb{CP}^1$-structures, that is, homeomorphisms $\Sigma \longrightarrow N$ where N is a $\mathbb{CP}^1$-manifold. Since $\mathsf{PSL}(2,\mathbb{C})$ acts holomorphicly, the $\mathbb{CP}^1$-atlas is a holomorphic atlas and N has a underlying Riemann surface M. The resulting Mod_Σ-equivariant map

$$\mathbb{CP}^1(\Sigma) \longrightarrow \mathfrak{T}_\Sigma$$

is a holomorphic affine bundle, whose underlying vector bundle is the holomorphic cotangent bundle of $\mathfrak{T}_\Sigma$. In particular Mod_Σ acts properly on $\mathbb{CP}^1(\Sigma)$ with quotient a holomorphic affine bundle over $\mathfrak{M}_\Sigma$.

The map which associates to a marked $\mathbb{CP}^1$-structure on Σ its holonomy representation is a local biholomorphism

$$\mathbb{CP}^1(\Sigma) \xrightarrow{\mathsf{hol}} \mathsf{Hom}(\pi,G)/G$$

which is known to be very complicated. Gallo-Kapovich-Marden [34] have shown that its image consists of all equivalence classes of representations ρ for which:

- ρ lifts to a representation $\pi \longrightarrow \mathsf{SL}(2,\mathbb{C})$;
- The image $\rho(\pi)$ is not precompact;
- The image $\rho(\pi)$ is not solvable.

The latter two conditions are equivalent to $\rho(\pi)$ not leaving invariant a finite subset of $\mathsf{H}^3 \cup \partial \mathsf{H}^3$. (The cardinality of this finite subset can be taken to be either 1 or 2.)

The holonomy map hol is Mod_Σ-equivariant. The action of Mod_Σ on $\mathbb{CP}^1(\Sigma)$ is proper, since it covers the action of Mod_Σ on $\mathfrak{T}_\Sigma$. The quotient $\mathbb{CP}^1(\Sigma)/\mathsf{Mod}_\Sigma$ affinely fibers over $\mathfrak{M}_\Sigma$. As before hol defines a foliation of $\mathbb{CP}^1(\Sigma)/\mathsf{Mod}_\Sigma$ orbit equivalent to the Mod_Σ-action on $\mathsf{Hom}(\pi,G)/G$. Thus hol may be regarded as a *resolution* of the Mod_Σ-action.

As a simple example, the Mod_Σ-action is proper on the quasi-Fuchsian subset $\mathcal{Q}_\Sigma \subset \mathsf{Hom}(\pi,G)/G$. As noted above, it is a maximal open set upon which Mod_Σ acts properly. Its restriction

$$\mathsf{hol}^{-1}(\mathcal{Q}_\Sigma) \longrightarrow \mathcal{Q}_\Sigma$$

is a covering space ([39]). However, the bumping phenomenon discovered by McMullen [76] implies that hol is not a covering space on any open neighborhood strictly containing $\mathcal{Q}_\Sigma$.

References

[1] Abikoff, W., "The Real Analytic Theory of Teichmüller Space," Lecture Notes in Mathematics **820**, Springer-Verlag, Berlin, 1980.

[2] Andersen, J., *Asymptotic faithfulness of the quantum $\mathsf{SU}(n)$-representations of the mapping class groups*, Ann. Math. (to appear), `math.QA/0204084`.

[3] Atiyah, M., and Bott, R., *The Yang-Mills equations over Riemann surfaces,* Phil. Trans. R. Soc. Lond. *A 308*

[4] Axelrod, S., Della Pietra, S., and Witten, E., *Geometric quantization of Chern-Simons gauge theory,* J. Diff. Geom. **33** (1991), 787–902. .

[5] Benedetto, R. and Goldman, W., *The Topology of the Relative Character Varieties of a Quadruply-Punctured, Sphere,* Experimental Mathematics (1999) **8:1**, 85 –104.

[6] Bergeron, N. and Gelander, T., *A note on local rigidity,* Geom. Ded. **107** (2004), 111–131.

[7] Bers, L., *Simultaneous uniformization,* Bull. Amer. Math. Soc. **66**, (1960), 94–97.

[8] Bers, L. and F. Gardiner, *Fricke spaces,* Adv. Math. **62** (1986), 249–284.

[9] Biswas, I., *On the mapping class group action on the cohomology of the representation space of a surface,* Proc. Amer. Math. Soc. **124** (1996), no. 6, 1959–1965.

[10] Bowditch, B. H., *Markoff triples and quasi-Fuchsian groups,* Proc. London Math. Soc. (3) **77** (1998), no. 3, 697–736.

[11] Bradlow, S., Garcia-Prada, O., and Gothen, P., *Surface group representations in $PU(p,q)$ and Higgs bundles,* J. Diff. Geo. **64** (2003), no. 1, 111-170.

[12] _____, *Surface group representations, Higgs bundles, and holomorphic triples,* math.AG/0206012.

[13] Brown, R., "Mapping class actions on the $SU(2)$-representation varieties of compact surfaces," Doctoral dissertation, University of Maryland (1996).

[14] _____, *Anosov mapping class actions on the $SU(2)$-representation variety of a punctured torus,* Ergod. Th. & Dynam. Sys. **18** (1998), 539–544.

[15] Burger, M., Iozzi, A., and Wienhard, A., *Surface group representations with maximal Toledo invariant,* C. R. Math. Acad. Sci. Paris, Ser. I **336** (2003), 387–390.

[16] *Maximal representations of surface groups: Symplectic Anosov structures,* Pure and Applied Mathematics Quarterly, Special Issue: In Memory of Armand Borel, **1** (2005), no. 2, 555-601. DG/0506079.

[17] Cappell, S., Lee, R., and Miller, E., *The Torelli group action on respresentation spaces,* in "Geometry and Topology: Aarhus 1998", 47–70, Contemp. Math. **258**, Amer. Math. Soc., Providence, RI, 2000.

[18] _____, *The action of the Torelli group on the homology of representation spaces is nontrivial,* Topology **39** (2000), no. 4, 851–871.

[19] Canary, R., Epstein, D. and Green, P. *Notes on notes of Thurston,* "Analytical and Geometrical Aspects of Hyperbolic Space (Coventry/Durham 1984)," London Math. Soc. Lecture Note Series **111**, Cambridge Univ. Press (1987), 3–92.

[20] Choi, S., and Goldman, W., *Convex real projective structures on closed surfaces are closed,* Proc. A.M.S. **118** No.2 (1993), 657–661.

[21] _____, *The Classification of Real Projective Structures on compact surfaces,* Bull. A.M.S. (New Series) **34**, No. 2, (1997), 161–170.

[22] Corlette, K., *Flat bundles with canonical metrics,* J. Diff. Geom. **28** (1988), 361–382.

[23] Donaldson, S., *Twisted harmonic maps and the self-duality equations,* Proc. London Math. Soc. (3)**55** (1987), 127–131.

[24] Earle, C.J. and Eels, J., *A fibre bundle description of Teichmüller theory,* J. Diff. Geom. **3** (1969), 19–43.

[25] D. Ebin, *The manifold of Riemannian metrics,* in "Global Analysis" (Proc. Sympos. Pure Math. Vol. XV, Berkeley, Calif.), 11–40, (1968), Amer. Math. Soc., Providence, R. I.

[26] Eels, J. and Sampson, J., *Harmonic mappings of Riemannian manifolds,* Amer. J. Math. **86** (1964), 109–160.

[27] Ehresmann, C., *Sur les espaces localement homogenes,* L'ens. Math. **35** (1936), 317–333

[28] _____, *Les connexions infinitesimales dans un fibré differentiable,* Extrait du "Colloque de Topologie," Bruxelles (1950), CBRM.

[29] Farb, B., and Margalit, D., "A primer on mapping class groups," (in preparation).

[30] Fock, V., and Goncharov,A., *Moduli spaces of local systems and higher Teichmüller theory,* math.AG/0311149 (submitted).

[31] _____, *Moduli spaces of convex projective structures on surfaces,* Adv. Math. (to appear) math.DG/0405348.

[32] Fock, V., and Rosly, A., *Poisson structure on moduli of flat connections on Riemann surfaces and r-matrix,.* Am. Math. Soc. Transl. **191** (1999) 67-86. math.QA/9802054.

[33] Fricke, R., and Klein, F., "Vorlesungen uber der Automorphen Funktionen", vol. I (1897)

[34] Gallo, D., Kapovich, M. and Marden, A., *The monodromy groups of Schwarzian equations on closed Riemann surfaces*, Ann. Math. **151** (2000), 625–704.
[35] Goldman, W., "Discontinuous groups and the Euler class," Doctoral dissertation, University of California, Berkeley (1980).
[36] _____, *The symplectic nature of fundamental groups of surfaces*, Adv. Math. Vol. 54 (1984), pp. 200–225.
[37] _____, *Representations of fundamental groups of surfaces*, in "Geometry and Topology, Proceedings, University of Maryland 1983–1984", J. Alexander and J. Harer (eds.), Lecture Notes in Mathematics **1167** 95–117, Springer-Verlag (1985).
[38] _____, *Invariant functions on Lie groups and Hamiltonian flows of surface group representations*, Inv. Math. **85** (1986), 1–40.
[39] _____, *Projective structures with Fuchsian holonomy*, J. Diff. Geom. **25** (1987), 297–326.
[40] _____, *Topological components of spaces of representations*, Inv. Math. **93** (3), (1988), 557–607.
[41] _____, *Geometric structures and varieties of representations*, in "The Geometry of Group Representations," Contemp. Math. **74** , Amer. Math. Soc. (1988), 169–198.
[42] _____, *Convex real projective structures on compact surfaces*, J. Diff. Geo. **31** (1990), 791–845.
[43] _____, *Ergodic theory on moduli spaces*, Ann. Math. **146** (1997), 1–33
[44] _____, *Action of the modular group on real SL(2)-characters of a one-holed torus*, Geometry and Topology **7** (2003), 443–486. mathDG/0305096.
[45] _____, *The complex symplectic geometry of $SL(2,C)$-characters over surfaces*, in "Proceedings of International Conference on Algebraic Groups and Arithmetic," December 17–22, 2001, TIFR, Mumbai. math.DG/0304307.
[46] _____, *An exposition of results of Vogt and Fricke*, math.GM/0402103 (submitted).
[47] _____, *An ergodic action of the outer automorphism group of a free group*, Geom. Func. Anal. (to appear). math.DG/0506401.
[48] _____, *The Mapping Class Group acts reducibly on $SU(n)$-character varieties*, to appear, "Knots and Primes: Proceedings of the 2004 Japan-American Mathematics Institute, Johns Hopkins University," Contemp. Math., Amer. Math. Soc. math.GT/0509115
[49] Goldman, W. and Millson, J., *Deformations of flat bundles over Kähler manifolds*, in "Geometry and Topology, Manifolds, Varieties, and Knots," C. McCrory and T.Shifrin (eds.), Lecture Notes in Pure and Applied Mathematics **105** (1987), 129–145 Marcel Dekker Inc., New York Basel.
[50] _____, *The deformation theory of representations of fundamental groups of Kähler manifolds*, Publ. Math. I. H. E. S. **67** (1988),43–96.
[51] Goldman, W. and Neumann, W., *Homological action of the modular group on some cubic moduli spaces*, Math. Research Letters (to appear).math.GT/0402039
[52] Goldman, W. and Stantchev, G., *Dynamics of the Automorphism Group of the $GL(2,R)$-Characters of a Once-punctured Torus*, math.DG/0309072 (submitted).
[53] Goldman, W. and Wentworth, R., *Energy of Twisted Harmonic Maps of Riemann Surfaces*, to appear in "Proceedings of the 2005 Alhfors-Bers Colloquium, University of Michigan," H. Masur and R. Canary, eds. Contemp. Math., Amer. Math. Soc. math.DG/0506212,
[54] Goldman, W. and Xia, E., *Rank one Higgs bundles on Riemann surfaces and representations of fundamental groups*, Memoirs A.M.S. (to appear). math.DG/0402429.
[55] Guruprasad, K., Huebschmann, J., Jeffrey, L., Weinstein, A., *Group systems, groupoids, and moduli spaces of parabolic bundles*, Duke Math. J. **89** No. 2, (1997), 377–412l.
[56] Harvey, W. J., *Spaces of Discrete Groups*, in "Discrete Groups and Automorphic Functions," Academic Press (1977), 295–347.
[57] Hitchin, N., *The self-duality equations on Riemann surfaces*, Proc. Lond. Math. Soc. **55** (1987), 59–126.
[58] _____, *Flat connections and geometric quantization*, Commun. Math. Phys. **131** (1990), 347–380.
[59] _____, *Lie groups and Teichmüller space*, Topology **31** (3), (1992), 449–473.
[60] Huebschmann, J., *Symplectic and Poisson structures of certain moduli spaces*, Duke Math J. **80** (1995) 737-756.
[61] Jeffrey,L. and Weitsman, J., *Bohr-Sommerfeld orbits and the Verlinde dimension formula*, Commun. Math. Phys. 150 (1992) 593-630

[62] _____, *Toric structures on the moduli space of flat connections on a Riemann surface: volumes and the moment map,* Advances in Mathematics **109**, 151-168 (1994).

[63] Jost, J. and Yau, S.T., *Harmonic maps and group representations,* Differential Geometry, H.B. Lawson and K. Tenenblat, eds., Longman, pp. 241–259.

[64] Kapovich, M., *On the dynamics of pseudo-Anosov homeomorphisms on representation varieties of surface groups,* Ann Acad. Sci. Fenn. Math. 23 (1998) no.1, 83-100.

[65] Keen, L., *On Fricke Moduli,* in "Advances in the Theory of Riemann Surfaces," Ann. Math. Studies **66** (1971), 205–224.

[66] Kleiner, B. and Leeb, B., *Rigidity of invariant convex sets in symmetric spaces,* math.DG/0412123.

[67] Labourie, F., *Existence d'applications harmoniques tordues à valeurs dans les variétés à courbure négative,* Proc. Amer. Math. Soc. **111** (1991), 877–882.

[68] _____, *$\mathbb{R}P^2$-structures et differentielles cubiques holomorphes,* Proc. of the GARC Conference in Differential Geometry, Seoul National University (1997).

[69] _____, *Anosov flows, surface group representations and curves in projective space,* Inv. Math. **165** no. 1, 51–114 (2006) math.DG/0401230.

[70] Lok, W., "Deformations of locally homogeneous spaces and Kleinian groups," Doctoral dissertation, Columbia University (1984).

[71] Loftin, J., *Applications of Affine Differential Geometry to $\mathbb{R}P^2$ Surfaces,* Doctoral dissertation, Harvard University (1999).

[72] _____, *Affine spheres and convex $\mathbf{R}P^n$-manifolds,* Amer. J. Math. **123** (2001), 255–274.

[73] _____, *The compactification of the moduli space of convex $\mathbb{R}P^2$-surfaces I,* J. Diff. Geo. **68**, No. 2, (2004), 233–276. math.DG/0311052

[74] Masur, H., *Interval exchange transformations and measured foliations,* Ann. Math. **115** (1982), no. 1, 169–200.

[75] McMullen, C., "Renormalization and 3-manifolds which fiber over the circle," Ann. Math. Studies **142**, Princeton University Press (1996).

[76] _____, *Complex earthquakes and Teichmüller theory,* J. Amer. Math. Soc. **11** (1998), 283–320.

[77] McOwen, R., *Prescribed curvature and singularities of conformal metrics on Riemann surfaces,* J. Math. Anal. Appl. **177** (1993), no. 1, 287–298.

[78] Narasimhan, M. S. and Seshadri, C. S., *Stable and unitary vector bundles over compact Riemann surfaces,* Ann. Math. **82** (1965), 540–567.

[79] Lerman, E., and Sjamaar, R. *Stratified spaces and reduction,* Ann. Math. (2) **134** (1991), no. 2, 375–422.

[80] Markovic, V., *Realization of the Mapping Class Group by Homeomorphisms,* preprint available from http://www.maths.warwick.ac.uk/ markovic/.

[81] Morita,S., *Characteristic classes of surface bundles,* Bull. Amer. Math. Soc. (N.S) **11** (1984), no. 2, 386–388.

[82] Pickrell, D. and Xia, E., *Ergodicity of Mapping Class Group Actions on Representation Varieties, I. Closed Surfaces,* Comment. Math. Helv. **77** (2001), 339–362.

[83] _____, *Ergodicity of Mapping Class Group Actions on Representation Varieties, II. Surfaces with Boundary,* Transformation Groups **8** (2003), no. 4, 397–402.

[84] Previte, J., and Xia, E., *Topological dynamics on moduli spaces I,* Pac. J. Math. **193** (200), no. 2, 397–417.

[85] _____, *Topological dynamics on moduli spaces II,* Trans. Amer. Math. Soc. **354** (2002), no. 6, 2475–2494.

[86] _____, *Exceptional discrete mapping class group orbits in moduli spaces,* Forum Math. **15** (2003), no. 6, 949–954.

[87] _____, *Dynamics of the Mapping Class Group on the Moduli of a Punctured Sphere with Rational Holonomy,* Geom. Ded. **112** (2005), 65–72.

[88] Raghnunathan, M. S.,. "Discrete Subgroups of Lie Groups" Springer-Verlag Ergebnisse (1972).

[89] Ramanathan, A., *Moduli for principal bundles over algebraic curves,* Proc. Indian Acad. Sci. Math. Sci. **106** (1996) I no.3, 301–326, II no.4, 421–449

[90] Sacks, J. and Uhlenbeck, K., *Minimal immersions of closed Riemann surfaces,* Trans. Amer. Math. Soc. **271** (1982), 639–652.

[91] Schoen, R. and Yau, S. T., *Existence of incompressible minimal surfaces and the topology of three-dimensional manifolds with non-negative scalar curvature,* Ann. Math. **110** (1979), 127–142.

[92] Simpson, C. T., *Nonabelian Hodge theory,* Proc. I.C.M., Kyoto 1990, Springer-Verlag (1991), 198–230.

[93] Souto, J., and Storm, P., *Dynamics of the mapping class group action on the variety of PSL(2,C) characters.* math.GT/0504474 (submitted).

[94] Stantchev, G., "Action of the modular group on GL(2,R)-characters on a once-punctured torus," Doctoral dissertation, University of Maryland (2003).

[95] Sullivan, D., *Quasiconformal homeomorphisms and dynamics. II. Structural stability implies hyperbolicity for Kleinian groups,* Acta Math. **155** (1985), no. 3-4, 243–260.

[96] Tan, S. P., *Branched $\mathbb{CP}^1$-structures with prescribed real holonomy,* Math. Ann. bf 300 (1994), 649–667.

[97] Tan, S. P., Wong, Y. L. and Zhang, Y., *Generalized Markoff maps and McShane's identity,* math.GT/0502464 (submitted).

[98] Thurston, W., "The geometry and topology of 3-manifolds," Princeton University lecture notes (1979) (unpublished).

[99] Tromba, A., "Teichmüller Theory in Riemannian Geometry," ETHZ Lectures in Mathematics, Birkhäuser (1992).

[100] Troyanov, M., *Prescribing curvature on compact surfaces with conical singularities,* Trans. Amer. Math. Soc. **324** (1991), no. 2, 793–821.

[101] Vogt, H., *Sur les invariants fondamentaux des equations différentielles linéaires du second ordre,* Ann. Sci. L'École Normale Supérieure, 3eme Série, Tome VI, (1889) Supplement 1 – 70.

[102] Weil, A. *On discrete subgroups of Lie groups II,* Ann. Math. **75** (1962), 578–602.

DEPARTMENT OF MATHEMATICS, UNIVERSITY OF MARYLAND, COLLEGE PARK, MD 20742
E-mail address: wmg@math.umd.edu

Geometric Properties of the Mapping Class Group

Ursula Hamenstädt

1. Introduction

Consider a compact oriented surface S of genus $g \geq 0$ from which $m \geq 0$ points, so-called *punctures*, have been deleted. The *mapping class group* $\mathcal{M}_{g,m}$ of S is the space of isotopy classes of orientation preserving homeomorphisms of S. It is a subgroup of index 2 in the *extended mapping class group* $\mathcal{M}_{g,m}^{\pm}$ defined as the group of isotopy classes of *all* homeomorphisms of S.

The extended mapping class group of the two-sphere S^2 is the group $\mathbb{Z}_2$ generated by the orientation-reversing involution $z \to -z$ (where we identify S^2 with the set $\{z \in \mathbb{R}^3 \mid \|z\| = 1\}$). In the case that S is a *closed* surface different from S^2 it was shown by Dehn, Nielsen and Baer that the extended mapping class group coincides with the group of outer automorphisms of the fundamental group $\pi_1(S)$ of S (see Section 2.9 of [**I**] for details and references). In particular, the mapping class group of the two-torus equals the group $SL(2,\mathbb{Z})$. Similarly, the extended mapping class group of a surface S with punctures coincides with the group of outer automorphisms of $\pi_1(S)$ which preserve the *peripheral structure*, i.e. the set of conjugacy classes of elements of $\pi_1(S)$ which can be represented by simple closed curves homotopic into a puncture. Thus the mapping class group $\mathcal{M}_{1,1}$ of the once punctured torus coincides with $SL(2,\mathbb{Z})$, and the mapping class group $\mathcal{M}_{0,4}$ of the four punctured sphere has a subgroup of finite index isomorphic to $SL(2,\mathbb{Z})$. We therefore only consider *non-exceptional* surfaces, i.e. we restrict to the case that $3g - 3 + m \geq 2$.

The mapping class groups have been intensively studied in the past. Many of the most important results known to date are described in the beautiful survey of Ivanov [**I**] which also contains an extensive list of references. The goal of this note is to present open problems about the mapping class groups of geometric nature. Our presentation includes the discussion of geometric results on the mapping class groups which either were obtained after the appearance of Ivanov's survey or can be understood with the present knowledge in a more consistent and unified way.

Our point of view will be the one of geometric group theory. This is possible because the mapping class group $\mathcal{M}_{g,m}$ is finitely generated (it is even finitely presented, see Section 4.3 of [**I**]). A finite symmetric set of generators $\mathcal{G}$ for $\mathcal{M}_{g,m}$ defines a *word norm* $\|\ \|$ on $\mathcal{M}_{g,m}$ where $\|g\|$ equals the smallest length of a word

---

Partially supported by Sonderforschungsbereich 611.

in the generators $\mathcal{G}$ which represents g. This word norm in turn induces a distance function on $\mathcal{M}_{g,m}$ which is invariant under the left action of $\mathcal{M}_{g,m}$ on itself by defining $d(g,h) = \|g^{-1}h\|$. The distance function depends on the choice of $\mathcal{G}$, but its large-scale properties are independent of this choice. Namely, changing the set of generators changes the metric d to an equivalent metric d' which means that the identity $(\mathcal{M}_{g,m}, d) \to (\mathcal{M}_{g,m}, d')$ is a bilipschitz map.

A basic idea of geometric group theory is to relate geometric properties of a finitely generated group Γ to global group theoretic properties of Γ. A fundamental and striking example for such an interplay between geometry and group theory is the following celebrated theorem of Gromov [**G**]. Define the *growth function* of a finitely generated group Γ as follows. For $R > 0$, let $m(R)$ be the number of elements in Γ whose distance to the identity element with respect to some fixed word norm is at most R. The group is called *of polynomial growth* if there is some $d > 0$ such that $m(R) < dR^d$ for all R. Note that this property is independent of the choice of generators. Gromov shows that a group is of polynomial growth if and only if it is *virtually nilpotent*, i.e. if it contains a nilpotent subgroup of finite index.

On the other hand, a group Γ which acts on a *geodesic* metric space X as a group of isometries inherits from X large-scale geometric properties provided that the action satisfies some discreteness assumptions. The most elementary result along this line is the theorem of Švarc and Milnor (see Chapter I.8 in [**BH**]) which can be stated as follows. Assume that a countable group Γ acts as a group of isometries on a proper geodesic metric space (X, d). If the action is proper and cocompact then Γ is finitely generated, and for any $x \in X$ the orbit map $g \in \Gamma \to gx \in X$ is a *quasi-isometry*. This means that if we denote by d_Γ any distance on Γ defined by a finite generating set, then there is a number $c > 0$ such that $d_\Gamma(g,h)/c - c \leq d(gx, hx) \leq cd_\Gamma(g,h) + c$ for all $g, h \in \Gamma$.

There are two natural metric graphs on which the mapping class group $\mathcal{M}_{g,m}$ acts by isometries. These graphs are the so-called *curve complex* (or, rather, its one-skeleton) and the *train track complex*. In Chapter 3 we give a description of the curve complex and its most relevant geometric properties. In Chapter 4 we discuss the action of $\mathcal{M}_{g,m}$ on the curve complex and some of its consequences for the structure of $\mathcal{M}_{g,m}$. In Chapter 5 we introduce the complex of train tracks and show how it can be used to study $\mathcal{M}_{g,m}$. In Chapter 2 we collect those properties of train tracks and geodesic laminations which are important for the later chapters. Chapters 3-5 also contain a collection of open problems.

2. Geodesic laminations and train tracks

A *geodesic lamination* for a complete hyperbolic structure of finite volume on S is a compact subset of S which is foliated into simple geodesics. Particular geodesic laminations are simple closed geodesics, i.e. laminations which consist of a single leaf. A geodesic lamination λ is called *minimal* if each of its half-leaves is dense in λ. Thus a simple closed geodesic is a minimal geodesic lamination. A minimal geodesic lamination with more than one leaf has uncountably many leaves. Every geodesic lamination λ is a disjoint union of finitely many minimal components and a finite number of non-compact isolated leaves. Each of the isolated leaves of λ either is an isolated closed geodesic and hence a minimal component, or it *spirals* about one or two minimal components ([**Bo1**], Theorem 4.2.8 of [**CEG**], [**O**]).

A geodesic lamination λ is *maximal* if all its complementary components are ideal triangles or once punctured monogons. A geodesic lamination is called *complete* if it is maximal and can be approximated in the Hausdorff topology for compact subsets of S by simple closed geodesics. Every minimal geodesic lamination λ is a *sublamination* of a complete geodesic lamination [**H1**], i.e. there is a complete geodesic lamination which contains λ as a closed subset. In particular, every simple closed geodesic c on S is a sublamination of a complete geodesic lamination. Such a lamination can be constructed as follows. Let P be a geodesic *pants decomposition* for S containing c; this means that P consists of a collection of $3g - 3 + m$ simple closed pairwise disjoint geodesics, and c is one of these. Then $S - P$ consists of $2g - 2 + m$ connected components, and each of these components is a *pairs of pants*, i.e. an oriented surface homeomorphic to a thrice punctured sphere. The metric completion of each such pair of pants is a bordered surface with one, two or three boundary circles depending on the number of punctures of S which it contains. For each such pair of pants S_0 choose a maximal collection of simple disjoint geodesics embedded in S_0 which spiral about the boundary circles of its metric completion $\overline{S}_0$. We also require that for every pair c, d of boundary components of S_0 there is a geodesic from the collection which spirals in one direction about c, in the other direction about d. In particular, there is at least one geodesic spiraling from each side of a curve from the collection P. We require that the spiraling directions from both sides of such a pants curve are opposite. The resulting lamination is then complete [**H1**].

A *measured geodesic lamination* on S is a geodesic lamination λ together with a translation invariant transverse measure supported in λ. Here a transverse measure for λ assigns to every smooth arc c on S with endpoints in the complement of λ and which intersects λ transversely a measure on c supported in $c \cap \lambda$. These measures transform in the natural way under homotopies of c by smooth arcs transverse to λ which move the endpoints of the arc c within fixed complementary components. The support of the measure is the smallest sublamination ν of λ such that the measure on any arc c which does not intersect ν is trivial. This support is necessarily a union of minimal components of λ. As an example, every simple closed geodesic γ naturally carries a transverse *counting measure* which associates to an arc c as above the sum of the Dirac masses at the intersection points between c and γ. If μ is any transverse measure for λ, then for every $a > 0$ the same is true for $a\mu$ and hence the group $(0, \infty)$ of positive reals naturally acts on the space $\mathcal{ML}$ of measured geodesic laminations. The space $\mathcal{ML}$ carries a natural topology, the so-called weak*-topology, which locally restricts to the usual weak*-topology for measures on transverse arcs. The action of $(0, \infty)$ is continuous with respect to the weak*-topology. The projectivization of $\mathcal{ML} - \{0\}$ is the space $\mathcal{PML}$ of *projective measured laminations* on S. Equipped with the quotient of the weak*-topology, $\mathcal{PML}$ is homeomorphic to a sphere of dimension $6g - 7 + 2m$; in particular, $\mathcal{PML}$ is compact (for all this, see [**FLP**] and [**PH**], in particular Theorem 3.1.4).

The *intersection number* $i(\gamma, \delta)$ between two simple closed curves $\gamma, \delta \in \mathcal{C}(S)$ equals the minimal number of intersection points between representatives of the free homotopy classes of γ, δ. This intersection number extends bilinearly to a continuous pairing for measured geodesic laminations on S.

A *train track* on the surface S is an embedded 1-complex $\tau \subset S$ whose edges (called *branches*) are smooth arcs with well-defined tangent vectors at the endpoints. At any vertex (called a *switch*) the incident edges are mutually tangent. Through each switch there is a path of class C^1 which is embedded in τ and contains the switch in its interior. In particular, the branches which are incident on a fixed switch are divided into "incoming" and "outgoing" branches according to their inward pointing tangent at the switch. Each closed curve component of τ has a unique bivalent switch, and all other switches are at least trivalent. The complementary regions of the train track have negative Euler characteristic, which means that they are different from discs with $0, 1$ or 2 cusps at the boundary and different from annuli and once-punctured discs with no cusps at the boundary. We always identify train tracks which are isotopic. A detailed account on train tracks can be found in [**PH**] and [**M2**].

A train track is called *generic* if all switches are at most trivalent. The train track τ is called *transversely recurrent* if every branch b of τ is intersected by an embedded simple closed curve $c = c(b) \subset S$ which intersects τ transversely and is such that $S - \tau - c$ does not contain an embedded *bigon*, i.e. a disc with two corners at the boundary. In this case we say that c *hits τ efficiently*.

A geodesic lamination or a train track λ is *carried* by a transversely recurrent train track τ if there is a map $F : S \to S$ of class C^1 which is isotopic to the identity and maps λ to τ in such a way that the restriction of its differential dF to every tangent line of λ is non-singular. Note that this makes sense since a train track has a tangent line everywhere. A train track τ is called *complete* if it is generic and transversely recurrent and if it carries a complete geodesic lamination [**H1**].

If c is a simple closed curve carried by τ with carrying map $F : c \to \tau$ then c defines a *counting measure* μ_c on τ. This counting measure is the non-negative weight function on the branches of τ which associates to an open branch b of τ the number of connected components of $F^{-1}(b)$. A counting measure is an example for a *transverse measure* on τ which is defined to be a nonnegative weight function μ on the branches of τ satisfying the *switch condition*: For every switch s of τ, the sum of the weights over all incoming branches at s is required to coincide with the sum of the weights over all outgoing branches at s. The set $V(\tau)$ of all transverse measures on τ is a closed convex cone in a linear space and hence topologically it is a closed cell. For every transverse measure μ on τ there is a measured geodesic lamination λ and a carrying map $F : \lambda \to \tau$ such that for every branch b of τ, the weight $\mu(b)$ is just the transverse measure of a compact arc transverse to λ which is mapped by F to a single point in the interior of b. A train track is called *recurrent* if it admits a transverse measure which is positive on every branch. For every recurrent train track τ, measures which are positive on every branch define the interior of the convex cone $V(\tau)$. A complete train track τ is recurrent [**H1**]. An arbitrary train track which is both recurrent and transversely recurrent is called *birecurrent*.

A half-branch $\tilde{b}$ in a generic train track τ incident on a switch v is called *large* if the switch v is trivalent and if every arc $\rho : (-\epsilon, \epsilon) \to \tau$ of class C^1 which passes through v meets the interior of $\tilde{b}$. A branch b in τ is called *large* if each of its two half-branches is large; in this case b is necessarily incident on two distinct switches (for all this, see [**PH**]).

There is a simple way to modify a complete train track τ to another complete train track. Namely, if e is a large branch of τ then we can perform a right or left *split* of τ at e as shown in Figure A below. The split τ' of a train track τ is carried by τ. If τ is complete and if the complete geodesic lamination λ is carried by τ, then for every large branch e of τ there is a unique choice of a right or left split of τ at e with the property that the split track τ' carries λ, and τ' is complete. In particular, a complete train track τ can always be split at any large branch e to a complete train track τ'; however there may be a choice of a right or left split at e such that the resulting track is not complete any more (compare p. 120 in [**PH**]).

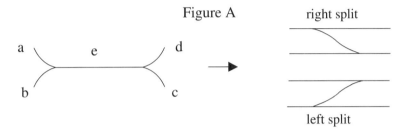

Figure A

In the sequel we denote by $\mathcal{TT}$ the collection of all isotopy classes of complete train tracks on S. A sequence $(\tau_i)_i \subset \mathcal{TT}$ of complete train tracks is called a *splitting sequence* if τ_{i+1} can be obtained from τ_i by a single split at some large branch e. Note that in this case for each i the train track τ_{i+1} is carried by τ_i.

3. The complex of curves

In [**Ha**], Harvey defined the *complex of curves* $\mathcal{C}(S)$ for S. The vertices of this complex are free homotopy classes of essential simple closed curves on S, i.e. simple closed curves which are not contractible nor homotopic into a puncture. For every fixed choice of a complete hyperbolic metric on S of finite volume, every such free homotopy class can be represented by a unique simple closed geodesic. The simplices in $\mathcal{C}(S)$ are spanned by collections of such curves which can be realized disjointly and hence the dimension of $\mathcal{C}(S)$ equals $3g - 3 + m - 1$ (recall that $3g - 3 + m$ is the number of curves in a pants decomposition of S). In the sequel we restrict our attention to the one-skeleton of $\mathcal{C}(S)$ which we denote again by $\mathcal{C}(S)$ by abuse of notation. Since $3g - 3 + m \geq 2$ by assumption, $\mathcal{C}(S)$ is a nontrivial graph which moreover is connected [**Ha**]. However, this graph is locally infinite. Namely, for every simple closed curve α on S the surface $S - \alpha$ which we obtain by cutting S open along α contains at least one connected component which is different from a thrice punctured sphere, and such a component contains infinitely many distinct free homotopy classes of simple closed curves which viewed as curves in S are disjoint from α.

Providing each edge in $\mathcal{C}(S)$ with the standard euclidean metric of diameter 1 equips the complex of curves with the structure of a geodesic metric space. Since $\mathcal{C}(S)$ is not locally finite, this metric space $(\mathcal{C}(S), d)$ is not locally compact. Nevertheless, its geometry can be understood quite explicitly. Namely, for some $\delta > 0$ a geodesic metric space is called δ-*hyperbolic in the sense of Gromov* if it satisfies the δ-*thin triangle condition*: For every geodesic triangle with sides a, b, c the side c is contained in the δ-neighborhood of $a \cup b$. The following important result is due to Masur and Minsky [**MM1**] (see also [**B1**] and [**H4**] for alternate proofs).

THEOREM 3.1 ([**MM1**]). *The complex of curves is hyperbolic.*

For some $q > 1$, a *q-quasi-geodesic* in $\mathcal{C}(S)$ is a curve $c : [a, b] \to \mathcal{C}(S)$ which satisfies
$$d(c(s), c(t))/q - q \leq |s - t| \leq qd(c(s), c(t)) + q \quad \text{for all } s, t \in [a, b].$$
Note that a quasi-geodesic does not have to be continuous. Call a curve $c : [0, m] \to \mathcal{C}(S)$ an *unparametrized q-quasigeodesic* if there is some $p > 0$ and a homeomorphism $\rho : [0, p] \to [0, m]$ such that the curve $c \circ \rho : [0, p] \to \mathcal{C}(S)$ is a q-quasi-geodesic. In a hyperbolic geodesic metric space, every q-quasi-geodesic is contained in a tubular neighborhood of fixed radius about any geodesic joining the same endpoints, so the δ-thin triangle condition also holds for triangles whose sides are uniform unparametrized quasi-geodesics (Theorem 1.7 in Chapter III.H of [**BH**]). Moreover, to understand the coarse geometric structure of $\mathcal{C}(S)$ it is enough to identify for a fixed $q > 1$ a collection of unparametrized q-quasi-geodesics connecting any pair of points in $\mathcal{C}(S)$.

To obtain such a system of curves we define a map from the set $\mathcal{T}\mathcal{T}$ of complete train tracks on S into $\mathcal{C}(S)$. Call a transverse measure μ on a birecurrent train track τ a *vertex cycle* if μ spans an extreme ray in the convex cone $V(\tau)$ of all transverse measures on τ. Up to scaling, every vertex cycle μ is a counting measure of a simple closed curve c which is carried by τ (p. 115 of [**MM1**]). A simple closed curve which is carried by τ, with carrying map $F : c \to \tau$, defines a vertex cycle on τ only if $F(c)$ passes through every branch of τ at most twice, with different orientation (Lemma 2.2 of [**H2**]). Thus if c is a vertex cycle for τ then its counting measure μ_c satisfies $\mu_c(b) \leq 2$ for every branch b of τ.

In the sequel we mean by a vertex cycle of a complete train track τ an *integral* transverse measure on τ which is the counting measure of a simple closed curve c on S carried by τ and which spans an extreme ray of $V(\tau)$; we also use the notion vertex cycle for the simple closed curve c. Since the number of branches of a complete train track on S only depends on the topological type of S, the number of vertex cycles for a complete train track on S is bounded by a universal constant (see [**MM1**] and [**H2**]). Moreover, there is a number $D_0 > 0$ with the property that for every train track $\tau \in \mathcal{T}\mathcal{T}$ the distance in $\mathcal{C}(S)$ between any two vertex cycles of τ is at most D_0 (see [**MM1**] and the discussion following Corollary 2.3 in [**H2**]).

Define a map $\Phi : \mathcal{T}\mathcal{T} \to \mathcal{C}(S)$ by assigning to a train track $\tau \in \mathcal{T}\mathcal{T}$ a vertex cycle $\Phi(\tau)$ for τ. By our above discussion, for any two choices Φ, Φ' of such a map we have $d(\Phi(\tau), \Phi'(\tau)) \leq D_0$ for all $\tau \in \mathcal{T}\mathcal{T}$. The following result is due to Masur and Minsky ([**MM3**], see also [**H2**] for an alternate proof).

THEOREM 3.2 ([**MM3**]). *There is a number $q > 0$ such that the image under Φ of an arbitrary splitting sequence in $\mathcal{T}\mathcal{T}$ is an unparametrized q-quasi-geodesic*

Theorem 3.2 can be used to construct for any pair α, β of points in $\mathcal{C}(S)$ an unparametrized q-quasi-geodesic connecting α to β. Namely, for a given $\alpha \in \mathcal{C}(S)$ choose a pants decomposition P containing α. Then every $\beta \in \mathcal{C}(S)$ is uniquely determined by the $3g - 3 + m$-tuple of intersection numbers between β and the pants curves of P and a $3g - 3 + m$-tuple of twist parameters with respect to a fixed system of *spanning arcs*. Such a system of spanning arcs consists of a choice of a point on each component of P and a maximal collection of disjoint simple pairwise not mutually homotopic arcs each embedded in a pair of pants and with endpoints at the distinguished points on the components of P (see [**FLP**]). Using the pants decomposition and the spanning arcs we can construct explicitly a complete train track τ which admits α as a vertex cycle and carries β (such a construction can be found in Section 2.6 of [**PH**]); this train track is then the initial point of a splitting sequence which connects a train track admitting α as a vertex cycle to a train track admitting β as a vertex cycle. For a suitable choice of the map Φ, the image under Φ of this splitting sequence is an unparametrized q-quasi-geodesic in $\mathcal{C}(S)$ connecting α to β.

However, it is also possible to construct explicitly for each pair of points $\alpha, \beta \in \mathcal{C}(S)$ a *geodesic* connecting α to β. Namely, building on the results of [**MM2**], Shackleton recently proved [**S**].

THEOREM 3.3 ([**S**]). *There is an algorithm which takes as input two curves $\alpha, \beta \in \mathcal{C}(S)$ and returns a geodesic between α and β.*

A hyperbolic geodesic metric space X admits a *Gromov boundary* which is defined as follows. Fix a point $p \in X$ and for two points $x, y \in X$ define the *Gromov product* $(x,y)_p = \frac{1}{2}(d(x,p) + d(y,p) - d(x,y))$. Call a sequence $(x_i) \subset X$ *admissible* if $(x_i, x_j)_p \to \infty$ $(i, j \to \infty)$. We define two admissible sequences $(x_i), (y_i) \subset X$ to be *equivalent* if $(x_i, y_i)_p \to \infty$. Since X is hyperbolic, this defines indeed an equivalence relation (see the discussion on p. 431 of [**BH**]). The Gromov boundary ∂X of X is the set of equivalence classes of admissible sequences $(x_i) \subset X$. It carries a natural Hausdorff topology. For the complex of curves, the Gromov boundary was determined by Klarreich [**K**] (see also [**H2**]).

For the formulation of Klarreich's result, we say that a geodesic lamination λ *fills up* S if every simple closed geodesic on S intersects λ transversely, i.e. if every complementary component of λ is an ideal polygon or a once punctured ideal polygon with geodesic boundary. For any geodesic lamination λ which fills up S, the number of geodesic laminations which contain λ as a sublamination is bounded by a universal constant only depending on the topological type of the surface S. Namely, each such lamination μ can be obtained from λ by successively subdividing complementary components P of λ which are different from an ideal triangle or a once punctured monogon by adding a simple geodesic line which either connects two non-adjacent cusps of P or goes around a puncture of S. Note that every leaf of μ which is not contained in λ is necessarily isolated in μ.

The space $\mathcal{L}$ of geodesic laminations on S can be equipped with the *Hausdorff topology* for compact subsets of S. With respect to this topology, $\mathcal{L}$ is compact and metrizable. We say that a sequence $(\lambda_i) \subset \mathcal{L}$ *converges in the coarse Hausdorff topology* to a minimal lamination μ which fills up S if every accumulation point of (λ_i) with respect to the Hausdorff topology contains μ as a sublamination. We equip the space $\mathcal{B}$ of minimal geodesic laminations which fill up S with the following topology. A set $A \subset \mathcal{B}$ is closed if and only if for every sequence $(\lambda_i) \subset A$ which converges in the coarse Hausdorff topology to a lamination $\lambda \in \mathcal{B}$ we have $\lambda \in A$. We call this topology on $\mathcal{B}$ the *coarse Hausdorff topology*. Using this terminology, Klarreich's result [**K**] can be formulated as follows.

THEOREM 3.4 ([**K**], [**H2**]). (1) *There is a natural homeomorphism Λ of $\mathcal{B}$ equipped with the coarse Hausdorff topology onto the Gromov boundary $\partial \mathcal{C}(S)$ of the complex of curves $\mathcal{C}(S)$ for S.*

(2) For $\mu \in \mathcal{B}$ a sequence $(c_i) \subset \mathcal{C}(S)$ is admissible and defines the point $\Lambda(\mu) \in \partial \mathcal{C}(S)$ if and only if (c_i) converges in the coarse Hausdorff topology to μ.

For every hyperbolic geodesic metric space X, the Gromov product $(\,,\,)_p$ based at a point $p \in X$ can be extended to a product on $X \cup \partial X$ by defining

$$(\xi, \eta)_p = \sup_{(x_i),(y_j)} \lim\inf_{i,j \to \infty} (x_i, y_j)_p$$

where the supremum is taken over all sequences representing the points ξ, η (i.e. if $\xi \in X$ the $x_i = \xi$ for all i). There is a natural topology on $X \cup \partial X$ which restricts to the given topology on X and on ∂X. For any given point $p \in X$ and every $\xi \in \partial X$, the family of *cones* based at p of the form $C_p(\xi, \delta) = \{y \in X \cup \partial X \mid (y, \xi)_p \geq -\log \delta\}$ ($\delta > 0$) define a neighborhood basis at ξ with respect to this topology.

The Gromov boundary ∂X of every Gromov hyperbolic geodesic metric space X carries a natural distance function δ defining its topology with the property that there are numbers $c > 0, \kappa > 0$ only depending on the hyperbolicity constant such that $ce^{-\kappa(\xi,\zeta)_p} \leq \delta(\xi, \zeta) \leq e^{-\kappa(\xi,\zeta)_p}$ for all $\xi, \zeta \in \partial X$ [**GH**]. If X is proper, then the metric δ is complete and $(\partial X, \delta)$ is compact. However, the metric space $\mathcal{C}(S)$ is not proper. Thus unlike in the case of proper hyperbolic metric spaces, (metrically) diverging sequences of points in $\mathcal{C}(S)$ may not have any accumulation point in $\partial \mathcal{C}(S)$.

PROBLEM 1. Is the Gromov boundary of $\mathcal{C}(S)$ locally connected? Relate the structure of this boundary to the geometry of $\mathcal{C}(S)$.

There is yet another way to construct unparametrized uniform quasi-geodesics in $\mathcal{C}(S)$. Namely, the *Teichmüller space* for S is the space $\mathcal{T}_{g,m}$ of marked isometry classes of complete hyperbolic metrics on S of finite volume. The Teichmüller space can naturally be identified with a domain in $\mathbb{C}^{3g-3+m}$ (see Chapter 6 of [**IT**]).

By a classical result of Bers (see [**Bu**]), there is a number $\chi > 0$ such that for every complete hyperbolic metric h on S there is a pants decomposition for S consisting of simple closed h-geodesics of length at most χ. Moreover, the diameter in $\mathcal{C}(S)$ of the set of simple closed curves on S of h-length at most χ is bounded from above by a universal constant $D > 0$. Thus we can define a map $\Psi : \mathcal{T}_{g,m} \to \mathcal{C}(S)$ by associating to a marked hyperbolic metric h a simple closed curve of h-length at most χ. For any two such maps Ψ, Ψ' we then have $\sup_{h \in \mathcal{T}_{g,m}} d(\Psi(h), \Psi'(h)) \leq D$.

The *Teichmüller metric* on $\mathcal{T}_{g,m}$ is a complete Finsler metric which is just the *Kobayashi metric* on the domain in $\mathbb{C}^{3g-3+m}$ representing $\mathcal{T}_{g,m}$ (see [**IT**]). Through any two distinct points in $\mathcal{T}_{g,m}$ passes a unique Teichmüller geodesic. Each such geodesic line in $\mathcal{T}_{g,m}$ is uniquely determined by its endpoints in the *Thurston boundary* of $\mathcal{T}_{g,m}$ which is just the space $\mathcal{PML}$ of projective measured laminations on S. The supports of the two measured laminations on S defining the endpoints of the geodesic together fill up the surface S, i.e. every simple closed curve on S intersects at least one of the two laminations transversely. These laminations then define a *holomorphic quadratic differential* (we refer to Section 4 of [**Ke**] for a discussion of this fact). The following result is implicitly contained in the paper [**MM1**] by Masur and Minsky; an explicit proof using a result of Rafi [**R**] can be found in Section 4 of [**H4**].

THEOREM 3.5 ([**MM1**]). *There is a universal constant $\tilde{q} > 0$ such that the image under Ψ of every Teichmüller geodesic is an unparametrized $\tilde{q}$-quasi-geodesic in $\mathcal{C}(S)$.*

The mapping class group acts properly discontinuously on $\mathcal{T}_{g,m}$, viewed as a domain in $\mathbb{C}^{3g-3+m}$, as a group of biholomorphic automorphisms. The quotient of $\mathcal{T}_{g,m}$ under this action is the *moduli space* $\text{Mod}(S)$ of S, a non-compact complex orbifold. A geodesic lamination λ on S is called *uniquely ergodic* if it supports up to scaling a *unique* transverse measure. Masur [**Mas**] showed that the endpoint in $\mathcal{PML}$ of a Teichmüller ray which projects to a *compact* subset of moduli space is uniquely ergodic. For a fixed number $R > 0$ there is a compact subset $K(R)$ of moduli space containing the projection of every Teichmüller geodesic γ which satisfies $d(\Psi(\gamma(s)), \Psi(\gamma(t))) \geq |s-t|/R - R$ for all $s,t \in \mathbb{R}$. Conversely, there is for a given compact set K in $\text{Mod}(S)$ a number $R = R(K) > 0$ such that $d(\Psi(\gamma(s)), \Psi(\gamma(t))) \geq |s-t|/R - R$ for every Teichmüller geodesic which projects to K [**H6**].

PROBLEM 2. Analyze the images in $\mathcal{C}(S)$ of geodesics in $\mathcal{T}_{g,m}$ determined by minimal geodesic laminations which fill up S and are *not* uniquely ergodic.

4. The action of $\mathcal{M}_{g,m}$ on the complex of curves

The relevance of the geometry of the complex of curves for the understanding of the geometry of the mapping class group comes from the obvious fact that the extended mapping class group $\mathcal{M}^{\pm}_{g,m}$ of S acts on the complex of curves as a group of simplicial automorphisms and hence isometries. Even more is true: If S is not a torus with 2 punctures then the extended mapping class group is *precisely* the group of isometries of $\mathcal{C}(S)$ (see Chapter 8 of [**I**] for references and a sketch of the proof). Thus the mapping class group inherits geometric properties from the complex of curves provided that the action satisfies some properness assumption. In this section we discuss a result of Bowditch [**B2**] who proved that this is indeed the case, and we derive some consequences for the group structure of $\mathcal{M}_{g,m}$.

To begin with, recall that the action of the mapping class group on $\mathcal{C}(S)$ is essentially transitive.

LEMMA 4.1. (1) *There are only finitely many orbits for the action of $\mathcal{M}_{g,m}$ on $\mathcal{C}(S)$.*
(2) *For any pair $\alpha, \beta \in \mathcal{C}(S)$ there is some $h \in \mathcal{M}_{g,m}$ with $d(h\alpha, \beta) \leq 2$.*

The *limit set* of a group Γ of isometries of a Gromov hyperbolic metric space X is the set of accumulation points in ∂X of a fixed Γ-orbit Γx on X. This limit set does not depend on x. The group Γ naturally acts as a group of homeomorphisms on the boundary ∂X, and the limit set $\Lambda(\Gamma)$ is invariant under this action. Moreover, if $\Lambda(\Gamma)$ contains at least 3 points then $\Lambda(\Gamma)$ is uncountable and it is the smallest nontrivial closed Γ-invariant subset of ∂X. This means in particular that for every $\xi \in \Lambda(\Gamma)$ the Γ-orbit of ξ is dense in $\Lambda(\Gamma)$.

The following is immediate from Lemma 4.1.

COROLLARY 4.2. *The limit set of the action of $\mathcal{M}_{g,m}$ on $\mathcal{C}(S)$ equals the whole boundary $\partial \mathcal{C}(S)$. In particular, every $\mathcal{M}_{g,m}$-orbit on $\partial \mathcal{C}(S)$ is dense.*

PROOF. By Lemma 4.1, the action of $\mathcal{M}_{g,m}$ on $\mathcal{C}(S)$ is essentially transitive. Thus for every fixed $\alpha \in \mathcal{C}(S)$ and every admissible sequence $(c_i) \subset \mathcal{C}(S)$ converging in $\mathcal{C}(S) \cup \partial \mathcal{C}(S)$ to some $\xi \in \partial \mathcal{C}(S)$ there is a sequence $(\varphi_i) \subset \mathcal{M}_{g,m}$ with $d(\varphi_i(\alpha), c_i) \leq 2$. Then $(\varphi_i(\alpha))$ converges to ξ, i.e. ξ is contained in the limit set of $\mathcal{M}_{g,m}$. This shows the corollary. $\square$

A *simple Dehn twist* about a simple closed essential curve c in S is an element of $\mathcal{M}_{g,m}$ which can be represented in the following form. Let $A \subset S$ be an embedded closed annulus with smooth boundary and core curve c. There is a diffeomorphism φ of A which preserves the boundary pointwise and maps an arc $\alpha : [0,1] \to A$ connecting two points on the two different boundary components and intersecting c in a single point $\alpha(t)$ to an arc with the same endpoints which is homotopic to the composition $\alpha[0,t] * c * \alpha[t,1]$. The homeomorphism $\tilde{\varphi}$ of S whose restriction to A equals φ and whose restriction to $S - A$ is the identity then represents a simple Dehn twist about c. Such a simple Dehn twist generates an infinite cyclic subgroup of $\mathcal{M}_{g,m}$ which just equals the center of the stabilizer of c in $\mathcal{M}_{g,m}$. It follows that the stabilizer of c in $\mathcal{M}_{g,m}$ is the direct product of the infinite cyclic group of Dehn twists about c and the mapping class group of the surface $S - c$.

As we saw so far, the action of $\mathcal{M}_{g,m}$ on $\mathcal{C}(S)$ is essentially transitive, and for every $\alpha \in \mathcal{C}(S)$ the stabilizer of α in $\mathcal{M}_{g,m}$ contains the mapping class group of the surface $S - \alpha$ as a subgroup of infinite index. Now if $\beta \in \mathcal{C}(S)$ is such that $d(\alpha, \beta) \geq 3$ then $\beta - \alpha$ consists of a collection of simple arcs which decompose $S - \alpha$ into discs and once punctured discs. Since β does not have self-intersections, the number of free homotopy classes relative to α of such components of $\beta - \alpha$ is bounded from above by a universal constant. However, the number of arcs in each free homotopy class is invariant under the action of the stabilizer of α in $\mathcal{M}_{g,m}$, indicating that this action is by no means transitive on the set of curves whose distance to α is a fixed constant. Indeed, Bowditch [**B2**] showed that the action of $\mathcal{M}_{g,m}$ on pairs of points in $\mathcal{C}(S)$ of sufficiently large distance is proper in a metric sense.

To explain his result, call an isometric action of a group Γ on a hyperbolic metric space X *acylindrical* if for every $m > 0$ there are numbers $R = R(m) > 0, c = c(m) > 0$ with the following property. Let $y, z \in X$ be such that $d(y, z) \geq R$; then there are at most $c(x, m)$ elements $\varphi \in \Gamma$ with $d(\varphi(y), y) \leq m, d(\varphi(x), x) \leq m$. This is a weak notion of properness in a metric sense for a group of isometries of a hyperbolic geodesic metric space. Using an indirect argument via the geometry and topology of 3-manifolds, Bowditch shows (Theorem 1.3 of [**B2**]).

THEOREM 4.3 ([**B2**]). *The action of* $\mathcal{M}_{g,m}$ *on* $\mathcal{C}(S)$ *is acylindrical.*

By Thurston's classification of elements of the mapping class group (see [**FLP**] and [**CB**]), $\mathcal{M}_{g,m}$ can be divided into three disjoint subsets. The first set contains all *periodic* elements $\varphi \in \mathcal{M}_{g,m}$, i.e. elements for which there is some $k > 1$ with $\varphi^k = Id$. Every orbit of the action on $\mathcal{C}(S)$ of the cyclic group generated by φ is finite and hence *bounded*. The second set contains all *reducible* elements in $\mathcal{M}_{g,m}$ which are not periodic. Such a reducible non-periodic element φ of $\mathcal{M}_{g,m}$ preserves a non-trivial *multi-curve*, i.e. a collection of pairwise disjoint mutually not freely homotopic simple closed essential curves on S. Then there is some $m \geq 1$ such that φ^m fixes an element of $\mathcal{C}(S)$ and once again, the orbits of the action on $\mathcal{C}(S)$

of the cyclic group generated by φ are bounded. The third set contains the so-called *pseudo-Anosov* elements. A pseudo-Anosov mapping class φ acts on $\mathcal{C}(S)$ as a *hyperbolic* isometry. The action of φ on the boundary $\partial\mathcal{C}(S)$ of $\mathcal{C}(S)$ has *north-south dynamics* with respect to a pair $\xi \neq \zeta$ of fixed points. By this we mean the following (compare the discussion in [**H3**]).

(1) For every neighborhood U of ξ and every neighborhood V of ζ there is some $m > 0$ such that $\varphi^m(\partial\mathcal{C}(S) - V) \subset U$ and $\varphi^{-m}(\partial\mathcal{C}(S) - U) \subset V$.
(2) There is a closed subset D of $\partial\mathcal{C}(S) - \{\xi, \zeta\}$ such that $\cup_i \varphi^i D = \partial\mathcal{C}(S) - \{\xi, \zeta\}$.

There is a number $p > 0$ only depending on the hyperbolicity constant for $\mathcal{C}(S)$ such that every pseudo-Anosov element preserves a p-quasi-geodesic connecting the two fixed points of φ on $\partial\mathcal{C}(S)$. Even more is true. Bowditch (Theorem 1.4 in [**B2**]) showed that for a pseudo-Anosov mapping class φ there is some $m > 0$ such that φ^m preserves a geodesic in $\mathcal{C}(S)$. Moreover, the *stable length* $\|\varphi\|$ of φ is positive (and moreover rational and bounded from below by a positive constant) where $\|\varphi\| = \lim_{n\to\infty} d(x, \varphi^n x)/n$ for an arbitrary point $x \in \mathcal{C}(S)$ (note that this limit always exists and is independent of x).

Let $\varphi \in \mathcal{M}_{g,m}$ be a pseudo-Anosov element; this element determines a point $a(\varphi)$ in the complement of the diagonal Δ of $\partial\mathcal{C}(S) \times \partial\mathcal{C}(S)$. By Bowditch's result, the set of points $a(\hat{\varphi})$ where $\hat{\varphi}$ ranges over all elements of $\mathcal{M}_{g,m}$ representing the conjugacy class of φ is a discrete subset of $\partial\mathcal{C}(S) \times \partial\mathcal{C}(S) - \Delta$ (this also follows from [**BFu**]). Imitating a construction for Riemannian manifolds of bounded negative curvature and hyperbolic groups (see [**Bo2**]), define a *geodesic current* for $\mathcal{M}_{g,m}$ to be a locally finite $\mathcal{M}_{g,m}$-invariant Borel measure on $\partial\mathcal{C}(S) \times \partial\mathcal{C}(S) - \Delta$. The sum of Dirac masses at the pairs of points corresponding to a pseudo-Anosov element of $\mathcal{M}_{g,m}$ is a such a geodesic current.

PROBLEM 3. Describe the space of geodesic currents for $\mathcal{M}_{g,m}$. Is the set of weighted sums of Dirac masses at the pairs of fixed points of pseudo-Anosov elements dense? Describe the geodesic currents μ which is *absolutely continuous*, i.e. such that there is a $\mathcal{M}_{g,m}$-invariant measure class μ_0 on $\partial\mathcal{C}(S)$ with the property that for every Borel subset A of $\partial\mathcal{C}(S)$ we have $\mu_0(A) = 0$ if and only if $\mu(A \times \partial\mathcal{C}(S) - \Delta) = 0$? Is there a distinguished absolutely continuous current such that the invariant measure class on $\partial\mathcal{C}(S)$ is determined by a Hausdorff measure with respect to one of the distance functions δ on $\partial\mathcal{C}(S)$?

A subgroup Γ of $\mathcal{M}_{g,m}$ is called *elementary* if its limit set contains at most 2 points. The next lemma follows from the work of McCarthy (compare [**MP**]).

LEMMA 4.4 ([**MP**]). *Let Γ be an elementary subgroup of $\mathcal{M}_{g,m}$. Then either Γ is virtually abelian or Γ contains a subgroup of finite index which stabilizes a nontrivial subsurface of S.*

PROOF. Our lemma relies on the following observation. Call a subgroup Γ of $\mathcal{M}_{g,m}$ *reducible* if there is a non-empty finite Γ-invariant family of disjoint simple closed curves on S. A reducible subgroup Γ of $\mathcal{M}_{g,m}$ has a finite orbit on $\mathcal{C}(S)$ and therefore its limit set is trivial. A subgroup Γ which is neither finite nor reducible contains a pseudo-Anosov element φ (this is claimed in Lemma 2.8 of [**MP**]) and hence its limit set contains at least the fixed points $\alpha \neq \beta$ of the action of φ on $\partial\mathcal{C}(S)$. Thus if Γ is elementary then the limit set of Γ coincides with the set $\{\alpha, \beta\}$

and therefore every element $\psi \in \Gamma$ preserves $\{\alpha, \beta\}$. Since the action of $\mathcal{M}_{g,m}$ on $\mathcal{C}(S)$ is weakly acylindrical the cyclic group generated by φ is of finite index in the subgroup of $\mathcal{M}_{g,m}$ which preserves α, β (this is claimed in Lemma 9.1 of [**Mc**] and also follows from the results in [**BFa**]). As a consequence, an elementary subgroup of $\mathcal{M}_{g,m}$ either is finite or reducible or virtually abelian. $\square$

Let Γ be a countable group and let V be a continuous Banach module for Γ. This means that V is a Banach space and that there is a representation of Γ into the group of linear isometries of V. We are only interested in the case when $V = \mathbb{R}$ with the trivial Γ-action or $V = \ell^p(\Gamma)$ for some $p \in [1, \infty)$ with the standard left action of Γ. The second bounded cohomology group $H_b^2(\Gamma, V)$ of Γ with coefficients in V is defined as the second cohomology group of the complex

$$0 \to L^\infty(\Gamma, V)^\Gamma \xrightarrow{d} L^\infty(\Gamma^2, V)^\Gamma \xrightarrow{d} \dots$$

with the usual homogeneous coboundary operator d and the twisted action of Γ. For $V = \mathbb{R}$ there is a natural homomorphism of $H_b^2(\Gamma, \mathbb{R})$ into the ordinary second cohomology group $H^2(\Gamma, \mathbb{R})$ of Γ which in general is neither injective nor surjective. Since the action of $\mathcal{M}_{g,m}$ on $\mathcal{C}(S)$ is weakly acylindrical we obtain the following [**BFu**], [**H3**].

THEOREM 4.5 ([**BFu**], [**H3**]). *Let $\Gamma < \mathcal{M}_{g,m}$ be any subgroup. If Γ is not virtually abelian then for every $p \in [1, \infty)$ the second bounded cohomology groups $H_b^2(\Gamma, \ell^p(\Gamma)), H_b^2(\Gamma, \mathbb{R})$ are infinitely generated.*

As a corollary, one obtains the following super-rigidity theorem for mapping class groups which was earlier shown by Farb and Masur [**FM**] building on the work of Kaimanovich and Masur [**KM**].

COROLLARY 4.6 ([**FM**], [**BFu**]). *Let G be a semi-simple Lie group without compact factors, with finite center and of rank at least 2. Let $\Gamma < G$ be an irreducible lattice and let $\rho : \Gamma \to \mathcal{M}_{g,m}$ be a homomorphism; then $\rho(\Gamma)$ is finite.*

PROOF. Burger and Monod [**BM**] observed that the second bounded cohomology group of an irreducible lattice in a semi-simple Lie group of higher rank as in the statement of the corollary is finite dimensional. On the other hand, by Margulis' normal subgroup theorem [**Ma**], for every homomorphism $\rho : \Gamma \to \mathcal{M}_{g,m}$ either the kernel of ρ is finite or the image of ρ is finite. If the kernel of ρ is finite then $\rho(\Gamma)$ is a subgroup of $\mathcal{M}_{g,m}$ which admits Γ as a finite extension. Since the second bounded cohomology group of a countable group coincides with the second bounded cohomology group of any finite extension, the second bounded cohomology group of $\rho(\Gamma)$ is finite dimensional. But $\rho(\Gamma)$ is not virtually abelian and hence this contradicts Theorem 4.5. $\square$

Theorem 4.3 and Theorem 4.5 can be viewed as structure theorems for subgroups of the mapping class group describing a *rank 1-phenomenon* (see also [**FLM**] for other results along this line). It indicates that a finitely generated infinite group whose geometry is incompatible with the geometry of a hyperbolic space (in a suitable sense) can not be a subgroup of $\mathcal{M}_{g,m}$.

However, the mapping class group has many interesting subgroups, for example free subgroups consisting of pseudo-Anosov elements. We conclude this section with a description of some families of subgroups with particularly simple geometric properties.

Let $\varphi, \eta \in \mathcal{M}_{g,m}$ be pseudo-Anosov elements. Then φ, η act as hyperbolic isometries on $\mathcal{C}(S)$, and they act with north-south dynamics on $\partial\mathcal{C}(S)$. Assume that the fixed point sets of φ, η on $\partial\mathcal{C}(S)$ are disjoint. By the classical ping-pong argument (see Chapter III.Γ.3 in [**BH**]), there are numbers $\ell > 0, k > 0$ such that the subgroup Γ of $\mathcal{M}_{g,m}$ generated by φ^ℓ, η^k is free and consists of pseudo-Anosov elements. We call such a group a *Schottky-group*. Clearly $\mathcal{M}_{g,m}$ contains infinitely many conjugacy classes of Schottky groups. These Schottky groups are *convex cocompact groups* in the sense of [**FMo**].

Now define a finitely generated subgroup Γ of $\mathcal{M}_{g,m}$ to be *convex cocompact* if for one (and hence every) $\alpha \in \mathcal{C}(S)$ the orbit map $\varphi \in \Gamma \to \varphi\alpha \in \mathcal{C}(S)$ is a quasi-isometry where Γ is equipped with the distance function defined by the word norm of a fixed symmetric set of generators. A convex cocompact subgroup of $\mathcal{M}_{g,m}$ is necessarily word hyperbolic. Schottky groups in $\mathcal{M}_{g,m}$ are convex cocompact in this sense. By [**KL, H6**] the above definition of a convex cocompact subgroup of $\mathcal{M}_{g,m}$ coincides with the definition given by Farb and Mosher in [**FMo**]. The natural extension of such a group Γ by the fundamental group $\pi_1(S)$ of S is word hyperbolic [**H6**].

PROBLEM 4. *Is there a convex cocompact subgroup of $\mathcal{M}_{g,m}$ which is not virtually free?*

A particular interesting class of subgroups of $\mathcal{M}_{g,m}$ arise from *Veech surfaces*. These surfaces are the projections to moduli space of the stabilizer of a *complex geodesic* in Teichmüller space (which is a maximal embedded complex disc in the Teichmüller space viewed as a bounded domain in $\mathbb{C}^{3g-3+m}$) with the additional property that this stabilizer is a lattice in $PSL(2,\mathbb{R})$. Veech surfaces are surfaces of finite type with isolated singularities embedded in moduli space; they are never closed [**V**]. Thus their corresponding subgroup of $\mathcal{M}_{g,m}$ contains a free group of finite index with a distinguished family of conjugacy classes corresponding to the cusps of the curve. Veech surfaces have many beautiful algebraic and geometric properties (see e.g. [**McM1**], [**McM2**]). Elementary constructions of such surfaces and their corresponding subgroups of $\mathcal{M}_{g,m}$ are for example discussed in [**L**]. Veech surfaces can also be used to construct explicit subgroups of mapping class groups with prescribed geometric properties. A particularly beautiful result along this line was recently obtained by Leininger and Reid [**LR**].

THEOREM 4.7 ([**LR**]). *For every $g \geq 2$ there exists a subgroup of $\mathcal{M}_{g,0}$ which is isomorphic to the fundamental group of a closed surface of genus $2g$ and such that all but one conjugacy class of its elements (up to powers) is pseudo-Anosov.*

PROBLEM 5. *Develop a theory of geometrically finite subgroups of $\mathcal{M}_{g,m}$ which include the groups defined by Veech surfaces.*

5. The train track complex

In Section 4 we indicated that it is possible to derive many large-scale geometric properties of the mapping class group from the fact that it admits an acylindrical action on a hyperbolic geodesic metric space. On the other hand, the mapping class group $\mathcal{M}_{g,m}$ of a non-exceptional surface is *not* hyperbolic except in the case when the surface is a twice punctured torus [**BFa**]. To get more precise informations on the geometry of the mapping class group we introduce now a geometric model.

Define a graph whose vertices are the isotopy classes of complete train tracks on S by connecting two such train tracks τ, τ' by a (directed) edge if τ' is obtained from τ by a single split at a large branch e. We call this graph the *train track complex*; it is locally finite and hence locally compact. The mapping class group acts on $\mathcal{TT}$ as a group of simplicial isometries. We have.

PROPOSITION 5.1 ([**H1**]). *$\mathcal{TT}$ is connected, and $\mathcal{M}_{g,m}$ acts on $\mathcal{TT}$ properly and cocompactly.*

As an immediate consequence of Proposition 5.1 we observe that the train track complex is $\mathcal{M}_{g,m}$-equivariantly quasi-isometric to the mapping class group and hence can be viewed as a geometric model for $\mathcal{M}_{g,m}$. The usefulness of this model comes from the fact that $\mathcal{TT}$ admits a natural family of uniform quasi-geodesics. Namely, we have [**M3**], [**H5**].

PROPOSITION 5.2. *Splitting sequences in $\mathcal{TT}$ are uniform quasi-geodesics.*

A finitely generated subgroup A of a finitely generate group Γ is called *undistorted* in Γ if the inclusion $\iota : A \to \Gamma$ satisfies $d(\iota g, \iota h) \geq cd(g,h)$ for some $c > 0$ and all $g, h \in A$ (note that the reverse estimate $d(\iota g, \iota h) \leq Cd(g,h)$ for a constant $C > 0$ is always satisfied). As an immediate consequence of Proposition 5.2 we conclude (see [**FLM**] which contains a proof of the first part of the corollary and [**MM2, H5**] for the second).

COROLLARY 5.3.
(1) *Any free abelian subgroup of $\mathcal{M}_{g,m}$ is undistorted in $\mathcal{M}_{g,m}$.*
(2) *Let $S' \subset S$ be a non-trivial connected subsurface. Then the mapping class group of S' as a subgroup of S is undistorted.*

PROOF. We begin with the proof of the second part of our lemma. Namely, let $S' \subset S$ be a subsurface bounded by some simple closed pairwise disjoint curves. Choose a pants decomposition P which contains this system of curves. For every geodesic lamination λ on the surface S' there is a train track $\tau \in \mathcal{TT}$ which carries λ and is adapted to P. As before, the mapping class group of S' is quasi-isometric to the subgraph G of $\mathcal{TT}$ which contains precisely all train tracks of this form and with a fixed intersection with $S - S'$. Since splitting sequences in $\mathcal{TT}$ which do not contain any split at a large branch which is *not* contained in S' are uniform quasi-geodesics in $\mathcal{TT}$ and define uniform quasi-geodesics in the mapping class group of S', the mapping class group of S' is an undistorted subgroup of $\mathcal{M}_{g,m}$ provided that the collection of pairs of vertices in the graph G which can be connected by a splitting sequence is κ-dense in $G \times G$ for some $\kappa > 0$. However, it was shown in [**H5**] that this is indeed the case.

To show the first part of the lemma it is enough to show that every infinite cyclic subgroup of $\mathcal{M}_{g,m}$ is undistorted in $\mathcal{M}_{g,m}$. For this let $\varphi \in \mathcal{M}_{g,m}$ be an element of infinite order. We may assume that φ either is a pseudo-Anosov element or it its reducible. In the first case φ preserves a pair of transverse minimal laminations which fill up S. The action of φ on the curve complex $\mathcal{C}(S)$ is hyperbolic and the stable length $\lim_{i \to \infty} \frac{1}{i} d(\varphi^i x, x)$ is positive. Since $\mathcal{TT}$ is quasi-isometric to $\mathcal{M}_{g,m}$ and the map $\Phi : \mathcal{TT} \to \mathcal{C}(S)$ introduced in Section 3 is uniformly Lipschitz we conclude that the cyclic group generated by φ is undistorted in $\mathcal{M}_{g,m}$.

If φ is reducible and not a Dehn twist then there is some $k > 0$ such that φ^k fixes a non-trivial subsurface of S and generates an undistorted subgroup of the mapping class group of this subsurface. Together with the second part of the corollary we conclude that the infinite cyclic subgroup of $\mathcal{M}_{g,m}$ generated by φ is undistorted. However, if φ is a Dehn-twist along a simple closed curve α then there is a splitting sequence $(\tau_i) \subset \mathcal{TT}$ issuing from a train track τ_0 which is adapted to a pants decomposition P containing α and such that $\varphi(\tau_{2i-2}) = \tau_{2i}$. Since splitting sequences are uniformly quasi-geodesic, our claim follows. $\square$

The *Torelli group* is the subgroup of $\mathcal{M}_{g,m}$ of all elements which act trivially on the first homology group of the surface S. For a closed surface of genus $g \geq 3$, the Torelli group is finitely generated [**J**]; however, this is not true for $g = 2$ [**MCM**].

PROBLEM 6. *For a closed surface of genus $g \geq 3$, is the Torelli subgroup of $\mathcal{M}_{g,m}$ undistorted? More generally, find a* distorted *finitely generated subgroup of $\mathcal{M}_{g,m}$.*

The splitting sequences on the complex of train tracks can be used to investigate the large-scale geometric behavior of the mapping class group. Note that such a splitting sequence (τ_i) is determined by an initial train track and for each i by a choice of a splitting move among a uniformly bounded number of possibilities which transforms the train track τ_i to the train track τ_{i+1}. In other words, it is possible to treat splitting sequences and hence uniform quasi-geodesics in $\mathcal{M}_{g,m}$ in an algorithmic way.

Algorithmic calculations in a finitely generated group Γ are very intimately related to two basis decision problems which go back to Dehn and can be formulated as follows (see Chapter III.Γ.1 in [**BH**]).

Word problem: A word w in a fixed system of generators for Γ is given. One is required to find a method to decide in a finite number of steps whether or not this word represents the identity in Γ.

Conjugacy problem: Two elements $g, h \in \Gamma$ are given. A method is sought to decide in a finite number of steps whether or not the elements g, h are conjugate, i.e. whether there is some $u \in \Gamma$ such that $h = ugu^{-1}$.

In the last decade of the twentith century, Epstein, Cannon, Holt, Levy, Paterson, Thurston [**E**] formulated a property for finitely generated groups which ensures that these problems can be solved in controlled time. Namely, a *biautomatic structure* for a finitely generated group Γ consists of a finite *alphabet* A, a (not necessarily injective) map $\pi : A \to \Gamma$ and a *regular language* L over the alphabet A with the following properties. The set $\pi(A)$ generates Γ, and there is an inversion $\iota : A \to A$ (i.e. $\iota^2 = \text{Id}$) with $\pi(\iota a) = \pi(a)^{-1}$ for all $a \in A$. In particular, $\pi(A)$ is a symmetric set of generators for Γ. Via concatenation, every word w in the alphabet A is mapped by π to a word in the generators $\pi(A)$ of Γ and hence it defines an element $\pi(w) \in \Gamma$. We require that the restriction of the map π to the set of all words from the language L maps L onto Γ. For all $x, y \in A$ and each word $w \in L$ of length $k \geq 0$, the word xwy defines via the projection π a path $s : [0, k+2] \to \Gamma$. By assumption, there is a word $w' \in L$ of length $\ell > 0$ with $\pi(w') = \pi(xwy)$. Let $s' : [0, \ell] \to \Gamma$ be the corresponding path in Γ; we require that the distance in Γ between $s(i)$ and $s'(i)$ is bounded by a universal constant which neither depends on i nor on the choice of x, y, w, w'.

Mosher [**M1**] showed:

THEOREM 5.4 ([**M1**]). *The mapping class group of a nonexceptional surface of finite type admits an automatic structure.*

Using the results of [**E, He**] one obtains as an immediate corollary.

COROLLARY 5.5. *Let $\mathcal{G}$ be a finite symmetric set of generators of $\mathcal{M}_{g,m}$ and let $\mathcal{F}(\mathcal{G})$ be the free group generated by $\mathcal{G}$.*
1. *There is a constant $\kappa_1 > 0$ such that a word w in $\mathcal{G}$ represents the identity in $\mathcal{M}_{g,m}$ if and only if in the free group $\mathcal{F}(\mathcal{G})$ we have $w = \prod_{i=1}^n x_i r_i x_i^{-1}$ where $n \leq \kappa_1 |w|^2$, r_i is a word in $\mathcal{G}$ of length at most κ_1 which represents the identity, and $|x_i| \leq \kappa_1 |w|$. Thus the word problem for $\mathcal{M}_{g,m}$ is solvable in quadratic time.*
2. *The conjugacy problem for $\mathcal{M}_{g,m}$ is solvable.*

From the automatic structure it is possible to deduce many of the other known properties of the mapping class group.

Call a finitely generated group Γ *linear* if it admits an injective homomorphism into $GL(n, \mathbb{C})$ for some $n > 0$. Consider for a moment the mapping class group of a *closed* surface of genus $g \geq 2$. This mapping class group contains a free abelian subgroup of dimension $3g - 3$ which is generated by Dehn twists about the curves of a pants decomposition for S. Since the Zariski closure of an abelian subgroup of a linear algebraic group is abelian, the existence of such a free abelian subgroup of $\mathcal{M}_{g,0}$ of dimension $3g-3$ can be used to show that there is no injective homomorphism of $\mathcal{M}_{g,0}$ into $GL(n, \mathbb{C})$ for $n < 2\sqrt{g-1}$ [**FLM**].

On the other hand, Krammer [**Kr**] recently showed that the *braid groups* are linear. There are many known similarities between the braid groups and the mapping class groups (see for example [**Bi**]). However, the following problem is open.

PROBLEM 7. Is the mapping class group linear?

A locally compact group Γ is said to satisfy the *Haagerup approximation property* or is *a-T-menable* if there exists a continuous, isometric action α of Γ on some affine Hilbert space $\mathcal{H}$ which is metrically proper. This means that for all bounded subsets B of $\mathcal{H}$, the set $\{g \in \Gamma \mid \alpha(g)B \cap B \neq \emptyset\}$ is relatively compact in Γ. There are other equivalent characterizations of this property (see [**CCJJV**]) which can be viewed as a strong negation of the (perhaps more widely know) property (T) of Kazhdan. The class of a-T-menable groups contains for example all amenable groups, Coxeter groups and the isometry groups of real and complex hyperbolic spaces. It is also known that for a-T-menable groups the Baum-Connes conjecture holds (see [**CCJJV**]). The strong Novikov conjecture for the mapping class group was established in [**H1, Ki**]

PROBLEM 8. Show that the mapping class group does not have property (T). Is the mapping class group a-T-menable?

References

[BFu] M. Bestvina, K. Fujiwara, *Bounded cohomology of subgroups of mapping class groups*, Geom. Topol. 6 (2002), 69-89.

[Bi] J. Birman, *Braids, links and mapping class groups*, Ann. Math. Stud. 82, Princeton Univ. Press 1975.

[Bo1] F. Bonahon, *Geodesic laminations on surfaces*, in "Laminations and foliations in dynamics, geometry and topology" (Stony Brook, NY, 1998), 1–37, Contemp. Math., 269, Amer. Math. Soc., Providence, RI, 2001.

[Bo2] F. Bonahon, *Geodesic currents on hyperbolic groups*, in "Arboretal group theory", Math. Sci. Res. Inst. Publ. 19, 143–168, Springer 1991.

[B1] B. Bowditch, *Intersection numbers and the hyperbolicity of the curve complex*, to appear in J. reine angew. Math.
[B2] B. Bowditch, *Tight geodesics in the curve complex*, preprint 2003.
[BH] M. Bridson, A. Haefliger, *Metric spaces of non-positive curvature*, Springer Grundlehren 319, Springer, Berlin 1999.
[BFa] J. Brock, B. Farb, *Curvature and rank of Teichmüller space*, Amer. J. Math. 128 (2006), 1-22.
[BM] M. Burger, N. Monod, *Bounded cohomology of lattices in higher rank Lie groups*, J. Eur. Math. Soc. 1 (1999), 199–235.
[Bu] P. Buser, *Geometry and spectra of compact Riemann surfaces*, Birkhäuser, Boston 1992.
[CEG] R. Canary, D. Epstein, P. Green, *Notes on notes of Thurston*, in "Analytical and geometric aspects of hyperbolic space", edited by D. Epstein, London Math. Soc. Lecture Notes 111, Cambridge University Press, Cambridge 1987.
[CB] A. Casson with S. Bleiler, *Automorphisms of surfaces after Nielsen and Thurston*, Cambridge University Press, Cambridge 1988.
[CCJJV] P. A. Cherix, M. Cowling, P. Jolissaint, P. Julg, A. Valette, *Groups with the Haagerup property*, Progress in Mathematics 197, Birkhäuser, Basel 2001.
[E] D. A. Epstein, with J. Cannon, D. Holt, S. Levy, M. Paterson, W. Thurston, *Word processing in groups*, Jones and Bartlett Publ., Boston 1992.
[FLM] B. Farb, A. Lubotzky, Y. Minsky, *Rank one phenomena for mapping class groups*, Duke Math. J. 106 (2001), 581–597.
[FM] B. Farb, H. Masur, *Superrigidity and mapping class groups*, Topology 37 (1998), 1169–1176.
[FMo] B. Farb, L. Mosher, *Convex cocompact subgroups of mapping class groups*, Geom. Topol. 6 (2002), 91-152.
[FLP] A. Fathi, F. Laudenbach, V. Poénaru, *Travaux de Thurston sur les surfaces*, Astérisque 1991.
[GH] E. Ghys, P. de la Harpe, *Sur les groupes hyperboliques d'après Mikhael Gromov*, Progr. Math. 83, Birkhäuser, Boston 1990.
[G] M. Gromov, *Groups of polynomial growth and expanding maps*, Publ. Math. IHES 53 (1981), 53–78.
[H1] U. Hamenstädt, *Geometry of the mapping class groups I: Boundary amenability*, arXiv: math.GR/0510116.
[H2] U. Hamenstädt, *Train tracks and the Gromov boundary of the complex of curves*, to appear in Proc. Newton Inst., arXiv: math.GT/0409611.
[H3] U. Hamenstädt, *Bounded cohomology and isometry groups of hyperbolic spaces*, arXiv: math.GR/0507097.
[H4] U. Hamenstädt, *Train tracks and the geometry of the complex of curves*, arXiv: math.GT/0502256.
[H5] U. Hamenstädt, *Geometry of the mapping class groups II: (Quasi)-geodesics*, arXiv: math.GR/0511349.
[H6] U. Hamenstädt, *Word hyperbolic extensions of surface groups*, arXiv: math.GT/0505244.
[Ha] W. J. Harvey, *Boundary structure of the modular group*, in "Riemann Surfaces and Related topics: Proceedings of the 1978 Stony Brook Conference" edited by I. Kra and B. Maskit, Ann. Math. Stud. 97, Princeton, 1981.
[He] G. Hemion, *On the classification of homeomorphisms of 2-manifolds and the classification of 3-manifolds*, Acta Math. 142 (1979), 123-155.
[IT] Y. Imayoshi, M. Taniguchi, *An introduction to Teichmüller spaces*, Springer, Tokyo 1992/1999.
[I] N. V. Ivanov, *Mapping class groups*, Chapter 12 in "Handbook of Geometric Topology", edited by R.J. Daverman and R.B. Sher, Elsevier Science (2002), 523–633.
[J] D. Johnson, *The structure of the Torelli group: A finite set of generators for $\mathcal{T}$*, Ann. of Math. 118 (1983), 423-442.
[KM] V. Kaimanovich, H. Masur, *The Poisson boundary of the mapping class group*, Invent. Math. 125 (1996), 221–264.
[KL] R. Kent, C. Leininger, *Shadows of mapping class groups: Capturing convex cocompactness*, arXiv: math.GT/0505114.

[Ke]   S. Kerckhoff, *The Nielsen realization problem*, Ann. Math. 117 (1983), 235–265.

[Ki]   Y. Kida, *The mapping class group from the viewpoint of measure equivalence theory*, to appear in Memoirs Math., AMS.

[K]   E. Klarreich, *The boundary at infinity of the curve complex and the relative Teichmüller space*, preprint 1999.

[Kr]   D. Krammer, *Braid groups are linear*, Ann. of Math. 155 (2002), 131–156.

[L]   C. Leininger, *On groups generated by two positive multi-twists: Teichmüller curves and Lehmer's number*, Geom. Topol. 8 (2004), 1301-1359.

[LR]   C. Leininger, A. Reid, *A combination theorem for Veech subgroups of the mapping class group*, arXiv: math.GT/0410041, to appear in Geom. Funct. Anal.

[Le]   G. Levitt, *Foliations and laminations on hyperbolic surfaces*, Topology 22 (1983), 119–135.

[Ma]   G. Margulis, *Discrete subgroups of semisimple Lie groups*, Springer Ergebnisse in Math. (3) 17, Springer Berlin Heidelberg 1991.

[Mas]   H. Masur, *Interval exchange transformations and measured foliations*, Ann. of Math. 115 (1982), 169–200.

[MM1]   H. Masur, Y. Minsky, *Geometry of the complex of curves I: Hyperbolicity*, Invent. Math. 138 (1999), 103–149.

[MM2]   H. Masur, Y. Minsky, *Geometry of the complex of curves II: Hierarchical structure*, GAFA 10 (2000), 902–974.

[MM3]   H. Masur, Y. Minsky, *Quasiconvexity in the curve complex*, in the tradition of Ahlfors and Bers III, 309–320, Contemp. Math. 355, AMS, 2004.

[Mc]   J. McCarthy, *A "Tits alternative" for subgroups of surface mapping class groups*, Trans. Amer. Math. Soc. 291 (1985), 583–612.

[MP]   J. McCarthy, A. Papadopoulos, *Dynamics on Thurston's sphere of projective measured foliations*, Comm. Math. Helv. 64 (1989), 133–166.

[MCM]   D. McCullough, A. Miller, *The genus 2 Torelli group is not finitely generated*, Topology Appl. 22 (1986), 43-49.

[McM1] C. McMullen, *Billiards and Teichmüller curves on Hilbert modular surfaces*, J. AMS 16 (2003), 857-885.

[McM2] C. McMullen, *Dynamics of $SL_2(\mathbb{R})$ over moduli space in genus 2*, to appear in Ann. Math.

[M1]   L. Mosher, *Mapping class groups are automatic*, Ann. Math. 142 (1995), 303-384.

[M2]   L. Mosher, *Train track expansions of measured foliations*, unpublished manuscript.

[M3]   L. Mosher, in preparation.

[O]   J. P. Otal, *Le Théorème d'hyperbolisation pour les variétés fibrées de dimension 3*, Astérisque 235, Soc. Math. Fr. 1996.

[PH]   R. Penner with J. Harer, *Combinatorics of train tracks*, Ann. Math. Studies 125, Princeton University Press, Princeton 1992.

[R]   K. Rafi, *A characterization of short curves of a Teichmüller geodesic*, Geom. Top. 9 (2005), 179–202.

[S]   K. Shackleton, *Tightness and computing distances in the curve complex*, arXiv: math.GT/0412078

[V]   W. Veech, *Teichmüller curves in moduli space, Eisenstein series and an application to triangular billiards*, Invent. Math. 97 (1989), 553-583.

Problems on Billiards, Flat Surfaces and Translation Surfaces

Pascal Hubert, Howard Masur, Thomas Schmidt, and Anton Zorich

Part of this list of problems grew out of a series of lectures given by the authors at the Luminy confrence in June 2003. These lectures appeared as survey articles in [**HuSt3**], [**Ma1**], see the related [**Es**] and [**Fo2**].

Additional general references for the background material for these problems are the survey articles [**MaTa**], [**Sm**], [**Gu1**], [**Gu2**], [**Zo4**] and the book [**St**].

Acknowledgments. The authors would like to thank the referee for numerous helpful comments.

1. Flat surfaces and billiards in polygons

The first set of problems concerns general flat surfaces with nontrivial holonomy and billiards in polygons.

Using any of various variations of a standard unfolding construction one can glue flat surfaces from several copies of the billiard table. When the resulting surface is folded back to the polygon, the geodesics on the surface are projected to billiard trajectories, so billiards in polygons and flat surfaces are closely related.

A quintessential example of a flat surface is given by the surface of the standard cube in real three-space. With its induced metric it is a flat sphere with eight singularities. Indeed, at each corner, three squares meet so that each corner has a neighborhood that is isometric to a Euclidean cone, with cone angle $3\pi/2$. These are indeed the only singularities of the flat metric; we call them *conical singularities* of the flat surface. Since the cone angle is not a multiple of 2π, parallel transport of a non-zero tangent vector about a simple closed curve around a corner will result in a distinct tangent vector; thus the holonomy is nontrivial. In general, a "flat surface" here refers to a surface of zero Gaussian curvature with isolated conical singularities.

Having a Riemannian metric it is natural to study geodesics. Away from singularities geodesics on a flat surface are (locally isometric to) straight lines. The geodesic flow on the unit tangent bundle is then also presumably well behaved. For simplicity, let "ergodic" here mean that a typical geodesic visits any region of the surface, and furthermore (under unit speed parametrization) spends a time in the region that is asymptotically proportional to the area of the region.

©2006 American Mathematical Society

PROBLEM 1 (Geodesics on general flat surfaces). Describe the behavior of geodesics on general flat surfaces. Prove (or disprove) the conjecture that the geodesic flow is ergodic on a typical (in any reasonable sense) flat surface. Does any (almost any) flat surface have at least one closed geodesic which does not pass through singular points?

If the answer is positive then one can ask for the asymptotics for the number of closed geodesics of bounded length as a function of the bound.

Note that typically a geodesic representative in a homotopy class of a simple closed curve is realized by a broken line containing many geodesic segments going from one conical singularity to the other. The counting problem for regular closed geodesics (ones which do not pass through singularities) is quite different from the counting problem for geodesics realized by broken lines.

The following questions treat billiards in arbitrary polygons in the plane.

PROBLEM 2 (Billiards in general polygons). Does every billiard table have at least one regular periodic trajectory? If the answer is affirmative, does this trajectory persist under deformations of the billiard table?

If a periodic trajectory exists, find the asymptotics for the number of periodic trajectories of bounded length as a function of the bound.

Describe the behavior of a generic regular billiard trajectory in a generic polygon; in particular, prove (or disprove) the assertion that the billiard flow is ergodic.[1]

We note that the case of triangles is already highly non-trivial. For recent work on billiards in obtuse triangles see [**Sc1**] and [**Sc2**].

In the case of triangles the notion of generic can be interpreted as follows. The space of triangles up to similarity can be parametrized as the set of triples $(\theta_1, \theta_2, \theta_3)$ with $\sum \theta_i = \pi$ and each $\theta_i > 0$. It is naturally an open simplex. Generic then refers to the natural Lebesgue measure.

To motivate the next problem we note that there is a close connection between the study of interval exchange transformations and billiards in rational polygons (defined below). An important technique in the study of interval exchange maps is that of renormalization. Given an interval exchange on the unit interval one can take the induced transformation on a subinterval. The resulting map is again an interval exchange map, and if one renormalizes so that the new interval has length one, then this gives a transformation on the space of unit interval exchange maps. This transformation is called the Rauzy-Veech induction ([**Ra**], [**Ve2**]), and it has proved to be of fundamental importance. There is a corresponding notion for the renormalization of translation surfaces given by the Teichmüller geodesic flow.

PROBLEM 3 (Renormalization of billiards in polygons). Is there a natural dynamical system acting on the space of billiards in polygons so as to allow a useful renormalization procedure?

2. Rational billiards, translation surfaces quadratic differentials and $SL(2, \mathbb{R})$ actions

An important special case of billiards is given by the *rational billiards* — billiards in polygonal tables whose vertex angles are rational multiples of π. There

---

[1]On behalf of the Center of Dynamics of Pennsylvania State University A. Katok promised a prize for a solution of this problem.

is a well-known procedure (see the surveys [**MaTa**], [**Gu1**]) which associates to a rational billiard an object called a translation surface. One labels the sides of the polygon and successively reflects the polygon across sides. The rationality assumption guarantees that after a certain number of reflections a labelled side appears parallel to itself. In that case the pair of sides with the same label is glued by a parallel translation. The result is a closed surface with conical singularities. The billiard flow on the polygon which involves reflection in the sides is replaced by a straight line flow on the glued surface; under the natural projection of the surface to the billiard table the geodesics are projected to billiard trajectories. It turns out that many results in rational billiards are found by studying more general translation surfaces.

A translation surface is defined by the following data:
- a finite collection of disjoint polygons $\Delta_1, \ldots, \Delta_n$ embedded into the oriented Euclidean plane;
- a pairing between the sides of the polygons: to each side s of any Δ_i is associated a unique side $s' \neq s$ of some Δ_j in such way that the two sides s, s' in each pair are parallel and have the same length $|s| = |s'|$. The pairing respects the induced orientation: gluing Δ_i to Δ_j by a parallel translation sending s to s' we get an oriented surface with boundary for any pair s, s';
- a choice of the positive vertical direction in the Euclidean plane.

A classical example is the square with opposite unit sides identified, giving the flat torus. This example arises from billiards in a square of side length $1/2$. Another example is a regular octagon with opposite sides identified. It arises by the unfolding process from billiards in a right triangle whose other angles are $\pi/8, 3\pi/8$. When translation surfaces arise from billiards the polygons in the gluings can be taken to be congruent, so translation surfaces arising from rational billiards always have extra symmetries not possessed by general translation surfaces. In this sense translation surfaces coming from rational billiards are always rather special.

Note that a translation surface is in particular a flat surface in the sense described before. It is locally Euclidean except possibly at the points corresponding to the vertices of the polygons. These points can be conical singularities, but the total angle around such a vertex — its cone angle — is always an integer multiple of 2π. For example, in the case of the regular octagon, the 8 vertices are identified to a single point with cone angle 6π.

Since the gluing maps are translations which are of course complex analytic, the underlying structure is that of a Riemann surface X. Moreover since translations preserve the form dz in each polygon, these forms dz fit together to give a holomorphic 1-form ω on X. Thus translation surfaces are often denoted by (X, ω). In this language a cone angle $2k\pi$ at a singularity corresponds to a zero of order $k-1$ of ω. The orders of the zeroes form a tuple $\alpha = (\alpha_1, \ldots, \alpha_n)$, where $\sum \alpha_i = 2g - 2$ and g is the genus of the surface. In the case of the regular octagon, there is a single zero of order 2 so $\alpha = (2)$. The set of all (X, ω) whose zeroes determine a fixed tuple α form a moduli space $\mathcal{H}(\alpha)$, called a stratum. We may think of the points of this moduli space as glued polygons where the vectors corresponding to the sides are allowed to vary. Since each (X, ω) has the underlying structure of a Riemann surface, remembering just the complex structure gives a projection from each stratum to the Riemann moduli space. One can introduce "markings" in order to get a well-defined projection map from spaces of marked Abelian differentials to Teichmüller space.

Studying translation surfaces from the different viewpoints of geometry and complex analysis has proved useful.

If we loosen our restrictions on the gluings so as to allow reflections in the origin as well as translations, then there is still an underlying Riemann surface; the resulting form is a quadratic differential. The structure is sometimes also called a half-translation surface. Now the cone angles are integer multiples of π. Thus each quadratic differential determines a set of zeroes whose orders again give a tuple $\beta = (\beta_1, \ldots, \beta_n)$ with $\sum \beta_i = 4g - 4$. We similarly have strata of quadratic differentials $\mathcal{Q}(\beta)$. If we fix a genus g, and introduce markings, then the union of the strata of quadratic differentials (including those that are naturally seen as the squares of Abelian differentials) of genus g fit together to form the cotangent bundle over Teichmüller space.

Much of the modern treatment of the subject arises from the study of the action of the group $SL(2, \mathbb{R})$ on each moduli space $\mathcal{H}(\alpha)$. Understanding the orbit of a translation surface allows one to understand much of the structure of the translation surface itself. For each (X, ω) realized as a union of glued polygons Δ_i, and $A \in SL(2, \mathbb{R})$, let A act on each Δ_i by the linear action on $\mathbb{R}^2$. Since A preserves parallel lines, this gives a map of (X, ω) to some $A \cdot (X, \omega)$. We have a similar action for $A \in GL^+(2, \mathbb{R})$. If we introduce markings then the projection of the orbit to Teichmüller space gives an isometric embedding of the hyperbolic plane into Teichmüller space equipped with the Teichmüller metric. The projection of the orbit of (X, ω) is called a Teichmüller disc. Similarly, we have Teichmüller discs for quadratic differentials. The image of the disc in the moduli space is typically dense. However there are (X, ω) whose orbit is closed in its moduli space $\mathcal{H}(\alpha)$. These are called Veech surfaces. We will discuss these in more detail in the next section.

2.1. Veech surfaces. This section discusses problems related to Veech surfaces and Veech groups. Given a translation surface (X, ω), or quadratic differential, one can discuss its affine diffeomorphism group; that is, the homomorphisms that are diffeomorphisms on the complement of the singularities, with constant Jacobian matrix (with respect to the flat metric). The group of Jacobians, $SL(X, \omega) \subset SL(2, \mathbb{R})$ can also be thought of as the stabilizer of (X, ω) in the moduli space under the action of $SL(2, \mathbb{R})$ on the moduli space of all translation surfaces. (The $SL(2, \mathbb{R})$ action is discussed in the next section). The Jacobians of orientation preserving affine diffeomorphisms form a discrete subgroup of $SL(2, \mathbb{R})$, also called a Fuchsian group. The image in $PSL(2, \mathbb{R})$ is the Veech group of the surface. One can also think of this group as a subgroup of the mapping class group of the surface. Hyperbolic elements of this group correspond to pseudo-Anosovs in the mapping class group; parabolic elements to reducible maps and elliptics to elements of finite order.

The surface (X, ω) is called a Veech surface if this group is a lattice (that is of cofinite volume) in $PSL(2, \mathbb{R})$. By a result of Smillie [**Ve4**] it is known that a surface is a Veech surface if and only if its $SL(2, \mathbb{R})$-orbit is closed in the corresponding stratum.

In genus 2 it is known [**McM1**] that if $SL(X, \omega)$ contains a hyperbolic element then in its action on the hyperbolic plane, it has as its limit set the entire circle at infinity. Consequently, it is either a lattice or infinitely generated. There are known examples of the latter, see [**HuSt2**] and [**McM1**].

PROBLEM 4 (Characterization of Veech surfaces). Characterize all Veech surfaces (for each stratum of each genus).

This problem is trivial in genus one; in genus two K. Calta [**Ca**] and C. Mc-Mullen [**McM2**] have provided solutions. In the papers [**KnSm**] and Puchta [**Pu**] the acute rational billiard triangles that give rise to Veech surfaces were classified. In [**SmWe**] there is a criterion for a surface (X, ω) to be a Veech surface that is given in terms of the areas of triangles embedded in (X, ω).

PROBLEM 5 (Fuchsian groups). Which Fuchsian groups are realized as Veech groups? Which subgroups of the mapping class group appear as Veech groups? This is equivalent to asking which subgroups are the stabilizers of a Teichmüller disc.

PROBLEM 6 (Purely cyclic). Is there a Veech group that is cyclic and generated by a single hyperbolic element? Equivalently, is there a pseudo-Anosov map such that its associated Teichmüller disk, is invariant only under powers of the pseudo-Anosov?

PROBLEM 7 (Algorithm for Veech groups). Is there an algorithm for determining the Veech group of a general translation surface or quadratic differential?

An interesting class of Veech surfaces are the square-tiled surfaces. These surfaces can be represented as a union of glued squares all of the same size, see [**Zo3**], [**HuLe**].

PROBLEM 8 (Orbits of square-tiled surfaces). Classify the $SL(2, \mathbb{R})$ orbits of square-tiled surfaces in any stratum. Describe their Teichmüller discs. A particular case of this problem is that of the stratum $\mathcal{H}(1, 1)$.

The above problem is solved only for the stratum $\mathcal{H}(2)$, see [**HuLe**], [**McM4**].

2.2. Minimal sets and analogue of Ratner's theorems. The next set of questions concern the $SL(2, \mathbb{R})$ action. They are motivated by trying to find an analogue of the $SL(2, \mathbb{R})$ action on the moduli spaces to Ratner's celebrated theorems on the actions of subgroups of a Lie group G on G/Γ where Γ is a lattice subgroup.

PROBLEM 9 (Orbit closures for moduli spaces). Determine the closures of the orbits for the $GL^+(2, \mathbb{R})$-action on $\mathcal{H}(\alpha)$ and $\mathcal{Q}(\beta)$. Are these closures always complex-analytic (complex-algebraic?) orbifolds? Characterize the closures geometrically.

Note that by a theorem of Kontsevich any $GL^+(2, \mathbb{R})$-invariant complex-analytic subvariety is represented by an affine subspace in period coordinates.

Consider the subset $\mathcal{H}_1(\alpha) \subset \mathcal{H}(\alpha)$ of translation surfaces of area one. It is a real codimension one subvariety in $\mathcal{H}(\alpha)$ invariant under the action of $SL(2, \mathbb{R})$. In the period coordinates it is defined by a quadratic equation (the Riemann bilinear relation). It is often called a *unit hyperboloid*. It is worth noting that it is a manifold locally modelled on a paraboloid. The invariant measure on $\mathcal{H}(\alpha)$ gives a natural invariant measure on the unit hyperboloid $\mathcal{H}_1(\alpha)$. Similarly one can define the unit hyperboloid $\mathcal{Q}_1(\beta)$. It was proved by Masur and Veech that the total measure of any $\mathcal{H}_1(\alpha)$, $\mathcal{Q}_1(\beta)$ is finite.

PROBLEM 10 (Ergodic measures). Classify the ergodic measures for the action of $\mathrm{SL}(2,\mathbb{R})$ on $\mathcal{H}_1(\alpha)$ and $\mathcal{Q}_1(\beta)$.

McMullen [**McM3**] has solved Problems 9 and 10 in the case of translation surfaces in genus 2.

A subset Ω is called minimal for the action of $\mathrm{SL}(2,\mathbb{R})$ if it is closed, invariant, and it has no proper closed invariant subsets. The $\mathrm{SL}(2,\mathbb{R})$ orbit of a Veech surface is an example of a minimal set.

PROBLEM 11 (Minimal sets). Describe the minimal sets for the $\mathrm{SL}(2,\mathbb{R})$-action on $\mathcal{H}_1(\alpha)$ and $\mathcal{Q}_1(\beta)$. Since Veech surfaces give rise to minimal sets, this problem generalizes the problem of characterizing Veech surfaces.

The problem below is particularly important for numerous applications. One application is to counting problems.

PROBLEM 12 (Analog of Ratner theorem). Classify the ergodic measures for the action of the unipotent subgroup $\begin{pmatrix} 1 & t \\ 0 & 1 \end{pmatrix}_{t \in \mathbb{R}}$ on $\mathcal{H}_1(\alpha)$ and $\mathcal{Q}_1(\beta)$. We note that a solution of Problem 10 does not imply a solution of this problem. In particular this problem is open even in genus 2.

Similarly classify the orbit closures on these moduli spaces.

There are some results in special cases on this problem, see [**EsMaSl**] and [**EsMkWt**].

K. Calta [**Ca**] and C. McMullen [**McM2**] have found unexpected closed $\mathrm{GL}^+(2,\mathbb{R})$-invariant sets in genus 2 which we now describe. One can form a family of translation surfaces from a given (X,ω) by varying the periods of the 1-form ω along cycles in the relative homology — those that join distinct zeroes — while keeping the "true" periods (that is, the absolute cohomology class of ω) fixed. One may also break up a zero of higher order into zeroes of lower order while keeping the absolute periods fixed. The resulting family of translation surfaces gives a leaf of the *kernel foliation* passing through (X,ω).

It follows from [**Ca**] and [**McM2**] that for *any* Veech surface $(X,\omega) \in \mathcal{H}(2)$, the union of the complex one-dimensional leaves of the kernel foliation passing through the $\mathrm{GL}^+(2,\mathbb{R})$-orbit of (X,ω) is a closed complex orbifold $\mathcal{N}$ of complex dimension 3. By construction $\mathcal{N}$ is $\mathrm{GL}^+(2,\mathbb{R})$-invariant. Note that the $\mathrm{GL}^+(2,\mathbb{R})$-orbit of the initial Veech surface (X,ω) is closed and has complex dimension 2, so what is surprising is that the union of the complex one-dimensional leaves passing through each point of the orbit $\mathrm{GL}^+(2,\mathbb{R}) \cdot (X,\omega)$ is again closed.

We may ask a similar question in higher genus. Let $\mathcal{O} \subset \mathcal{H}(\alpha_1,\ldots,\alpha_m)$ be a $\mathrm{GL}^+(2,\mathbb{R})$-invariant submanifold (suborbifold) on translation surfaces of genus g. Let $\mathcal{H}(\alpha'_1,\ldots,\alpha'_n)$ be a stratum of surfaces of genus g that is adjacent, in that each α_i is the sum of corresponding α'_j. The complex dimension of the leaves of the kernel foliation in $\mathcal{H}(\alpha'_1,\ldots,\alpha'_n)$ is $n - m$.

Consider the closure of the union of leaves of the kernel foliation in the stratum $\mathcal{H}(\alpha'_1,\ldots,\alpha'_n)$ passing through $\mathcal{O}$; this is a closed $\mathrm{GL}^+(2,\mathbb{R})$-invariant subset $\mathcal{N} \subset \mathcal{H}(\alpha'_1,\ldots,\alpha'_n)$ of dimension at least $\dim_{\mathbb{C}} \mathcal{O} + n - m$.

PROBLEM 13 (Kernel foliation). Is $\mathcal{N}$ a complex-analytic (complex-algebraic) orbifold? When is $\dim_{\mathbb{C}} \mathcal{N} = \dim_{\mathbb{C}} \mathcal{O} + n - m$? On the other hand when does $\mathcal{N}$ coincide with the entire connected component of the enveloping stratum $\mathcal{H}(\alpha'_1, \ldots, \alpha'_n)$?

One of the key properties used in [**McM3**] for the classification of the closures of orbits of $\mathrm{GL}(2,\mathbb{R})$ in each of $\mathcal{H}(1,1)$ and $\mathcal{H}(2)$ was the knowledge that on *any* translation surface of either stratum, one can find a pair of homologous saddle connections.

For example, cutting a surface (X, ω) in $\mathcal{H}(1,1)$ along two homologous saddle connections joining distinct zeroes decomposes the surface into two tori, allowing one to apply the machinery of Ratner's Theorem.

PROBLEM 14 (Decomposition of surfaces). Given a connected component of the stratum $\mathcal{H}(\alpha)$ of Abelian differentials (or of quadratic differentials $\mathcal{Q}(\beta)$ find those configurations of homologous saddle connections (or homologous closed geodesics), which are present on every surface in the stratum.

For quadratic differentials the notion of homologous saddle connections (homologous closed geodesics) should be understood in terms of homology with local coefficients, see [**MaZo**].

The last two problems in this section concern the Teichmüller geodesic flow on the moduli spaces. This is the flow defined by the 1-parameter subgroup $\begin{pmatrix} e^t & 0 \\ 0 & e^{-t} \end{pmatrix}_{t \in \mathbb{R}}$ on $\mathcal{H}(\alpha_1, \ldots, \alpha_m)$.

In any smooth dynamical system the Lyapunov exponents (see [**BaPe**], [**Fo2**]) are important. Recently, A. Avila and M. Viana [**AvVi**] have shown the simplicity of the spectrum for the cocycle related to the Teichmüller geodesic flow (strengthening the earlier result of Forni on positivity of the smallest Lyapunov exponent).

PROBLEM 15 (Lyapunov exponents). Study *individual* Lyapunov exponents of the Teichmüller geodesic flow:
– for all known $\mathrm{SL}(2;\mathbb{R})$-invariant subvarieties;
– for strata;
– for strata of large genera as the genus tends to infinity.

Are they related to characteristic numbers of any natural bundles over appropriate compactifications of the strata?

The motivation for this problem is a beautiful formula of Kontsevich [**Ko**] representing the sum of the first g Lyapunov exponents in differential-geometric terms.

It follows from the Calabi Theorem [**Cb**] that given a *real* closed 1-form ω_0 with isolated zeroes Σ (satisfying some natural conditions) on a smooth surface S of real dimension two, one can find a complex structure on S and a holomorphic 1-form ω such that the ω_0 is the real part of ω. Consider the resulting point (X, ω) in the corresponding stratum. For generic (X, ω) the cocycle related to the Teichmüller geodesic flow acting on $H^1(X, \mathbb{R})$ defines a pair of transverse Lagrangian subspaces $H^1(X, \mathbb{R}) = L_0 \oplus L_1$ by means of the Oseledets Theorem ([**BaPe**]). These subspaces correspond to contracting and to expanding directions.

Though the pair (X, ω) is not uniquely determined by ω_0, the subspace $L_0 \subset H^1(X, \mathbb{R})$ does not depend on (X, ω) for a given ω_0. Moreover, L_0 does not change under small deformations of ω_0 that preserve the cohomology class $[\omega] \in H^1(S, \Sigma; \mathbb{R})$. We get a topological object L_0 defined in implicit dynamical terms.

PROBLEM 16 (Dynamical Hodge decomposition). Study properties of distributions of the Lagrangian subspaces in $H^1(S;\mathbb{R})$ defined by the Teichmüller geodesic flow, in particular, their continuity. Is there any topological or geometric way to define them?

The Lagrangian subspaces are an interesting structure relating topology and geometry.

2.3. Geometry of individual flat surfaces. Let (X,ω) be a translation surface (resp. quadratic differential). Fix a direction $0 \le \theta < 2\pi$ and consider a vector field (resp. line field) on each polygon of unit vectors in direction θ. Since the gluings are by translations (or by rotations by π about the origin followed by translations) which preserve this vector field, there is a well-defined vector field on (X,ω) (resp. line field) defined except at the zeroes. There is a corresponding flow ϕ_t^θ (resp. foliation). A basic question is to understand the dynamics of this flow or foliation. In the case of a flat torus this is classical. For any direction either every orbit in that direction is closed or every orbit is dense and uniformly distributed on the surface. This property is called unique ergodicity.

For a general translation surface a saddle connection is defined to be a leaf joining a pair of conical singularities. There are only countably many saddle connections in all possible directions. For any direction which does not have a saddle connection, the flow or foliation is minimal, which means that for any point, if the orbit in either the forward or backward direction does not hit a conical singularity then it is dense. Veech [**Ve3**] showed that as in the case of a flat torus, every Veech surface satisfies the dichotomy that for any θ, the flow or foliation has the property that either:

- Every leaf which does not pass through a singularity is closed. This implies that the surface decomposes into a union of cylinders of parallel closed leaves. The boundary of each cylinder is made up of saddle connections. The directional flow is said to be *completely periodic* if it has this property.
- The foliation is minimal and uniquely ergodic.

PROBLEM 17 (Converse to dichotomy). Characterize translation surfaces for which

(1) the set of minimal directions coincides with the set of uniquely ergodic directions;

(2) the set of completely periodic directions coincides with the set of non-uniquely ergodic directions.

Note that Property (2) implies Property (1).

In genus $g = 2$ it is known that for every translation surface which is not a Veech surface there is a direction θ which is minimal and not uniquely ergodic [**ChMa**]. On the other hand, using work of Hubert–Schmidt [**HuSt2**], J. Smillie and B. Weiss [**SmWe**] have given an example of a surface which is not a Veech surface and yet for which Property (2) holds. (This surface is obtained as a ramified covering over a Veech surface with a single ramification point.)

As mentioned above, a closed orbit avoiding the conical singularities of the flat surface determines a cylinder of parallel lines, all of the same length. It is of interest to find the asymptotics for the number of cylinders (in all possible directions) of lengths less than a given number. In the case of the standard flat torus the number

of cylinders of length at most L is asymptotic to
$$\frac{1}{\zeta(2)}\pi L^2.$$
Each Veech surface also has quadratic asymptotics [**Ve3**] and the same is true for generic surfaces in each stratum [**EsMa**].

PROBLEM 18 (Quadratic asymptotics for any surface). Is it true that *every* translation surface or quadratic differential has exact quadratic asymptotics for the number of saddle connections and for the number of regular closed geodesics?

PROBLEM 19 (Error term for counting functions). What can be said about the error term in the quadratic asymptotics for counting functions
$$N((X,\omega), L) \sim c \cdot L^2$$
on a generic translation surface (X,ω)? In particular, is it true that
$$\limsup_{L\to\infty} \frac{\log |N(S,L) - c \cdot L^2|}{\log L} < 2?$$
Is the lim sup the same for almost all flat surfaces in a given connected component of a stratum?

The classical Circle Problem gives an estimate for the error term in the case of the torus.

Veech proved that for Veech surfaces the limsup in the error term is actually a limit, see [**Ve3**]. However, nothing is known about the value of this limit. One may ask whether there is a uniform bound for this limit for Veech surfaces in a given stratum or even for the square-tiled surfaces in a given stratum.

2.4. Topological and geometric properties of strata.

PROBLEM 20 (Topology of strata). Is it true that the connected components of the strata $\mathcal{H}(\alpha)$ and of the strata $\mathcal{Q}(\beta)$ are $K(\pi,1)$-spaces (i.e. their universal covers are contractible)?

It is known [**KoZo**], [**La**] that the strata $\mathcal{H}(\alpha)$ and $\mathcal{Q}(\beta)$ need not be connected. With the exception of the four strata listed below, there are intrinsic invariants that allow one to tell which component a given translation surface or quadratic differential belongs to.

PROBLEM 21 (Exceptional strata). Find a geometric invariant which distinguishes different connected components of the four exceptional strata $\mathcal{Q}(-1,9)$, $\mathcal{Q}(-1,3,6)$, $\mathcal{Q}(-1,3,3,3)$ and $\mathcal{Q}(12)$.

At the moment the known invariant (called the extended Rauzy class) distinguishing connected components is given in combinatorial and not geometric terms. ([**La**])

References

[AvVi]   A. Avila, M. Viana: Simplicity of Lyapunov Spectra: Proof of the Kontsevich—Zorich Conjecture. Preprint (2005)

[BaPe]   L. Barreira and Ya. Pesin: Lyapunov Exponents and Smooth Ergodic Theory. Univ. Lect. Series 23, Amer. Math. Soc. (2002)

[Cb]   E. Calabi: An intrinsic characterization of harmonic 1-forms, Global Analysis. In: D. C. Spencer and S. Iyanaga (ed) Papers in Honor of K. Kodaira. 101–117 (1969)

[Ca]   K. Calta: Veech surfaces and complete periodicity in genus two. J. Amer. Math. Soc., **17** 871-908 (2004)

[ChMa]   Y. Cheung, H. Masur, Minimal nonergodic directions on genus two translation surfaces, Ergodic Theory and Dynamical Systems **26** (2006), 341–351; Eprint in math.DS/0501296

[Es]   A. Eskin: Counting problems in moduli space. In: B. Hasselblatt and A. Katok (ed) Handbook of Dynamical Systems, Vol. 1B. Elsevier Science B.V. (2005)

[EsMa]   A. Eskin, H. Masur: Asymptotic formulas on flat surfaces. Ergodic Theory and Dynamical Systems, **21:2**, 443–478 (2001)

[EsMaSl]   A. Eskin, H. Masur, M. Schmoll; Billiards in rectangles with barriers, Duke Math. J., **118** (3), 427–463 (2003)

[EsMkWt]   A. Eskin, J. Marklof, D. Witte Morris; Unipotent flows on the space of branched covers of Veech surfaces, Ergodic Theory and Dynamical Systems, **26** (2006), 129-162; Eprint in math.DS/0408090

[EsMaZo]   A. Eskin, H. Masur, A. Zorich: Moduli spaces of Abelian differentials: the principal boundary, counting problems, and the Siegel–Veech constants. Publications de l'IHES, **97:1**, pp. 61–179 (2003)

[Fo1]   G. Forni: Deviation of ergodic averages for area-preserving flows on surfaces of higher genus. Annals of Math., **155**, no. 1, 1–103 (2002)

[Fo2]   G. Forni: On the Lyapunov exponents of the Kontsevich–Zorich cocycle. In: B. Hasselblatt and A. Katok (ed) Handbook of Dynamical Systems, Vol. 1B. Elsevier Science B.V. (2005)

[Gu1]   E. Gutkin: Billiards in polygons: survey of recent results. J. Stat. Phys., **83**, 7–26 (1996)

[Gu2]   E. Gutkin: Billiard dynamics: a survey with the emphasis on open problems. Regul. Chaotic Dyn. **8** no. 1, 1–13 (2003)

[HuLe]   P. Hubert, S. Lelièvre: Square-tiled surfaces in $\mathcal{H}(2)$. Israel Journal of Math. **151** (2006), 281–321; Eprint in math.GT/0401056

[HuSt1]   P. Hubert, T. A. Schmidt: Veech groups and polygonal coverings. J. Geom. and Phys. **35**, 75–91 (2000)

[HuSt2]   P. Hubert, T. A. Schmidt: Infinitely generated Veech groups. Duke Math. J. **123**, 49–69 (2004)

[HuSt3]   P. Hubert and T. Schmidt: Affine diffeomorphisms and the Veech dichotomy. In: B. Hasselblatt and A. Katok (ed) Handbook of Dynamical Systems, Vol. 1B. Elsevier Science B.V. (2005)

[KnSm]   R. Kenyon and J. Smillie: Billiards in rational-angled triangles, Comment. Mathem. Helv. **75**, 65–108 (2000)

[KfMaSm]   S. Kerckhoff, H. Masur, and J. Smillie: Ergodicity of billiard flows and quadratic differentials. Annals of Math., **124**, 293–311 (1986)

[Ko]   M. Kontsevich: Lyapunov exponents and Hodge theory. "The mathematical beauty of physics" (Saclay, 1996), (in Honor of C. Itzykson) 318–332, Adv. Ser. Math. Phys., 24. World Sci. Publishing, River Edge, NJ (1997)

[KoZo]   M. Kontsevich, A. Zorich: Connected components of the moduli spaces of Abelian differentials. Invent. Math., **153:3**, 631–678 (2003)

[La]   E. Lanneau: Connected components of the moduli spaces of quadratic differentials. Preprint (2003)

[Ma1]   H. Masur: Ergodic theory of flat surfaces. In: B. Hasselblatt and A. Katok (ed) Handbook of Dynamical Systems, Vol. 1B Elsevier Science B.V. (2005)

[MaTa]   H. Masur and S. Tabachnikov: Rational Billiards and Flat Structures. In: B. Hasselblatt and A. Katok (ed) Handbook of Dynamical Systems, Vol. 1A, 1015–1089. Elsevier Science B.V. (2002)

[MaZo]   H. Masur and A. Zorich: Multiple Saddle Connections on Flat Surfaces and Principal Boundary of the Moduli Spaces of Quadratic Differentials, Preprint 73 pp. (2004); Eprint in math.GT/0402197

[McM1]   C. McMullen: Teichmüller geodesics of infinite complexity, Acta Math. **191**, 191–223 (2003)

[McM2]   C. McMullen: Billiards and Teichmüller curves on Hilbert modular surfaces. J. Amer. Math. Soc., **16**, no. 4, 857–885 (2003)
[McM3]   C. McMullen: Dynamics of $SL_2(\mathbb{R})$ over moduli space in genus two, to appear in Annals of Math.
[McM4]   C. McMullen: Teichmüller curves in genus two: Discriminant and spin, Math. Ann. **333** (2005), 87–130.
[McM5]   C. McMullen: Teichmüller curves in genus two: The decagon and beyond. J. Reine Angew. Math. **582** (2005), 173–200.
[Pu]   J.-Ch. Puchta: On triangular billiards, Comment. Mathem. Helv. **76**, 501–505 (2001)
[Ra]   G. Rauzy. Echanges d'intervalles et transformations induites. Acta Arithmetica. XXXIV 315-328 (1979)
[Sn]   G. Schmithüsen: An algorithm for finding the Veech group of an origami. Experimental Mathematics **13:4**, 459–472 (2004)
[Sc1]   R. Schwartz: Obtuse triangular billiards I: near the (2,3,6) triangle, preprint, math.umd.edu/ res/papers.html
[Sc2]   R. Schwartz: Obtuse triangular billiards II: near the degenerate (2,2,infinity) triangle, preprint, math.umd.edu/ res/papers.html
[Sm]   J. Smillie: Dynamics of billiard flow in rational polygons. In: Ya. G. Sinai (ed) Dynamical Systems. Ecyclopedia of Math. Sciences. Vol. 100. Math. Physics 1. Springer Verlag (2000)
[SmWe]   J. Smillie, B. Weiss, Isolation theorems for quadratic differentials and lattice surfaces, in preparation
[St]   K. Strebel: Quadratic Differentials. Springer-Verlag (1984)
[Ve1]   W. A. Veech: Strict ergodicity in zero dimensional dynamical systems and the Kronecker-Weyl theorem mod 2. Trans. Amer. Math. Soc. **140**, 1–33 (1969)
[Ve2]   W. A. Veech: Gauss measures for transformations on the space of interval exchange maps. Annals of Mathematics. **115**, 201-242 (1982)
[Ve3]   W. A. Veech: Teichmüller curves in modular space, Eisenstein series, and an application to triangular billiards, Inv. Math. **97**, 553–583 (1989)
[Ve4]   W. A. Veech: Geometric realization of hyperelliptic curves. Chaos, Dynamics and Fractals. Plenum (1995)
[Zo1]   A. Zorich: Finite Gauss measure on the space of interval exchange transformations. Lyapunov exponents. Annales de l'Institut Fourier, **46:2**, 325–370 (1996)
[Zo2]   A. Zorich: Deviation for interval exchange transformations. Ergodic Theory and Dynamical Systems, **17**, 1477–1499 (1997)
[Zo3]   A. Zorich: Square tiled surfaces and Teichmüller volumes of the moduli spaces of Abelian differentials. In collection "Rigidity in Dynamics and Geometry", M. Burger, A. Iozzi (Editors), Springer Verlag, 459–471 (2002)
[Zo4]   A. Zorich: Flat surfaces. In collection "Frontiers in Number Theory, Physics and Geometry. Volume 1: On random matrices, zeta functions and dynamical systems", 145 pp; Ecole de physique des Houches, France, March 9-21 2003, Springer-Verlag, Berlin, 2006.

INSTITUT DE MATHÉMATIQUES DE LUMINY, 163 AV. DE LUMINY, CASE 907, 13288 MARSEILLE CEDEX 09 FRANCE
E-mail address: hubert@iml.univ-mrs.fr

UNIVERSITY OF ILLINOIS AT CHICAGO, CHICAGO, ILLINOIS 60607
E-mail address: masur@math.uic.edu

OREGON STATE UNIVERSITY, CORVALLIS, OREGON 97331
E-mail address: toms@math.orst.edu

IRMAR, UNIVERSITÉ RENNES-1, CAMPUS DE BEAULIEU, 35402 RENNES, CEDEX FRANCE
E-mail address: zorich@math.univ-rennes1.fr

Problems in the Geometry of Surface Group Extensions

Lee Mosher

1. Surface group extensions

A *surface group extension* is a short exact sequence of the form

(1.1) $$1 \to \pi_1 S \to \Gamma \to G \to 1$$

where S is a closed, oriented surface of genus $g \geq 2$. The canonical example is the sequence

(1.2) $$1 \to \pi_1 S \xrightarrow{i} \operatorname{Aut}(\pi_1 S) \xrightarrow{q} \operatorname{Out}(\pi_1 S) \to 1$$

where $\operatorname{Aut}(\pi_1 S)$ is the automorphism group of $\pi_1 S$, the injection $\pi_1(\pi_1 S) \xrightarrow{i} \operatorname{Aut}(\pi_1 S)$ is given by inner automorphisms, $g \to i_g$ where $i_g(h) = g^{-1}hg$, and the quotient $\operatorname{Out}(\pi_1 S)$ is the *outer automorphism group*. This short exact sequence is universal for surface group extensions, in the sense that for any extension as in (1.1) above, there exists a commutative diagram

(1.3)
$$\begin{array}{ccccccccc}
1 & \longrightarrow & \pi_1(S) & \longrightarrow & \Gamma & \longrightarrow & G & \longrightarrow & 1 \\
& & \downarrow & & \downarrow & & \downarrow \alpha & & \\
1 & \longrightarrow & \pi_1(S) & \xrightarrow{i} & \operatorname{Aut}(\pi_1(S)) & \xrightarrow{q} & \operatorname{Out}(\pi_1(S)) & \longrightarrow & 1
\end{array}$$

This shows that Γ is identified with the pushout group

$$\Gamma_\alpha = \{(\phi, \gamma) \in \operatorname{Aut}(\pi_1(S)) \times G \mid q(\phi) = \alpha(\gamma)\}$$

and the homomorphisms $\Gamma \to G$ and $\Gamma \to \operatorname{Aut}(\pi_1 S)$ are identified with the projection homomorphisms of the pushout group.

We are interested here only in the case that α is the inclusion map of a subgroup $G < \operatorname{Out}(\pi_1 S)$, in which case we shall write Γ_G instead of Γ_α, and so $\Gamma_G \approx q^{-1}(G)$.

The short exact sequence (1.2) has a more topological interpretation due to Dehn, Nielsen, Epstein, and Baer. Let $\mathcal{MCG}(S) = \operatorname{Homeo}(S)/\operatorname{Homeo}_0(S)$ denote the mapping class group of S, where Homeo denotes the homeomorphism group and $\operatorname{Homeo}_0(S)$ denotes the component of the identity, consisting of those homeomorphisms isotopic to the identity. The notation $\pi_1(S)$ assumes, implicitly, that a base point has been chosen; denoting this base point by p we have, more explicitly, $\pi_1(S) = \pi_1(S, p)$. Let $\mathcal{MCG}(S, p) = \operatorname{Homeo}(S, p)/\operatorname{Homeo}_0(S, p)$ denote the mapping class group of S punctured at p. There is an isomorphism of short exact sequences:

$$\begin{array}{ccccccccc}
1 & \longrightarrow & \pi_1(S,p) & \longrightarrow & \mathcal{MCG}(S,p) & \longrightarrow & \mathcal{MCG}(S) & \longrightarrow & 1 \\
& & \| & & \wr & & \wr & & \\
1 & \longrightarrow & \pi_1(S,p) & \longrightarrow & \mathrm{Aut}(\pi_1(S,p)) & \longrightarrow & \mathrm{Out}(\pi_1(S,p)) & \longrightarrow & 1
\end{array}$$

where the isomorphism $\mathcal{MCG}(S,p) \approx \mathrm{Aut}(\pi_1(S,p))$ is just the induced homomorphism in the fundamental group functor. Using this isomorphism, to each subgroup $G < \mathcal{MCG}(S)$ there is an associated surface group extension $\Gamma_G < \mathcal{MCG}(S,p)$.

Here is a broad umbrella which covers the problems we will explore:

PROBLEM 1. Given a subgroup $G < \mathcal{MCG}(S)$, how is the geometry of Γ_G related to the dynamics of the action of G on Thurston's compactification of Teichmüller space $\overline{\mathcal{T}}(S) = \mathcal{T}(S) \cup \partial \mathcal{T}(S)$?

Several of the problems that we propose in Sections 2—4 are taken from the author's joint paper with Benson Farb [**FM02a**].

2. Convex cocompact groups: Background

Given a finitely generated subgroup $G < \mathcal{MCG}(S)$, word hyperbolicity of the surface group extension Γ_G is related to the dynamics of G acting on $\overline{\mathcal{T}}(S)$ as follows. We say that G is a *convex cocompact subgroup* of $\mathcal{MCG}(S)$ if G is a word hyperbolic group and there exists an G-equivariant continuous map

$$f \colon G \cup \partial G \to \mathcal{T}(S) \cup \partial \mathcal{T}(S)$$

from the Gromov topology to the Thurston topology, taking Γ to $\mathcal{T}(S)$ and taking ∂G injectively to $\partial \mathcal{T}(S)$, such that any two distinct points $\xi, \eta \in \partial G$ are endpoints of a unique convergent Teichmüller geodesic $\overline{\xi \eta}$, and the union of these geodesics has finite Hausdorff distance from $f(G)$. We use the terminology "convergent" to refer to a Teichmüller geodesic whose ends each converge to a point in $\partial \mathcal{T}(S)$.

THEOREM 1 ([**FM02a**]). *Given a finitely generated subgroup $G \subset \mathcal{MCG}(S)$, if Γ_G is word hyperbolic then G is convex cocompact. If G is a free group then the converse holds as well.*

Every infinite order element in a convex cocompact subgroup of $\mathcal{MCG}(S)$ is pseudo-Anosov. Free, convex cocompact subgroups of $\mathcal{MCG}(S)$ are called *Schottky subgroups*, and they consist entirely of pseudo-Anosov elements except for the identity.

PROBLEM 2. Does the converse hold in the above theorem without the assumption that G is free?

The gist of Problem 2 is to find an extension of the Bestvina-Feighn combination theorem beyond the setting of groups acting on trees, to the wider setting of groups acting on Gromov hyperbolic cell complexes. In particular, in the context of Problem 2, the group Γ_G acts on the Rips complex of G, and one wants to use this information to decide about word hyperbolicity of Γ_G.

Unfortunately there are at present no examples on which to test Problem 2, but we will propose some problems along these lines in Section 4.

3. Schottky subgroups of $\mathcal{MCG}(S)$

Schottky subgroups exist in abundance, as a consequence of the following result. Recall that a pseudo-Anosov element of $\mathcal{MCG}(S)$ acts on $\partial \mathcal{T}(S)$ with "source–sink" dynamics, having one repelling fixed point $\xi_-(\phi)$, one attracting fixed point $\xi_+(\phi)$, and every other orbit tending to $\xi_-(\phi)$ under backward iteration and to $\xi_+(\phi)$ under forward iteration. Two pseudo-Anosov elements $\phi, \psi \in \mathcal{MCG}(S)$ are *independent* if $\{\xi_\pm(\phi)\} \cap \{\xi_\pm(\psi)\} = \emptyset$, equivalently, if $\{\xi_\pm(\phi)\} \neq \{\xi_\pm(\psi)\}$.

THEOREM 2 ([**Mos97**]). *Let $\phi_1, \ldots, \phi_n \in \mathcal{MCG}(S)$ be pairwise independent pseudo-Anosov elements. Then there exists $M > 0$ such that for integers $\alpha_1, \ldots, \alpha_n \geq M$, the mapping classes $\phi_1^{\alpha_1}, \ldots, \phi_n^{\alpha_n}$ freely generate a subgroup H such that Γ_H is word hyperbolic.*

Applying Theorem 1 we obtain:

THEOREM 3 ([**FM02a**]). *In Theorem 2, it follows that H is a Schottky subgroup of $\mathcal{MCG}(S)$.*

It is unclear how to use Theorem 2 to write down a single example of a Schottky subgroup. It would be interesting to make the theorem more effective:

PROBLEM 3. Does there exist an algorithm which produces the integer M in Theorem 2, given $\phi_1, \ldots, \phi_n$?

One approach to this problem is to use train track expansions of pseudo-Anosov homeomorphisms [**Mos03c**].

Hamidi-Tehrani's thesis [**HT97a**] (see also the preprint [**HT97b**]) contains a related algorithm which produces an integer M such that if $\alpha_1, \ldots, \alpha_n \geq M$ then the mapping classes $\phi_1^{\alpha_1}, \ldots, \phi_n^{\alpha_n}$ freely generate a subgroup of $\mathcal{MCG}(S)$.

Concrete examples of Schottky subgroups of $\mathcal{MCG}(S)$. We will construct examples in stabilizers of Teichmüller discs in $\mathcal{T}(S)$, maximal holomorphically embedded copies of the Poincaré disc in $\mathcal{T}(S)$. Recall Royden's Theorem [**Roy71**] that the restriction of the Teichmüller metric to a Teichmüller disc D equals the hyperbolic metric on D. Also, the embedding $D \hookrightarrow \mathcal{T}(S)$ extends to a proper embedding $D \cup \partial D \to \mathcal{T}(S) \cup \partial \mathcal{T}(S)$ taking ∂D to $\mathcal{T}(S)$. If $H < \text{Stab}(D)$ is a finitely generated, torsion free subgroup with no parabolics, then H acts properly on D because $\mathcal{MCG}(S)$ acts properly on $\mathcal{T}(S)$, and the quotient hyperbolic surface D/H has finite topological type and each end is nonparabolic, so the convex hull of D/H is a compact hyperbolic surface with totally geodesic boundary. It follows that H is a free subgroup acting convex cocompactly on D — a Schottky group in the hyperbolic plane — which immediately implies that H is a Schottky subgroup of $\mathcal{MCG}(S)$.

Thurston [**Thu88**] first constructed Teichmüller discs $D \subset \mathcal{T}(S)$ with interesting stabilizer groups $\text{Stab}(D) < \mathcal{MCG}(S)$, including examples in which the action of $\text{Stab}(D)$ on D has finite co-area. After Veech [**Vee89**] initiated an in-depth study of this phenomenon the groups $\text{Stab}(D)$ have come to be known as *Veech subgroups* of $\mathcal{MCG}(S)$. This study has blossomed beautifully; some recent papers, which can be consulted for a fuller bibliography, are [**HS04**], [**MT02**], [**McM03**].

We shall follow Thurston's original construction of infinite co-area examples, producing an explicit example of a Schottky subgroup of $\mathcal{MCG}(S)$, but first we give some background on Teichmüller discs.

Consider a conformal structure σ on S and a quadratic differential q on σ. Away from the singularities of q there are *regular* local coordinate charts with complex parameter z in which $q = dz^2$, and such charts are unique up to transformations of the type $z \mapsto \pm z + c$ for complex constants c. Given $A \in \mathrm{SL}(2, \mathbf{R})$, the action of A in a regular local coordinate $z = x + iy \in \mathbf{C} \approx \mathbf{R}^2$ transforms z by the ordinary linear action of A on $\mathbf{R}^2$, producing a new quadratic differential and a new conformal structure denoted $A(\sigma)$. If A is post-composed with an element of $\mathrm{SL}(2, \mathbf{R})$ that fixes the point $A(i)$ in the upper half plane, then the resulting element of $\mathrm{SL}(2, \mathbf{R})$ fixes $A(\sigma)$. The map $\sigma \mapsto A(\sigma)$ therefore induces a map $\mathbf{H}^2 = \mathrm{SL}(2, \mathbf{R})/\mathrm{SO}(2, \mathbf{R}) \to \mathcal{T}(S)$ whose image is a Teichmüller disc denoted D_q.

Now consider $c, d \subset S$ two closed curves intersecting transversely such that each component of $S - (c \cup d)$ is a $2n$-sided polygon with $n \geq 2$, where the sides alternate between segments of c and segments of d. Let $c \cap d = \{x_1, \ldots, x_m\}$. There is a quadratic differential q on S whose nonsingular horizontal trajectories are all isotopic to c and form an open dense annulus A_c, and whose nonsingular vertical trajectories are isotopic to d and form an open dense annulus A_d. On each component of $A_c \cap A_d$ the differential q is isometric to the restriction of dz to a unit square in the complex plane. The positive Dehn twist τ_d preserves D_q, acting like the fractional linear transformation $M_d = \left(\begin{smallmatrix} 1 & m \\ 0 & 1 \end{smallmatrix} \right)$, and the negative Dehn twist τ_c preserves D_q acting like $M_c = \left(\begin{smallmatrix} 1 & 0 \\ m & 1 \end{smallmatrix} \right)$. Consider the action of $\mathrm{PSL}(2, \mathbf{R})$ on $D_q \approx \mathbf{H}^2$ extending to $\partial \mathbf{H}^2 = \mathbf{R}\mathcal{P}^1$. Consider the following four closed intervals in $\mathbf{R}\mathcal{P}^1 \approx \mathbf{R} \cup \{\infty\}$:

$$I_0^- = \left[-\frac{1}{m-1}, -\frac{1}{m+1} \right] \qquad I_0^+ = [m-1, m+1]$$
$$I_1^- = \left[\frac{1}{m+1}, \frac{1}{m-1} \right] \qquad I_1^+ = [-(m+1), -(m-1)].$$

The matrix $M_d M_c = \left(\begin{smallmatrix} m^2+1 & m \\ m & 1 \end{smallmatrix} \right)$ takes the inside of I_0^- in $\mathbf{R}\mathcal{P}^1$ to the outside of I_0^+ and $M_c M_d = \left(\begin{smallmatrix} 1 & m \\ m & m^2+1 \end{smallmatrix} \right)$ takes the inside of I_1^- to the outside of I_1^+. This can be verified analytically, or by drawing the standard modular diagram depicting the action of $\mathrm{SL}(2, \mathbf{Z})$ on the hyperbolic plane. If $m \geq 3$ then the four closed intervals $I_0^-, I_0^+, I_1^-, I_1^+$ are disjoint, and so $M_d M_c$ and $M_c M_d$ generate a Schottky group of rank 2 acting on D_q. But D_q is isometrically embedded in $\mathcal{T}(S)$ and so this gives a Schottky subgroup of $\mathcal{MCG}(S)$. Even if $m = 2$, where $I_1^- \cap I_0^+ = \{1\}$ and $I_0^- \cap I_1^+ = \{-1\}$, by working a bit harder one can show that $M_d M_c$ and $M_c M_d$ generate a rank 2 Schottky subgroup.

Questions about Schottky subgroups. As noted above, every discrete, finite rank, free subgroup of $\mathrm{Isom}(\mathbf{H}^2)$ with no parabolics acts as a Schottky group on $\mathbf{H}^2$. A much deeper result says that if M is a closed hyperbolic 3-manifold, then every finite rank, free subgroup of $\pi_1(M)$ acts as a Schottky group on $\mathbf{H}^3$; as noted by Canary in [**Can94**], this is a consequence of tameness of every complete hyperbolic 3-manifold with finitely generated fundamental group, recently proved by Agol [**Ago04**] and by Calegari and Gabai [**CG04**]. These results motivate the following:

PROBLEM 4. *If $H \subset \mathcal{MCG}(S)$ is finite rank free subgroup whose nonidentity elements are pseudo-Anosov, is H a Schottky group?*

For specific examples on which to test the above problem, Whittlesey [**Whi00b**] answered a question of Penner by producing a normal subgroup of $\mathcal{MCG}(S)$, S of genus 2, whose nonidentity elements are entirely pseudo-Anosov. Moreover, this subgroup is an infinite rank free subgroup [**Whi00a**].

PROBLEM 5. Is every finite rank subgroup of Whittlesey's group a Schottky subgroup of $\mathcal{MCG}(S)$?

As a consequence of Theorem 1, if $H < \mathcal{MCG}(S)$ has a finite index subgroup which is Schottky, that is, if H is a virtual Schottky subgroup, then Γ_H has a finite index word hyperbolic subgroup and so Γ_H is itself word hyperbolic.

PROBLEM 6. Give examples and constructions of virtual Schottky subgroups of $\mathcal{MCG}(S)$.

One such construction is due to Honglin Min, currently a doctoral candidate at Rutgers University, Newark:

THEOREM 4 (H. Min). *If $A, B \subset \mathcal{MCG}(S)$ are finite subgroups, if $\phi \in \mathcal{MCG}(S)$ is pseudo-Anosov, and if A, B each have trivial intersection with the normalizer of ϕ in $\mathcal{MCG}(S)$, then for sufficiently high n the subgroup of $\mathcal{MCG}(S)$ generated by the finite subgroups $A, \phi^n B \phi^{-n}$ is the free product of these subgroups, and is a virtual Schottky subgroup of $\mathcal{MCG}(S)$.*

4. Convex cocompact groups: Questions

One of the most intriguing questions around is the following:

PROBLEM 7. Do there exist two surfaces S, S', closed and of genus ≥ 2, such that $\mathcal{MCG}(S)$ contains a subgroup isomorphic to $\pi_1(S')$ all of whose nontrivial elements are pseudo-Anosov?

The closest attempt so far is an example of Leininger and Reid [**LR04**] in which all closed curves on S' give pseudo-Anosov elements of $\mathcal{MCG}(S)$ with the exception of a single simple closed curve and its iterates.

This problem invites the following refinement:

PROBLEM 8. Do there exist two surfaces S, S', closed and of genus ≥ 2, and a subgroup $G < \mathcal{MCG}(S)$ isomorphic to $\pi_1(S')$, so that Γ_G is word hyperbolic? Or so that G is convex cocompact?

The equivalence of the two questions in Problem 8 is still open, as indicated in Problem 2.

PROBLEM 9. Does there exist *any* non-virtually free, finitely generated subgroup $G < \mathcal{MCG}(S)$ whose nontorsion elements are all pseudo-Anosov? Does G exist so that Γ_G is word hyperbolic? So that G is convex cocompact?

A speculation about reflection groups. Here's a speculative approach to constructing $\mathbf{H}^2$ polygon reflection groups in $\mathcal{MCG}(S)$ on which to test Problem 9. If successful this would produce examples on which to test Problem 8 by passing to a finite index subgroup. In order to see analogues of this construction in more familiar territory we formulate it for lattices in $\text{Isom}(\mathbf{H}^n)$ as well as for $\mathcal{MCG}(S)$.

Consider a group Γ, which the reader may imagine as either $\mathcal{MCG}(S)$, or as a discrete, cocompact subgroup of $\text{Isom}(\mathbf{H}^n)$. A *cycle of dihedral subgroups* of Γ is a cyclic sequence of the form

(4.1) $$D_0, Z_1, D_1, Z_2, D_2, \ldots, Z_n, D_n = D_0$$

where each D_i is a finite dihedral subgroup of order $2k_i$, each Z_i is a subgroup of order 2, $D_{i-1} \cap D_i = Z_i$, and the respective generators z_i, z_{i+1} of Z_i, Z_{i+1} form the generating set of a standard presentation

$$D_i = \langle z_i, z_{i+1} \mid z_i^2 = z_{i+1}^2 = (z_i z_{i+1})^{k_i} = 1 \rangle.$$

From the cycle (4.1) one obtains a presentation of a polygon reflection group

$$P = \langle z_0, z_1, \ldots, z_n = z_0 \mid z_i^2 = (z_i z_{i+1})^{k_i} = 1 \quad \text{for } i = 1, \ldots, n \rangle$$

and a homomorphism $P \to \Gamma$. A necessary condition for injectivity of this homomorphism is that the cycle (4.1) is *simple*, meaning that the subgroups Z_i have pairwise trivial intersection, and the subgroup $D_i \cap D_j$ is trivial if $|i - j| > 1$. In general simplicity of the cycle (4.1) is *not* sufficient for injectivity of $P \to \Gamma$.

For example, consider the case that Γ is a discrete, cocompact subgroup of $\text{Isom}(\mathbf{H}^3)$, and we assume that Γ contains dihedral subgroups so that there is somewhere to start. Consider the quotient orbifold $O = \mathbf{H}^3/\Gamma$. Suppose that we have in our hands a 2-dimensional suborbifold $E \subset O$ whose underlying 2-manifold is a disc, and we assume that E is a hyperbolic 2-orbifold. Then E determines a cycle of dihedral subgroups of Γ whose associated reflection group P can be identified with $\pi_1 E$, and the homomorphism $P \to \Gamma$ is induced by the inclusion $E \hookrightarrow O$. It is well known that, as a consequence of the Meeks-Yau equivariant loop theorem [**MY82**], the homomorphism $P \to \Gamma$ is injective if and only if E is an essential 2-dimensional suborbifold of the 3-orbifold O (see [**CHK00**] for the precise definition of essential in this context).

Consider now the case that $\Gamma = \mathcal{MCG}(S)$. There are certainly many finite dihedral subgroups to work with: for instance, take a dihedral group acting on $\mathbf{R}^3$ fixing the points on the z-axis, and take any surface that is invariant under this group. It might be worthwhile setting a computer to search for simple cycles of dihedral subgroups in $\mathcal{MCG}(S)$. If one is found, who knows? Maybe the resulting homomorphism $P \to \mathcal{MCG}(S)$ will be injective, convex cocompact, and have a word hyperbolic extension group Γ_P.

To summarize:

PROBLEM 10. *Does there exist a simple cycle of dihedral subgroups of $\mathcal{MCG}(S)$ so that the associated reflection group P injects in $\mathcal{MCG}(S)$? So that the image of P in $\mathcal{MCG}(S)$ answers any of the questions in Problem 9?*

5. Quasi-isometric rigidity problems

A *quasi-isometric embedding* between metric spaces is a map $f: X \to Y$ such that for some constants $K \geq 1$, $C \geq 0$ we have

$$\frac{1}{K} d(x, x') - C \leq d(fx, fx') \leq K d(x, x') + C \quad \text{for all} \quad x, x' \in X.$$

We say that f is a *quasi-isometry* if it also holds, after possibly rechoosing C, that for all $y \in Y$ there exists $x \in X$ such that $d(fx, y) \leq C$. Two metric spaces are *quasi-isometric* if there exists a quasi-isometry between them; this is an equivalence relation on metric spaces. If Γ is a finitely generated group, and if X is any proper geodesic metric space on which Γ acts properly and cocompactly, then X is quasi-isometric to Γ with its word metric, and so all such X are quasi-isometric to each other.

Gromov proposed the problem of classifying finitely generated groups up to quasi-isometry of their word metrics. Loosely speaking, a group Γ is "quasi-isometrically rigid" if all the groups in its quasi-isometry class are closely related to Γ; the precise sense of "closely related" varies widely for different groups Γ.

One particularly strong notion of quasi-isometric rigidity goes like this. Given a metric space X, such as a finitely generated group with the word metric, the set of self quasi-isometries of X with the operation of composition forms a group after identifying any two quasi-isometries which have finite distance in the sup norm; this group is denoted QI(X). The action of G on itself by left translation induces a natural homomorphism $G \to$ QI(G). One of the strongest quasi-isometric properties occurs when this homomorphism has finite kernel and finite index image; when this happens then *every* group H quasi-isometric to G has a homomorphism $H \to$ QI(G) with finite kernel and finite index image. In such a case, the group QI(G) itself is a finitely generated group in the quasi-isometry class of G, and in some sense it is the canonical representative of this class.

Among the class of groups Γ_G for finitely generated $G < \mathcal{MCG}(S)$, here are some quasi-isometric rigidity theorems:

THEOREM 5 (Farb–M. [**FM02b**]). *If $G < \mathcal{MCG}(S)$ is a Schottky subgroup then the natural homomorphism $\Gamma_G \to$ QI(Γ_G) is an injection with finite index image.*

THEOREM 6 (M.–Whyte [**Mos03b**]). *The natural homomorphism from a once-punctured mapping class group $\mathcal{MCG}(S,p) = \Gamma_{\mathcal{MCG}(S)}$ to its quasi-isometry group is an isomorphism.*

The last theorem is the first settled case of a well known conjecture that if S is an oriented, finite type surface with $\chi(S) < 0$ (not the once-punctured torus nor the four punctured sphere), then the homomorphism $\mathcal{MCG}(S) \to$ QI($\mathcal{MCG}(S)$) is an isomorphism.

The proof of Theorem 6, described in [**Mos03b**], combines a machine developed by the author in [**Mos03a**] for investigating quasi-isometric rigidity of more general groups of the form Γ_G, together with homological methods developed by Whyte. The machine of [**Mos03a**] grew out of the proof of Theorem 5 in [**FM02b**]. We describe here some of the details of this machine, with proposals to apply the machine to Γ_G for specific examples of G, particularly Veech groups.

Suppose that $G < \mathcal{MCG}(S)$ is free but not necessarily Schottky. Let $S \to \mathcal{O}_G$ be the orbifold covering map of largest degree such that G descends to a subgroup of $\mathcal{MCG}(\mathcal{O}_G)$, denoted G'. Let $\mathcal{C}$ be the relative commensurator of G' in $\mathcal{MCG}(\mathcal{O}_G)$, that is, the group of all $\phi \in \mathcal{MCG}(\mathcal{O}_G)$ for which $\phi^{-1} G' \phi \cap G'$ has finite index in both $\phi^{-1} G' \phi$ and G'. We obtain an extension group

$$1 \to \pi_1(\mathcal{O}_G) \to \Gamma_\mathcal{C} \to \mathcal{C} \to 1$$

and a homomorphism $\Gamma_\mathcal{C} \to$ QI(Γ_G).

Recall that a subgroup $G < \mathcal{MCG}(S)$ is *irreducible* if there does not exist a pairwise disjoint, pairwise nonisotopic family C of simple closed curves on S such that each element of G preserves C up to isotopy. By Ivanov's Theorem [**Iva92**],

G is infinite and irreducible if and only if it contains at least one pseudo-Anosov element.

THEOREM 7 ([**Mos03a**]). *If $G \subset \mathcal{MCG}(S)$ is finite rank free and irreducible then the homomorphism $\Gamma_{\mathcal{C}} \to \mathrm{QI}(\Gamma_G)$ is an isomorphism.*

When G is Schottky, the results of [**FM02b**] involve an explicit calculation of $\Gamma_{\mathcal{C}}$, with the outcome that the natural injection $\Gamma_G \to \mathrm{QI}(\Gamma_G) \approx \Gamma_{\mathcal{C}}$ has finite index image.

The next step is to investigate what happens in cases when G is not Schottky:

PROBLEM 11. Suppose that $G < \mathcal{MCG}(S)$ is a finite co-area Veech subgroup. What can one say about $\Gamma_{\mathcal{C}}$? In particular, does it contain Γ_G with finite index?

PROBLEM 12. Explore $\Gamma_{\mathcal{C}}$ for other free subgroups $G < \mathcal{MCG}(S)$, for example free subgroups generated by high powers of Dehn twists about a pair of filling curves.

If G is not free then [**Mos03a**] contains a result more general than Theorem 7, with $\mathrm{QI}(\Gamma_G)$ replaced by a group $\mathrm{QI}_f(\Gamma_G)$ constructed from quasi-isometries $\Gamma_G \to \Gamma_G$ which coarsely preserve the family of cosets of $\pi_1(S) < \Gamma_G$. The point of G being free is that in this case one can use methods of coarse algebraic topology to prove that $\mathrm{QI}_f(\Gamma_G) = \mathrm{QI}(\Gamma_G)$; see e.g. [**FM00**]. In the case that $G = \mathcal{MCG}(S)$ covered in Theorem 6, Whyte's homological methods are used to prove that $\mathrm{QI}_f(\Gamma_G) = \mathrm{QI}(\Gamma_G)$.

One might be able to use Whyte's homological methods for other subgroups $G < \mathcal{MCG}(S)$, particularly when G is a Bieri-Eckmann duality group which is not a Poincaré duality group, such as $G = \mathcal{MCG}(S)$ [**Har86**].

Unfortunately, none of these methods will work when G is a Leininger-Reid group, because in this case, G is a 2-dimensional Poincaré duality group, which rules out the methods of [**FM00**] as well as Whyte's homological methods. We do not have any idea of how to overcome this obstacle. Nonetheless it should still be interesting to calculate $\Gamma_{\mathcal{C}} = \mathrm{QI}_f(\Gamma_G)$ for a Leininger-Reid subgroup G.

6. Geometrically finite subgroups of $\mathcal{MCG}(S)$

In studying finitely generated discrete subgroups of $\mathrm{Isom}(\mathbf{H}^n)$, the best behaved examples are the convex cocompact groups, and the next best are the geometrically finite groups. Guided by Problem 1, and keeping in mind a comparison between Problem 11 and Theorem 1, it might be interesting to develop a definition and a theory of "geometrically finite subgroups of $\mathcal{MCG}(S)$". In order to provoke thought we shall propose such a definition here. We will see that it holds for any finitely generated Veech subgroup, and it would be interesting to know if it held for, say, a Leininger-Reid subgroup. The definition we offer will be tailored *only* for the case where the cusp groups are infinite cyclic; see "Remarks on the definition" below.

First we review the theory of relatively hyperbolic groups following Farb [**Far98**] and Bowditch [**Bow**]. Consider a finitely generated group G and finitely generated subgroups $H_1, \ldots, H_n$. We say that G is *hyperbolic relative to* $H_1, \ldots, H_n$ if G satisfies Farb's relative hyperbolicity plus bounded coset penetration, relative to the subgroups $H_1, \ldots, H_n$. As Bowditch observed [**Bow**], this is equivalent to saying that G has a proper, isometric action on some proper hyperbolic metric space X, so that:

- For each point $x \in \partial X$, either x is a conical limit point of the action of G, or the subgroup of G stabilizing x acts by parabolic isometries, restricting to a proper cocompact action on some horosphere based at x.
- The subgroups $H_1, \ldots, H_n$ are stabilizers of parabolic points $x_1, \ldots, x_n$ respectively, and $x_1, \ldots, x_n$ are representatives of the orbits of G acting on the parabolic points.

Consider now a subgroup $G < \mathcal{MCG}(S)$ and subgroups $H_1, \ldots, H_n < G$. We propose to say that G is a *geometrically finite subgroup of* $\mathcal{MCG}(S)$, *with cusp groups* $H_1, \ldots, H_n$, if G is hyperbolic relative to $H_1, \ldots, H_n$ and, letting G act on X as above, there are G-equivariant continuous maps

$$f \colon X \to \mathcal{T}(S)$$
$$\partial f \colon \partial X \to \partial \mathcal{T}(S)$$

such that f is a quasi-isometric embedding and ∂f is a topological embedding, for any $\xi \neq \eta \in \partial X$ the points $f(\xi), f(\eta) \in \partial \mathcal{T}(S)$ are a pair of filling measured foliations in S which therefore determine a $\mathcal{T}(S)$ geodesic $\overline{f(\xi)f(\eta)}$, the image of a geodesic in X with ideal endpoints ξ, η is a fellow traveller of the geodesic $\overline{f(\xi)f(\eta)}$, and the image $f(X)$ has finite Hausdorff distance from the union of the $\mathcal{T}(S)$ geodesics $\overline{f(\xi)f(\eta)}$ for $\xi \neq \eta \in \partial X$.

REMARK. If, in the above definition, one replaces $\mathcal{T}(S)$ with $\mathbf{H}^n$ and $\mathcal{MCG}(S)$ with $\mathrm{Isom}(\mathbf{H}^n)$ then the result, with the additional proviso that G be discrete, becomes equivalent to the usual definition of a geometrically finite discrete group acting on $\mathbf{H}^n$.

EXAMPLE. Suppose that G is a Veech subgroup acting with co-finite area on a Teichmüller disc D_q. In this case one can take $X = D_q$ and f to be the identity map. The image of ∂f is an embedded circle, namely the horizontal measured foliations of the quadratic differentials θq parameterized by the unit complex numbers θ; this parameterization is a double covering of the image of ∂f. It follows that G is geometrically finite. In fact if G is any finitely generated group stabilizing a Teichmüller disc D_q then, taking X to be the convex hull of the limit set of G in the circle at infinity of D_q, we see that G is a geometrically finite subgroup of $\mathcal{MCG}(S)$.

The next obvious test case is:

PROBLEM 13. Are the Leininger–Reid subgroups geometrically finite, with cusp groups the reducible cyclic subgroups?

Note that a Leininger–Reid subgroup is isomorphic to $\pi_1(S')$ for some closed surface group S' of genus ≥ 2, and the reducible cyclic subgroups correspond to a pairwise disjoint family of simple closed curves. Relative hyperbolicity certainly holds in this case, as can be seen by constructing X as follows. Consider the universal covering $\widetilde{S'} \approx \mathbf{H}^2$. The axes of the reducible cyclic subgroups form a pairwise disjoint family of geodesics Γ in $\widetilde{S'}$. Attach to each of these axes a copy of a horodisc in $\mathbf{H}^2$; one might visualize these horodiscs as sticking out of $\widetilde{S'}$ in a third dimension. It is not too hard to see that the space X is Gromov hyperbolic, and that the reducible cyclic subgroups have the required action. Chris Leininger suggested an alternate geometry on X, which is obviously $\mathrm{CAT}(-1)$, and which closely mimics the natural geometry of a Leininger–Reid subgroup: leave the attached horodiscs as they are, but for each component M of $\widetilde{S}' - \Gamma$, instead of giving M a hyperbolic structure with totally geodesic boundary, give M the metric obtained from the corresponding Teichmüller disc D by equivariantly truncating horodiscs at the cusp points of D.

REMARKS ON THE DEFINITION. I thank Chris Leininger for comments and observations leading to the following remarks.

(1) In the concluding remarks of [**LR04**] it is noted that the inclusion map $D_q \hookrightarrow \mathcal{T}(S)$ *need not* extend continuously to a map from the Gromov compactification of the hyperbolic plane D_q taking the circle at infinity to the image of ∂f. This is a consequence of Masur's proof [**Mas82**] that a Teichmüller ray defined by a measured foliation f whose trajectories are all compact converges not necessarily to the projective class of f but to the class of the measured foliation obtained from f by putting equal transverse measure on the annuli comprising f. For this reason, after initially writing the definition to require $f \cup \partial f$ to be continuous, we have dropped this requirement; however, one might want to impose conditions that mimic the results of [**Mas82**] regarding almost everywhere continuity.

(2) In our proposed definition of geometrically finite, the fixed point set of the cusp group H_i in image(∂f) is a single point. This fits well with examples where H_i is an infinite cyclic Dehn twist group. However, if H_i is an abelian subgroup of rank $r \geq 2$ generated by Dehn twists about r disjoint curves, then the fixed point set of H_i in $\partial \mathcal{T}(S)$ is a simplex of dimension $r - 1 \geq 1$, and an ambiguity arises in defining the map ∂f on the point of ∂X fixed by H_i. For this reason, our definition of geometric finiteness is inadequate in the broader context of higher rank cusp groups.

(3) For understanding geometric finiteness with higher rank cusp subgroups, here is a good example to ponder. Let τ_1, τ_2 be Dehn twists about disjoint, nonisotopic simple closed curves in S, let ϕ be a pseudo-Anosov mapping class on S, and let G be the free product of the $\mathbf{Z}^2$ subgroup $H = \langle \tau_1, \tau_2 \rangle$ and the cyclic group $K = \langle \phi^n \rangle$ for some large n. For n sufficiently large the group G is likely to be the free product of H and K, and then G will be hyperbolic relative to H. Whatever "geometrically finite" might mean in a broader context, the group G ought to be geometrically finite with cusp subgroup H.

As these remarks show, the reader should beware that any definition of geometrically finite subgroups of $\mathcal{MCG}(S)$ needs to be carefully tested against examples. For instance, the requirement that f be a quasi-isometric embedding may be too strong in general, even though it works perfectly well for cofinite area Veech suroups.

Here is an entirely open-ended speculation, related to Problems 11 and 12:

PROBLEM 14. If $G < \mathcal{MCG}(S)$ is geometrically finite with cusp groups $H_1, \ldots, H_n$, what can be said about the geometric properties of the group Γ_G? Does it have useful large scale geometric properties? Does it satisfy some form of quasi-isometric rigidity?

Note that Γ_G is *not* hyperbolic relative to its subgroups $\Gamma_{H_1}, \ldots, \Gamma_{H_n}$, because each of these subgroups and their conjugates contains the kernel $\pi_1(S)$, whereas relative hyperbolicity implies that the intersection of any two of these conjugates is finite.

References

[Ago04] I. Agol, *Tameness of hyperbolic 3-manifolds*, preprint, ARXIV:MATH.GT/0405568, 2004.

[Bow] B. Bowditch, *Relatively hyperbolic groups*, Preprint, University of Southampton http://www.maths.soton.ac.uk.

[Can94] R. Canary, *Covering theorems for hyperbolic 3-manifolds*, Low-dimensional topology (Knoxville, TN, 1992), Conf. Proc. Lecture Notes Geom. Topology, III, Internat. Press, Cambridge, MA, 1994, pp. 21–30.

[CG04] D. Calegari and D. Gabai, *Shrinkwrapping and the taming of hyperbolic 3-manifolds*, preprint, ARXIV:MATH.GT/0417161, 2004.

[CHK00] D. Cooper, C. D. Hodgson, and S. P. Kerckhoff, *Three-dimensional orbifolds and cone-manifolds*, MSJ Memoirs, vol. 5, Mathematical Society of Japan, 2000.

[Far98] B. Farb, *Relatively hyperbolic groups*, Geom. Funct. Anal. **8** (1998), no. 5, 810–840.

[FM00] B. Farb and L. Mosher, *On the asymptotic geometry of abelian-by-cyclic groups*, Acta Math. **184** (2000), no. 2, 145–202.

[FM02a] _____, *Convex cocompact subgroups of mapping class groups*, Geometry and Topology **6** (2002), 91–152.

[FM02b] _____, *The geometry of surface-by-free groups*, Geom. Funct. Anal. **12** (2002), 915–963, Preprint, ARXIV:MATH.GR/0008215.

[Har86] J. L. Harer, *The virtual cohomological dimension of the mapping class group of an orientable surface*, Invent. Math. **84** (1986), 157–176.

[HS04] P. Hubert and T. Schmidt, *Infinitely generated Veech groups*, Duke Math. J. **123** (2004), no. 1, 49–69.

[HT97a] H. Hamidi-Tehrani, *Algorithms in the surface mapping class groups*, Ph.D. thesis, Columbia University, 1997.

[HT97b] _____, *On free subgroups of the mapping class groups*, preprint, 1997.

[Iva92] N. V. Ivanov, *Subgroups of Teichmüller modular groups*, Translations of Mathematical Monographs, vol. 115, Amer. Math. Soc., 1992.

[LR04] C. Leininger and A. Reid, *A combination theorem for Veech subgroups of the mapping class group*, preprint, arXiv:math.GT/0410041, 2004.

[Mas82] H. Masur, *Two boundaries of Teichmüller space*, Duke Math. J. **49** (1982), no. 1, 183–190.

[McM03] C. McMullen, *Teichmüller geodesics of infinite complexity*, Acta Math. **191** (2003), no. 2, 191–223.

[Mos97] L. Mosher, *A hyperbolic-by-hyperbolic hyperbolic group*, Proc. AMS **125** (1997), no. 12, 3447–3455.

[Mos03a] _____, *Fiber respecting quasi-isometries of surface group extensions*, Preprint, ARXIV:MATH.GR/0308067, 2003.

[Mos03b] _____, *Homology and dynamics in quasi-isometric rigidity of once-punctured mapping class groups*, Lecture Notes from "LMS Durham Symposium: Geometry and Cohomology in Group Theory, University of Durham, UK (July 2003)". Preprint, ARXIV:MATH.GR/0308065, 2003.

[Mos03c] _____, *Train track expansions of measured foliations*, preprint, version 2, http://newark.rutgers.edu:80/ mosher/, 2003.

[MT02] H. Masur and S. Tabachnikov, *Rational billiards and flat structures*, Handbook of dynamical systems, Vol. 1A, North-Holland, Amsterdam, 2002, pp. 1015–1089.

[MY82] W. H. Meeks and S. T. Yau, *The equivariant Dehn's lemma and loop theorem*, Comment. Math. Helv. **56** (1982), 225–239.

[Roy71] H. Royden, *Automorphisms and isometries of Teichmüller space*, Advances in the Theory of Riemann Surfaces (Proc. Conf., Stony Brook, N.Y., 1969), Ann. of Math. Studies, vol. 66, 1971, pp. 369–383.

[Thu88] W. P. Thurston, *On the geometry and dynamics of diffeomorphisms of surfaces*, Bull. AMS **19** (1988), 417–431.

[Vee89] W. A. Veech, *Teichmüller curves in moduli space, Eisenstein series and an application to triangular billiards*, Invent. Math. **97** (1989), no. 3, 553–583, Erratum: volume 103 number 2, 1991, page 447.

[Whi00a] K. Whittlesey, 2000, private correspondence.
[Whi00b] _____, *Normal all pseudo-Anosov subgroups of mapping class groups*, Geometry and Topology **4** (2000), 293–307.

DEPARTMENT OF MATHEMATICS, RUTGERS UNIVERSITY, NEWARK, NEWARK, NJ 07102
E-mail address: mosher@andromeda.rutgers.edu

Surface Subgroups of Mapping Class Groups

Alan W. Reid

1. Introduction

Let Σ be a compact oriented surface, possibly with boundary, the *Mapping Class group* of Σ is

$$\mathrm{Mod}(\Sigma) = \mathrm{Homeo}^+(\Sigma)/\mathrm{Homeo}_0(\Sigma),$$

where, $\mathrm{Homeo}^+(\Sigma)$ is the group of orientation preserving homeomorphisms of Σ and $\mathrm{Homeo}_0(\Sigma)$ are those homeomorphisms isotopic to the identity. When Σ is a closed surface of genus $g \geq 1$ then we denote $\mathrm{Mod}(\Sigma)$ by Γ_g. In this case it is well-known that Γ_g is isomorphic to a subgroup of index 2 in $\mathrm{Out}(\pi_1(\Sigma))$.

When $g = 1$, the subgroup structure of Γ_1 is well-understood since it is simply the group $\mathrm{SL}(2, \mathbf{Z})$. When $g \geq 2$, attempts to understand the subgroup structure of Γ_g have been made by exploiting the many analogies between Γ_g and non-uniform lattices in Lie groups. In particular, both the finite and infinite index subgroup structure of Γ_g has many parallels in the theory of lattices. For example, the question of whether Property T holds for $\Gamma_g \geq 3$ (it fails for $g = 2$ by [30]), and whether Γ_g has a version of the Congruence Subgroup Property, or towards the other extreme, whether there are finite index subgroups of Γ_g that surject onto $\mathbf{Z}$ are questions that have received much attention recently (see for example [11], [16], [18] and [30] for more on these directions). The discussion and questions raised in this paper are motivated by analogies between the subgroup structure of Γ_g and non-cocompact but finite co-volume Kleinian groups. To that end, we are particularly interested in the nature of surface subgroups of Γ_g. For an exploration of other analogies between Γ_g and Kleinian groups see [13], [19] and [34].

Throughout, the terms *surface (sub)group* will be reserved for $\pi_1(\Sigma_g)$ where $g \geq 2$, and a subgroup of Γ_g is said to be *purely pseudo-Anosov* if all non-trivial elements are pseudo-Anosov. Simply put our motivation is the following question.

QUESTION 1.1. *For $g \geq 2$, does Γ_g contain a purely pseudo-Anosov surface subgroup?*

The paper is organized as follows. In §2, we discuss the existence of surface subgroups in Kleinian groups of finite co-volume. In §3, we discuss surface subgroups of Γ_g, and contrast and compare with §2. In §4 we discuss some related topics;

This work was partially supported by an N.S.F. grant.

for example we discuss the connection of Question 1.1 with some conjectures in 4-manifold topology.

Acknowledgments. Much of the content of this paper is related to joint work of the author and Chris Leininger [**23**], and the author gratefully acknowledges his contributions to this work. In addition, some of the questions mentioned here also appear in [**23**]. The author would also like to thank Ian Agol, Richard Kent and Darren Long for useful conversations on some of the topics discussed here.

2. Surface subgroups of Kleinian groups

We begin by discussing the case of surface subgroups in Kleinian groups; see [**26**], [**38**], or [**33**] for terminolgy.

2.1. A *Kleinian group* Γ is a discrete subgroup of $\text{PSL}(2, \mathbf{C})$, and as such, Γ acts discontinuously on $\mathbf{H}^3$ and the quotient $\mathbf{H}^3/\Gamma$ is a hyperbolic 3-orbifold. When Γ is torsion-free, $\mathbf{H}^3/\Gamma$ is a hyperbolic 3-manifold. Of interest to us is the case when $\mathbf{H}^3/\Gamma$ is closed or finite volume.

Let Σ be a closed orientable surface of genus at least 2, $M = \mathbf{H}^3/\Gamma$ an orientable finite volume hyperbolic 3-manifold and $f : \Sigma \to M$ a map. We shall call $f(\Sigma)$ (or by abuse simply Σ) an *essential surface* if $f_* : \pi_1(\Sigma) \to \Gamma$ is injective. This is non-standard terminology, but will be convenient for our purposes. If M contains an essential surface then Γ contains a surface subgroup. The converse is also true, and given this, an important question in 3-manifold topology is:

QUESTION 2.1. *Let $M = \mathbf{H}^3/\Gamma$ be a finite volume hyperbolic 3-manifold, does Γ contain a surface subgroup.*

This was answered in [**8**] for *non-compact* but finite volume manifolds, and so the remaining cases of Question 2.1 are the closed manifolds, and this seems far from resolution at present. We will discuss [**8**] in more detail below.

2.2. Given an essential surface Σ in a finite volume hyperbolic 3-manifold M, one can attempt to understand the surface in terms of how the hyperbolic metric on M restricts to the surface Σ. In a non-compact finite volume hyperbolic 3-manifold there are two possibilities for the geometry of a closed essential surface Σ. Σ is either *quasi-Fuchsian* or it is said *to contain accidental parabolics*. In the latter case, as suggested by the name, these surface groups contain parabolic elements, whilst in the former case, all non-trivial elements are hyperbolic. Both these surfaces are *geometrically finite*. The methods of [**8**] only provide surfaces containing accidental parabolics, and a comment on the construction in [**8**] will be informative for our discussion of Γ_g (see also [**7**]).

Suppose that $M = \mathbf{H}^3/\Gamma$ is a non-compact finite volume hyperbolic 3-manifold. We can assume that M is orientable on passing to a double cover if necessary, and so M is the interior of a compact 3-manifold with boundary consisting of a disjoint union of tori. Using residual finiteness of Γ there is a finite cover, $\hat{M}$ of M, which has at least 3 boundary components. Standard 3-manifold topology shows that the first betti number of $\hat{M}$ is at least 3, and that one can find an embedded orientable essential surface F with non-empty boundary in $\hat{M}$ missing at least one of the boundary components of $\hat{M}$. This surface can then be used to build a finite cyclic cover of $\hat{M}$ containing a closed embedded essential surface Σ of genus at least 2. Indeed, the construction is explicit, the surface Σ is constructed by taking two copies of F "tubed together" along their boundary. Σ pushes down to M to provide an essential surface in M and so the desired surface subgroup. By construction, these "tubed surfaces" contain accidental parabolic elements corresponding to the peripheral elements from ∂F.

The question of existence of closed quasi-Fuchsian surfaces in M as above remains open. However recent work of Masters and Zhang [27] makes some progress on this.

2.3. The surface subgroups built in [8] can also be viewed as being built by repeated applications of the Maskit combination theorems, a version of which is given below (see [23] and [26] Chapter VII for more details).

Suppose $G_0, G_1, G_2 < \Gamma$ with $G_1 \cap G_2 = G_0 < \Gamma$. If Γ acts on a set X, then we say that a pair of subsets $\Theta_1, \Theta_2 \subset X$ is a *proper interactive pair* for G_1, G_2 if

(1) $\Theta_i \neq \emptyset$ for each $i = 1, 2$,
(2) $\Theta_1 \cap \Theta_2 = \emptyset$,
(3) G_0 leaves Θ_i invariant for each $i = 1, 2$,
(4) for every $\phi_1 \in G_1 \setminus G_0$ we have $\phi_1(\Theta_2) \subset \Theta_1$ and for every $\phi_2 \in G_2 \setminus G_0$ we have $\phi_2(\Theta_1) \subset \Theta_2$, and
(5) for $i = 1, 2$, there exists $\theta_i \in \Theta_i$, such that for every $\phi_i \in G_i \setminus G_0$, $\theta_i \notin \phi_i(\Theta_{i'})$ for $i' \neq i$.

With this notation, we can state the following combination theorem.

THEOREM 2.2. *Suppose G_0, G_1, G_2, Γ, X are as above and $\Theta_1, \Theta_2 \subset X$ is a proper interactive pair for G_1, G_2. Then*
$$G = G_1 *_{G_0} G_2 \hookrightarrow \Gamma$$
is an injection.

The construction of essential surfaces in non-compact finite volume hyperbolic 3-manifolds discussed in §2.2 can also be described using Theorem 2.2 (and an HNN version of it) applied to the two copies of the surface F. In the particular case when the surface F has only one boundary component, the closed surface DF constructed by the above has the property that all non-trivial elements are hyperbolic except those conjugate into the cyclic subgroup generated by the parabolic element $\alpha = [\partial F]$ in $\pi_1(DF)$.

2.4. The dichotomy of closed quasi-Fuchsian versus those that contain an accidental parabolic also manifests itself in the structure of surface subgroups of a Kleinian group Γ. Using the compactness of pleated surfaces, Thurston shows in [38] that for a fixed genus g there are only finitely many Γ-conjugacy classes of quasi-Fuchsian surface subgroups of Γ of genus g. On the otherhand, given an essential surface Σ of genus g in M with an accidental parabolic it is easy to construct infinitely many conjugacy classes of surface subgroups of genus g (see [38] Chapter 8).

3. Surface subgroups in Γ_g

Throughout this section we assume that Σ is a closed oriented surface of genus $g \geq 2$. We will denote the Teichmüller space of Σ by $\mathcal{T}(\Sigma)$, the space of compactly supported measured laminations by $\mathcal{ML}_0(\Sigma)$, and $\mathbb{P}\mathcal{ML}_0(\Sigma)$ the space of projective measured laminations on Σ.

3.1. Given any element $\phi \in \Gamma_g$, one of the following holds; ϕ has finite order, ϕ is reducible, or ϕ is pseudo-Anosov. An infinite order element $\phi \in \Gamma_g \setminus \{1\}$ is pseudo-Anosov if there exists a pair of measured laminations λ_s, λ_u which bind Σ and such that $\{[\lambda_s],[\lambda_u]\}$ is invariant by ϕ. An element $\phi \in \Gamma_g \setminus \{1\}$ is reducible if there exists a multi-curve (ie a finite union of essential simple closed curves on Σ which are pairwise disjoint and parallel) invariant by ϕ.

If q is holomorphic quadratic differential on Σ, then for any constant, nonzero 1-form $adx + bdy$ on $\mathbf{C}$ ($a, b \in \mathbf{R}$), the measured foliation $|adx + bdy|$ is invariant under the transition functions for the atlas of q-coordinates, and so pulls back to a measured foliation on the complement of the cone points. This defines a singular measured foliation on Σ and the space of all such measured foliations is denoted $\mathcal{ML}_0(q)$.

3.2. Although an answer to Question 1.1 remains unknown at present, there appears to be an interesting analogue to the discussion in §2. It had been known for some time that Γ_g contained surface subgroups (see [2] and [15] for example), however a more uniform treatment of constructions of surface subgroups is given in [23]. In some sense this can be viewed as an analogue of the results in [8]. To describe this more fully, we require some terminology.

Veech subgroups of Γ_g arise from stabilizers of Teichmüller discs in the Teichmüller space and have been objects of some interest of late (see [22], [32] and [41] for more on related topics). Briefly, these arise as follows.

Any holomorphic quadratic differential q on Σ defines a holomorphic totally geodesic embedding of the hyperbolic plane

$$f_q : \mathbf{H}^2 \to \mathbf{H}_q \subset \mathcal{T}(\Sigma).$$

The stabilizer $\mathrm{Stab}_{\Gamma_g}(\mathbf{H}_q)$ acts on $\mathbf{H}_q$ and this action can be conjugated back to $\mathbf{H}^2$, via f_q, thus defining a homomorphism D to $\mathrm{PSL}(2,\mathbf{R})$. Subgroups of $\mathrm{Stab}_{\Gamma_g}(\mathbf{H}_q)$ are called Veech groups. If $\mathrm{PSL}(q)$ denotes the image of $\mathrm{Stab}_{\Gamma_g}(\mathbf{H}_q)$ under D, then in the special case when $\mathbf{H}^2/\mathrm{PSL}(q)$ has finite area, this quotient is called a Teichmüller curve and this immerses into the moduli space. This is our analogue of an immersed essential non-compact surface in a finite volume hyperbolic 3-manifold.

An example we will make use of is the following due to Veech [41]. Let Δ_g be the non-convex polygon obtained as the union of two regular $2g+1$-gons in the Euclidean plane which meet along an edge and have disjoint interiors. Let R_g denote the closed surface of genus g obtained by gluing opposite sides of Δ_g by translations. The Euclidean metric on the interior of Δ_g is the restriction of a Euclidean cone metric on R_g, and we can find local coordinates defining a quadratic differential on R_g compatible with this metric. We denote this quadratic differential ξ_g, and $F(\xi_g)$ the associated Veech group. Veech showed that $F(\xi_g)$ is isomorphic to a triangle group of type $(2, 2g+1, \infty)$, where the single primitive parabolic conjugacy class in the triangle group corresponds to the conjugacy class of a reducible element δ. It is easy to see that the above triangle group contains a subgroup of index $2(2g+1)$ which is the fundamental group of a 1-punctured surface of genus g. Hence this determins a subgroup $G(\xi_g)$ of $F(\xi_g)$ isomorphic to the fundamental group of a 1-punctured surface of genus g.

A particular case of the main construction of [23], using the Combination Theorem (Theorem 2.2) is the following, and which produces examples that are "closest to purely pseudo-Anosov" obtained thus far (cf. the discussion after Theorem 2.2).

THEOREM 3.1. *For every $g \geq 2$, there exist subgroups of Γ_g isomorphic to the fundamental group of a closed surface of genus $2g$. These are obtained as the amalgamated free product of two copies of $G(\xi_g)$ along the infinite cyclic subgroup $<\delta>$. In addition, all but one conjugacy class of non-trivial elements (up to powers) is pseudo-Anosov.*

As in the 3-manifold setting, it can also be shown that there are infinitely many distinct conjugacy classes of surface subgroups of genus $2g$ (see [**23**]). Motivated by Thurston's result discussed in §2.4 we pose:

QUESTION 3.2. *Are there are only finitely many Γ_g-conjugacy classes of purely pseudo-Anosov surface subgroups of any fixed genus?*

Of course given that Question 1.1 is open, the answer to Question 3.2 could be zero.

3.3. As mentioned in §2.2, quasi-Fuchsian surface subgroups and those with accidental parabolics are geometrically finite. Furthermore, the combination theorem in the Kleinian group setting shows that combining two geometrically finite groups (with some assumption on the amalgamating subgroup) gives a geometrically finite group. Quasi-Fuchsian groups form part of a special subclass of geometrically finite Kleinian groups; namely they are convex cocompact subgroups. Another way to describe when a surface subgroup of $\mathrm{PSL}(2, \mathbf{C})$ is quasi-Fuchsian is that an orbit of a quasi-Fuchsian group acting on $\mathbf{H}^3$ is quasi-convex.

In [**13**] the notion of convex cocompactness of subgroups of Γ_g is defined, and an attempt to extend this to geometrically finite is proposed in [**34**]. Following [**13**], a finitely generated subgroup G of Γ_g is *convex cocompact* if some orbit of G acting on the Teichmüller space of Σ is quasi-convex. Now Teichmüller space is not hyperbolic in any reasonable sense, but the curve complex is by [**29**], and the analogy with the hyperbolic setting is made more precise in [**19**], where it is shown that a finitely generated subgroup G of Γ_g is convex cocompact if and only if sending G to an orbit in the curve complex of Σ defines a quasi-isometric embedding. Thus convex cocompact surface subgroups of Γ_g are the natural analogue of quasi-Fuchsian subgroups of Kleinian groups.

One interest of [**13**] in this property is its connections with when the extension of $\pi_1(\Sigma)$ by G is word hyperbolic (see §4 for a discussion of a related topological question).

The notion of geometrical finiteness is somewhat more delicate in the setting of subgroups of Γ_g, and we will not discuss this here. We refer the reader to the article by Mosher in this volume for more on this. However, we finish this section by commenting that (given the discussion at the start of the subsection) a class of groups that should be geometrically finite (by analogy with the Kleinian case) are the groups in Theorem 3.1 as well as others constructed in [**23**].

3.4. Taking the analogy with hyperbolic spaces further, we recall that Thurston's compactification of Teichmüller space is obtained by adding $\mathbb{PML}_0(\Sigma)$ at infinity to obtain

$$\overline{\mathcal{T}}(\Sigma) = \mathcal{T}(\Sigma) \cup \mathbb{PML}_0(\Sigma) \cong B^{6g-6}$$

where B^{6g-6} is the closed ball of dimension $6g - 6$ and $\mathbb{PML}_0(\Sigma)$ is identified with the boundary. Moreover, $\mathbb{PML}_0(\Sigma)$ has a natural piecewise projective structure and the action of Γ_g on $\mathcal{T}(\Sigma)$ and $\mathbb{PML}_0(\Sigma)$ fit together to give a well defined action on $\overline{\mathcal{T}}(\Sigma)$ which is holomorphic on the interior and piecewise projective on the boundary.

There is a natural identification of $\mathbb{PML}_0(q)$ with the boundary at infinity $\partial_\infty \mathbf{H}^2$. In this way, the inclusion of $\mathbb{PML}_0(q)$ into $\mathbb{PML}_0(\Sigma)$ can be thought of as an extension $\partial_\infty f_q$ of f_q to infinity. Indeed, the natural projective structure on $\mathbf{RP}^1 = \partial_\infty \mathbf{H}^2 = \mathbb{PML}_0(q)$ makes $\partial_\infty f_q$ into a piecewise projective embedding, equivariant with respect to the $\mathrm{Stab}_{\Gamma_g}(\mathbf{H}_q)$ action.

Moreover, $\partial_\infty f_q$ sends the limit set $\Lambda(\mathrm{PSL}(q)) \subset \partial_\infty \mathbf{H}^2 = \mathbb{PML}_0(q)$ homeomorphically and $\mathrm{Stab}_{\Gamma_g}(\mathbf{H}_q)$-equivariantly to the limit set $\Lambda(\mathrm{Stab}_{\Gamma_g}(\mathbf{H}_q)) \subset \mathbb{PML}_0(\Sigma)$ as defined by McCarthy and Papadopoulos [31].

The map
$$\overline{f}_q = f_q \cup \partial_\infty f_q : \overline{\mathbf{H}}^2 = \mathbf{H}^2 \cup \mathbb{PML}_0(q) \to \overline{\mathcal{T}}(\Sigma)$$
is continuous for every $p \in \mathbb{H}^2$ and almost every $p \in \mathbb{PML}_0(q)$ by a theorem of Masur [28]. However, Masur's theorem implies that this is in general not continuous at every point of $\mathbb{PML}_0(q)$.

This now leads us to the following natural question, the analog of which is true in the setting of Kleinian groups.

QUESTION 3.3. *Let $G \cong \pi_1(S_{2g}) \to \Gamma_g$ be the injection given by Theorem 3.1. Consider $\partial_\infty(G)$ which can be canonically identified with the circle at infinity of the universal cover $\widetilde{S}_{2g} \cong \mathbb{H}^2$ of S_{2g}. Does there exist a continuous G-equivariant map*
$$\partial_\infty(G) \to \mathbb{PML}_0(\Sigma)?$$

3.5. In the context of 3-manifold topology, given an essential surface Σ in a finite volume hyperbolic 3-manifold M, a natural question is whether there is a finite cover of M to which Σ lifts to an embedded surface. A group theoretic property that is closely related to this question is *LERF*, which now define.

If Γ is a group, and H a subgroup of Γ, then Γ is called H-separable if for every $g \in G \setminus H$, there is a subgroup K of finite index in Γ such that $H \subset K$ but $g \notin K$. Γ is called LERF or subgroup separable if Γ is H-separable for all finitely generated subgroups H. This has been widely studied in the setting of low-dimensional topology (see [1] and [36] for example). Indeed, it is often the case that one does not need the full power of LERF for applications to hyperbolic manifolds, separating geometrically finite subgroups often suffices; this led to the property of GFERF, that is separable on all geometrically finite subgroups.

Now it is known that the groups Γ_g are not LERF whenever $g \geq 2$ since they contain $F_2 \times F_2$, however in analogy with the case for hyperbolic manifolds we pose (for $g = 1$ it is known that Γ_1 is LERF since it is virtually free):

QUESTION 3.4. *Is Γ_g GFERF for $g \geq 2$?*

QUESTION 3.5. *Let H be a convex cocompact subgroup of Γ_g. Is Γ_g H-separable?*

Just focusing on surface subgroups, we can ask:

QUESTION 3.6. *Let H be a surface subgroup of Γ_g. Is Γ_g H-separable?*

For recent progress on various classes of subgroups of Γ_g that are separable, we refer the reader to [24].

4. Related topics

Although it is widely believed that if $M = \mathbf{H}^3/\Gamma$ is non-compact and finite volume, then Γ contains a closed quasi-Fuchsian surface subgroup, it is unclear how Question 1.1 will be resolved at present. We next discuss some related questions, that may help shed light on it.

4.1. One motivation for Question 1.1 is the following question from 4-manifold topology.

QUESTION 4.1. *Does there exist a closed hyperbolic 4-manifold X that is the total space of a smooth fiber bundle $\Sigma_g \to X \to \Sigma_h$?*

We will call such an X a *surface bundle over a surface*. The following is well-known.

THEOREM 4.2. *A postive answer to Question 4.1 implies a positive answer to Question 1.1*

PROOF. Let X be a closed hyperbolic 4-manifold that is a surface bundle over a surface. Then the long homotopy exact sequence defines $\pi_1(X)$ as a short exact sequence:
$$1 \to \pi_1(\Sigma_g) \to \pi_1(X) \to \pi_1(\Sigma_h) \to 1$$
and therefore defines a homomorphism $\phi : \pi_1(\Sigma_h) \to \Gamma_g$. Indeed, since X is hyperbolic, it follows that ϕ is injective. Furthermore the image of ϕ must be purely pseudo-Anosov. For if not, since $\pi_1(X)$ is torsion-free, there is a reducible element ψ in the image. Let M_ψ denote the 3-manifold fibered over the circle with fiber Σ_g and monodromy ψ. From the injection of $\pi_1(\Sigma_h)$ into Γ_g it follows that $\pi_1(M_\psi)$ injects into $\pi_1(X)$. Now reducibility implies that $\pi_1(M_\psi)$ contains a copy of $\mathbf{Z} \oplus \mathbf{Z}$, but X is a closed hyperbolic 4-manifold and $\pi_1(X)$ contains no such subgroup. □

At present there seems little about the geometry and topology of hyperbolic 4-manifolds that can be brought to bear on this question. However, two trivial remarks are:

1. *With X as above then the signature of X is 0.*

In fact this holds more generally for any closed hyperbolic 4-manifold; since any hyperbolic 4-manifold is conformally flat, so the Hirzebruch signature theorem implies that the signature is zero.

2. *The hyperbolic volume of X is*
$$\mathrm{Vol}(X) = \frac{4\pi^2}{3}\chi(\Sigma_g)\chi(\Sigma_h).$$

This follows from the observations that for any finite volume hyperbolic 4-manifold M, $\mathrm{Vol}(M) = \frac{4\pi^2}{3}\chi(M)$, and for X a surface bundle over a surface with base Σ_h and fiber Σ_g, we have $\chi(X) = \chi(\Sigma_g)\chi(\Sigma_h)$.

Note that the volume formula shows quickly that the genus of both the base and fiber in this setting is at least 2. Also, using Wang's finiteness result [42] on the finiteness of the number of isometry classes of hyperbolic manifolds of a fixed volume, it follows from this second remark that for fixed $g, h \geq 2$, there are only finitely many manifolds X as above with fiber of genus g and base of genus h.

Given that the smallest known Euler characteristic of a closed orientable hyperbolic 4-manifold is 16 [6], a natural warm-up question might be.

QUESTION 4.3. *Does there exist a closed hyperbolic 4-manifold that is a surface bundle over a surface where the genus of the fiber and base is 2?*

4.2. Some evidence for a negative answer to Question 4.1 is given in [21] where the following conjecture is stated (this is a special case of a more general conjecture on vanishing of Seiberg-Witten invariants). We refer to [37] and [14] for definitions.

CONJECTURE 4.4. *Let M be a closed hyperbolic 4-manifold. Then all the Seiberg-Witten invariants of M vanish.*

The relevance of this is given in the following proposition. We give a proof for hyperbolic manifolds that makes use only of Taubes celebrated paper [37] when $b_2^+ > 1$, and hence avoid complexities that arise when $b_2^+ = 1$.

PROPOSITION 4.5. *A postive answer to Conjecture 4.4 implies that a closed hyperbolic 4-manifold M cannot be symplectic.*

PROOF. This follows automatically from [37] in the case when $b_2^+ > 1$, since [37] shows that a compact oriented symplectic 4-manifold with $b_2^+ > 1$ has non-vanishing Seiberg-Witten invariants. Thus assume that $b_2^+(M) = 1$, and M is symplectic (if $b_2(M) = 0$ there is nothing to prove). From the discussion above, since M is hyperbolic, the signature of M is 0, and so $b_2(M) = 2$. Also M being hyperbolic implies $\pi_1(M)$ is residually finite and so M has many finite covers. Let $p : M_1 \to M$ be a cover of degree $d > 1$. Since M is symplectic, M_1 will be symplectic using the pullback of the symplectic form on M. We claim that $b_2^+(M_1) > 1$, and so we can apply [37] to get a contradiction.

For if $b_2(M_1) = 2$, then (from the volume formula above) since the Euler characteristic satisfies $\chi(M_1) > 0$, it follows that $b_1(M_1) < 2$. However, $\chi(M_1) = d\chi(M)$ and this shows $b_1(M_1) = 2 - d(2 - b_1(M))$. These remarks yield the desired contradiction. □

On the other hand, any X that has the description of a surface bundle over a surface where both the base and fiber have genus ≥ 2 is symplectic by an old argument of Thurston (see [39] or [14] Theorem 10.2.17).

Motivated by this discussion, one can ask the following generalization of Question 4.1, which is an analogue of the virtual fibering question in dimension 3.

QUESTION 4.6. *Does there exist a closed hyperbolic 4-manifold X for which no finite cover admits a symplectic structure? (i.e. X is not virtually symplectic.)*

We note that it is easy to construct non-symplectic hyperbolic 4-manifolds. For example it is easy to see that the Davis manifold D (see [9]) admits no symplectic structure using the following simple parity rule (see [14] Corollary 10.1.10):

Suppose that (M, ω) is a closed symplectic 4-manifold. Then $1 - b_1(M) + b_2^+(M)$ is even.

For the Davis manifold we have from [35] that $b_1(D) = 24$ and $b_2(D) = 72$. Since the signature of D is zero, $b_2^+(D) = 36$, and so the parity condition fails.

4.3. A natural weakening of Question 4.1 is the following.

QUESTION 4.7. *For $g, h \geq 2$, does there exist a short exact sequence:*
$$1 \to \pi_1(\Sigma_g) \to \Gamma \to \pi_1(\Sigma_h) \to 1$$
for which Γ is a word hyperbolic group?

Arguing as in the proof of Theorem 4.2, it follows that such an extension defines a purely pseudo-Anosov surface subgroup of Γ_g. However, even in this case, little is known. We make two comments in this regard.

1. One result is that Γ cannot be the fundamental group of a closed complex hyperbolic surface. As described in [20], this follows from [25], on showing that if Γ is as decribed, then there is a non-singular holomorphic fibration $X = \mathbf{H}_\mathbf{C}^2/\Gamma \to \Sigma_h$ that induces the short exact sequence (see also [17]).

2. The following idea to construct purely pseudo-Anosov surface subgroups was described to me by Ian Agol. In [40], Thurston proves the following result (see [40] for terminology and further details).

THEOREM 4.8. *Let $n \geq 3$ and $k_1, k_2 \ldots k_n \in (0, 2\pi)$ whose sum is 4π. Then the set of Euclidean cone metrics on S^2 with cone points of curvature k_i and of total area 1 forms a complex hyperbolic manifold whose metric completion is a complex hyperbolic cone manifold of finite volume.*

In addition, Thurston also gives conditions when such a cone manifold is an orbifold, and shows that certain of these completions give rise to a cocompact arithmetic lattices arising from Hermitian forms. These arithmetic lattices also arise in work of Mostow, and go back to Picard (see [10]). We will consider the example in [38] when $n = 5$ and denote the lattice that Thurston constructs by Δ_5. Since these arithmetic groups arise from Hermitian forms, it is well-known that they contain many cocompact Fuchsian subgroups; these can be **R**-Fuchsian, in the sense that they are subgroups of a group $SO(2,1)$ or **C**-Fuchsian in the sense that they are subgroups of $SU(1,1)$.

Now if $\mathcal{M}_{0,n}$ denotes the moduli space of the n-times punctured sphere, then Thurston shows that $\mathbf{H}_\mathbf{C}^2/\Delta_5$ is a compact orbifold corresponding to a compactification of $\mathcal{M}_{0,5}$. Agol has informed me that this compactification is the Mumford compactification. In terms of Thurston's description, the compactification locus is the locus at which pairs of cone points collide.

QUESTION 4.9. *Does there exist a cocompact Fuchsian subgroup of Δ_5 that misses the compactification locus?*

Given such a Fuchsian subgroup $F < \Delta_5$, Agol produces a purely pseudo-Anosov subgroup in some Γ_g using a branched cover construction.

4.4. One generalization of looking for surface subgroups of Γ_g is to look for injections of (cocompact) lattices in Lie groups into Γ_g. If the lattices are superrigid the image of such a lattice in Γ_g is necessarily finite (see [12] and [43]), and so it follows (cf. Theorem 2 of [43]), that the only lattices that can admit a faithful representation (or even an infinite representation) into Γ_g are lattices in $\mathrm{SO}(m,1)$, $m \geq 2$ or $\mathrm{SU}(q,1)$, $q \geq 1$. Indeed, since solvable subgroups of Γ_g are virtually abelian [4], this observation also excludes non-cocompact lattices of $\mathrm{SU}(q,1)$ for $q \geq 2$ from injecting. In addition, there is a simple obstruction to injecting certain of these lattices, or indeed for any group. Namely if a finitely generated group G admits an injection into Γ_g, then $\mathrm{vcd}(G) \leq \mathrm{vcd}(\Gamma_g)$ (see [5]). If the vcd of a group G satisfies the above inequality, then we call G *admissable*. The vcd's of the groups Γ_g are known to be $4g - 5$ when $g \geq 2$ [16]. Motivated by this discussion we pose.

QUESTION 4.10. *Let Γ be a lattice in $\mathrm{SO}(m,1)$, $m \geq 3$ or $\mathrm{SU}(q,1)$, $q \geq 2$ which is admissable for Γ_g. Does Γ inject in Γ_g? Can there be purely pseudo-Anosov representations?*

If such an injection exists does there exist a continuous Γ-equivariant map

$$\partial_\infty(\Gamma) \to \mathbb{PML}_0(\Sigma)?$$

More generally, for a fixed Σ and hence fixed vcd, a further natural generalization of the discussion here is:

QUESTION 4.11. *Which 1-ended admissable word hyperbolic groups G inject in Γ_g (as purely pseudo-Anosov subgroups)?*

If such an injection exists does there exist a continuous G-equivariant map

$$\partial_\infty(G) \to \mathbb{PML}_0(\Sigma)?$$

REMARKS. **1.** As discussed in [34], no example of a purely pseudo-Anosov non-free subgroup of Γ_g is known at present.

2. It is a conjecture of Gromov that every 1-ended word hyperbolic group contains a surface subgroup (see [3]). Assuming this conjecture holds, then if there were **any** purely pseudo-Anosov injection of a 1-ended word hyperbolic group into Γ_g, this would produce a purely pseudo-Anosov surface subgroup of Γ_g.

3. As discussed in the Introduction, there are many analogies between non-uniform lattices in Lie groups and Γ_g. An analogy for purely pseduo-Anosov surface subgroup of Γ_g would be a purely semisimple surface subgroup of a non-uniform lattice. In this regard, the groups $\mathrm{SL}(n, \mathbf{Z})$ for $n \geq 3$ all contain such surface groups. This can be seen be realizing certain arithmetic Fuchsian groups as subgroups of finite index in groups $\mathrm{SO}(f; \mathbf{Z}) \subset \mathrm{SL}(n, \mathbf{Z})$, where f is an indefinite ternary quadratic form with coefficients in $\mathbf{Z}$.

References

[1] I. Agol, D. D. Long and A. W. Reid, *The Bianchi groups are separable on geometrically finite subgroups*, Ann. of Math. **153** (2001), 599–621.

[2] M. F. Atiyah, *The signature of fiber bundles*, In Global Analysis (Papers dedicated to K. Kodaira), Princeton University Press, Princeton (1969), 73–84.

[3] M. Bestvina, *Questions in geometric group theory*, available at www.math.utah.edu/~bestvina.

[4] J. Birman, A. Lubotzky and J. McCarthy, *Abelian and solvable subgroups of the mapping class groups*, Duke Math. J. **50** (1983), 1107–1120.

[5] K. Brown, *Cohomology of Groups*, Graduate Texts in Math. **87** Springer-Verlag (1982).

[6] M. Conder and C. Maclachlan, *Compact hyperbolic 4-manifolds of small volume*, to appear Proc. A. M. S.

[7] D. Cooper and D. D. Long, *Some surface subgroups survive surgery*, Geom. Topol. **5** (2001), 347–367.

[8] D. Cooper, D. D. Long and A. W. Reid, *Essential surfaces in bounded 3-manifolds*, J. A. M. S. **10** (1997), 553–563.

[9] M. W. Davis, *A hyperbolic 4-manifold*, Proc. A. M. S. **93** (1985), 325–328.

[10] P. Deligne and G. D. Mostow, *Commensurabilities among lattices in* $PU(1,n)$, Annals of Math. Studies, **132** Princeton University Press, Princeton, (1993).

[11] B. Farb, A. Lubotzky, and Y. Minsky, *Rank-1 phenomena for mapping class groups*, Duke Math. J. **106** (2001), 581–597.

[12] B. Farb and H. Masur, *Superrigidity and mapping class groups*, Topology **37** (1998), 1169–1176.

[13] B. Farb and L. Mosher, *Convex cocompact subgroups of mapping class groups*, Geom. Topol. **6** (2002), 91–152.

[14] R. E. Gompf and A. I. Stipsicz, *4-Manifolds and Kirby Calculus*, Graduate Studies in Mathematics, **20** American Mathematical Society, (1999).

[15] G. Gonzalez-Diez and W. J. Harvey, *Surface groups in Mapping Class groups*, Topology **38** (1999), 57–69.

[16] J. L. Harer, *The cohomology of the moduli space of curves*, Theory of moduli (Montecatini Terme, 1985), 138–221, Lecture Notes in Math., **1337**, Springer-Verlag (1988).

[17] J. A. Hillman, *Complex surfaces which are fibre bundles*, Topology and its Appl. **100** (2000), 187–191.

[18] N. V. Ivanov, *Subgroups of Teichmüller modular groups*, Transl. Math. Monogr., **115**, Am. Math. Soc., Providence, RI, 1992.

[19] R. P. Kent and C. J. Leininger, *Shadows of Mapping Class groups: capturing convex cocompactness*, preprint.

[20] M. Kapovich, *On normal subgroups in the fundamental groups of complex surfaces*, preprint.

[21] C. LeBrun, *Hyperbolic manifolds, harmonic forms and Seiberg-Witten invariants*, Proceedings of the Euroconference on Partial Differential Equations and their Applications to Geometry and Physics (Castelvecchio Pascoli, 2000). Geom. Dedicata **91** (2002), 137–154.

[22] C. J. Leininger, *On groups generated by two positive multi-twists: Teichmüller curves and Lehmer's number*, Geometry and Topology **8** (2004), 1301–1359.

[23] C. J. Leininger and A. W. Reid, *A combination theorem for Veech subgroups of the mapping class group*, to appear in G.A.F.A., arXiv:math.GT/0410041.

[24] C. J. Leininger and D. B. McReynolds, *Separable subgroups of the Mapping class group*, to appear in Topology and its Applications.

[25] K. Liu, *Geometric height inequalities*, Math. Research Letters **3** (1996), 693–702.

[26] B. Maskit, *Kleinian groups*, Springer-Verlag Berlin Heidelberg, 1988.

[27] J. D. Masters and X. Zhang, *Closed quasi-Fuchsian surfaces in hyperbolic knot complements*, preprint arXiv:math/GT/0601445

[28] H. Masur, *Two boundaries of Teichmüller space*, Duke Math. J. **49** (1982), 183–190.

[29] H. Masur and Y. Minsky, *Geometry of the complex of curves I: Hyperbolicity*, Invent. Math. **138** (1999), 103–149.

[30] J. D. McCarthy, *On the first cohomology of cofinite subgroups in surface mapping class groups*, Topology **40** (2001), 401–418.

[31] J. McCarthy and A. Papadopoulos, *Dynamics on Thurston's sphere of projective measured foliations*, Comment. Math. Helv. **64** (1989), 133–166.

[32] C. T. McMullen, *Billiards and Teichmüller curves on Hilbert Modular surfaces*, J. A. M. S. **16** (2003), 857–885.

[33] J. W. Morgan, *On Thurston's uniformization theorem for three-dimensional manifolds*, The Smith conjecture (New York, 1979), 37–125, Pure Appl. Math., **112**, Academic Press, (1984).

[34] L. Mosher, *Problems in the geometry of surface group extensions*, preprint.

[35] J. Ratcliffe and S. Tschantz, *On the Davis hyperbolic 4-manifold*, Topology and its Appl. **111** (2001), 327–342.

[36] G. P. Scott, *Subgroups of surface groups are almost geometric*, J. London Math. Soc. **17** (1978), 555 - 565. See also *ibid Correction*: J. London Math. Soc. **32** (1985), 217–220.

[37] C. H. Taubes, *The Seiberg-Witten invariants and symplectic forms*, Math. Research Letters **1** (1994), 809–822.

[38] W. P. Thurston, *The Geometry and Topology of Three-Manifolds*, Princeton University course notes, available at http://www.msri.org/publications/books/gt3m/ (1980).

[39] W. P. Thurston, *Some simple examples of symplectic manifolds*, Proc. A. M. S. **55** (1976), 467–468.

[40] W. P. Thurston, *Shapes of polyhedra and triangulations of the sphere*, Geometry and Topology Monographs, **1** The Epstein Birthday Schrift, eds. I. Rivin, C. Rourke, and C. Series 511–549 (1998).

[41] W. A. Veech, *Teichmüller curves in moduli space, Eisenstein series and an application to triangular billiards*, Invent. Math. **97** (1989), 553–583.

[42] H. C. Wang, *Topics in totally discontinuous groups*, in Symmetric Spaces, eds. W. M. Boothby and G. L. Weiss, Pure and Appl. Math **8**, Marcel Dekker (1972), pp 459–487.

[43] S-K. Yeung, *Representations of semisimple lattices in Mapping class groups*, International Math. Research Notices **31** (2003), 1677-1686.

DEPARTMENT OF MATHEMATICS, UNIVERSITY OF TEXAS, AUSTIN, TX 78712

Weil-Petersson Perspectives

Scott A. Wolpert

1. Introduction

We highlight recent progress in the study of the Weil-Petersson (WP) geometry of finite dimensional Teichmüller spaces. For recent progress on and the understanding of infinite dimensional Teichmüller spaces the reader is directed to [**TT04a, TT04b**]. As part of the highlight, we also present possible directions for future investigations. Recent works on WP geometry involve new techniques which present new opportunities.

We begin with background highlights. The reader should see [**Ahl61, Ber74, Nag88, Roy75**], as well as [**Wol03**, Sec. 1 and 2] for particulars of our setup and notation for the augmented Teichmüller space $\overline{\mathcal{T}}$, the mapping class group MCG, Fenchel-Nielsen (FN) coordinates and the complex of curves $C(F)$ for a surface F. We describe the main elements.

For a reference topological surface F of genus g with n punctures and negative Euler characteristic, a Riemann surface R (homeomorphic to F) with complete hyperbolic metric, a marked Riemann/hyperbolic surface is the equivalence class of a pair $\{(R, f)\}$ for $f : F \to R$ an orientation preserving homeomorphism; equivalence for post composition with a conformal homeomorphism. Teichmüller space $\mathcal{T}$ is the space of equivalence classes, [**Ahl61, IT92, Nag88**]. The mapping class group $MCG = Homeo^+(F)/Homeo_0(F)$ acts properly discontinuously on $\mathcal{T}$ by taking $\{(R, f)\}$ to $\{(R, f \circ h^{-1})\}$ for a homeomorphism h of F.

$\mathcal{T}$ is a complex manifold with the MCG acting by biholomorphisms. The cotangent space at $\{(R, f)\}$ is $Q(R)$, the space of integrable holomorphic quadratic differentials, [**Ahl61, Nag88**]. For $\phi, \psi \in Q(R)$ the Teichmüller (Finsler metric) conorm of ϕ is $\int_R |\phi|$, while the WP dual Hermitian pairing is $\int_R \phi\overline{\psi}(ds^2)^{-1}$, where ds^2 is the R complete hyperbolic metric. The WP metric is Kähler, not complete, with negative sectional curvature and the MCG acts by isometries (see Section 3 below), [**Ahl61, Mas76, Roy75, Tro86, Wol86**]. The *complex of curves* $C(F)$ is defined as follows. The vertices of $C(F)$ are free homotopy classes of homotopically nontrivial, nonperipheral, simple closed curves on F. A k-simplex of $C(F)$ consists of $k+1$ distinct homotopy classes of mutually disjoint simple closed curves. A maximal set of mutually disjoint simple closed curves, a *pants decomposition*, has $3g - 3 + n$ elements for F a genus g, n punctured surface.

A free homotopy class γ on F determines the geodesic-length function ℓ_γ on $\mathcal{T}$; $\ell_\gamma(\{R, f,)\})$ is defined to be the hyperbolic length of the unique geodesic freely homotopic to $f(\gamma)$. The geodesic-length functions are convex along WP geodesics, [**Wol87b, Wol**]. On a hyperbolic surface the geodesics for a pants decomposition decompose the surface into geometric pairs of pants: subsurfaces homeomorphic to spheres with a combination of three discs or points removed, [**Abi80, IT92**]. The hyperbolic geometric structure on a pair of pants is uniquely determined by its boundary geodesic-lengths. Two pants boundaries with a common length can be abutted to form a new hyperbolic surface (a complete hyperbolic structure with possible further geodesic boundaries.) The common length ℓ for the joined boundary and the offset, or *twist* τ, for adjoining the boundaries combine to provide parameters (ℓ, τ) for the construction. The twist τ is measured as displacement along boundaries. The Fenchel-Nielsen (FN) parameters $(\ell_{\gamma_1}, \tau_{\gamma_1}, \ldots, \ell_{\gamma_{3g-3+n}}, \tau_{\gamma_{3g-3+n}})$ valued in $(\mathbb{R}_+ \times \mathbb{R})^{3g-3+n}$ provide global real-analytic coordinates for $\mathcal{T}$, [**Abi80, Mas01, Wol82**]. Each pants decomposition determines a global coordinate.

A bordification of $\mathcal{T}$, the *augmented Teichmüller space*, is introduced by extending the range of the parameters. For an ℓ_γ equal to zero, the twist is not defined and in place of the geodesic for γ there appears a pair of cusps. Following Bers [**Ber74**] the extended FN parameters describe marked (possibly) noded Riemann surfaces (marked stable curves.) An equivalence relation is defined for marked noded Riemann surfaces and a construction is provided for adjoining to $\mathcal{T}$ frontier spaces (where subsets of geodesic-lengths vanish) to obtain the augmented Teichmüller space $\overline{\mathcal{T}}$, [**Abi77, Abi80**]. $\overline{\mathcal{T}}$ is not locally compact since in a neighborhood of ℓ_γ vanishing the FN angle $\theta_\gamma = 2\pi\tau_\gamma/\ell_\gamma$ has values filling $\mathbb{R}$. The leading-term expansion for the WP metric is provided in [**DW03, Yam01, Wol03**]. Following Masur the WP metric extends to a complete metric on $\overline{\mathcal{T}}$, [**Mas76**].

The group MCG acts (not properly discontinuously) as a group of homeomorphisms of $\overline{\mathcal{T}}$ and $\overline{\mathcal{T}}/MCG$ is topologically the Deligne-Mumford compactified moduli space of stable curves $\overline{\mathcal{M}}$. We will include the empty set as a -1-simplex of $C(F)$. The complex is partially ordered by inclusion of simplices. For a marked noded Riemann surface $\Lambda(\{(R, f)\}) \in C(F)$ is the simplex of free homotopy classes on F mapped to nodes of R. The level sets of Λ are the strata of $\overline{\mathcal{T}}$.

$(\overline{\mathcal{T}}, d_{WP})$ is a $CAT(0)$ metric space; see [**DW03, Yam01, Wol03**] and the attribution to Benson Farb in [**MW02**]. The general structure of $CAT(0)$ spaces is described in detail in [**BH99**]. $CAT(0)$ spaces are complete metric spaces of *nonpositive curvature*. In particular for $p, q \in \overline{\mathcal{T}}$ there is a unique length-minimizing path $\widehat{pq}$ connecting p and q. An additional property is that WP geodesics do not *refract*; an open WP geodesic segment $\widehat{pq} - \{p, q\}$ is contained in the (open) stratum $\Lambda(p) \cap \Lambda(q)$ [**DW03, Yam01, Wol03**]; an open segment is a solution of the WP geodesic differential equation on a product of Teichmüller spaces. The stratum of $\overline{\mathcal{T}}$ (the level sets of Λ) are totally geodesic complete subspaces. Strata also have a metric-intrinsic description: a stratum is the union of all open length-minimizing paths containing a given point, [**Wol03**, Thrm. 13].

A classification of *flats*, the locally Euclidean isometrically embedded subspaces, of $\overline{\mathcal{T}}$ is given in [**Wol03**, Prop. 16]. Each flat is contained in a proper substratum and is the Cartesian product of geodesics from component Teichmüller spaces. A structure theorem characterizes limits of WP geodesics on $\overline{\mathcal{T}}$ modulo the action of the mapping class group, [**Wol03**, Sec. 7]. Modulo the action the general limit is the unique length-minimizing piecewise geodesic path connecting the initial point, a sequence of strata and the final point. An application is the construction of axes in $\overline{\mathcal{T}}$ for elements of the MCG, [**DW03, Wol03**]. The geodesic limit behavior is suggested by studying the sequence $\widehat{pT_\gamma^n p}$ for a Dehn twist $T_\gamma \in MCG$: modulo MCG the limit is two copies of the length-minimizing path from p to $\{\ell_\gamma = 0\}$.

Jeffrey Brock established a collection of important results on WP synthetic geometry, [**Bro03, Bro02**]. Brock introduced an approach for approximating WP geodesics. A result is that the geodesic rays from a point of $\mathcal{T}$ to the maximally noded Riemann surfaces have initial tangents dense in the initial tangent space. We used the approach in [**Wol03**, Coro. 19] to find that $\overline{\mathcal{T}}$ is the closed convex hull of the discrete subset of marked maximally noded Riemann surfaces. We also showed that the geodesics connecting the maximally noded Riemann surfaces have tangents dense in the tangent bundle of $\mathcal{T}$.

Howard Masur and Michael Wolf established the WP counterpart to H. Royden's celebrated theorem: each WP isometry of $\mathcal{T}$ is induced by an element of the extended MCG, [**MW02**]. A simplified proof of the Masur-Wolf result is given as follows: [**Wol03**, Thrm. 20], the elements of $Isom_{WP}(\mathcal{T})$ extend to isometries of $\overline{\mathcal{T}}$; the extensions preserve the metric-intrinsic stratum of $\overline{\mathcal{T}}$ and so correspond to simplicial automorphisms of $C(F)$; the simplicial automorphisms of $C(F)$ are in general induced by the elements of the MCG from the work of N. Ivanov, M. Korkmaz and F. Luo, [**Iva97, Kor99, Luo00**]; and finally $\overline{\mathcal{T}} - \mathcal{T}$ is a uniqueness set for WP isometries. Brock and Dan Margalit have recently extended available techniques to include the special (g,n) types of $(1,1), (1,2)$ and $(2,0)$, [**BM04**].

I would like to take the opportunity to thank Benson Farb, Richard Wentworth and Howard Weiss for their assistance.

2. Classical metrics for $\mathcal{T}$

A puzzle of Teichmüller space geometry is that the classical Teichmüller and Weil-Petersson metrics in combination have desired properties: Kähler, complete, finite volume, negative curvature, and a suitable sphere at infinity, yet individually the metrics already lack the basic properties of completeness and non-positive curvature. In fact N. Ivanov showed that $\mathcal{M}_{g,n}$, $3g - 3 + n \geq 2$, admits no complete Riemannian metric of pinched negative sectional curvature, [**Iva88**]. Recently authors have considered asymptotics of the WP metric and curvature with application to introducing modifications to obtain *designer metrics*.

Curt McMullen considered adding a small multiple of the Hessian of the *short geodesic log length sum* to the WP Kähler form

$$\omega_{1/\ell} = \omega_{WP} + ic \sum_{\ell_\gamma < \epsilon} \partial\overline{\partial} Log \, \ell_\gamma$$

(for $Log \approx \min\{\log, 0\}$) to obtain the Kähler form for a modified metric, [**McM00**]. The constructed metric is *Kähler hyperbolic*: on the universal cover the Kähler form $\omega_{1/\ell}$ is the exterior derivative of a bounded 1-form and the injectivity radius is positive; on the moduli space $\mathcal{M}_{g,n}$ the metric $g_{1/\ell}$ is: complete, finite volume, of bounded sectional curvature. Of primary interest is that the constructed metric $g_{1/\ell}$ is comparable to the Teichmüller metric. The result provides that the Teichmüller metric qualitatively has the properties of a Kähler hyperbolic metric, [**McM00**].

As applications McMullen establishes a positive lower bound for the Teichmüller Rayleigh-Ritz quotient and a Teichmüller metric isoperimetric inequality for complex manifolds. McMullen also uses the metric to give a simple derivation that the sign of the orbifold Euler characteristic alternates with the parity of the complex dimension. The techniques of [**Wol**] can be applied to show that in a neighborhood of the compactification divisor $\mathcal{D} = \overline{\mathcal{M}_g} - \mathcal{M}_g \subset \overline{\mathcal{M}_g}$ the McMullen modification to WP is principally in the directions transverse to the divisor.

Lipman Bers promoted the question of understanding the equivalence of the classical metrics on Teichmüller spaces in 1972, [**Ber72**]. A collection of authors have now achieved major breakthroughs. Kefeng Liu, Xiaofeng Sun and Shing-Tung Yau have analyzed WP curvature, applied Yau's Schwarz lemma, and considered the Bers embedding to compare and study the Kähler-Einstein metric for Teichmüller space and also the classical metrics, [**LSY04**]. Sai-Kee Yeung considered the Bers embedding and applied Schwarz type lemmas to compare classical metrics, [**Yeu04**]. Yeung's approach is based on his analysis of convexity properties of fractional powers of geodesic-length functions, [**Yeu03**]. Bo-Yong Chen considered McMullen's metric, the WP Kähler potential of Takhtajan-Teo, and L^2-estimates to establish equivalence of the Bergman and Teichmüller metrics, [**BY04**].

In earlier work, Cheng-Yau [**CY80**] and Mok-Yau [**MY83**] had established existence and completeness for a Kähler-Einstein metric for Teichmüller space. Liu-Sun-Yau study the negative WP Ricci form $-Ricci_{WP}$; the form defines a complete Kähler metric on the Teichmüller space. The authors find that the metric is equivalent to the *asymptotic Poincaré metric* and that asymptotically its holomorphic sectional curvature has negative upper and lower bounds. The authors develop the asymptotics for the curvature of the metric, [**LSY04**].

To overcome difficulty with controlling interior curvatures Liu-Sun-Yau further introduce the modification

$$g_{LSY} = -Ricci_{WP} + c\, g_{WP}, \ c > 0, \ [\mathbf{LSY04}].$$

The authors find that the LSY metric is complete with bounded negative holomorphic sectional curvature, bounded negative Ricci curvature and bounded sectional curvature. The authors apply their techniques and results of others to establish comparability of the seven classical metrics

LSY $\sim$ asymptotic Poincaré $\sim$ Kähler-Einstein $\sim$ Teichmüller-Kobayashi $\sim$ McMullen $\sim$ Bergman $\sim$ Carathéodory.

Yeung established the comparability of the last five listed classical metrics. Now the classical metrics are expected to have the same qualitative behavior. In combination the metrics have a substantial list of properties. Liu-Sun-Yau use their understanding of the metrics in a neighborhood of the compactification divisor $\mathcal{D} \subset \overline{\mathcal{M}_g}$ to show that the logarithmic cotangent bundle (Bers' bundle of *holomorphic 2-differentials*) of the compactified moduli space is stable in the sense of Mumford. The authors also show that the bundle is stable with respect to its first Chern class. The authors further find that the Kähler-Einstein metric has bounded geometry in a strong sense.

> *The comparison of metrics now provides a research opportunity to combine different approaches for studying Teichmüller geometry.*

3. WP synthetic geometry

A collection of authors including Jeffrey Brock, Sumio Yamada and Richard Wentworth have recently studied WP synthetic geometry. A range of new techniques have been developed. Brock established a collection of very interesting results on the large-scale behavior of WP distance. Brock considered the *pants graph* $C_{\mathbf{P}}(F) \subset C(F)$ having vertices the distinct pants decompositions of F and joining edges of unit-length for pants decompositions differing by an elementary move, [**Bro03**]. He showed that the 0-skeleton of $C_{\mathbf{P}}(F)$ with the edge-metric is quasi-isometric to $\mathcal{T}$ with the WP metric. He further showed for $p, q \in \mathcal{T}$ that the corresponding quasifuchsian hyperbolic three-manifold has convex core volume comparable to $d_{WP}(p, q)$. At large-scale, WP distance and convex core volume are approximately combinatorially determined.

Brock's approach begins with Bers' observation that there is a constant L depending only on the topological type of F, such that a surface R has a pants decomposition with lengths $\ell_\gamma(R) < L$. For $\mathcal{P}$ a pants decomposition Brock associates the sublevel set

$$V(\mathcal{P}) = \{R \mid \gamma \in \mathcal{P}, \ell_\gamma < L\}.$$

By Bers' theorem the sublevel sets $V(\mathcal{P})$ cover $\mathcal{T}$, [**Ber74**]. Brock finds that the configuration of the sublevel sets in terms of WP distance on $\mathcal{T}$ is coarsely approximated by the metric space $C_{\mathbf{P}}(F)$. He finds for $R, S, \in \mathcal{T}$ with $R \in V(\mathcal{P}_R)$, $S \in V(\mathcal{P}_S)$ that

$$d_{\mathbf{P}}(\mathcal{P}_R, \mathcal{P}_S) \asymp d_{WP}(R, S)$$

for $d_{\mathbf{P}}$ the $C_{\mathbf{P}}(F)$ edge-metric, [**Bro03**]. Brock's underlying idea is that $C_{\mathbf{P}}(F)$ is the 1-skeleton for the nerve of the covering $\{V(\mathcal{P})\}$ and that the minimal count of *elementary moves* in $C_{\mathbf{P}}(F)$ approximates WP distance. Paths in the pants graph $C_{\mathbf{P}}(F)$ can be encoded as sequences of elementary moves.

> *Can sequences of elementary moves be used to study the bi-infinite WP geodesics? Can $C_{\mathbf{P}}(F)$ be used to investigate the large g, n behavior of $\operatorname{diam}_{WP}(\mathcal{M}_{g,n})$?*

Brock's model can be used to study and compare the volume growth of $C_{\mathbf{P}}(F)$ and $\mathcal{T}$. A growth function for $C_{\mathbf{P}}(F)$ is the number of vertices in an edge-metric-ball of a given radius about a base point. A growth function for $\mathcal{T}$ is the WP volume of a ball of a given radius about a base point.

> *A research question is to determine the growth function for $C_{\mathbf{P}}(F)$ and to effect the comparison to $\mathcal{T}$.*

Questions regarding WP geodesics include behavior in-the-small, as well as in-the-large. Understanding the WP metric tangent cones of $\overline{\mathcal{T}}$, as well as the metric tangent cone bundle over $\overline{\mathcal{T}}$ is a basic matter. Recall the definition of the Alexandrov angle for unit-speed geodesics $\alpha(t)$ and $\beta(t)$ emanating from a common point $\alpha(0) = \beta(0) = o$ of a metric space (X, d), [**BH99**, pg. 184]. The Alexandrov angle between α and β is defined by

$$\angle(\alpha, \beta) = \lim_{t \to 0} \frac{1}{2t} d(\alpha(t), \beta(t)).$$

Basic properties of the angle are provided in the opening and closing sections of [**BH99**, Chap. II.3]. Angle zero provides an equivalence relation on the space of germs of unit-speed geodesics beginning at o. For the model space $(\mathbb{H}; ds^2 =$

$dr^2 + r^6 d\theta^2, r > 0$) for the WP metric transverse to a stratum: the geodesics $\{\theta = \text{constant}\}$ are at angle zero at the *origin point* $\{r = 0\}$. The WP metric tangent cone at $p \in \overline{\mathcal{T}}$ is defined as the space of germs of constant-speed geodesics beginning at p modulo the relation of Alexandrov angle zero. Another matter is behavior of WP geodesics in a neighborhood of a stratum. In particular for $\mathcal{S}$ a stratum defined by the vanishing of a geodesic-length function ℓ_γ and θ_γ the corresponding Fenchel-Nielsen angle, is θ_γ bounded along each geodesic ending on $\mathcal{S}$? (See the next section for results of Wentworth precluding spiraling for harmonic maps from 2-dimensional domains.)

> *Research questions on geodesic behavior in-the-small include the following. Define and determine the basic properties of the metric tangent cone bundle of $\overline{\mathcal{T}}$. Determine if WP geodesics can spiral to a stratum. Determine the behavior of the Fenchel-Nielsen angles along finite geodesics.*

There are questions regarding in-the-large behavior of WP geodesics. A basic invariant of hyperbolic metrics, the relative systole $sys_{rel}(R)$ of a surface R is the length of the shortest closed geodesic. In [**Wol03**] we note that the WP injectivity radius inj_{WP} of the moduli space $\mathcal{M}$ is comparable to $(sys_{rel})^{1/2}$. A basic question is to understand the relative systole sys_{rel} along infinite WP geodesics in $\mathcal{T}$. It would also be interesting to understand the combinatorics of the sequence of short geodesics, and any limits of the (weighted) collections of simple closed geodesics for which the Bers decomposition is realized. Brock, Masur and Minsky have preliminary results on such limits in the space of measured geodesic laminations $\mathcal{MGL}$. The authors are studying the family of hyperbolic metrics $\{R_t\}$ over a WP geodesic. A geometric model M for a 3-manifold is constructed from the family; the authors seek to show that the model is biLipschitz to the *thick part* of a hyperbolic 3-dimensional structure. To construct M, the authors remove the *thin parts* of the fibers and then use the hyperbolic metric along fibers and rescalings of the WP metric in the parameter to define a 3-dimensional metric.

> *Two research questions on geodesic behavior in-the-large involve bi-infinite geodesics. Determine the behaviors of sys_{rel} along WP geodesics. Understand the combinatorics along WP geodesics of the sequence of short geodesics (pants decompositions) on surfaces and any limits of the weighted collections of short geodesics.*

The divergence of a geodesic and a geodesic/stratum is a measure of the curvature of the join. Although the WP metric on $\mathcal{T}$ has negative sectional curvature there are directions of almost vanishing curvature near the bordification $\overline{\mathcal{T}} - \mathcal{T}$, as well as flats in the bordification. To codify, Zheng Huang has shown that the sectional curvatures on $\mathcal{T}$ are pointwise bounded as $-c'(sys_{rel})^{-1} < sec_{WP} < -c'' sys_{rel}$ for positive constants, [**Hua04**]. In particular, a complex parameter for the formation of a node determines a tangent 2-plane with large negative curvature realizing the lower curvature bound. The two parameters for a pair of independent formations of nodes determine a tangent 2-plane with almost vanishing curvature. Huang has also found for sys_{rel} bounded away from zero that the sectional curvatures are bounded independent of the genus, [**Hua06**].

A research matter is to analyze the rate of divergence of pairs consisting of an infinite WP geodesic and either a second infinite WP geodesic or a stratum.

A third matter proposed by Jeffrey Brock is to investigate the dynamics of the WP geodesic flow on the tangent bundle $T\mathcal{M}$. The $CAT(0)$ geometry can be used to show that the *finite and semi-finite* WP geodesics form a subset of smaller Hausdorff dimension, [**Wol03**, Sec. 6].

Are there WP geodesics with lifts dense in $T\mathcal{T}$? Are the lifts of axes of (pseudo Anosov) elements of the MCG dense in $T\mathcal{T}$? What is the growth rate of the minimal translation lengths for the conjugacy classes of the pseudo Anosov elements? Is WP geodesic flow (restricted to the bi-infinite directions) ergodic?

Jason Behrstock-Yair Minsky [**BM06**] and Ursula Hamenstädt [**Ham06**] have verified the Brock-Farb conjecture for the maximal dimension of a quasi-flat as $\lfloor \frac{3g+n-2}{2} \rfloor$ (the lower bound is straightforward [**BF01, Wol03**].) The maximal dimension of a quasi-flat is the rank in the sense of Gromov. The rank is important for understanding the global WP geometry, as well as for understanding mapping class groups. Brock-Farb, Jason Behrstock and Javier Aramayona had already found that certain low dimensional Teichmüller spaces are Gromov-hyperbolic and necessarily rank one, [**Ara04, Beh05, BF01**].

A research matter is to understand the behavior of quasi-geodesics and especially quasi-flats, the quasi-isometric embeddings of Euclidean space into $\mathcal{T}$.

4. Harmonic maps to $\overline{\mathcal{T}}$

A collection of authors including Georgios Daskalopoulos, Richard Wentworth and Sumio Yamada have considered harmonic maps from (finite volume) Riemannian domains Ω into the $CAT(0)$ space $\overline{\mathcal{T}}$, [**DKW00, DW03, Wen04, Yam01**]. The authors apply the Sobolev theory of Korevaar-Schoen, [**KS93**], to study maps energy-minimizing for prescribed boundary values. Possible applications are rigidity results for homomorphisms of lattices in Lie groups to mapping class groups and existence results for harmonic representatives of the classifying maps associated to symplectic Lefschetz pencils, [**DW03**].

The behavior of a harmonic map to a neighborhood of a lower stratum of $\overline{\mathcal{T}}$ is a basic matter. Again this is especially important since $\overline{\mathcal{T}}$ is not locally compact in a neighborhood of a lower stratum; in each neighborhood the corresponding Fenchel-Nielson angles surject to $\mathbb{R}$. A harmonic map or even a WP geodesic may *spiral* with unbounded Fenchel-Nielsen angles. With the non-refraction of WP geodesics at lower stratum [**DW03, Yam01, Wol03**], a question is to consider analogs of *non-refraction* for harmonic mappings. In particular for the complex of curves $C(F)$ and $\Lambda : \overline{\mathcal{T}} \to C(F)$ the labeling function and $u : \Omega \to \overline{\mathcal{T}}$ a harmonic map, what is the behavior of the composition $\Lambda \circ u$? For 2-dimensional domains Wentworth has already provided important results on *non-spiraling* and on $\Lambda \circ u$ being constant on the interior of Ω, [**Wen04**]. A second general matter is to develop an encompassing approach (including treating regularity and singularity behavior) for approximation of maps harmonic to a neighborhood of a stratum by harmonic maps to model spaces, in particular for: $(\mathbb{H}; ds^2 = dr^2 + r^6 d\theta^2, r > 0)$.

Two research matters on harmonic maps are as follows. Understand the behavior of harmonic maps to neighborhoods of lower strata in $\overline{\mathcal{T}}$. Develop an encompassing approach for approximation of maps harmonic to a neighborhood of a stratum in terms of harmonic maps to model spaces.

5. Characteristic classes and the WP Kähler form

Andrew McIntyre in joint work with Leon Takhtajan [**MT04**], Maryam Mirzakhani [**Mir04a, Mir04b, Mir**], and Lin Weng in part in joint work with Wing-Keung To [**Wen01**] have been studying the questions in algebraic geometry involving the WP Kähler form. The authors' considerations are guided by the study of the Quillen and Arakelov metrics, and in particular calculations from *string theory*. We sketch aspects of their work and use the opportunity to describe research with Kunio Obitsu.

Andrew McIntyre and Leon Takhtajan have extended the work of Peter Zograf and provided a new *holomorphic factorization formula* for the regularized determinant $\det' \Delta_k$ of the hyperbolic Laplace operator acting on (smooth) symmetric-tensor k-differentials for compact Riemann surfaces, [**MT04**]. Alexey Kokotov and Dmitry Korotkin have also provided a new *holomorphic factorization formula* for the regularized determinant $\det' \Delta_0$ of the hyperbolic Laplace operator acting on functions, [**KK04**]. Each factorization involves the exponential of an *action integral*: in the first formula for a Schottky uniformization, and in the second formula for a branched covering of $\mathbb{C}P^1$. The McIntyre-Takhtajan formula provides a holomorphic factorization for the $\partial\overline{\partial}$ *antiderivative* for the celebrated families local index theorem

$$\overline{\partial}\partial \log \frac{\det N_k}{\det' \Delta_k} = \frac{6k^2 - 6k + 1}{6\pi i} \omega_{WP}$$

for N_k the Gram matrix of the natural basis for holomorphic k-differentials relative to the Petersson product, [**MT04**]. The McIntyre-Takhtajan formula for $\frac{\det N_k}{\det' \Delta_k}$ gives rise to an isometry between the determinant bundle for holomorphic k-differentials with Quillen metric and with a metric defined from the Liouville action. Recall that the quotient $\frac{\det N_k}{\det' \Delta_k}$ is the Quillen norm of the natural frame for $\det N_k$. The formulas represent progress in the ongoing study of the behavior of the Quillen metric, the hyperbolic regularized determinant $\det' \Delta_k$ and positive integral values of the Selberg zeta function, since $\det' \Delta_k = c_{g,k} Z(k), k > 2$, [**DP86, Sar87**]. The above formulas provide another approach for studying the degeneration of $\det' \Delta_k$; see [**Hej90, Wol87a**].

Mumford's tautological class κ_1, the pushdown of the square of the relative dualizing sheaf from the universal curve, is represented by the WP class $\frac{1}{\pi^2} \omega_{WP}$, [**Wol90**]. The top self-intersection number of κ_1 on $\overline{\mathcal{M}_{g,n}}$ is the WP volume. From effective estimates for intersections of divisors on the moduli space, Georg Schumacher and Stefano Trapani [**ST01**] have given lower bounds for $vol_{WP}(\overline{\mathcal{M}_{g,n}})$. From Robert Penner's [**Pen92**] decorated Teichmüller theory and a combinatorial description of the moduli space, Samuel Grushevsky [**Gru01**] has given upper bounds with the same leading growth order.

In a series of innovative papers Maryam Mirzakhani has presented a collection of new results on hyperbolic geometry and calculations of WP integrals. The asymptotics for the count of the number of simple closed geodesics on a hyperbolic

surface of at most a given length is presented in [**Mir04a**]. She establishes the asymptotic
$$\#\{\gamma \mid \ell_\gamma(R) \leq L\} \sim c_R L^{6g-6+2n}.$$
A recursive method for calculating the WP volumes of moduli spaces of bordered hyperbolic surfaces with prescribed boundary lengths is presented in [**Mir04b**]. And a proof of the Witten-Kontsevich formula for the tautological classes on $\overline{\mathcal{M}}_{g,n}$ is presented in [**Mir**].

Central to Mirzakhani's considerations is a recursive scheme for evaluating WP integrals over the moduli space of bordered hyperbolic surfaces with prescribed boundary lengths. Her approach is based on recognizing an integration role for McShane's length sum identity. Greg McShane discovered a remarkable identity for geodesic-lengths of simple closed curves on punctured hyperbolic surfaces, [**McS98**]. To illustrate the recursive scheme for evaluation of integrals we sketch the consideration for $(g, n) = (1, 1)$.

For the $(1, 1)$ case the identity provides that
$$\sum_{\gamma \; scg} \frac{1}{1+e^{\ell_\gamma}} = \frac{1}{2}$$
for the sum over simple closed geodesics (scg's). The identity corresponds to a decomposition of a horocycle about the puncture; the identity arises from classifying the behavior of simple complete geodesics emanating from the puncture. Mirzakhani's insight is that the identity can be combined with the $d\tau \wedge d\ell$ formula for ω_{WP}, [**Wol85**], to give an *unfolding* of the $\mathcal{M}_{1,1}$ volume integral. For a MCG-fundamental domain $\mathcal{F}_{1,1} \subset \mathcal{T}_{1,1}$, she observes that
$$\int_{\mathcal{M}_{1,1}} \frac{1}{2}\omega_{WP} = \int_{\mathcal{F}_{1,1}} \sum_{h \in MCG/Stab_\gamma} \frac{1}{1+e^{\ell_{h(\gamma)}}} d\tau \wedge d\ell =$$
$$\sum_{h \in MCG/Stab_\gamma} \int_{h^{-1}(\mathcal{F}_{1,1})} \frac{1}{1+e^\ell} d\tau \wedge d\ell = \int_{\mathcal{T}_{1,1}/Stab_\gamma} \frac{1}{1+e^\ell} d\tau \wedge d\ell$$
(using that $\ell_\gamma \circ h^{-1} = \ell_{h(\gamma)}$) with the last integral elementary since
$$\mathcal{T}_{1,1}/Stab_\gamma = \{\ell > 0, 0 < \tau < \ell\}.$$

Mirzakhani established a general identity for bordered hyperbolic surfaces that generalizes McShane's identity, [**Mir04b**, Sec. 4]. Both identities are based on a sum over configurations of simple closed curves (*sub-partitions*) which with a fixed boundary bound a pair of pants. Mirzakhani uses the identity to *unfold* WP integrals to sums of product integrals over lower dimensional moduli spaces for surfaces with boundaries. Her overall approach provides a recursive scheme for determining WP volumes $vol_{WP}(\mathcal{M}_g(b_1, \ldots, b_n))$ for the (real analytic) moduli spaces of hyperbolic surfaces with prescribed boundary lengths $(b_1, \ldots, b_n)$. As an instance it is shown that
$$vol_{WP}(\mathcal{M}_1(b)) = \frac{\pi^2}{6} + \frac{b^2}{24}$$
and
$$vol_{WP}(\mathcal{M}_1(b_1, b_2)) = \frac{1}{384}(4\pi^2 + b_1^2 + b_2^2)(12\pi^2 + b_1^2 + b_2^2).$$

In complete generality Mirzakhani found that the WP volume is a polynomial
$$vol_g(b) = \sum_{|\alpha| \leq 3g-3+n} c_\alpha b^{2\alpha}, \ c_\alpha > 0, \ c_\alpha \in \pi^{6g-6+2n-2}\mathbb{Q},$$
for b the vector of boundary lengths and α a multi index, [**Mir04b**]. An easy application is an expansion for the volume of the tube $\mathcal{N}_\epsilon(\mathcal{D}) \subset \overline{\mathcal{M}_g}$ about the compactification divisor. Recently Ser Peow Tan, Yan Loi Wong and Ying Zhang have also generalized McShane's identity for conical hyperbolic surfaces, [**TWZ04**]. A general identity is obtained by studying gaps formed by simple normal geodesics emanating from a distinguished cone point, cusp or boundary geodesic.

Mirzakhani separately established that the volumes $vol_{WP}(\mathcal{M}_g(b_1,\ldots,b_n))$ are determined from the intersection numbers of tautological characteristic classes on $\overline{\mathcal{M}_{g,n}}$. A point of $\overline{\mathcal{M}_{g,n}}$ describes a Riemann surface R possibly with nodes with distinct points $x_1,\ldots,x_n$. The line bundle $\mathcal{L}_i$ on $\overline{\mathcal{M}_{g,n}}$ is the unique line bundle whose fiber over $(R; x_1,\ldots,x_n)$ is the cotangent space of R at x_i; write $\psi_i = c_1(\mathcal{L}_i)$ for the Chern class. Mirzakhani showed using symplectic reduction (for an S^1 quasi-free action following Guillemin-Sternberg) that
$$vol_g(b) = \sum_{|\alpha| \leq N} \frac{b_1^{2\alpha_1} \cdots b_n^{2\alpha_n}}{2^{|\alpha|}\alpha!(N-|\alpha|)!} \int_{\overline{\mathcal{M}_{g,n}}} \psi_1^{\alpha_1} \cdots \psi_n^{\alpha_n} \omega_{WP}^{N-|\alpha|}$$
for $N = \dim_\mathbb{C} \overline{\mathcal{M}_{g,n}}$, [**Mir04b**]. She then combined the above formula and her recursive integration scheme to find that the collection of integrals
$$\langle \tau_{k_1} \cdots \tau_{k_n} \rangle = \int_{\overline{\mathcal{M}_{g,n}}} \psi_1^{k_1} \cdots \psi_n^{k_n}, \ \sum k_i = \dim_\mathbb{C} \overline{\mathcal{M}_{g,n}},$$
satisfy the recursion for the *string equation* and the *dilaton equation*, [**Mir**, Sec. 6]. The recursion is the Witten-Kontsevich conjecture. She also found that the volume intergrands satisfy a recursive scheme. Mirzakhani's results represent major progress for the study of volumes and intersection numbers; her results clearly raise the prospect of further insights.

> *Can Mirzakhani's approach be applied for other integrals? Can the considerations of Grushevsky, Schumacher-Trapani be extended to give further effective intersection estimates?*

More than a decade ago Leon Takhtajan and Peter Zograf studied the local index theorem for a family of $\overline{\partial}$-operators on Riemann surfaces of type (g,n), [**TZ91**]. The authors calculated the first Chern form of the determinant line bundle provided with Quillen's metric. For a Riemann surface with punctures there are several candidates for $\det' \Delta$ (a renormalization is necessary since punctures give rise to *continuous spectrum* for Δ). The authors considered the Selberg zeta function $Z(s)$ and set $\det' \Delta_k = c_{g,n} Z(k), k \geq 2$, [**TZ91**]. For $\lambda_k = \det N_k$, the determinant line bundle of the bundle of holomorphic k-differentials (with poles allowed at punctures), the authors found for the Chern form
$$c_1(\lambda_k) = \frac{6k^2 - 6k + 1}{12\pi^2} \omega_{WP} - \frac{1}{9}\omega_*$$
with the 2-form ω_* on holomorphic quadratic differentials $\phi, \psi \in Q(R)$
$$\omega_*(\phi,\psi) = \sum_{i=1}^n \Im \int_R \phi\overline{\psi} E_i(z;2) \, (ds_{hyp}^2)^{-1}$$

with $E_i(z;2)$, the Eisenstein-Maass series at $s=2$ for the cusp x_i. By construction ω_* is a closed $(1,1)$ form; the associated Hermitian pairing (absent the imaginary part) is a MCG-invariant Kähler metric for $\mathcal{T}$. The Takhtajan-Zograf (TZ) metric g_{TZ} has been studied in several works, [**Obi99, Obi01, Wen01, Wol06**]. Kunio Obitsu showed that the metric is not complete [**Obi01**] and is now studying the degeneration of the metric. His estimates provide a local comparison for the TZ and WP metrics. Although beyond our exposition, we cite the important work of Lin Weng; he has been pursuing an Arakelov theory for punctured Riemann surfaces and has obtained a global comparison for the WP and TZ metrized determinant line bundles $\Delta_{WP}^{\otimes n^2} \leq \Delta_{TZ}^{\otimes((2g-2+n)^2)}$, [**Wen01**].

In joint work with Obitsu we are considering the expansion for the *tangential to the compactification divisor* $\mathcal{D} \subset \overline{\mathcal{M}_g}$ component g_{WP}^{tgt} of the WP metric, [**OW**]. The tangential component is the (orthogonal) complement to Yamada's normal form $dr^2 + r^6 d\theta^2$ for the transversal component. The tangential component expansion is given for a neighborhood of $\mathcal{D} \subset \overline{\mathcal{M}_g}$ for g_{WP} restricted to subspaces parallel to $\mathcal{D}$. For a family $\{R_\ell\}$ of hyperbolic surfaces given by *pinching* short geodesics all with common length ℓ, we find

$$g_{WP}^{tgt}(\ell) = g_{WP}^{tgt}(0) + \frac{\ell^2}{3} g_{TZ}(0) + O(\ell^3).$$

The formula establishes a direct relationship between the WP and TZ metrics. There is also a relationship with the work of Mirzakhani. The formula is based on an explicit form of the earlier 2-term expansion for the degeneration of hyperbolic metrics, [**Wol90**, Exp. 4.2].

What is the consequence of the above expansion for the determinant line bundle? Explore the properties of the TZ metric and its relationship to WP geometry.

References

[Abi77] William Abikoff. Degenerating families of Riemann surfaces. *Ann. of Math. (2)*, 105(1):29–44, 1977.

[Abi80] William Abikoff. *The real analytic theory of Teichmüller space*. Springer, Berlin, 1980.

[Ahl61] Lars V. Ahlfors. Some remarks on Teichmüller's space of Riemann surfaces. *Ann. of Math. (2)*, 74:171–191, 1961.

[Ara04] Javier Aramayona. The Weil-Petersson geometry of the five-times punctured sphere. preprint, 2004.

[Beh05] Jason A. Behrstock. Asymptotic geometry of of the mapping class group and Teichmüller space. preprint, 2005.

[Ber72] Lipman Bers. Uniformization, moduli, and Kleinian groups. *Bull. London Math. Soc.*, 4:257–300, 1972.

[Ber74] Lipman Bers. Spaces of degenerating Riemann surfaces. In *Discontinuous groups and Riemann surfaces (Proc. Conf., Univ. Maryland, College Park, Md., 1973)*, pages 43–55. Ann. of Math. Studies, No. 79. Princeton Univ. Press, Princeton, N.J., 1974.

[BF01] Jeffrey F. Brock and Benson Farb. Curvature and rank of Teichmüller space. preprint, 2001.

[BH99] Martin R. Bridson and André Haefliger. *Metric spaces of non-positive curvature*. Springer-Verlag, Berlin, 1999.

[BM04] Jeffrey Brock and Dan Margalit. Weil-Petersson isometries via the pants complex. preprint, 2004.

[BM06] Jason Behrstock and Yair Minsky. Dimension and rank for mapping class groups. preprint, 2006.

[Bro02] Jeffrey F. Brock. The Weil-Petersson visual sphere. preprint, 2002.

[Bro03] Jeffrey F. Brock. The Weil-Petersson metric and volumes of 3-dimensional hyperbolic convex cores. *J. Amer. Math. Soc.*, 16(3):495–535 (electronic), 2003.
[BY04] Chen Bo-Yong. Equivalence of the Bergman and Teichmüller metrics on Teichmüller spaces. preprint, 2004.
[CY80] Shiu Yuen Cheng and Shing Tung Yau. On the existence of a complete Kähler metric on noncompact complex manifolds and the regularity of Fefferman's equation. *Comm. Pure Appl. Math.*, 33(4):507–544, 1980.
[DKW00] Georgios Daskalopoulos, Ludmil Katzarkov, and Richard Wentworth. Harmonic maps to Teichmüller space. *Math. Res. Lett.*, 7(1):133–146, 2000.
[DP86] Eric D'Hoker and D. H. Phong. On determinants of Laplacians on Riemann surfaces. *Comm. Math. Phys.*, 104(4):537–545, 1986.
[DW03] Georgios Daskalopoulos and Richard Wentworth. Classification of Weil-Petersson isometries. *Amer. J. Math.*, 125(4):941–975, 2003.
[Gru01] Samuel Grushevsky. An explicit upper bound for Weil-Petersson volumes of the moduli spaces of punctured Riemann surfaces. *Math. Ann.*, 321(1):1–13, 2001.
[Ham06] Ursula Hamenstädt. Geometry of the mapping class groups III: Geometric rank. preprint, 2006.
[Hej90] Dennis A. Hejhal. Regular b-groups, degenerating Riemann surfaces, and spectral theory. *Mem. Amer. Math. Soc.*, 88(437):iv+138, 1990.
[Hua04] Zheng Huang. On asymptotic Weil-Petersson geometry of Teichmüller Space of Riemann surfaces. preprint, 2004.
[Hua06] Zheng Huang. The Weil-Petersson geometry on the thick part of the moduli space of Riemann surfaces. preprint, 2006.
[IT92] Y. Imayoshi and M. Taniguchi. *An introduction to Teichmüller spaces*. Springer-Verlag, Tokyo, 1992. Translated and revised from the Japanese by the authors.
[Iva88] N. V. Ivanov. Teichmüller modular groups and arithmetic groups. *Zap. Nauchn. Sem. Leningrad. Otdel. Mat. Inst. Steklov. (LOMI)*, 167(Issled. Topol. 6):95–110, 190–191, 1988.
[Iva97] Nikolai V. Ivanov. Automorphisms of complexes of curves and of Teichmüller spaces. In *Progress in knot theory and related topics*, volume 56 of *Travaux en Cours*, pages 113–120. Hermann, Paris, 1997.
[KK04] Alexey Kokotov and Dmitry Korotkin. Bergmann tau-function on Hurwitz spaces and its applications. preprint, 2004.
[Kor99] Mustafa Korkmaz. Automorphisms of complexes of curves on punctured spheres and on punctured tori. *Topology Appl.*, 95(2):85–111, 1999.
[KS93] Nicholas J. Korevaar and Richard M. Schoen. Sobolev spaces and harmonic maps for metric space targets. *Comm. Anal. Geom.*, 1(3-4):561–659, 1993.
[LSY04] Kefeng Liu, Sun Sun, Xiaofeng, and Shing-Tung Yau. Canonical metrics on the moduli space of Riemann surfaces, I, II. preprints, 2004.
[Luo00] Feng Luo. Automorphisms of the complex of curves. *Topology*, 39(2):283–298, 2000.
[Mas76] Howard Masur. Extension of the Weil-Petersson metric to the boundary of Teichmuller space. *Duke Math. J.*, 43(3):623–635, 1976.
[Mas01] Bernard Maskit. Matrices for Fenchel-Nielsen coordinates. *Ann. Acad. Sci. Fenn. Math.*, 26(2):267–304, 2001.
[McM00] Curtis T. McMullen. The moduli space of Riemann surfaces is Kähler hyperbolic. *Ann. of Math. (2)*, 151(1):327–357, 2000.
[McS98] Greg McShane. Simple geodesics and a series constant over Teichmuller space. *Invent. Math.*, 132(3):607–632, 1998.
[Mir] Maryam Mirzakhani. Weil-Petersson volumes and intersection theory on the moduli space of curves. to appear.
[Mir04a] Maryam Mirzakhani. Growth of the number of simple closed geodesics on hyperbolic surfaces. to appear, 2004.
[Mir04b] Maryam Mirzakhani. Simple geodesics and Weil-Petersson volumes of moduli spaces of bordered Riemann surfaces. preprint, 2004.
[MT04] Andrew McIntyre and Leon A. Takhtajan. Holomorphic factorization of determinants of Laplacians on Riemann surfaces and a higher genus generalization of Kronecker's first limit formula. preprint, 2004.

[MW02] Howard Masur and Michael Wolf. The Weil-Petersson isometry group. *Geom. Dedicata*, 93:177–190, 2002.

[MY83] Ngaiming Mok and Shing-Tung Yau. Completeness of the Kähler-Einstein metric on bounded domains and the characterization of domains of holomorphy by curvature conditions. In *The mathematical heritage of Henri Poincaré, Part 1 (Bloomington, Ind., 1980)*, volume 39 of *Proc. Sympos. Pure Math.*, pages 41–59. Amer. Math. Soc., Providence, RI, 1983.

[Nag88] Subhashis Nag. *The complex analytic theory of Teichmüller spaces*. Canadian Mathematical Society Series of Monographs and Advanced Texts. John Wiley & Sons Inc., New York, 1988. A Wiley-Interscience Publication.

[Obi99] Kunio Obitsu. Non-completeness of Zograf-Takhtajan's Kähler metric for Teichmüller space of punctured Riemann surfaces. *Comm. Math. Phys.*, 205(2):405–420, 1999.

[Obi01] Kunio Obitsu. The asymptotic behavior of Eisenstein series and a comparison of the Weil-Petersson and the Zograf-Takhtajan metrics. *Publ. Res. Inst. Math. Sci.*, 37(3):459–478, 2001.

[OW] Kunio Obitsu and Scott A. Wolpert. in preparation.

[Pen92] R. C. Penner. Weil-Petersson volumes. *J. Differential Geom.*, 35(3):559–608, 1992.

[Roy75] H. L. Royden. Intrinsic metrics on Teichmüller space. In *Proceedings of the International Congress of Mathematicians (Vancouver, B. C., 1974), Vol. 2*, pages 217–221. Canad. Math. Congress, Montreal, Que., 1975.

[Sar87] Peter Sarnak. Determinants of Laplacians. *Comm. Math. Phys.*, 110(1):113–120, 1987.

[ST01] Georg Schumacher and Stefano Trapani. Estimates of Weil-Petersson volumes via effective divisors. *Comm. Math. Phys.*, 222(1):1–7, 2001.

[Tro86] A. J. Tromba. On a natural algebraic affine connection on the space of almost complex structures and the curvature of Teichmüller space with respect to its Weil-Petersson metric. *Manuscripta Math.*, 56(4):475–497, 1986.

[TT04a] Lee-Peng Teo and Leon A. Takhtajan. Weil-Petersson metric on the universal Teichmüller space i: curvature properties and Chern forms. preprint, 2004.

[TT04b] Lee-Peng Teo and Leon A. Takhtajan. Weil-Petersson metric on the universal Teichmüller space ii: Kähler potential and period mapping. preprint, 2004.

[TWZ04] Ser Peow Tan, Yan Loi Wong, and Ying Zhang. Generalizations of McShane's identity to hyperbolic cone-surfaces. preprint, 2004.

[TZ91] L. A. Takhtajan and P. G. Zograf. A local index theorem for families of $\overline{\partial}$-operators on punctured Riemann surfaces and a new Kähler metric on their moduli spaces. *Comm. Math. Phys.*, 137(2):399–426, 1991.

[Wen01] Lin Weng. Ω-admissible theory. II. Deligne pairings over moduli spaces of punctured Riemann surfaces. *Math. Ann.*, 320(2):239–283, 2001.

[Wen04] Richard A. Wentworth. Regularity of harmonic maps from riemann surfaces to the Weil-Petersson completion of Teichüller space. preprint, 2004.

[Wol] Scott A. Wolpert. Convexity of geodesic-length functions: a reprise. In *Spaces of Kleinian Groups*, Lec. Notes. Cambridge Univ. Press. to appear.

[Wol82] Scott A. Wolpert. The Fenchel-Nielsen deformation. *Ann. of Math. (2)*, 115(3):501–528, 1982.

[Wol85] Scott A. Wolpert. On the Weil-Petersson geometry of the moduli space of curves. *Amer. J. Math.*, 107(4):969–997, 1985.

[Wol86] Scott A. Wolpert. Chern forms and the Riemann tensor for the moduli space of curves. *Invent. Math.*, 85(1):119–145, 1986.

[Wol87a] Scott A. Wolpert. Asymptotics of the spectrum and the Selberg zeta function on the space of Riemann surfaces. *Comm. Math. Phys.*, 112(2):283–315, 1987.

[Wol87b] Scott A. Wolpert. Geodesic length functions and the Nielsen problem. *J. Differential Geom.*, 25(2):275–296, 1987.

[Wol90] Scott A. Wolpert. The hyperbolic metric and the geometry of the universal curve. *J. Differential Geom.*, 31(2):417–472, 1990.

[Wol03] Scott A. Wolpert. Geometry of the Weil-Petersson completion of Teichmüller space. In *Surveys in Differential Geometry VIII: Papers in Honor of Calabi, Lawson, Siu and Uhlenbeck*, pages 357–393. Intl. Press, Cambridge, MA, 2003.

[Wol06] Scott A. Wolpert. Cusps and the family hyperbolic metric. preprint, 2006.

[Yam01] Sumio Yamada. Weil-Petersson Completion of Teichmüller Spaces and Mapping Class Group Actions. preprint, 2001.
[Yeu03] Sai-Kee Yeung. Bounded smooth strictly plurisubharmonic exhaustion functions on Teichmüller spaces. *Math. Res. Lett.*, 10(2-3):391–400, 2003.
[Yeu04] Sai-Kee Yeung. Quasi-isometry of metrics on Teichmüller spaces. preprint, 2004.

Part IV

Braid Groups, $\mathrm{Out}(F_n)$ and other Related Groups

Braid Groups and Iwahori-Hecke Algebras

Stephen Bigelow

ABSTRACT. The braid group B_n is the mapping class group of an n-times punctured disk. The Iwahori-Hecke algebra $\mathcal{H}_n$ is a quotient of the braid group algebra of B_n by a quadratic relation in the standard generators. We discuss how to use $\mathcal{H}_n$ to define the Jones polynomial of a knot or link. We also summarize the classification of the irreducible representations of $\mathcal{H}_n$. We conclude with some directions for future research that would apply mapping class group techniques to questions related to $\mathcal{H}_n$.

1. Introduction

The braid group B_n is the mapping class group of an n-times punctured disk. It can also be defined using certain kinds of arrangements of strings in space, or certain kinds of diagrams in the plane. Our main interest in the braid group B_n will be in relation to the Iwahori-Hecke algebra $\mathcal{H}_n$, which is a certain quotient of the group algebra of B_n. The exact definition will be given in Section 3.

The Iwahori-Hecke algebra plays an important role in representation theory. It first came to the widespread attention of topologists when Jones used it to define the knot invariant now called the Jones polynomial [**Jon85**]. This came as a huge surprise, since it brought together two subjects that were previously unrelated. It has given knot theorists a host of new knot invariants, and intriguing connections to other areas of mathematics to explore. It has also helped to promote the use of pictures and topological thinking in representation theory.

As far as I know, no major results related to the Iwahori-Hecke algebra have yet been proved using the fact that B_n is a mapping class group. I think the time is ripe for such a result. Unfortunately this paper will use the diagrammatic definition of the braid group almost exclusively. I hope it will at least help to provide a basic grounding for someone who wants to pursue the connection to mapping class groups in the future.

The outline of this paper is as follows. In Sections 2 and 3 we introduce the braid group and Iwahori-Hecke algebra. In Sections 4 and 5 we give a basis for the Iwahori-Hecke algebra and for its module of trace functions. In Section 6 we explain how one such trace function leads to the definition of the Jones polynomial of a knot or a link. In Section 7 we briefly summarize the work of Dipper and James [**DJ86**]

Partially supported by NSF grant DMS 0307235 and the Sloan Foundation.

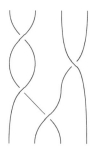

FIGURE 1. A braid with four strands.

classifying the irreducible representations of the Iwahori-Hecke algebra. In Section 8 we conclude with some speculation on possible directions for future research. Open problems will be scattered throughout the paper.

2. The braid group

Like most important mathematical objects, the braid group B_n has several equivalent definitions. Of greatest relevance to this volume is its definition as a mapping class group. Let D be a closed disk, let $p_1, \ldots, p_n$ be distinct points in the interior of D, and let $D_n = D \setminus \{p_1, \ldots, p_n\}$. The braid group B_n is the mapping class group of D_n. Thus a braid is the equivalence class of a homeomorphism from D_n to itself that acts as the identity of the boundary of the disk.

Artin's original definition of B_n was in terms of *geometric braids*. A geometric braid is a disjoint union of n edges, called *strands*, in $D \times I$, where I is the interval $[0, 1]$. The set of endpoints of the strands is required to be $\{p_1, \ldots, p_n\} \times \{0, 1\}$, and each strand is required to intersect each disk cross-section exactly once. Two geometric braids are said to be equivalent if it is possible to deform one to the other through a continuous family of geometric braids. The elements of B_n are equivalence classes of geometric braids.

We will need some terminology to refer to directions in a geometric braid. Take D to be the unit disk centered at 0 in the complex plane. Take $p_1, \ldots, p_n$ to be real numbers with $-1 < p_1 < \cdots < p_n < 1$. The *top* and *bottom* of the braid are $D \times \{1\}$ and $D \times \{0\}$, respectively. In a disk cross-section, the *left* and *right* are the directions of decreasing and increasing real part, respectively, while the *front* and *back* are the directions of decreasing and increasing imaginary part, respectively.

Multiplication in B_n is defined as follows. If a and b are geometric braids with n strands then the product ab is obtained by stacking a on top of b and then rescaling vertically to the correct height. This can be shown to give a well-defined product of equivalence classes, and to satisfy the axioms of a group.

A geometric braid can be drawn in the plane using a projection from $D \times I$ to $[-1, 1] \times I$. An example is shown in Figure 1. The projection map is given by $(x + iy, t) \mapsto (x, t)$. Note that this sends each strand to an embedded edge. We also require that the braid be in *general position* in the sense that the images of the strands intersect each other transversely, with only two edges meeting at each point of intersection. The points of intersection are called *crossings*. At each crossing, we record which of the two strands passed in front of the other at the corresponding disk cross-section of the geometric braid. This is usually represented pictorially by

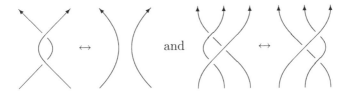

FIGURE 2. Reidemeister moves of types two and three.

a small break in the segment that passes behind. The image of a geometric braid under a projection in general position, together with this crossing information, is called a *braid diagram*.

Let us fix some terminology related to braid diagrams. The directions left, right, top, and bottom are the images of these same directions in the geometric braid, so for example the point $(-1, 1)$ is the top left of the braid diagram. We say a strand makes an *overcrossing* or an *undercrossing* when it passes respectively in front of or behind another strand at a crossing. A crossing is called *positive* if the strand making the overcrossing goes from the bottom left to the top right of the crossing, otherwise it is called *negative*. The endpoints of strands are called *nodes*.

Two braid diagrams represent the same braid if and only if they are related by an isotopy of the plane and a sequence of Reidemeister moves of types two and three. These are moves in which the diagram remains unchanged except in a small disk, where it changes as shown in Figure 2. (There is also a Reidemeister move of type one, which is relevant to knots but not to braids.)

For $i = 1, \ldots, n-1$, let σ_i be the braid diagram with one crossing, which is a positive crossing between strands i and $i+1$. The braid group B_n is generated by $\sigma_1, \ldots, \sigma_{n-1}$, with defining relations

- $\sigma_i \sigma_j = \sigma_j \sigma_i$ if $|i - j| > 1$,
- $\sigma_i \sigma_j \sigma_i = \sigma_j \sigma_i \sigma_j$ if $|i - j| = 1$.

There is an imprecise but vivid physical description of the correspondence between a geometric braid and a mapping class of the n-times punctured disk D_n. Imagine a braid made of inflexible wires and a disk made of flexible rubber. Press the disk onto the top of the braid, puncturing the disk at n points in its interior. Now hold the disk by its boundary and push it down. As the wires of the braid twist around each other, the punctures of the disk will twist around and the rubber will be stretched and distorted to accommodate this. The mapping class corresponding to the geometric braid is represented by the function taking each point on D_n to its image in D_n after the disk has been pushed all the way to the bottom of the braid. (With our conventions, this description gives the group of mapping classes acting on the right.)

See [**Bir74**], or [**BB**], for proofs that these and other definitions of B_n are all equivalent. This paper will primarily use the definition of a braid group as a braid diagram. This is in some sense the least elegant choice since it involves an arbitrary projection and a loss of the true three-dimensional character of the geometric braid. The main goal is to provide an introduction that may inspire someone to apply mapping class group techniques to problems that have previously been studied algebraically and combinatorially.

FIGURE 3. The skein relation

3. The Iwahori-Hecke algebra

Let n be a positive integer and let q_1 and q_2 be units in a domain R. The *Iwahori-Hecke algebra* $\mathcal{H}_n(q_1, q_2)$, or simply $\mathcal{H}_n$, is the associative R-algebra given by generators $T_1, \ldots, T_{n-1}$ and relations

- $T_i T_j = T_j T_i$ if $|i - j| > 1$,
- $T_i T_j T_i = T_j T_i T_j$ if $|i - j| = 1$,
- $(T_i - q_1)(T_i - q_2) = 0$.

The usual definition of the Iwahori-Hecke algebra uses only one parameter q. It corresponds to $\mathcal{H}_n(-1, q)$, or in some texts to $\mathcal{H}_n(1, -q)$. There is no loss of generality because there is an isomorphism from $\mathcal{H}_n(q_1, q_2)$ to $\mathcal{H}_n(-1, -q_2/q_1)$ given by $T_i \mapsto -q_1 T_i$. It will be convenient for us to keep two parameters.

We now explore some of the basic properties of the Iwahori-Hecke algebra. Since q_1 and q_2 are units of R, the generators T_i are units of $\mathcal{H}_n$, with

$$T_i^{-1} = (T_i - q_1 - q_2)/(q_1 q_2).$$

Thus there is a well-defined homomorphism from B_n to the group of units in $\mathcal{H}_n$ given by

$$\sigma_i \mapsto T_i.$$

The following is a major open question.

QUESTION 1. *If $R = \mathbf{Q}(q_1, q_2)$, is the above map from B_n to $\mathcal{H}_n(q_1, q_2)$ injective?*

For $n = 3$, the answer is yes. For $n = 4$, the answer is yes if and only if the Burau representation of B_4 is injective, or *faithful*. The Burau representation is one of the irreducible summands of the Iwahori-Hecke algebra over $\mathbf{Q}(q_1, q_2)$. By a result of Long [**Lon86**], the map from B_n to $\mathcal{H}_n$ is injective if and only if at least one of these irreducible summands is faithful. For $n = 4$, they are all easily shown to be unfaithful except for the Burau representation, which remains unknown. For $n \geq 5$, the Burau representation is unfaithful [**Big99**], but there are other summands whose status remains unknown.

One can also ask Question 1 for other choices of ring R and parameters q_1 and q_2. If the map from B_n to $\mathcal{H}_n$ is injective for any such choice then it is injective when $R = \mathbf{Q}(q_1, q_2)$. A non-trivial case when the map is not injective is when $n = 4$ and $R = k[q_1^{\pm 1}, q_2^{\pm 1}]$, where k is a field of characteristic 2 [**CL97**] or 3 [**CL98**]. Another is when $n = 4$, $R = \mathbf{Q}$ and $q_2/q_1 = -2$ [**Big02**].

Using the map from B_n to $\mathcal{H}_n$, we can represent any element of $\mathcal{H}_n$ by a linear combination of braid diagrams. The quadratic relation is equivalent to the *skein relation* shown in Figure 3. Here, an instance of the skein relation is a relation involving three diagrams that are identical except inside a small disk where they are as shown in the figure.

One motivation for studying the Iwahori-Hecke algebra is its connection with the representation theory of the braid groups. The representations of $\mathcal{H}_n$ are precisely those representations of B_n for which the image of the generators satisfy a quadratic relation. The study of these representations led Jones to the discovery of his knot invariant, which we define in Section 6

Another reason for interest is the connection between the Iwahori-Hecke algebra and the symmetric group. There is an isomorphism from $\mathcal{H}_n(1,-1)$ to the group algebra $R\mathfrak{S}_n$ taking T_i to the transposition $(i, i+1)$. Thus $\mathcal{H}_n(q_1, q_2)$ can be thought of as a *deformation* of $R\mathfrak{S}_n$. The Iwahori-Hecke algebra plays a role in the representation theory of the general linear group over a finite field that is analogous to the role of the symmetric group in the representation theory of the general linear group over the real numbers. See for example [**Dip85**].

This process of realizing a classical algebraic object as the case $q = 1$ in a family of algebraic objects parametrized by q is part of a large circle of ideas called *quantum mathematics*, or *q-mathematics*. The exact nature and significance of any connection to quantum mechanics not clear at present. One example is [**Bar03**], in which Barrett uses quantum mathematics to analyze quantum gravity in a universe with no matter and three space-time dimensions.

4. A basis

The aim of this section is to show that $\mathcal{H}_n$ is a free R-module of rank $n!$, and to give an explicit basis.

Let $\phi\colon B_n \to \mathfrak{S}_n$ be the map such that $\phi(\sigma_i)$ is the transposition $(i, i+1)$. Thus in any braid b, the strand with lower endpoint at node number i has upper endpoint at node number $\phi(b)(i)$.

For $w \in \mathfrak{S}_n$, let T_w be a braid diagram with the minimal number of crossings such that every crossing is positive and $\phi(T_w) = w$. Such a braid can be thought of as "layered" in the following sense. In the front layer is a strand connecting node 1 at the bottom to node $w(1)$ at the top. Behind that is a strand connecting node 2 at the bottom to node $w(2)$ at the top. This continues until the back layer, in which a strand connects node n at the bottom to node $w(n)$ at the top. From this description it is clear that our definition of T_w specifies a unique braid in B_n. By abuse of notation, let T_w denote the image of this braid in $\mathcal{H}_n$. For example, if w is a transposition $(i, i+1)$ then T_w is the generator T_i.

THEOREM 4.1. *The set of T_w for $w \in \mathfrak{S}_n$ forms a basis for $\mathcal{H}_n$.*

To prove this, we first describe an algorithm that will input a linear combination of braid diagrams and output a linear combination of basis elements T_w that represents the same element of $\mathcal{H}_n$. By linearity, it suffices to describe how to apply the algorithm to a single braid diagram v.

A crossing in v will be called *bad* if the strand that makes the overcrossing is the one whose lower endpoint is farther to the right. If v has no bad crossings, stop here.

Suppose v has at least one bad crossing. Let the *worst* crossing be a bad crossing whose undercrossing strand has lower endpoint farthest to the left. If there is more than one such bad crossing, let the worst be the one that is closest to the bottom of the diagram.

Use the skein relation to rewrite v as a linear combination of v' and v_0, where v' is the result of changing the sign of the worst crossing and v_0 is the result of removing it. Now recursively apply this procedure to v' and v_0.

Note that any bad crossings in v' and v_0 are "better" than the worst crossing of v in the sense that either the lower endpoint of their undercrossing strand is farther to the right or they have the same undercrossing strand and are closer to the top of the diagram. Thus the above algorithm must eventually terminate with a linear combination of diagrams that have no bad crossings. Any such diagram must equal T_w for some $w \in \mathfrak{S}_n$.

This algorithm shows that the T_w span $\mathcal{H}_n$. It remains to show that they are linearly independent. Note that if the algorithm is given as input a linear combination of diagrams of the form T_w, then its output will be the same linear combination. Thus it suffices to show that the output of the algorithm does not depend on the initial choice of linear combination of braid diagrams to represent a given element of $\mathcal{H}_n$. We prove this in three claims, which show that the output of the algorithm is invariant under the skein relation and Reidemeister moves of types two and three.

CLAIM 4.2. *Suppose v_+, v_- and v_0 are three braid diagrams that are identical except in a small disk where v_+ has a positive crossing, v_- has a negative crossing, and v_0 has no crossing. Then the algorithm gives the same output for both sides of the skein relation $v_+ + q_1 q_2 v_- = (q_1 + q_2) v_0$.*

PROOF. For exactly one of v_+ and v_-, the crossing inside the small disk is a bad crossing. For convenience assume it is v_+, since it makes no difference to the argument.

Suppose the worst crossing for v_+ is the crossing in the small disk. Applying the next step of the algorithm to v_+ results in a linear combination of v_- and v_0 which, by design, will exactly cancel the other two terms in the skein relation.

Now suppose the worst crossing for v_+ is not inside the small disk. Then it must be the same as the worst crossing for v_- and for v_0. Thus the next step of the algorithm has the same effect on v_+, v_- and v_0. The claim now follows by induction. □

CLAIM 4.3. *If u and v are diagrams that differ by a Reidemeister move of type two then the algorithm gives the same output for u as for v.*

PROOF. As in the proof of the previous claim, we can reduce to the case where a worst crossing lies inside the small disk affected by the Reidemeister move. The claim now follows by computing the result of applying the algorithm inside the small disk. Alternatively, observe that this computation amounts to checking the case $n = 2$ of Theorem 4.1, which follows easily from the presentation of $\mathcal{H}_2$. □

CLAIM 4.4. *Suppose u and v are diagrams that differ by a Reidemeister move of type three. Then the algorithm gives the same output for u as for v.*

PROOF. Once again, one solution involves a brute force computation of the algorithm. Here we describe a somewhat more comprehensible approach.

Label the three strands in the small disk in each of u and v the *front, back,* and *middle* strands, where the front strand makes two overcrossings, the back strand makes two undercrossings, and the middle strand makes one overcrossing and one undercrossing.

Let u' and v' be the result of changing the sign of the crossings between the front and middle strands of u and v respectively. Note that eliminating these crossings results in identical braid diagrams. Thus by Claim 4.2, the algorithm gives the same output for u as for v if and only if it gives the same output for u' as for v'.

Relabel the three strands in u' and v' so that once again the front strand makes two overcrossings, the back strand makes two undercrossings, and the middle strand makes one overcrossing and one undercrossing. Now let u'' and v'' be the output of changing the sign of the crossing between the middle and back strands of u' and v' respectively. Note that eliminating these crossings results in braid diagrams that differ by Reidemeister moves of type two. Thus by Claims 4.2 and 4.3, the algorithm gives the same output for u' as for v' if and only if it gives the same output for u'' as for v''.

We can continue in this way, alternately changing crossings between front and middle, and middle and back strands. We obtain six different versions of the Reidemeister move of type three. Each is obtained from the original by some crossing changes, and corresponds to one of the six permutations of the roles of front, middle, and back strands.

The algorithm gives the same output for u as for v if and only if it gives the same output when the relevant disks in u and v are changed to represent any one of the six versions of the Reidemeister move of type three. Thus we can choose a version to suit our convenience. In particular we can always choose the front and back strands to be the ones with lower endpoints farthest to the left and right respectively. That way there will be no bad crossings inside the small disk, and the algorithm will proceed identically for the diagrams on either side of the move.

This completes the proof of the claim, and hence of the theorem. $\square$

5. Trace functions

A trace function on $\mathcal{H}_n$ is a linear function tr: $\mathcal{H}_n \to R$ such that $\text{tr}(ab) = \text{tr}(ba)$ for all $a, b \in \mathcal{H}_n$. Let V be the quotient of $\mathcal{H}_n$ by the vector subspace spanned by elements of the form $ab - ba$ for $a, b \in \mathcal{H}_n$. Then the trace functions of $\mathcal{H}_n$ correspond to the linear maps from V to R.

The aim of this section is to find a basis for V, and hence classify all trace functions of $\mathcal{H}_n$. This has been done by Turaev [**Tur88**] and independently by Hoste and Kidwell [**HK90**]. They actually consider a larger algebra in which the strands can have arbitrary orientations, but the result is very similar.

We define a *closed n-braid* to be a disjoint union of circles in $D \times S^1$ that intersects each disk cross-section at a total of n points. We say two closed braids are equivalent if one can be deformed to the other through a continuous family of closed braids. The *closure* of a geometric braid in $D \times I$ is the result of identifying $D \times \{0\}$ to $D \times \{1\}$. It is not difficult to show that two braids have equivalent closures if and only if they are conjugate in B_n.

Define a *diagram* of a closed braid to be a projection onto the annulus $I \times S^1$ in general position, together with crossing information, similar to the diagram of a braid. Then V is the vector space of formal linear combination of closed n-braid diagrams modulo the skein relation and Reidemeister moves of types two and three. This is an example of a *skein algebra* of the annulus.

A *partition of* n is a sequence $\lambda = (\lambda_1, \ldots, \lambda_k)$ of integers such that $\lambda_1 \geq \cdots \geq \lambda_k > 0$ and $\lambda_1 + \cdots + \lambda_k = n$. For any $m > 0$, let $b_{(m)}$ be the braid $\sigma_{m-1} \ldots \sigma_2 \sigma_1$.

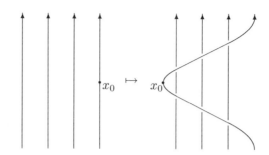

FIGURE 4. Pulling the basepoint toward $\{0\} \times S^1$

If $\lambda = (\lambda_1, \ldots, \lambda_k)$ is a partition of n, let b_λ be the braid with diagram consisting of a disjoint union of diagrams of the braids $b_{(\lambda_i)}$, in order from left to right. Let v_λ be the closure of b_λ.

THEOREM 5.1. *The set of v_λ for partitions λ of n forms a basis for V.*

We start by defining an algorithm similar to that of Theorem 4.1. There are some added complications because a strand can circle around and cross itself, and there is no "bottom" of the closed braid to use as a starting point. Therefore the first step is to choose a basepoint x_0 on the diagram that is not a crossing point. Pull x_0 in front of the other strands so that x_0 becomes the closest point to the boundary component $\{0\} \times S^1$, as suggested by Figure 4.

Consider the oriented edge that begins at x_0 and proceeds in the positive direction around the annulus. Call a crossing *bad* if this edge makes an undercrossing on the first (or only) time it passes through that crossing. If there is a bad crossing, use the skein relation to eliminate the first bad crossing the edge encounters. This process will eventually terminate with a linear combination of diagrams that have no bad crossings.

If there are no bad crossings then the loop through x_0 passes in front of every other loop in the closed braid. Furthermore, we can assume that its distance toward the front of the diagram steadily decreases as it progresses in the positive direction from x_0 until just before it closes up again at x_0. Since x_0 is the closest point to $\{0\} \times S^1$, this implies that the loop through x_0 is isotopic to $v_{(m)}$ for some positive integer m.

Isotope this loop, keeping it in front of all other strands, toward the boundary component $\{1\} \times S^1$, until its projection is disjoint from that of all other loops. Now ignore this loop and repeat the above procedure to the remainder of the closed braid diagram. This process must eventually terminate with a linear combination of closed braids of the form v_λ.

To show that the v_λ are linearly independent, it suffices to show that output is invariant under the skein relation, Reidemeister moves of types two and three, and the choices of basepoint. By induction on n we can assume that the output of the algorithm does not depend on choices of basepoint made after the first loop has been made disjoint from the other loops. Thus the algorithm produces a unique output given a diagram of a closed n-braid together with a single choice of initial basepoint x_0.

CLAIM 5.2. *The output of the algorithm is invariant under the skein relation, and under any Reidemeister move for which the basepoint does not lie in the disk affected by the move.*

PROOF. The proofs of Claims 4.2, 4.3 and 4.4 go through unchanged. □

It remains only to prove the following.

CLAIM 5.3. *For a given diagram of a closed n-braid, the output of the algorithm does not depend on the choice of basepoint.*

PROOF. First we show that the output of the algorithm is not affected by moving the basepoint over an overcrossing. Recall that the first step of the algorithm is to pull the basepoint in front of the other strands as in Figure 4. If the basepoint is moved over an overcrossing, the output of this first step will be altered by a Reidemeister move of type two. Furthermore, the basepoint does not lie in the disk affected by this Reidemeister move. Thus the output of the algorithm is unchanged.

Now fix a diagram v of a closed n-braid. By induction, assume that the claim is true for any diagram with fewer crossings than v. By the skein relation, if v' is the result of changing the sign of one of the crossings of v then the claim is true for v if and only if it is true for v'. Thus we are free to change the signs of the crossings in v to suit our convenience.

By changing crossings and moving the basepoint past overcrossings we can move the basepoint to any other point on the same loop in v. Now suppose basepoints x_0 and y_0 lie on two distinct loops in v. We can choose each to be the closest point on its loop to the boundary component $\{0\} \times S^1$. Assume, without loss of generality, that x_0 is at least as close as to $\{0\} \times S^1$ as y_0 is. By changing the signs of crossings, we can assume that the loop through y_0 has no bad crossings. Applying the algorithm using y_0 as the basepoint then has the affect of isotoping the corresponding loop toward $\{1\} \times S^1$, keeping it in front of all other strands. This can be achieved by a sequence of Reidemeister moves of types two and three. No point in this loop is closer to $\{0\} \times S^1$ than x_0, so x_0 does not lie in the disk affected by any of these Reidemeister moves. Thus the algorithm will give the same output using the basepoint x_0 as it does using basepoint y_0.

This completes the proof of the claim, and hence of the theorem. □

6. The Jones polynomial

The Jones polynomial is an invariant of knots and links, first defined by Jones [**Jon85**]. Jones arrived at his definition as an outgrowth of his work on operator algebras, as opposed to knot theory. To this day the topological meaning of his polynomial seems somewhat mysterious, and it has a very different flavor to classical knot invariants such as the Alexander polynomial.

After the discovery of the Jones polynomial, several people independently realized that it could be generalized to a two-variable polynomial now called the HOMFLY or HOMFLYPT polynomial. The names are acronyms of the authors of [**FYH**+**85**], where the polynomial was defined, and of [**PT88**], where related results were discovered independently.

The aim of this section is to show how to use the Iwahori-Hecke algebra to define a polynomial invariant of knots and links called the HOMFLY or HOMFLYPT polynomial.

Given a geometric braid b, we can obtain a closed braid in the solid torus by identifying the top and the bottom of b. Now embed this solid torus into S^3 in a standard unknotted fashion. The resulting knot or link in S^3 is called the *closure* of b. It is a classical theorem of Alexander that any knot or link in S^3 can be obtained in this way.

Let B_∞ be the disjoint union of the braid groups B_n for $n \geq 1$. For every $n \geq 1$, let
$$\iota\colon B_n \to B_{n+1}$$
be the inclusion map that adds a single straight strand to the right of any n-braid. The *Markov moves* are as follows.

- $ab \leftrightarrow ba$,
- $b \leftrightarrow \sigma_n \iota(b)$,
- $b \leftrightarrow \sigma_n^{-1} \iota(b)$,

for any $a, b \in B_n$.

THEOREM 6.1 (Markov's theorem). *Two braids have the same closure if and only if they are connected by a sequence of Markov moves.*

An R-valued link invariant is thus equivalent to a function from B_∞ to R that is invariant under the Markov moves. We now look for such a function that factors through the maps $B_n \to \mathcal{H}_n$.

Let $\iota\colon \mathcal{H}_n \to \mathcal{H}_{n+1}$ be the inclusion map $T_i \mapsto T_i$. A family of linear maps $\mathrm{tr}\colon \mathcal{H}_n \to R$ defines a link invariant if and only if it satisfies the following.

- $\mathrm{tr}(ab) = \mathrm{tr}(ba)$,
- $\mathrm{tr}(b) = \mathrm{tr}(T_n \iota(b))$,
- $\mathrm{tr}(b) = \mathrm{tr}(T_n^{-1} \iota(b))$,

for every $n \geq 1$ and $a, b \in \mathcal{H}_n$. We will call such a family of maps a *normalized Markov trace*. The usual definition of Markov trace is slightly different, but can easily be rescaled to satisfy the above conditions.

By the skein relation, the third condition on a normalized Markov trace is equivalent to

$$(1) \qquad (1 + q_1 q_2)\,\mathrm{tr}(b) = (q_1 + q_2)\,\mathrm{tr}(\iota(b)),$$

for every $n \geq 1$ and $b \in \mathcal{H}_n$. To obtain an interesting invariant, we assume from now on that $q_1 + q_2$ is a unit of R.

Let λ be a partition of n and let b_λ be the corresponding braid diagram as defined in the previous section. Let k be the number of components of b_λ, that is, the number of nonzero entries in λ. Using the Markov moves and Equation (1), it is easy to show that any normalized Markov trace must satisfy

$$(2) \qquad \mathrm{tr}(b_\lambda) = \left(\frac{1 + q_1 q_2}{q_1 + q_2}\right)^{k-1} \mathrm{tr}(\mathrm{id}_1),$$

where id_1 is the identity element of $\mathcal{H}_1$.

We now show that this equation defines a normalized Markov trace. Let b be a braid diagram. Let v be the closed braid diagram obtained by identifying the top and the bottom of b. Apply the algorithm from the previous section to write v as a linear combination of v_λ. Let $\mathrm{tr}(b)$ be as given by Equation (2) with $\mathrm{tr}(\mathrm{id}_1) = 1$.

Let v_0 be the closed braid diagram obtained by identifying the top and the bottom of $\iota(b)$. This is obtained from v by adding a disjoint loop. Now apply the algorithm to write v_0 as a linear combination of basis elements v_λ. The added loop in v_0 remains unchanged throughout the algorithm. Thus it has the effect of adding an extra component to each term of the resulting linear combination of basis elements. This shows that tr satisfies Equation (1).

Now let v_+ be the closed braid diagram obtained by identifying the top and the bottom of $\sigma_n \iota(b)$. then v_+ is obtained from v by adding an extra "kink" in one of the strands. Now apply the algorithm to write v_+ as a linear combination of basis elements v_λ. We can assume that we never choose a basepoint that lies on the added kink. Then the added kink remains unchanged throughout the algorithm. It has no effect on the number of components in each term of the resulting linear combination of basis elements. This shows that tr is invariant under the second Markov move.

This completes the proof that tr is a normalized Markov trace. Any other normalized Markov trace must be a scalar multiple of tr. The HOMFLYPT polynomial $P_L(q_1, q_2)$ of a link L is defined to be $\text{tr}(b)$ for any braid b whose closure is L. There are many different definitions of P_L in the literature, each of which can be obtained from any other by a change of variables. They are usually specified by giving the coefficients of the three terms of the skein relation in Figure 3. As far as I know, mine is yet another addition to the collection of possible choices that appear in the literature.

The Jones polynomial V_L is given by

$$V_L(t) = P_L(-t^{\frac{1}{2}}, t^{\frac{3}{2}}).$$

If L is a knot, V_L turns out to involve only integer powers of t. This polynomial was originally defined as a trace function of the Temperley-Lieb algebra, which is a certain quotient of the Iwahori-Hecke algebra.

A somewhat tangential question is worth mentioning here. In its most open-ended form, it is as follows.

QUESTION 2. *What are the equivalence classes of braids modulo the moves*
- $ab \leftrightarrow ba$, *and*
- $b \leftrightarrow \sigma_n \iota(b)$?

In other words, what happens if the Markov move

$$b \leftrightarrow \sigma_n^{-1} \iota(b)$$

is omitted? This question was shown in [**OS03**] to be equivalent to the important problem in contact geometry of classifying transversal links up to transversal isotopy.

The *Bennequin number* of a braid $b \in B_n$ is $e - n$, where e is the sum of the exponents in a word in the generators σ_i representing b. The Bennequin number is invariant under the moves in Question 2. Thus it can be used to show that there are braids that are related by Markov moves, but not by the moves in Question 2.

Birman and Menasco [**BM**] and Etnyre and Honda [**EH**] have independently found pairs of braids that are related by Markov moves and have the same Bennequin invariant, but are not related by the moves in Question 2. Their proofs are quite complicated, and it would be nice to have a new invariant that could distinguish their pairs of braids.

7. Representations of $\mathcal{H}_n$

A *representation* of $\mathcal{H}_n$ is simply a $\mathcal{H}_n$-module. If R is a field then an *irreducible* representation of $\mathcal{H}_n$ is a nonzero $\mathcal{H}_n$-module with no nonzero proper submodules. In [**DJ86**], Dipper and James gave a complete list of the irreducible representations of $\mathcal{H}_n$. The aim of this section is to summarize their results. Our approach comes from the theory of *cellular algebras*, as defined in [**GL96**]. For convenience we will take
$$\mathcal{H}_n = \mathcal{H}_n(-1, q)$$
from now on.

Let λ be a partition of n. The *Young subgroup* $\mathfrak{S}_\lambda$ of $\mathfrak{S}_n$ is the image of the obvious embedding
$$\mathfrak{S}_{\lambda_1} \times \cdots \times \mathfrak{S}_{\lambda_k} \to \mathfrak{S}_n.$$
More precisely, it is the set of permutations of $\{1, \ldots, n\}$ that fix setwise each set of the form $\{k_i + 1, \ldots, k_i + \lambda_i\}$ where $k_i = \lambda_1 + \cdots + \lambda_{i-1}$. Let
$$m_\lambda = \sum_{w \in \mathfrak{S}_\lambda} T_w.$$

Let λ be a partition of n. Let M^λ be the left-ideal $\mathcal{H}_n m_\lambda$. We say a partition μ of n *dominates* λ if $\sum_{i=1}^j \mu_i \leq \sum_{i=1}^j$ for all $j \geq 1$. Let I^λ be the two-sided ideal of $\mathcal{H}_n$ generated by m_μ for all partitions μ of n that dominate λ. The *Specht module* is the quotient
$$S^\lambda = M^\lambda / (M^\lambda \cap I^\lambda).$$
Let $\operatorname{rad} S^\lambda$ be the set of $v \in S^\lambda$ such that $m_\lambda h v = 0$ for all $h \in \mathcal{H}_n$. Let
$$D^\lambda = S^\lambda / \operatorname{rad} S^\lambda.$$

THEOREM 7.1. *Suppose R is a field. Then every irreducible representation of $\mathcal{H}_n$ is of the form D^λ for some partition λ of n. If λ and μ are distinct partitions of n then D^λ and D^μ are either distinct or both zero.*

The definition of D^λ can be better motivated by defining a bilinear form on S^λ. Let $\star \colon \mathcal{H}_n \to \mathcal{H}_n$ be the antiautomorphism given by $T_w^* = T_{w^{-1}}$ for all $w \in \mathfrak{S}_n$. Note that $m_\lambda^* = m_\lambda$. The following lemma is due to Murphy [**Mur92**].

LEMMA 7.2. *If $h \in \mathcal{H}_n$ then $m_\lambda h m_\lambda = r m_\lambda$ modulo I^λ, for some $r \in R$.*

Thus we can define a bilinear form
$$\langle \cdot, \cdot \rangle \colon S^\lambda \times S^\lambda \to R$$
by
$$(h_1 m_\lambda)^* (h_2 m_\lambda) = \langle h_1 m_\lambda, h_2 m_\lambda \rangle m_\lambda.$$
Then $\operatorname{rad} S^\lambda$ is the set of $y \in S^\lambda$ such that $\langle x, y \rangle = 0$ for all $x \in \mathcal{H}_n$.

Dipper and James also determined which values of λ give a nonzero D^λ. Let e be the smallest positive integer such that $1 + q + \cdots + q^{e-1} = 0$, or infinity if there is no such integer.

THEOREM 7.3. *$D^\lambda \neq 0$ if and only if $\lambda_i - \lambda_{i+1} < e$ for all $i \geq 1$.*

Thus the work of Dipper and James completely characterizes the irreducible representations of $\mathcal{H}_n$. However, understanding these irreducible representations remains an active area of research to this day. An example of a major open-ended question in the area is the following.

QUESTION 3. *What can be said about the dimensions of D^λ?*

8. The future

My hope for the future is that the definition of B_n as a mapping class group will provide solutions to problems related to the representation theory of the Iwahori-Hecke algebra. One reason for optimism is the mechanism described in [**Big04**] to obtain representations of the $\mathcal{H}_n$ from the induced action of B_n on homology modules of configuration spaces in D_n. There I conjectured that all irreducible representations D^λ can be obtained in this way. The inner product we defined on D^λ would presumably correspond to the intersection form on homology.

Another direction for future research is to look at other quotient algebras of RB_n. After the Iwahori-Hecke algebra, the next obvious candidate is the Birman-Wenzl-Murakami algebra. A somewhat non-standard presentation of this algebra is as follows.

Let X be the following element of the braid group algebra RB_n.

$$X = q\bar{\sigma}_1 + 1 - q - \sigma_1.$$

The Birman-Wenzl-Murakami algebra is the quotient of RB_n by the following relations.

- $(q^2\sigma_1^{-1}\sigma_2^{-1} - \sigma_1\sigma_2)X = 0$,
- $(q\sigma_2^{-1} + 1 - q - \sigma_2)X = (q\sigma_1^{-1}\sigma_2^{-1} - \sigma_1\sigma_2)X$,
- $\sigma_1 X = tX$.

This algebra has an interesting history. After the discovery of the Jones polynomial, Kauffman [**Kau90**] discovered a new knot invariant which he defined directly using the knot or link diagram and a skein relation. The Birman-Wenzl-Murakami algebra was then constructed in [**BW89**], and independently in [**Mur87**], so as to give the Kauffman polynomial via a trace function. Thus the history of the Birman-Wenzl-Murakami algebra traces the history of the Jones polynomial in reverse.

QUESTION 4. *How much of this paper can be generalized to the Birman-Wenzl-Murakami algebra?*

In this direction, John Enyang [**Eny04**] has shown that the Birman-Murakami-Wenzl algebra is a cellular algebra, and used this to give a definition of its irreducible representations similar to the approach in Section 7.

Next we would like to generalize [**Big04**] to the Birman-Wenzl-Murakami algebra.

QUESTION 5. *Is there a homological definition of representations of the Birman-Wenzl-Murakami algebra?*

I believe the answer to this is yes. Furthermore, the homological construction suggests a new algebra Z_n, which would further generalize the Iwahori-Hecke and Birman-Wenzl-Murakami algebras. I will conclude this paper with a definition of Z_n and some related open questions. I hope these might be amenable to some combinatorial computations, even without the homological motivation, which is currently unclear and unpublished.

We use the notation $\sigma_{i_1\ldots i_k}$ as shorthand for $\sigma_{i_1}\ldots\sigma_{i_k}$, and $\bar{\sigma}_{i_1\ldots i_k}$ for $\sigma_{i_1\ldots i_k}^{-1}$. Define the following elements of RB_n:

$$X_2 = q\bar{\sigma}_1 + 1 - q - \sigma_1$$
$$X_3 = (q^2\bar{\sigma}_{21} - \sigma_{12})X_2$$
$$X_4 = (q^3\bar{\sigma}_{321} - \sigma_{123})X_3$$
$$\vdots$$
$$X_n = (q^{n-1}\bar{\sigma}_{(n-1)\ldots 1} - \sigma_{1\ldots(n-1)})X_{n-1}.$$

Then Z_n is the algebra RB_n modulo the following relations:

$$(q\bar{\sigma}_2 + 1 - q - \sigma_2)X_2 = (q\bar{\sigma}_{21} - \sigma_{12})X_2,$$
$$(q^2\bar{\sigma}_{32} - \sigma_{23})X_3 = (q^2\bar{\sigma}_{321} - \sigma_{123})X_3,$$
$$(q^3\bar{\sigma}_{432} - \sigma_{234})X_4 = (q^3\bar{\sigma}_{4321} - \sigma_{1234})X_4,$$
$$\vdots$$
$$(q^{n-1}\bar{\sigma}_{n\ldots 2} - \sigma_{2\ldots n})X_n = (q^{n-1}\bar{\sigma}_{n\ldots 1} - \sigma_{1\ldots n})X_n.$$

Note that $\mathcal{H}_n(1,-q)$ is the quotient of Z_n by the relation $X_2 = 0$. Also the Birman-Wenzl-Murakami algebra is the quotient of Z_n by the relations $X_3 = 0$ and $\sigma_1 X_2 = tX_2$. The following basically asks if Z_n is bigger than the Birman-Wenzl-Murakami algebra.

QUESTION 6. *Does X_3 equal 0 in Z_n?*

Presumably some extra relations should be added to Z_n, such as $\sigma_1 X_2 = tX_2$, or something more general.

QUESTION 7. *What extra relations should be added to Z_n to make it finite-dimensional?*

QUESTION 8. *How much of this paper can be generalized to Z_n?*

It might be easier to first study these questions for the quotient of Z_n by the relation $X_4 = 0$.

References

[Bar03] John W. Barrett, *Geometrical measurements in three-dimensional quantum gravity*, Proceedings of the Tenth Oporto Meeting on Geometry, Topology and Physics (2001), vol. 18, 2003, pp. 97–113. MR2029691 (2005b:83034)

[BB] Joan S. Birman and Tara E. Brendle, *Braids: A survey*, Handbook of knot theory, 19–103, Elsevier B. V., Amsterdam, 2005. MR2179260

[Big99] Stephen Bigelow, *The Burau representation is not faithful for $n = 5$*, Geom. Topol. **3** (1999), 397–404 (electronic). MR1725480 (2001j:20055)

[Big02] _____, *Does the Jones polynomial detect the unknot?*, J. Knot Theory Ramifications **11** (2002), no. 4, 493–505, Knots 2000 Korea, Vol. 2 (Yongpyong). MR1915491 (2003c:57010)

[Big04] _____, *Homological representations of the Iwahori-Hecke algebra*, Geometry and Topology Monographs **7** (2004), 493–507.

[Bir74] Joan S. Birman, *Braids, links, and mapping class groups*, Princeton University Press, Princeton, N.J., 1974, Annals of Mathematics Studies, No. 82.

[BM] Joan S. Birman and William W. Menasco, *Stabilization in the braid groups-II: Transversal simplicity of knots.*

[BW89]   Joan S. Birman and Hans Wenzl, *Braids, link polynomials and a new algebra*, Trans. Amer. Math. Soc. **313** (1989), no. 1, 249–273.
[CL97]   D. Cooper and D. D. Long, *A presentation for the image of* Burau(4) $\otimes$ Z_2, Invent. Math. **127** (1997), no. 3, 535–570. MR1431138 (97m:20050)
[CL98]   _____, *On the Burau representation modulo a small prime*, The Epstein birthday schrift, Geom. Topol. Monogr., vol. 1, Geom. Topol. Publ., Coventry, 1998, pp. 127–138 (electronic). MR1668343 (99k:20077)
[Dip85]   Richard Dipper, *On the decomposition numbers of the finite general linear groups. II*, Trans. Amer. Math. Soc. **292** (1985), no. 1, 123–133. MR805956 (87c:20028)
[DJ86]   Richard Dipper and Gordon James, *Representations of Hecke algebras of general linear groups*, Proc. London Math. Soc. (3) **52** (1986), no. 1, 20–52. MR88b:20065
[EH]   John B. Etnyre and Ko Honda, *Cabling and transverse simplicity*, Ann. of Math. (2) **162** (2005), no. 3, 1305–1333. MR2179731
[Eny04]   John Enyang, *Cellular bases for the Brauer and Birman-Murakami-Wenzl algebras*, J. Algebra **281** (2004), no. 2, 413–449. MR2098377
[FYH+85]   P. Freyd, D. Yetter, J. Hoste, W. B. R. Lickorish, K. Millett, and A. Ocneanu, *A new polynomial invariant of knots and links*, Bull. Amer. Math. Soc. (N.S.) **12** (1985), no. 2, 239–246.
[GL96]   J. J. Graham and G. I. Lehrer, *Cellular algebras*, Invent. Math. **123** (1996), no. 1, 1–34. MR1376244 (97h:20016)
[HK90]   Jim Hoste and Mark E. Kidwell, *Dichromatic link invariants*, Trans. Amer. Math. Soc. **321** (1990), no. 1, 197–229. MR961623 (90m:57007)
[Jon85]   Vaughan F. R. Jones, *A polynomial invariant for knots via von Neumann algebras*, Bull. Amer. Math. Soc. (N.S.) **12** (1985), no. 1, 103–111. MR86e:57006
[Kau90]   Louis H. Kauffman, *An invariant of regular isotopy*, Trans. Amer. Math. Soc. **318** (1990), no. 2, 417–471. MR958895 (90g:57007)
[Lon86]   D. D. Long, *A note on the normal subgroups of mapping class groups*, Math. Proc. Cambridge Philos. Soc. **99** (1986), no. 1, 79–87. MR809501 (87c:57009)
[Mur87]   Jun Murakami, *The Kauffman polynomial of links and representation theory*, Osaka J. Math. **24** (1987), no. 4, 745–758.
[Mur92]   G. E. Murphy, *On the representation theory of the symmetric groups and associated Hecke algebras*, J. Algebra **152** (1992), no. 2, 492–513. MR1194316 (94c:17031)
[OS03]   S. Yu. Orevkov and V. V. Shevchishin, *Markov theorem for transversal links*, J. Knot Theory Ramifications **12** (2003), no. 7, 905–913. MR2017961 (2004j:57011)
[PT88]   Józef H. Przytycki and Paweł Traczyk, *Invariants of links of Conway type*, Kobe J. Math. **4** (1988), no. 2, 115–139. MR945888 (89h:57006)
[Tur88]   V. G. Turaev, *The Conway and Kauffman modules of a solid torus*, Zap. Nauchn. Sem. Leningrad. Otdel. Mat. Inst. Steklov. (LOMI) **167** (1988), no. Issled. Topol. 6, 79–89, 190. MR964255 (90f:57012)

DEPARTMENT OF MATHEMATICS, UNIVERSITY OF CALIFORNIA AT SANTA BARBARA, CALIFORNIA 93106, USA

E-mail address: `bigelow@math.ucsb.edu`

Automorphism Groups of Free Groups, Surface Groups and Free Abelian Groups

Martin R. Bridson and Karen Vogtmann

The group of 2×2 matrices with integer entries and determinant ± 1 can be identified either with the group of outer automorphisms of a rank two free group or with the group of isotopy classes of homeomorphisms of a 2-dimensional torus. Thus this group is the beginning of three natural sequences of groups, namely the general linear groups $GL(n, \mathbb{Z})$, the groups $\text{Out}(F_n)$ of outer automorphisms of free groups of rank $n \geq 2$, and the mapping class groups $\text{Mod}^\pm(S_g)$ of orientable surfaces of genus $g \geq 1$. Much of the work on mapping class groups and automorphisms of free groups is motivated by the idea that these sequences of groups are strongly analogous, and should have many properties in common. This program is occasionally derailed by uncooperative facts but has in general proved to be a successful strategy, leading to fundamental discoveries about the structure of these groups. In this article we will highlight a few of the most striking similarities and differences between these series of groups and present some open problems motivated by this philosophy.

Similarities among the groups $\text{Out}(F_n)$, $GL(n, \mathbb{Z})$ and $\text{Mod}^\pm(S_g)$ begin with the fact that these are the outer automorphism groups of the most primitive types of torsion-free discrete groups, namely free groups, free abelian groups and the fundamental groups of closed orientable surfaces $\pi_1 S_g$. In the case of $\text{Out}(F_n)$ and $GL(n, \mathbb{Z})$ this is obvious, in the case of $\text{Mod}^\pm(S_g)$ it is a classical theorem of Nielsen. In all cases there is a *determinant* homomorphism to $\mathbb{Z}/2$; the kernel of this map is the group of "orientation-preserving" or "special" automorphisms, and is denoted $\text{SOut}(F_n)$, $SL(n, Z)$ or $\text{Mod}(S_g)$ respectively.

1. Geometric and topological models

A natural geometric context for studying the global structure of $GL(n, \mathbb{Z})$ is provided by the symmetric space X of positive-definite, real symmetric matrices of determinant 1 (see [**78**] for a nice introduction to this subject). This is a non-positively curved manifold diffeomorphic to $\mathbb{R}^d$, where $d = \frac{1}{2}n(n+1) - 1$. $GL(n, \mathbb{Z})$ acts properly by isometries on X with a quotient of finite volume.

Each $A \in X$ defines an inner product on $\mathbb{R}^n$ and hence a Riemannian metric ν of constant curvature and volume 1 on the n-torus $T^n = \mathbb{R}^n/\mathbb{Z}^n$. One can recover A from the metric ν and an ordered basis for $\pi_1 T^n$. Thus X is homeomorphic to the space of equivalence classes of *marked* Euclidean tori (T^n, ν) of volume 1, where a *marking* is a homotopy class of homeomorphisms $\rho : T^n \to (T^n, \nu)$ and two marked tori are considered equivalent if there is an isometry $i : (T_1^n, \nu_1) \to (T_2^n, \nu_2)$ such that $\rho_2^{-1} \circ i \circ \rho_1$ is homotopic to the identity. The natural action of $\mathrm{GL}(n, \mathbb{Z}) = \mathrm{Out}(\mathbb{Z}^n)$ on $T^n = K(\mathbb{Z}^n, 1)$ twists the markings on tori, and when one traces through the identifications this is the standard action on X.

If one replaces T^n by S_g and follows exactly this formalism with marked metrics of constant curvature[1] and fixed volume, then one arrives at the definition of *Teichmüller space* and the natural action of $\mathrm{Mod}^{\pm}(S_g) = \mathrm{Out}(\pi_1 S_g)$ on it. Teichmüller space is again homeomorphic to a Euclidean space, this time $\mathbb{R}^{6g-6}$.

In the case of $\mathrm{Out}(F_n)$ there is no canonical choice of classifying space $K(F_n, 1)$ but rather a finite collection of natural models, namely the finite graphs of genus n with no vertices of valence less than 3. Nevertheless, one can proceed in essentially the same way: one considers metrics of fixed volume (sum of the lengths of edges =1) on the various models for $K(F_n, 1)$, each equipped with a marking, and one makes the obvious identifications as the homeomorphism type of a graph changes with a sequence of metrics that shrink an edge to length zero. The space of marked metric structures obtained in this case is Culler and Vogtmann's Outer space [27], which is stratified by manifold subspaces corresponding to the different homeomorphism types of graphs that arise. This space is not a manifold, but it is contractible and its local homotopical structure is a natural generalization of that for a manifold (cf. [80]).

One can also learn a great deal about the group $\mathrm{GL}(n, \mathbb{Z})$ by examining its actions on the Borel-Serre bordification of the symmetric space X and on the spherical Tits building, which encodes the asymptotic geometry of X. Teichmüller space and Outer space both admit useful bordifications that are closely analogous to the Borel-Serre bordification [44, 53, 2]. And in place of the spherical Tits building for $\mathrm{GL}(n, \mathbb{Z})$ one has the complex of curves [46] for $\mathrm{Mod}^{\pm}(S_g)$, which has played an important role in recent advances concerning the large scale geometry of $\mathrm{Mod}^{\pm}(S_g)$. For the moment this complex has no well-established counterpart in the context of $\mathrm{Out}(F_n)$.

These closely parallel descriptions of geometries for the three families of groups have led mathematicians to try to push the analogies further, both for the geometry and topology of the "symmetric spaces" and for purely group-theoretic properties that are most naturally proved using the geometry of the symmetric space. For example, the symmetric space for $\mathrm{GL}(n, \mathbb{Z})$ admits a natural equivariant deformation retraction onto an $n(n-1)/2$-dimensional cocompact subspace, the *well-rounded retract* [1]. Similarly, both Outer space and the Teichmüller space of a punctured or bounded orientable surface retract equivariantly onto cocompact simplicial spines [27, 44]. In all these cases, the retracts have dimension equal to the virtual cohomological dimension of the relevant group. For closed surfaces, however, the question remains open:

QUESTION 1. *Does the Teichmüller space for S_g admit an equivariant deformation retraction onto a cocompact spine whose dimension is equal to $4g - 5$, the virtual cohomological dimension of $\mathrm{Mod}^{\pm}(S_g)$?*

Further questions of a similar nature are discussed in (2.1).

---

[1]If $g \geq 2$, then the curvature will be negative.

The issues involved in using these symmetric space analogs to prove purely group theoretic properties are illustrated in the proof of the Tits alternative, which holds for all three classes of groups. A group Γ is said to satisfy the Tits alternative if each of its subgroups either contains a non-abelian free group or else is virtually solvable. The strategy for proving this is similar in each of the three families that we are considering: inspired by Tits's original proof for linear groups (such as $\mathrm{GL}(n,\mathbb{Z})$), one attempts to use a ping-pong argument on a suitable boundary at infinity of the symmetric space. This strategy ultimately succeeds but the details vary enormously between the three contexts, and in the case of $\mathrm{Out}(F_n)$ they are particularly intricate ([**4, 3**] versus [**9**]). One finds that this is often the case: analogies between the three classes of groups can be carried through to theorems, and the architecture of the expected proof is often a good guide, but at a more detailed level the techniques required vary in essential ways from one class to the next and can be of completely different orders of difficulty.

Let us return to problems more directly phrased in terms of the geometry of the symmetric spaces. The symmetric space for $\mathrm{GL}(n,\mathbb{Z})$ has a left-invariant metric of non-positive curvature, the geometry of which is relevant to many areas of mathematics beyond geometric group theory. Teichmüller space has two natural metrics, the Teichmüller metric and the Weyl-Petersen metric, and again the study of each is a rich subject. In contrast, the metric theory of Outer space has not been developed, and in fact there is no obvious candidate for a natural metric. Thus, the following question has been left deliberately vague:

QUESTION 2. *Develop a metric theory of Outer space.*

The elements of infinite order in $\mathrm{GL}(n,\mathbb{Z})$ that are diagonalizable over $\mathbb{C}$ act as loxodromic isometries of X. When $n = 2$, these elements are the hyperbolic matrices; each fixes two points at infinity in $X = \mathbb{H}^2$, one a source and one a sink. The analogous type of element in $\mathrm{Mod}^{\pm}(S_g)$ is a pseudo-Anosov, and in $\mathrm{Out}(F_n)$ it is an *iwip* (irreducible with irreducible powers). In both cases, such elements have two fixed points at infinity (i.e. in the natural boundary of the symmetric space analog), and the action of the cyclic subgroup generated by the element exhibits the north-south dynamics familiar from the action of hyperbolic matrices on the closure of the Poincaré disc [**62**], [**54**]. In the case of $\mathrm{Mod}^{\pm}(S_g)$ this cyclic subgroup leaves invariant a unique geodesic line in Teichmüller space, i.e. pseudo-Anosov's are axial like the semi-simple elements of infinite order in $\mathrm{GL}(n,\mathbb{Z})$. Initial work of Handel and Mosher [**43**] shows that in the case of iwips one cannot hope to have a unique axis in the same metric sense, but leaves open the possibility that there may be a reasonable notion of axis in a weaker sense. (We highlighted this problem in an earlier version of the current article.) In a more recent preprint [**42**] they have addressed this last point directly, defining an *axis bundle* associated to any iwip, cf. [**63**]. Nevertheless, many interesting questions remain (some of which are highlighted by Handel and Mosher). Thus we retain a modified version of our original question:

QUESTION 3. *Describe the geometry of the axis bundle (and associated objects) for an iwip acting on Outer Space.*

2. Actions of $\mathrm{Aut}(F_n)$ and $\mathrm{Out}(F_n)$ on other spaces

Some of the questions that we shall present are more naturally stated in terms of $\mathrm{Aut}(F_n)$ rather than $\mathrm{Out}(F_n)$, while some are natural for both. To avoid redundancy, we shall state only one form of each question.

2.1. Baum-Connes and Novikov conjectures.

Two famous conjectures relating topology, geometry and functional analysis are the Novikov and Baum-Connes conjectures. The Novikov conjecture for closed oriented manifolds with fundamental group Γ says that certain *higher signatures* coming from $H^*(\Gamma; \mathbb{Q})$ are homotopy invariants. It is implied by the Baum-Connes conjecture, which says that a certain *assembly map* between two K-theoretic objects associated to Γ is an isomorphism. Kasparov [57] proved the Novikov conjecture for $\mathrm{GL}(n, \mathbb{Z})$, and Guenther, Higson and Weinberger proved it for all linear groups [40]. The Baum-Connes conjecture for $\mathrm{GL}(n, \mathbb{Z})$ is open when $n \geq 4$ (cf. [61]).

Recently Storm [79] pointed out that the Novikov conjecture for mapping class groups follows from results that have been announced by Hamenstädt [41] and Kato [59], leaving open the following:

QUESTION 4. *Do mapping class groups or* $\mathrm{Out}(F_n)$ *satisfy the Baum-Connes conjecture? Does* $\mathrm{Out}(F_n)$ *satisfy the Novikov conjecture?*

An approach to proving these conjectures is given by work of Rosenthal [75], generalizing results of Carlsson and Pedersen [23]. A contractible space on which a group Γ acts properly and for which the fixed point sets of finite subgroups are contractible is called an $\underline{E}\Gamma$. Rosenthal's theorem says that the Baum-Connes map for Γ is split injective if there is a cocompact $\underline{E}\Gamma = E$ that admits a compactification X, such that

(1) the Γ-action extends to X;
(2) X is metrizable;
(3) X^G is contractible for every finite subgroup G of Γ
(4) E^G is dense in X^G for every finite subgroup G of Γ
(5) compact subsets of E become small near $Y = X \smallsetminus E$ under the Γ-action: for every compact $K \subset E$ and every neighborhood $U \subset X$ of $y \in Y$, there exists a neighborhood $V \subset X$ of y such that $\gamma K \cap V \neq \emptyset$ implies $\gamma K \subset U$.

The existence of such a space E also implies the Novikov conjecture for Γ.

For $\mathrm{Out}(F_n)$ the spine of Outer space mentioned in the previous section is a reasonable candidate for the required $\underline{E}\Gamma$, and there is a similarly defined candidate for $\mathrm{Aut}(F_n)$. For mapping class groups of punctured surfaces the complex of arc systems which fill up the surface is a good candidate (note that this can be identified with a subcomplex of Outer space, as in [47], section 5).

QUESTION 5. *Does there exist a compactification of the spine of Outer space satisfying Rosenthal's conditions? Same question for the complex of arc systems filling a punctured surface.*

In all of the cases mentioned above, the candidate space E has dimension equal to the virtual cohomological dimension of the group. G. Mislin [68] has constructed a cocompact $\underline{E}G$ for the mapping class group of a closed surface, but it has much higher dimension, equal to the dimension of the Teichmüller space. This leads us to a slight variation on Question 1.

QUESTION 6. *Can one construct a cocompact $\underline{E}G$ with dimension equal to the virtual cohomological dimension of the mapping class group of a closed surface?*

2.2. Properties (T) and FA.

A group has Kazdhan's Property (T) if any action of the group by isometries on a Hilbert space has fixed vectors. Kazdhan proved that $\mathrm{GL}(n, \mathbb{Z})$ has property (T) for $n \geq 3$.

QUESTION 7. *For $n > 3$, does $\mathrm{Aut}(F_n)$ have property (T)?*

The corresponding question for mapping class groups is also open. If $\mathrm{Aut}(F_n)$ were to have Property (T), then an argument of Lubotzky and Pak [64] would provide a conceptual explanation of the apparently-unreasonable effectiveness of certain algorithms in computer science, specifically the Product Replacement Algorithm of Leedham-Green *et al.*

If a group has Property (T) then it has Serre's property FA: every action of the group on an $\mathbb{R}$-tree has a fixed point. When $n \geq 3$, $\mathrm{GL}(n, \mathbb{Z})$ has property FA, as do $\mathrm{Aut}(F_n)$ and $\mathrm{Out}(F_n)$, and mapping class groups in genus ≥ 3 (see [28]). In contrast, McCool [67] has shown that $\mathrm{Aut}(F_3)$ has a subgroup of finite-index with positive first betti number, i.e. a subgroup which maps onto $\mathbb{Z}$. In particular this subgroup acts by translations on the line and therefore does not have property FA or (T). Since property (T) passes to finite-index subgroups, it follows that $\mathrm{Aut}(F_3)$ does not have property (T).

QUESTION 8. *For $n > 3$, does $\mathrm{Aut}(F_n)$ have a subgroup of finite index with positive first betti number?*

Another finite-index subgroup of $\mathrm{Aut}(F_3)$ mapping onto $\mathbb{Z}$ was constructed by Alex Lubotzky, and was explained to us by Andrew Casson. Regard F_3 as the fundamental group of a graph R with one vertex. The single-edge loops provide a basis $\{a, b, c\}$ for F_3. Consider the 2-sheeted covering $\hat{R} \to R$ with fundamental group $\langle a, b, c^2, cac^{-1}, cbc^{-1}\rangle$ and let $G \subset \mathrm{Aut}(F_3)$ be the stabilizer of this subgroup. G acts on $H_1(\hat{R}, \mathbb{Q})$ leaving invariant the eigenspaces of the involution that generates the Galois group of the covering. The eigenspace corresponding to the eigenvalue -1 is two dimensional with basis $\{a - cac^{-1}, b - cbc^{-1}\}$. The action of G with respect to this basis gives an epimorphism $G \to \mathrm{GL}(2, \mathbb{Z})$. Since $\mathrm{GL}(2, \mathbb{Z})$ has a free subgroup of finite-index, we obtain a subgroup of finite index in $\mathrm{Aut}(F_3)$ that maps onto a non-abelian free group.

One can imitate the essential features of this construction with various other finite-index subgroups of F_n, thus producing subgroups of finite index in $\mathrm{Aut}(F_n)$ that map onto $\mathrm{GL}(m, \mathbb{Z})$. In each case one finds that $m \geq n - 1$.

QUESTION 9. *If there is a homomorphism from a subgroup of finite index in $\mathrm{Aut}(F_n)$ onto a subgroup of finite index in $\mathrm{GL}(m, \mathbb{Z})$, then must $m \geq n - 1$?*

Indeed one might ask:

QUESTION 10. *If $m < n - 1$ and $H \subset \mathrm{Aut}(F_n)$ is a subgroup of finite index, then does every homomorphism $H \to \mathrm{GL}(m, \mathbb{Z})$ have finite image?*

Similar questions are interesting for the other groups in our families (cf. section 3). For example, if $m < n - 1$ and $H \subset \mathrm{Aut}(F_n)$ is a subgroup of finite index, then does every homomorphism $H \to \mathrm{Aut}(F_m)$ have finite image?

A positive answer to the following question would answer Question 8; a negative answer would show that $\mathrm{Aut}(F_n)$ does not have property (T).

QUESTION 11. *For $n \geq 4$, do subgroups of finite index in $\mathrm{Aut}(F_n)$ have Property FA?*

A promising approach to this last question breaks down because we do not know the answer to the following question.

QUESTION 12. *Fix a basis for F_n and let $A_{n-1} \subset \operatorname{Aut}(F_n)$ be the copy of $\operatorname{Aut}(F_{n-1})$ corresponding to the first $n-1$ basis elements. Let $\phi : \operatorname{Aut}(F_n) \to G$ be a homomorphism of groups. If $\phi(A_{n-1})$ is finite, must the image of ϕ be finite?*

Note that the obvious analog of this question for $\operatorname{GL}(n, \mathbb{Z})$ has a positive answer and plays a role in the foundations of algebraic K-theory.

A different approach to establishing Property (T) was developed by Zuk [85]. He established a combinatorial criterion on the links of vertices in a simply connected G-complex which, if satisfied, implies that G has property (T): one must show that the smallest positive eigenvalue of the discrete Laplacian on links is sufficiently large. One might hope to apply this criterion to one of the natural complexes on which $\operatorname{Aut}(F_n)$ and $\operatorname{Out}(F_n)$ act, such as the spine of Outer space. But David Fisher has pointed out to us that the results of Izeki and Natayani [55] (alternatively, Schoen and Wang – unpublished) imply that such a strategy cannot succeed.

2.3. Actions on CAT(0) spaces.

An $\mathbb{R}$-tree may be defined as a complete CAT(0) space of dimension[2] 1. Thus one might generalize property FA by asking, for each $d \in \mathbb{N}$, which groups must fix a point whenever they act by isometries on a complete CAT(0) space of dimension $\leq d$.

QUESTION 13. *What is the least integer δ such that $\operatorname{Out}(F_n)$ acts without a global fixed point on a complete CAT(0) space of dimension δ? And what is the least dimension for the mapping class group $\operatorname{Mod}^{\pm}(S_g)$?*

The action of $\operatorname{Out}(F_n)$ on the first homology of F_n defines a map from $\operatorname{Out}(F_n)$ to $\operatorname{GL}(n,\mathbb{Z})$ and hence an action of $\operatorname{Out}(F_n)$ on the symmetric space of dimension $\frac{1}{2}n(n+1) - 1$. This action does not have a global fixed point and hence we obtain an upper bound on δ. On the other hand, since $\operatorname{Out}(F_n)$ has property FA, $\delta \geq 2$. In fact, motivated by work of Farb on $\operatorname{GL}(n,\mathbb{Z})$, Bridson [14] has shown that using a Helly-type theorem and the structure of finite subgroups in $\operatorname{Out}(F_n)$, one can obtain a lower bound on δ that grows as a linear function of n. Note that a lower bound of $3n - 3$ on δ would imply that Outer Space did not support a complete $\operatorname{Out}(F_n)$-equivariant metric of non-positive curvature.

If X is a CAT(0) polyhedral complex with only finitely many isometry types of cells (e.g. a finite dimensional cube complex), then each isometry of X is either elliptic (fixes a point) or hyperbolic (has an axis of translation) [15]. If $n \geq 4$ then a variation on an argument of Gersten [36] shows that in any action of $\operatorname{Out}(F_n)$ on X, no Nielsen generator can act as a hyperbolic isometry.

QUESTION 14. *If $n \geq 4$, then can $\operatorname{Out}(F_n)$ act without a global fixed point on a finite-dimensional CAT(0) cube complex?*

2.4. Linearity.

Formanek and Procesi [33] proved that $\operatorname{Aut}(F_n)$ is not linear for $n \geq 3$ by showing that $\operatorname{Aut}(F_3)$ contains a "poison subgroup", i.e. a subgroup which has no faithful linear representation.

Since $\operatorname{Aut}(F_n)$ embeds in $\operatorname{Out}(F_{n+1})$, this settles the question of linearity for $\operatorname{Out}(F_n)$ as well, except when $n = 3$.

QUESTION 15. *Does $\operatorname{Out}(F_3)$ have a faithful representation into $\operatorname{GL}(m, \mathbb{C})$ for some $m \in \mathbb{N}$?*

---
[2] topological covering dimension

Note that braid groups are linear [8] but it is unknown if mapping class groups of closed surfaces are. Brendle and Hamidi-Tehrani [13] showed that the approach of Formanek and Procesi cannot be adapted directly to the mapping class groups. More precisely, they prove that the type of "poison subgroup" described above does not arise in mapping class groups.

The fact that the above question remains open is an indication that $\mathrm{Out}(F_3)$ can behave differently from $\mathrm{Out}(F_n)$ for n large; the existence of finite index subgroups mapping onto $\mathbb{Z}$ was another instance of this, and we shall see another in our discussion of automatic structures and isoperimetric inequalities.

3. Maps to and from $\mathrm{Out}(F_n)$

A particularly intriguing aspect of the analogy between $\mathrm{GL}(n,\mathbb{Z})$ and the two other classes of groups is the extent to which the celebrated rigidity phenomena for lattices in higher rank semisimple groups transfer to mapping class groups and $\mathrm{Out}(F_n)$. Many of the questions in this section concern aspects of this rigidity; questions 9 to 11 should also be viewed in this light.

Bridson and Vogtmann [21] showed that any homomorphism from $\mathrm{Aut}(F_n)$ to a group G has finite image if G does not contain the symmetric group Σ_{n+1}; in particular, any homomorphism $\mathrm{Aut}(F_n) \to \mathrm{Aut}(F_{n-1})$ has image of order at most 2.

QUESTION 16. *If $n \geq 4$ and $g \geq 1$, does every homomorphism from $\mathrm{Aut}(F_n)$ to $\mathrm{Mod}^{\pm}(S_g)$ have finite image?*

By [21], one cannot obtain homomorphisms with infinite image unless $\mathrm{Mod}^{\pm}(S_g)$ contains the symmetric group Σ_{n+1}. For large enough genus, you can realize any symmetric group; but the order of a finite group of symmetries is at most 84g-6, so here one needs $84g - 6 \geq (n+1)!$.

There are no *injective* maps from $\mathrm{Aut}(F_n)$ to mapping class groups. This follows from the result of Brendle and Hamidi-Tehrani that we quoted earlier. For certain g one can construct homomorphisms $\mathrm{Aut}(F_3) \to \mathrm{Mod}^{\pm}(S_g)$ with infinite image, but we do not know the minimal such g.

QUESTION 17. *Let Γ be an irreducible lattice in a semisimple Lie group of $\mathbb{R}$-rank at least 2. Does every homomorphism from Γ to $\mathrm{Out}(F_n)$ have finite image?*

This is known for non-uniform lattices (see [16]; it follows easily from the Kazdhan-Margulis finiteness theorem and the fact that solvable subgroups of $\mathrm{Out}(F_n)$ are virtually abelian [5]). Farb and Masur provided a positive answer to the analogous question for maps to mapping class groups [32]. The proof of their theorem was based on results of Kaimanovich and Masur [56] concerning random walks on Teichmüller space. (See [54] and, for an alternative approach, [6].)

QUESTION 18. *Is there a theory of random walks on Outer space similar to that of Kaimanovich and Masur for Teichmüller space?*

Perhaps the most promising approach to Question 17 is via bounded cohomology, following the template of Bestvina and Fujiwara's work on subgroups of the mapping class group [6].

QUESTION 19. *If a subgroup $G \subset \mathrm{Out}(F_n)$ is not virtually abelian, then is $H_b^2(G;\mathbb{R})$ infinite dimensional?*

If $m \geq n$ then there are obvious embeddings $\mathrm{GL}(n,\mathbb{Z}) \to \mathrm{GL}(m,\mathbb{Z})$ and $\mathrm{Aut}(F_n) \to \mathrm{Aut}(F_m)$, but there are no obvious embeddings $\mathrm{Out}(F_n) \to \mathrm{Out}(F_m)$. Bogopolski and Puga [10] have shown that, for $m = 1 + (n-1)kn$, where k is an arbitrary natural number coprime to $n-1$, there is in fact an embedding, by restricting automorphisms to a suitable characteristic subgroup of F_m.

QUESTION 20. *For which values of m does $\mathrm{Out}(F_n)$ embed in $\mathrm{Out}(F_m)$? What is the minimal such m, and is it true for all sufficiently large m?*

It has been shown that when n is sufficiently large with respect to i, the homology group $H_i(\mathrm{Out}(F_n),\mathbb{Z})$ is independent of n [50, 51].

QUESTION 21. *Is there a map $\mathrm{Out}(F_n) \to \mathrm{Out}(F_m)$ that induces an isomorphism on homology in the stable range?*

A number of the questions in this section and (2.2) ask whether certain quotients of $\mathrm{Out}(F_n)$ or $\mathrm{Aut}(F_n)$ are necessarily finite. The following quotients arise naturally in this setting: define $Q(n,m)$ to be the quotient of $\mathrm{Aut}(F_n)$ by the normal closure of λ^m, where λ is the Nielsen move defined on a basis $\{a_1,\ldots,a_n\}$ by $a_1 \mapsto a_2 a_1$. (All such Nielsen moves are conjugate in $\mathrm{Aut}(F_n)$, so the choice of basis does not alter the quotient.)

The image of a Nielsen move in $\mathrm{GL}(n,\mathbb{Z})$ is an elementary matrix and the quotient of $\mathrm{GL}(n,\mathbb{Z})$ by the normal subgroup generated by the m-th powers of the elementary matrices is the finite group $\mathrm{GL}(n,\mathbb{Z}/m)$. But Bridson and Vogtmann [21] showed that if m is sufficiently large then $Q(n,m)$ is infinite because it has a quotient that contains a copy of the free Burnside group $B(n-1,m)$. Some further information can be gained by replacing $B(n-1,m)$ with the quotients of F_n considered in subsection 39.3 of A.Yu. Ol'shanskii's book [73]. But we know very little about the groups $Q(n,m)$. For example:

QUESTION 22. *For which values of n and m is $Q(n,m)$ infinite? Is $Q(3,5)$ infinite?*

QUESTION 23. *Can $Q(n,m)$ have infinitely many finite quotients? Is it residually finite?*

4. Individual elements and mapping tori

Individual elements $\alpha \in \mathrm{GL}(n,\mathbb{Z})$ can be realized as diffeomorphisms $\hat{\alpha}$ of the n-torus, while individual elements $\psi \in \mathrm{Mod}^{\pm}(S_g)$ can be realized as diffeomorphisms $\hat{\psi}$ of the surface S_g. Thus one can study α via the geometry of the torus bundle over $\mathbb{S}^1$ with holonomy $\hat{\alpha}$ and one can study ψ via the geometry of the 3-manifold that fibres over $\mathbb{S}^1$ with holonomy $\hat{\psi}$. (In each case the manifold depends only on the conjugacy class of the element.)

The situation for $\mathrm{Aut}(F_n)$ and $\mathrm{Out}(F_n)$ is more complicated: the natural choices of classifying space $Y = K(F_n,1)$ are finite graphs of genus n, and no element of infinite order $\phi \in \mathrm{Out}(F_n)$ is induced by the action on $\pi_1(Y)$ of a homeomorphism of Y. Thus the best that one can hope for in this situation is to identify a graph Y_ϕ that admits a homotopy equivalence inducing ϕ and that has additional structure well-adapted to ϕ. One would then form the mapping torus of this homotopy equivalence to get a good classifying space for the algebraic mapping torus $F_n \rtimes_\phi \mathbb{Z}$.

The *train track technology* of Bestvina, Feighn and Handel [**7, 4, 3**] is a major piece of work that derives suitable graphs Y_ϕ with additional structure encoding key properties of ϕ. This results in a decomposition theory for elements of $\operatorname{Out}(F_n)$ that is closely analogous to (but more complicated than) the Nielsen-Thurston theory for surface automorphisms. Many of the results mentioned in this section are premised on a detailed knowledge of this technology and one expects that a resolution of the questions will be too.

There are several natural ways to define the *growth* of an automorphism ϕ of a group G with finite generating set A; in the case of free, free-abelian, and surface groups these are all asymptotically equivalent. The most easily defined growth function is $\gamma_\phi(k)$ where $\gamma_\phi(k) := \max\{d(1, \phi^k(a)) \mid a \in A\}$. If $G = \mathbb{Z}^n$ then $\gamma_\phi(k) \simeq k^d$ for some integer $d \leq n-1$, or else $\gamma_\phi(k)$ grows exponentially. If G is a surface group, the Nielsen-Thurston theory shows that only bounded, linear and exponential growth can occur. If $G = F_n$ and $\phi \in \operatorname{Aut}(F_n)$ then, as in the abelian case, $\gamma_\phi(k) \simeq k^d$ for some integer $d \leq n-1$ or else $\gamma_\phi(k)$ grows exponentially.

QUESTION 24. *Can one detect the growth of a surface or free-group homomorphism by its action on the homology of a characteristic subgroup of finite index?*

Notice that one has to pass to a subgroup of finite index in order to have any hope because automorphisms of exponential growth can act trivially on homology. A. Piggott [**74**] has answered the above question for free-group automorphisms of polynomial growth, and linear-growth automorphisms of surfaces are easily dealt with, but the exponential case remains open in both settings.

Finer questions concerning growth are addressed in the on-going work of Handel and Mosher [**43**]. They explore, for example, the implications of the following contrast in behaviour between surface automorphisms and free-group automorphisms: in the surface case the exponential growth rate of a pseudo-Anosov automorphism is the same as that of its inverse, but this is not the case for iwip free-group automorphisms.

For mapping tori of automorphisms of free abelian groups $G = \mathbb{Z}^n \rtimes_\phi \mathbb{Z}$, the following conditions are equivalent (see [**17**]): G is automatic; G is a CAT(0) group[3]; G satisfies a quadratic isoperimetric inequality. In the case of mapping tori of surface automorphisms, all mapping tori satisfy the first and last of these conditions and one understands exactly which $S_g \rtimes \mathbb{Z}$ are CAT(0) groups.

Brady, Bridson and Reeves [**12**] show that there exist mapping tori of free-group automorphisms $F \rtimes \mathbb{Z}$ that are not automatic, and Gersten showed that some are not CAT(0) groups [**36**]. On the other hand, many such groups do have these properties, and they all satisfy a quadratic isoperimetric inequality [**18**].

QUESTION 25. *Classify those $\phi \in \operatorname{Aut}(F_n)$ for which $F_n \rtimes_\phi \mathbb{Z}$ is automatic and those for which it is CAT(0).*

Of central importance in trying to understand mapping tori is:

QUESTION 26. *Is there an alogrithm to decide isomorphism among groups of the form $F \rtimes \mathbb{Z}$.*

In the purest form of this question one is given the groups as finite presentations, so one has to address issues of how to find the decomposition $F \rtimes \mathbb{Z}$ and one has to

---

[3]This means that G acts properly and cocompactly by isometries on a CAT(0) space.

combat the fact that this decomposition may not be unique. But the heart of any solution should be an answer to:

QUESTION 27. *Is the conjugacy problem solvable in* $\operatorname{Out}(F_n)$?

Martin Lustig posted a detailed outline of a solution to this problem on his web page some years ago [65], but neither this proof nor any other has been accepted for publication. This problem is of central importance to the field and a clear, compelling solution would be of great interest. The conjugacy problem for mapping class groups was shown to be solvable by Hemion [52], and an effective algorithm for determining conjugacy, at least for pseudo-Anosov mapping classes, was given by Mosher [70]. The isomorphism problem for groups of the form $S_g \rtimes \mathbb{Z}$ can be viewed as a particular case of the solution to the isomorphism problem for fundamental groups of geometrizable 3-manifolds [76]. The solvability of the conjugacy problem for $\operatorname{GL}(n, \mathbb{Z})$ is due to Grunewald [39]

5. Cohomology

In each of the series of groups $\{\Gamma_n\}$ we are considering, the ith homology of Γ_n has been shown to be independent of n for n sufficiently large. For $\operatorname{GL}(n, \mathbb{Z})$ this is due to Charney [24], for mapping class groups to Harer [45], for $\operatorname{Aut}(F_n)$ and $\operatorname{Out}(F_n)$ to Hatcher and Vogtmann [48, 50], though for $\operatorname{Out}(F_n)$ this requires an erratum by Hatcher, Vogtmann and Wahl [51]. With trivial rational coefficients, the stable cohomology of $\operatorname{GL}(n, \mathbb{Z})$ was computed in the 1970's by Borel [11], and the stable rational cohomology of the mapping class group computed by Madsen and Weiss in 2002 [66]. The stable rational cohomology of $\operatorname{Aut}(F_n)$ (and $\operatorname{Out}(F_n)$) was very recently determined by S. Galatius [34] to be trivial.

The exact stable range for trivial rational coefficients is known for $\operatorname{GL}(n, \mathbb{Z})$ and for mapping class groups of punctured surfaces. For $\operatorname{Aut}(F_n)$ the best known result is that the ith homology is independent of n for $n > 5i/4$ [49], but the exact range is unknown:

QUESTION 28. *Where precisely does the rational homology of* $\operatorname{Aut}(F_n)$ *stabilize? And for* $\operatorname{Out}(F_n)$?

There are only two known non-trivial classes in the (unstable) rational homology of $\operatorname{Out}(F_n)$ [49, 26]. However, Morita [69] has defined an infinite series of cycles, using work of Kontsevich which identifies the homology of $\operatorname{Out}(F_n)$ with the cohomology of a certain infinite-dimensional Lie algebra. The first of these cycles is the generator of $H_4(\operatorname{Out}(F_4); \mathbb{Q}) \cong \mathbb{Q}$, and Conant and Vogtmann showed that the second also gives a non-trivial class, in $H_8(\operatorname{Out}(F_6); \mathbb{Q})$ [26]. Both Morita and Conant-Vogtmann also defined more general cycles, parametrized by odd-valent graphs.

QUESTION 29. *Are Morita's original cycles non-trivial in homology? Are the generalizations due to Morita and to Conant and Vogtmann non-trivial in homology?*

No other classes have been found to date in the homology of $\operatorname{Out}(F_n)$, leading naturally to the question of whether these give all of the rational homology.

QUESTION 30. *Do the Morita classes generate all of the rational homology of* $\operatorname{Out}(F_n)$?

The maximum dimension of a Morita class is about $4n/3$. Morita's cycles lift naturally to $\mathrm{Aut}(F_n)$, and again the first two are non-trivial in homology. By Galatius' result, all of these cycles must eventually disappear under the stabilization map $\mathrm{Aut}(F_n) \to \mathrm{Aut}(F_{n+1})$. Conant and Vogtmann show that in fact they disappear immediately after they appear, i.e. one application of the stabilization map kills them [**25**]. If it is true that the Morita classes generate all of the rational homology of $\mathrm{Out}(F_n)$ then this implies that the stable range is significantly lower than the current bound.

We note that Morita has identified several conjectural relationships between his cycles and various other interesting objects, including the image of the Johnson homomorphism, the group of homology cobordism classes of homology cylinders, and the motivic Lie algebra associated to the algebraic mapping class group (see Morita's article in this volume).

Since the stable rational homology of $\mathrm{Out}(F_n)$ is trivial, the natural maps from mapping class groups to $\mathrm{Out}(F_n)$ and from $\mathrm{Out}(F_n)$ to $\mathrm{GL}(n,\mathbb{Z})$ are of course zero. However, the unstable homology of all three classes of groups remains largely unkown and in the unstable range these maps might well be nontrivial. In particular, we note that $H_8(\mathrm{GL}(6,\mathbb{Z});\mathbb{Q}) \cong \mathbb{Q}$ [**30**]; this leads naturally to the question

QUESTION 31. *Is the image of the second Morita class in $H_8(\mathrm{GL}(6,\mathbb{Z});\mathbb{Q}))$ non-trivial?*

For further discussion of the cohomology of $\mathrm{Aut}(F_n)$ and $\mathrm{Out}(F_n)$ we refer to [**81**].

6. Generators and Relations

The groups we are considering are all finitely generated. In each case, the most natural set of generators consists of a single orientation-reversing generator of order two, together with a collection of simple infinite-order special automorphisms. For $\mathrm{Out}(F_n)$, these special automorphisms are the Nielsen automorphisms, which multiply one generator of F_n by another and leave the rest of the generators fixed; for $\mathrm{GL}(n,\mathbb{Z})$ these are the elementary matrices; and for mapping class groups they are Dehn twists around a small set of non-separating simple closed curves.

These generating sets have a number of important features in common. First, implicit in the description of each is a choice of generating set for the group B on which Γ is acting. In the case of $\mathrm{Mod}^{\pm}(S_g)$ this "basis" can be taken to consist of $2g+1$ simple closed curves representing the standard generators $a_1, b_1, a_2, b_2, \ldots, a_g, b_g$, of $\pi_1(S_g)$ together with $z = a_2^{-1} b_3 a_3 b_3^{-1}$. In the case of $\mathrm{Out}(F_n)$ and $\mathrm{GL}(n,\mathbb{Z})$, the generating set is a basis for F_n and $\mathbb{Z}^n$ respectively.

Note that in the cases $\Gamma = \mathrm{Out}(F_n)$ or $\mathrm{GL}(n,\mathbb{Z})$, the universal property of the underlying free objects $B = F_n$ or $\mathbb{Z}^n$ ensures that Γ acts transitively on the set of preferred generating sets (bases). In the case $B = \pi_1 S_g$, the corresponding result is that any two collections of simple closed curves with the same pattern of intersection numbers and complementary regions are related by a homeomorphism of the surface, hence (at the level of π_1) by the action of Γ.

If we identify $\mathbb{Z}^n$ with the abelianization of F_n and choose bases accordingly, then the action of $\mathrm{Out}(F_n)$ on the abelianization induces a homomorphism $\mathrm{Out}(F_n) \to \mathrm{GL}(n,\mathbb{Z})$ that sends each Nielsen move to the corresponding elementary matrix (and hence is surjective). Correspondingly, the action $\mathrm{Mod}^{\pm}(S_g)$ on the abelianization of $\pi_1 S_g$ yields a homomorphism onto the symplectic group $Sp(2g,\mathbb{Z})$ sending the generators of $\text{Mod}^\pm(S_g)$ given by Dehn twists around the a_i and b_i to transvections. Another common feature of these generating sets is that they all have linear growth (see section 4).

Smaller (but less transparent) generating sets exist in each case. Indeed, B.H. Neumann [**72**] proved that $\text{Aut}(F_n)$ (hence its quotients $\text{Out}(F_n)$ and $\text{GL}(n,\mathbb{Z})$) is generated by just 2 elements when $n \geq 4$. Wajnryb [**83**] proved that this is also true of mapping class groups.

In each case one can also find generating sets consisting of finite order elements, involutions in fact. Zucca showed that $\text{Aut}(F_n)$ can be generated by 3 involutions two of which commute [**84**], and Kassabov, building on work of Farb and Brendle, showed that mapping class groups of large enough genus can be generated by 4 involutions [**58**].

Our groups are also all finitely presented. For $\text{GL}(n,\mathbb{Z})$, or more precisely for $\text{SL}(n,\mathbb{Z})$, there are the classical Steinberg relations, which involve commutators of the elementary matrices. For the special automorphisms $\text{SAut}(F_n)$, Gersten gave a presentation in terms of corresponding commutator relations of the Nielsen generators [**35**]. Finite presentations of the mapping class groups are more complicated. The first was given by Hatcher and Thurston, and worked out explicitly by Wajnryb [**82**].

QUESTION 32. *Is there a set of simple Steinberg-type relations for the mapping class group?*

There is also a presentation of $\text{Aut}(F_n)$ coming from the action of $\text{Aut}(F_n)$ on the subcomplex of Auter space spanned by graphs of degree at most 2. This is simply-connected by [**48**], so Brown's method [**22**] can be used to write down a presentation. The vertex groups are stabilizers of marked graphs, and the edge groups are the stabilizers of pairs consisting of a marked graph and a forest in the graph. The quotient of the subcomplex modulo $\text{Aut}(F_n)$ can be computed explicitly, and one finds that $\text{Aut}(F_n)$ is generated by the (finite) stabilizers of seven specific marked graphs. In addition, all of the relations except two come from the natural inclusions of edge stabilizers into vertex stabilizers, i.e. either including the stabilizer of a pair (graph, forest) into the stabilizer of the graph, or into the stabilizer of the quotient of the graph modulo the forest. Thus the whole group is almost (but not quite) a pushout of these finite subgroups. In the terminology of Haefliger (see [**19**], II.12), the complex of groups is not simple.

QUESTION 33. *Can $\text{Out}(F_n)$ and $\text{Mod}^\pm(S_g)$ be obtained as a pushout of a finite subsystem of their finite subgroups, i.e. is either the fundamental group of a developable simple complex of finite groups on a 1-connected base?*

6.1. IA automorphisms. We conclude with a well-known problem about the kernel $\text{IA}(n)$ of the map from $\text{Out}(F_n)$ to $\text{GL}(n,Z)$. The notation "IA" stands for *identity on the abelianization*; these are (outer) automorphisms of F_n which are the identity on the abelianization Z^n of F_n. Magnus showed that this kernel is finitely generated, and for $n=3$ Krstic and McCool showed that it is not finitely presentable [**60**]. It is also known that in some dimension the homology is not finitely generated [**77**]. But that is the extent of our knowledge of basic finiteness properties.

QUESTION 34. *Establish finiteness properties of the kernel* IA(n) *of the map from* Out(F_n) *to* GL(n, $\mathbb{Z}$). *In particular, determine whether* IA(n) *is finitely presentable for* n > 3.

The subgroup IA(n) is analogous to the Torelli subgroup of the mapping class group of a surface, which also remains quite mysterious in spite of having been extensively studied.

7. Automaticity and Isoperimetric Inequalities

In the foundational text on automatic groups [31], Epstein gives a detailed account of Thurston's proof that if $n \geq 3$ then GL(n, $\mathbb{Z}$) is not automatic. The argument uses the geometry of the symmetric space to obtain an exponential lower bound on the $(n-1)$-dimensional isoperimetric function of GL(n, $\mathbb{Z}$); in particular the Dehn function of GL(3, $\mathbb{Z}$) is shown to be exponential.

Bridson and Vogtmann [20], building on this last result, proved that the Dehn functions of Aut(F_3) and Out(F_3) are exponential. They also proved that for all $n \geq 3$, neither Aut(F_n) nor Out(F_n) is biautomatic. In contrast, Mosher proved that mapping class groups are automatic [71] and Hamenstädt [41] proved that they are biautomatic; in particular these groups have quadratic Dehn functions and satisfy a polynomial isoperimetric inequality in every dimension. Hatcher and Vogtmann [47] obtain an exponential upper bound on the isoperimetric function of Aut(F_n) and Out(F_n) in every dimension.

An argument sketched by Thurston and expanded upon by Gromov [37], [38] (cf. [29]) indicates that the Dehn function of GL(n, $\mathbb{Z}$) is quadratic when $n \geq 4$. More generally, the isoperimetric functions of GL(n, $\mathbb{Z}$) should parallel those of Euclidean space in dimensions $m \leq n/2$.

QUESTION 35. *What are the Dehn functions of* Aut(F_n) *and* Out(F_n) *for* n > 3?

QUESTION 36. *What are the higher-dimensional isoperimetric functions of* GL(n, $\mathbb{Z}$), Aut(F_n) *and* Out(F_n)?

QUESTION 37. *Is* Aut(F_n) *automatic for* n > 3?

References

[1] A. ASH, *Small-dimensional classifying spaces for arithmetic subgroups of general linear groups*, Duke Math. J., 51 (1984), pp. 459–468.

[2] M. BESTVINA AND M. FEIGHN, *The topology at infinity of* Out(F_n), Invent. Math., 140 (2000), pp. 651–692.

[3] M. BESTVINA, M. FEIGHN, AND M. HANDEL, *The Tits alternative for* Out(F_n) *II: A Kolchin type theorem.* arXiv:math.GT/9712218.

[4] ———, *The Tits alternative for* Out(F_n). *I. Dynamics of exponentially-growing automorphisms*, Ann. of Math. (2), 151 (2000), pp. 517–623.

[5] ———, *Solvable subgroups of* Out(F_n) *are virtually Abelian*, Geom. Dedicata, 104 (2004), pp. 71–96.

[6] M. BESTVINA AND K. FUJIWARA, *Bounded cohomology of subgroups of mapping class groups*, Geom. Topol., 6 (2002), pp. 69–89 (electronic).

[7] M. BESTVINA AND M. HANDEL, *Train tracks and automorphisms of free groups*, Ann. of Math. (2), 135 (1992), pp. 1–51.

[8] S. J. BIGELOW, *Braid groups are linear*, J. Amer. Math. Soc., 14 (2001), pp. 471–486 (electronic).

[9] J. S. BIRMAN, A. LUBOTZKY, AND J. MCCARTHY, *Abelian and solvable subgroups of the mapping class groups*, Duke Math. J., 50 (1983), pp. 1107–1120.

[10] O. BOGOPOLSKI AND D. V. PUGA, *Embedding the outer automorphism group* Out(F_n) *of a free group of rank n in the group* Out(F_m) *for* $m > n$. preprint, 2004.

[11] A. BOREL, *Stable real cohomology of arithmetic groups*, Ann. Sci. École Norm. Sup. (4), 7 (1974), pp. 235–272 (1975).

[12] N. BRADY, M. R. BRIDSON, AND L. REEVES, *Free-by-cyclic groups that are not automatic.* preprint 2005.

[13] T. E. BRENDLE AND H. HAMIDI-TEHRANI, *On the linearity problem for mapping class groups*, Algebr. Geom. Topol., 1 (2001), pp. 445–468 (electronic).

[14] M. R. BRIDSON, *Helly's theorem and actions of the automorphism group of a free group.* In preparation.

[15] ———, *On the semisimplicity of polyhedral isometries*, Proc. Amer. Math. Soc., 127 (1999), pp. 2143–2146.

[16] M. R. BRIDSON AND B. FARB, *A remark about actions of lattices on free groups*, Topology Appl., 110 (2001), pp. 21–24. Geometric topology and geometric group theory (Milwaukee, WI, 1997).

[17] M. R. BRIDSON AND S. M. GERSTEN, *The optimal isoperimetric inequality for torus bundles over the circle*, Quart. J. Math. Oxford Ser. (2), 47 (1996), pp. 1–23.

[18] M. R. BRIDSON AND D. GROVES, *Free-group automorphisms, train tracks and the beaded decomposition.* arXiv:math.GR/0507589.

[19] M. R. BRIDSON AND A. HAEFLIGER, *Metric spaces of non-positive curvature*, vol. 319 of Grundlehren der Mathematischen Wissenschaften [Fundamental Principles of Mathematical Sciences], Springer-Verlag, Berlin, 1999.

[20] M. R. BRIDSON AND K. VOGTMANN, *On the geometry of the automorphism group of a free group*, Bull. London Math. Soc., 27 (1995), pp. 544–552.

[21] ———, *Homomorphisms from automorphism groups of free groups*, Bull. London Math. Soc., 35 (2003), pp. 785–792.

[22] K. S. BROWN, *Presentations for groups acting on simply-connected complexes*, J. Pure Appl. Algebra, 32 (1984), pp. 1–10.

[23] G. CARLSSON AND E. K. PEDERSEN, *Controlled algebra and the Novikov conjectures for K- and L-theory*, Topology, 34 (1995), pp. 731–758.

[24] R. M. CHARNEY, *Homology stability of* GL_n *of a Dedekind domain*, Bull. Amer. Math. Soc. (N.S.), 1 (1979), pp. 428–431.

[25] J. CONANT AND K. VOGTMANN, *The Morita classes are stably trivial.* preprint, 2006.

[26] ———, *Morita classes in the homology of automorphism groups of free groups*, Geom. Topol., 8 (2004), pp. 1471–1499 (electronic).

[27] M. CULLER AND K. VOGTMANN, *Moduli of graphs and automorphisms of free groups*, Invent. Math., 84 (1986), pp. 91–119.

[28] ———, *A group-theoretic criterion for property FA*, Proc. Amer. Math. Soc., 124 (1996), pp. 677–683.

[29] C. DRUŢU, *Filling in solvable groups and in lattices in semisimple groups*, Topology, 43 (2004), pp. 983–1033.

[30] P. ELBAZ-VINCENT, H. GANGL, AND C. SOULÉ, *Quelques calculs de la cohomologie de* $GL_N(\mathbb{Z})$ *et de la K-théorie de* $\mathbb{Z}$, C. R. Math. Acad. Sci. Paris, 335 (2002), pp. 321–324.

[31] D. B. A. EPSTEIN, J. W. CANNON, D. F. HOLT, S. V. F. LEVY, M. S. PATERSON, AND W. P. THURSTON, *Word processing in groups*, Jones and Bartlett Publishers, Boston, MA, 1992.

[32] B. FARB AND H. MASUR, *Superrigidity and mapping class groups*, Topology, 37 (1998), pp. 1169–1176.

[33] E. FORMANEK AND C. PROCESI, *The automorphism group of a free group is not linear*, J. Algebra, 149 (1992), pp. 494–499.

[34] S. GALATIUS. in preparation.

[35] S. M. GERSTEN, *A presentation for the special automorphism group of a free group*, J. Pure Appl. Algebra, 33 (1984), pp. 269–279.

[36] ———, *The automorphism group of a free group is not a CAT(0) group*, Proc. Amer. Math. Soc., 121 (1994), pp. 999–1002.

[37] M. GROMOV, *Asymptotic invariants of infinite groups*, in Geometric group theory, Vol. 2 (Sussex, 1991), vol. 182 of London Math. Soc. Lecture Note Ser., Cambridge Univ. Press, Cambridge, 1993, pp. 1–295.

[38] M. GROMOV, *Metric structures for Riemannian and non-Riemannian spaces*, vol. 152 of Progress in Mathematics, Birkhäuser Boston Inc., Boston, MA, 1999. Based on the 1981 French original [MR0682063 (85e:53051)], With appendices by M. Katz, P. Pansu and S. Semmes, Translated from the French by Sean Michael Bates.

[39] F. J. GRUNEWALD, *Solution of the conjugacy problem in certain arithmetic groups*, in Word problems, II (Conf. on Decision Problems in Algebra, Oxford, 1976), vol. 95 of Stud. Logic Foundations Math., North-Holland, Amsterdam, 1980, pp. 101–139.

[40] E. GUENTNER, N. HIGSON, AND S. WEINBURGER, *The Novikov conjecture for linear groups*. preprint, 2003.

[41] U. HAMMENSTADT, *Train tracks and mapping class groups I*. preprint, available at http://www.math.uni-bonn.de/people/ursula/papers.html.

[42] M. HANDEL AND L. MOSHER, *Axes in Outer Space*. arXiv:math.GR/0605355.

[43] ———, *The expansion factors of an outer automorphism and its inverse*. arXiv:math.GR/0410015.

[44] J. L. HARER, *The cohomology of the moduli space of curves*, in Theory of moduli (Montecatini Terme, 1985), vol. 1337 of Lecture Notes in Math., Springer, Berlin, 1988, pp. 138–221.

[45] ———, *Stability of the homology of the moduli spaces of Riemann surfaces with spin structure*, Math. Ann., 287 (1990), pp. 323–334.

[46] W. J. HARVEY, *Boundary structure of the modular group*, in Riemann surfaces and related topics: Proceedings of the 1978 Stony Brook Conference (State Univ. New York, Stony Brook, N.Y., 1978), vol. 97 of Ann. of Math. Stud., Princeton, N.J., 1981, Princeton Univ. Press, pp. 245–251.

[47] A. HATCHER AND K. VOGTMANN, *Isoperimetric inequalities for automorphism groups of free groups*, Pacific J. Math., 173 (1996), pp. 425–441.

[48] ———, *Cerf theory for graphs*, J. London Math. Soc. (2), 58 (1998), pp. 633–655.

[49] ———, *Rational homology of* $\mathrm{Aut}(F_n)$, Math. Res. Lett., 5 (1998), pp. 759–780.

[50] ———, *Homology stability for outer automorphism groups of free groups*, Algebr. Geom. Topol., 4 (2004), pp. 1253–1272 (electronic).

[51] A. HATCHER, K. VOGTMANN, AND N. WAHL, *Erratum to: Homology stability for outer automorphism groups of free groups*, 2006. arXiv math.GR/0603577.

[52] G. HEMION, *On the classification of homeomorphisms of 2-manifolds and the classification of 3-manifolds*, Acta Math., 142 (1979), pp. 123–155.

[53] N. V. IVANOV, *Complexes of curves and Teichmüller modular groups*, Uspekhi Mat. Nauk, 42 (1987), pp. 49–91, 255.

[54] ———, *Mapping class groups*, in Handbook of geometric topology, North-Holland, Amsterdam, 2002, pp. 523–633.

[55] H. IZEKI AND S. NAYATANI, *Combinatorial harmonic maps and discrete-group actions on Hadamard spaces*.

[56] V. A. KAIMANOVICH AND H. MASUR, *The Poisson boundary of the mapping class group*, Invent. Math., 125 (1996), pp. 221–264.

[57] G. G. KASPAROV, *Equivariant kk-theory and the novikov conjecture*, Invent. Math., 91 (1988), pp. 147–201.

[58] M. KASSABOV, *Generating Mapping Class Groups by Involutions*. arXiv:math.GT/0311455.

[59] T. KATO, *Asymptotic Lipschitz maps, combable groups and higher signatures*, Geom. Funct. Anal., 10 (2000), pp. 51–110.

[60] S. KRSTIĆ AND J. MCCOOL, *The non-finite presentability of* $\mathrm{IA}(F_3)$ *and* $\mathrm{GL}_2(\mathbf{Z}[t,t^{-1}])$, Invent. Math., 129 (1997), pp. 595–606.

[61] V. LAFFORGUE, *Une démonstration de la conjecture de Baum-Connes pour les groupes réductifs sur un corps p-adique et pour certains groupes discrets possédant la propriété (T)*, C. R. Acad. Sci. Paris Sér. I Math., 327 (1998), pp. 439–444.

[62] G. LEVITT AND M. LUSTIG, *Irreducible automorphisms of* F_n *have north-south dynamics on compactified outer space*, J. Inst. Math. Jussieu, 2 (2003), pp. 59–72.

[63] J. LOS AND M. LUSTIG, *The set of train track representatives of an irreducible free group automorphism is contractible*. preprint, December 2004.

[64] A. LUBOTZKY AND I. PAK, *The product replacement algorithm and Kazhdan's property (T)*, J. Amer. Math. Soc., 14 (2001), pp. 347–363 (electronic).

[65] M. LUSTIG, *Structure and conjugacy for automorphisms of free groups*. available at http://junon.u-3mrs.fr/lustig/.

[66] I. MADSEN AND M. S. WEISS, *The stable moduli space of Riemann surfaces: Mumford's conjecture.* arXiv:math.AT/0212321.

[67] J. MCCOOL, *A faithful polynomial representation of* Out F_3, Math. Proc. Cambridge Philos. Soc., 106 (1989), pp. 207–213.

[68] G. MISLIN, *An EG for the mapping class group.* Workshop on moduli spaces, Munster 2004.

[69] S. MORITA, *Structure of the mapping class groups of surfaces: a survey and a prospect,* in Proceedings of the Kirbyfest (Berkeley, CA, 1998), vol. 2 of Geom. Topol. Monogr., Geom. Topol. Publ., Coventry, 1999, pp. 349–406 (electronic).

[70] L. MOSHER, *The classification of pseudo-Anosovs,* in Low-dimensional topology and Kleinian groups (Coventry/Durham, 1984), vol. 112 of London Math. Soc. Lecture Note Ser., Cambridge Univ. Press, Cambridge, 1986, pp. 13–75.

[71] ———, *Mapping class groups are automatic,* Ann. of Math. (2), 142 (1995), pp. 303–384.

[72] B. NEUMANN, *Die automorphismengruppe der freien gruppen,* Math. Ann., 107 (1932), pp. 367–386.

[73] A. Y. OL'SHANSKIĬ, *Geometry of defining relations in groups,* vol. 70 of Mathematics and its Applications (Soviet Series), Kluwer Academic Publishers Group, Dordrecht, 1991. Translated from the 1989 Russian original by Yu. A. Bakhturin.

[74] A. PIGGOTT, *Detecting the growth of free group automorphisms by their action on the homology of subgroups of finite index.* arXiv:math.GR/0409319.

[75] D. ROSENTHAL, *Split Injectivity of the Baum-Connes Assembly Map.* arXiv:math.AT/0312047.

[76] Z. SELA, *The isomorphism problem for hyperbolic groups. I,* Ann. of Math. (2), 141 (1995), pp. 217–283.

[77] J. SMILLIE AND K. VOGTMANN, *A generating function for the Euler characteristic of* Out(F_n), J. Pure Appl. Algebra, 44 (1987), pp. 329–348.

[78] C. SOULE, *An introduction to arithmetic groups,* 2004. cours aux Houches 2003, "Number theory, Physics and Geometry", Preprint IHES, arxiv:math.math.GR/0403390.

[79] P. STORM, *The Novikov conjecture for mapping class groups as a corollary of Hamenstadt's theorem'.* arXiv:math.GT/0504248.

[80] K. VOGTMANN, *Local structure of some* Out(F_n)-*complexes,* Proc. Edinburgh Math. Soc. (2), 33 (1990), pp. 367–379.

[81] ———, *The cohomology of automorphism groups of free groups,* in Proceedings of the International Congress of Mathematicians (Madrid, 2006), 2006.

[82] B. WAJNRYB, *A simple presentation for the mapping class group of an orientable surface,* Israel J. Math., 45 (1983), pp. 157–174.

[83] ———, *Mapping class group of a surface is generated by two elements,* Topology, 35 (1996), pp. 377–383.

[84] P. ZUCCA, *On the* $(2, 2 \times 2)$-*generation of the automorphism groups of free groups,* Istit. Lombardo Accad. Sci. Lett. Rend. A, 131 (1997), pp. 179–188 (1998).

[85] A. ŻUK, *La propriété (T) de Kazhdan pour les groupes agissant sur les polyèdres,* C. R. Acad. Sci. Paris Sér. I Math., 323 (1996), pp. 453–458.

MATHEMATICS. HUXLEY BUILDING, IMPERIAL COLLEGE LONDON, LONDON SW7 2AZ
E-mail address: m.bridson@imperial.ac.uk

MATHEMATICS DEPARTMENT, 555 MALOTT HALL, CORNELL UNIVERSITY, ITHACA, NY 14850
E-mail address: vogtmann@math.cornell.edu

Problems: Braid Groups, Homotopy, Cohomology, and Representations

F. R. Cohen

ABSTRACT. This article is a list of problems concerning Artin's braid groups. These problems overlap with several subjects with motivation for these overlaps also given below.

1. Introduction

This article has five sections of related problems. The problems themselves are at the confluence of Artin's braid groups [6], homotopy theory, representations, Lie algebras, group cohomology, modular forms, and classical knot theory. The problems appear natural, and naive, perhaps too naive. A table of contents follows.

(1) Introduction
(2) Braid groups and homotopy groups
(3) Braid groups and modular forms
(4) Representation spaces for braid groups and fundamental groups of complements of complex hyperplane arrangements
(5) Representations and the Lie algebra associated to the descending central series of a discrete group
(6) Embedding spaces, and loop spaces of configuration spaces

2. Braid groups and homotopy groups

The section addresses connections between the structure of braid groups, Vassiliev invariants, Lie algebras, and the homotopy groups of spheres. These problems arise in joint work with Jon Berrick, Yan Loi Wong, and Jie Wu [15, 5], as well as [36].

Let P_n denote Artin's n-stranded pure braid group with B_n the full n-stranded braid group. These braids will be seen to assemble into a topological space which "contains" the loop space of the 2-sphere as described in the next few paragraphs. The relevant structure arises in the context of a simplicial group, that is a collection of groups $\{\Gamma_0, \Gamma_1, \Gamma_2, \cdots\}$ denoted Γ_* with face operations $d_i : \Gamma_n \to \Gamma_{n-1}$, $0 \leq i \leq n$, and degeneracy operations $s_j : \Gamma_n \to \Gamma_{n+1}$, $0 \leq j \leq n$, which satisfy the simplicial identities.

The author was partially supported by the NSF.

©2006 American Mathematical Society

One basic example of a simplicial group, denoted AP_* below, is given by the collection of groups
$$\Gamma_n = P_{n+1}$$
for $n = 0, 1, 2, 3, \cdots$.

(1) There are $n+1$ homomorphisms
$$d_i : P_{n+1} \to P_n$$
for $0 \leq i \leq n$ obtained by deleting the $(i+1)$-st strand.

(2) In addition, there are $n+1$ homomorphisms
$$s_i : P_{n+1} \to P_{n+2}$$
obtained by "doubling" the $(i+1)$-st strand.

One property of these maps given explicitly in [**15, 5**] is that these homomorphisms satisfy the classical simplicial identities. The resulting simplicial group is denoted AP_*.

A second example of a simplicial group arises from an elementary "cabling" operation associated to the pure braid groups which is pictured as follows:

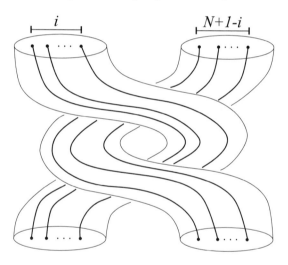

FIGURE 2.1. The braid x_i in P_{N+1}.

The braid x_1 with $N = 1 = i$ in Figure 2.1 is Artin's generator $A_{1,2}$ of P_2. The braids x_i for $1 \leq i \leq N$ in Figure 2.1 yield homomorphisms from a free group on N letters F_N to P_{N+1},
$$\Theta_N : F_N \to P_{N+1}.$$
The homomorphism Θ_N is defined on generators y_i in F_N by the formula
$$\Theta_N(y_i) = x_i.$$
The maps Θ_N are the subject of [**15**] where it is shown that
$$\Theta_n : F_n \to P_{n+1}$$
is faithful for all n. A second Theorem in [**15**] implies that the subgroups F_n of P_{n+1} generated by the elements x_i for $1 \leq i \leq n$ for all $n \geq 0$ admit the structure of a simplicial group which is isomorphic to $F[S^1]$ Milnor's free group construction for

the simplicial circle S^1: the geometric realization of $F[S^1]$ is homotopy equivalent to the loop space of the 2-sphere [29]. More precisely, the smallest simplicial group in AP_* which contains Artin's generator $A_{1,2}$ in $\Gamma_1 = P_2$ is isomorphic to $F[S^1]$.

It is easy to check that $\Theta_n : F_n \to P_{n+1}$ is faithful in case n is 1, or 2. If $n > 2$, the method of proof in [15] is by a comparison of Lie algebras, the structure which gives the Vassiliev invariants of pure braids as analyzed in work of Toshitake Kohno [25, 26]. Related work appears in [24], and [18].

That is, consider the homomorphism $\Theta_n : F_n \to P_{n+1}$ on the level of associated graded Lie algebras obtained by filtering via the descending central series

$$E_0^*(\Theta_n) : E_0^*(F_n) \to E_0^*(P_{n+1})$$

for which F_n is $F[S^1]$ in degree n. Using the structure of the Lie algebra $E_0^*(P_{n+1})$, it is shown in [15] that $E_0^*(\Theta_n)$, and thus Θ_n are injections.

Simplicial groups admit homotopy groups which are analogous to the classical homotopy groups of spaces, and are defined as follows. The n-th homotopy group of Γ_* is defined to be the quotient group

$$\pi_n(\Gamma_*) = Z_n/B_n$$

for which

$$Z_n = \cap_{0 \leq i \leq n} ker(d_i : \Gamma_n \to \Gamma_{n-1}),$$

and

$$B_n = d_o(\cap_{1 \leq i \leq n+1} ker(d_i : \Gamma_{n+1} \to \Gamma_n)).$$

In this special case given by AP_* for which $\Gamma_n = P_{n+1}$, Z_n is the group of $(n+1)$-stranded Brunnian braids

$$Brun_{n+1},$$

those braids which are trivial after deleting any strand. A group analogous to the group of Brunnian braids is the "almost Brunnian" $(n+2)$-stranded braid group

$$QBrun_{n+2} = \cap_{1 \leq i \leq n+1} ker(d_i : P_{n+2} \to P_{n+1}).$$

The subgroup $QBrun_{n+2}$ of P_{n+2} consists of those braids which are trivial after deleting any one of the strands $2, 3, \cdots, n+2$, but not necessarily the first. Observe that the map $d_0 : QBrun_{n+2} \to Brun_{n+1}$ is a surjection. By the results in [15], the restriction of d_0 to the map $d_0 : F_{n+1} \cap QBrun_{n+2} \to F_n \cap Brun_{n+1}$ has cokernel given by $\pi_{n+1}S^2$.

A natural variation of a simplicial group is a Δ-group, a collection of groups

$$\Lambda_* = \{\Lambda_0, \Lambda_1, \Lambda_2, \cdots\}$$

together with homomorphisms $d_i : \Lambda_n \to \Lambda_{n-1}$ which (1) satisfy the standard commutation formulas for face operations in a simplicial group, but (2) are not necessarily equipped with degeneracy operations. Homotopy sets for Δ-groups are defined in the analogous way as in the case of simplicial groups:

$$\pi_n(\Lambda_*) = Z_n/B_n$$

the set of left cosets for which

$$Z_n = \cap_{0 \leq i \leq n} ker(d_i : \Lambda_n \to \Lambda_{n-1}),$$

and

$$B_n = d_o(\cap_{1 \leq i \leq n+1} ker(d_i : \Lambda_{n+1} \to \Lambda_n)).$$

These homotopy sets are not necessarily groups.

An example of a Δ-group is obtained from any simplicial group by retaining the face operations d_i, but omitting the degeneracies. Examples of Δ-groups, $\Lambda_*(M_g)$, which are not necessarily simplicial groups are given by

$$\Lambda_n(M_g) = \pi_1(Conf(M_g, n+1)),$$

the fundamental group of the configuration space $Conf(M_g, n+1)$ of ordered, $(n+1)$-tuples of distinct points in a closed orientable surface M_g of genus $g \geq 0$. If $n+1 \geq 4$, and $M_g = M_0 = S^2$ is of genus 0, then the associated homotopy sets are groups and are given by the $(n+1)$-st homotopy group of the 2-sphere [5]. That is, if $n+1 \geq 4$, then there are isomorphisms

$$\pi_{n+1}S^2 \to \pi_{n+1}\Lambda_*(M_0)).$$

Similar results apply to other spheres as well as simply-connected CW-complexes [15].

Some natural problems which occur here are as follows.

Q 2.1: The Vassiliev invariants of pure braids distinguish all pure braids [**25, 26**]. Give weaker versions of Vassiliev invariants which distinguish elements in homotopy groups.

Q 2.2: Describe properties of the mod-p analogue of Vassiliev invariants obtained from the Lie algebra arising from the mod-p descending central series for the pure braid groups. Are there natural interpretations of this Lie algebra in terms of iterated integrals analogous to those in [**25, 26**]?

In this case, the filtration quotients associated to the map $\Theta_n : F_n \to P_{n+1}$ on the level of the mod-p descending central series also induces a monomorphism of associated graded Lie algebras [**15**]. On the other-hand, the associated graded Lie algebra for F_n gives the E^0-term of the mod-p unstable Adams spectral sequence for the 2-sphere [**16**]. Is there an informative interpretation of the mod-p analogue of Vassiliev invariants?

Q 2.3: The map $d_0 : QBrun_{n+2} \to Brun_{n+1}$ is a surjection, however, the restriction

$$d_0 : QBrun_{n+2} \cap F_{n+1} \to Brun_{n+1} \cap F_n$$

is not a surjection. Describe features of the cokernel in terms of Brunnian braids.

Q 2.4: Do the known elements in the homotopy groups of the 2-sphere impact the structure of pure braids through the structures above?

Q 2.5: The Lie algebras associated to the descending central series for $Brun_{n+1}$ and $Brun_{n+1} \cap F_n$ are bigraded, and finitely generated free abelian groups in each bidegree [**15**]. Give properties of the two variable generating functions listing these ranks. For example, do these series satisfy any natural invariance properties (such as is the case of the one variable generating function associated to the classical partition function)?

Q 2.6: The groups $QBrun_{n+2} \cap F_{n+1}$, and $Brun_{n+1} \cap F_n$ are free. Give combinatorial descriptions of the map

$$d_0 : QBrun_{n+2} \cap F_{n+1} \to Brun_{n+1} \cap F_n$$

on the level of abelianizations $H_1(QBrun_{n+2} \cap F_{n+1})$, and $H_1(Brun_{n+1} \cap F_n)$.

Q 2.7: Prove that $\Theta_n : F_n \to P_{n+1}$ is a monomorphism via other means such as the "ping-pong" lemma given in [**21**], Lemma II.24 or [**3**].

3. Braid groups and modular forms

There is a natural connection between the above structures, and the cohomology of the mapping class group of an orientable surface of genus g, $\mathcal{M}_g$, via classical work of Eichler, and Shimura [**17, 33, 19, 30, 31**]. This section gives some related problems.

Consider maps of the braid group to the mapping class group $\mathcal{M}_g$. For example, there is a classical map $h : B_{2g+2} \to \mathcal{M}_g$ which is a surjection in the cases $g = 1$, or $g = 2$ [**6**]. Then consider the composition with the natural map $\mathcal{M}_g \to Sp(2g, \mathbb{Z})$ to obtain a $2g$-dimensional representation

$$\alpha : B_{2g+2} \to Sp(2g, \mathbb{Z})$$

labelled V_{2g}.

A classical connection to modular forms is via the Eichler-Shimura isomorphism which identifies the ring of modular forms based on the action of $SL(2, \mathbb{Z})$ (respectively certain subgroups Γ of $SL(2, \mathbb{Z})$) by fractional linear transformations on the upper 1/2-plane. This identification is given by the real cohomology of $SL(2, \mathbb{Z})$ (respectively Γ) with coefficients in the k-fold symmetric powers of V_2, $Sym^k(V_2)$, for $k \geq 0$. This structure was used in work of Furusawa, Tezuka, and Yagita [**19**] to work out the cohomology of $BDiff^+(S^1 \times S^1)$. The analogous structure was also used in [**10**] to work out the cohomology for the genus 1 mapping class group with marked points. Connections to elliptic cohomology are given in [**27, 30, 31**].

Q 3.1: The Eichler-Shimura isomorphism implies that cohomology groups

$$H^*(B_3; \oplus_{k \geq 0} Sym^k(V_2 \otimes \mathbb{R}))$$

are given by the

$$E[u] \otimes \mathbb{M}_*(SL(2, \mathbb{Z}))$$

for which $E[u]$ denotes an exterior algebra on one generator of degree 1, and $\mathbb{M}_*(SL(2, \mathbb{Z}))$ denotes the ring of modular forms associated to $SL(2, \mathbb{Z})$. What is the cohomology of B_{2g+2}, or $\mathcal{M}_g$ with coefficients in the symmetric powers of V_{2g}?

Q 3.2: A similar question arises with symmetric powers replaced by tensor products of these representations with both exterior powers, and tensor powers. For example, what is the cohomology of the braid groups with coefficients in tensor powers of V_{2g}?

As an aside, the cohomology groups of the braid groups with certain other choices of local coefficient systems are given in [**13**]: in these cases, cohomology groups are considered with coefficients in any graded permutation representation. The answers are given in terms of free Lie algebras.

Q 3.3: Give an analytic interpretation of the cohomology of the braid groups with coefficients in the symmetric powers of the natural symplectic representations analogous to that of Shimura [**33**]. What are analogues of Eisenstein series for B_{2g+2}?

Q 3.4: A map $\Theta_2 : F_2 \to P_3$ was described in section 2 addressing a connection between braid groups, and homotopy groups. Furthermore, the classical surjection $B_3 \to SL(2,\mathbb{Z})$ restricts to give a composite of the following maps

$$F_2 \xrightarrow{\Theta_2} P_3 \xrightarrow{\alpha|_{P_3}} SL(2,\mathbb{Z}) \longrightarrow PSL(2,\mathbb{Z}).$$

A direct computation gives that this map is an embedding, and that the image is the principal congruence subgroup of level 2 in $PSL(2,\mathbb{Z})$. The cohomology of F_2 with coefficients in this representation is classical, and is given in terms of the modular forms associated to the principal congruence subgroup of level 2. For example, $H^1(F_2; Sym^2(V_2 \otimes \mathbb{R}))$ is of rank 3, and there are 3 linearly independent forms of weight 4 (via Shimura's weight convention).

Determine the cohomology of F_n, and P_{n+1} for $n+1 \leq 2g+2$ with coefficients in $Sym^k(V_{2g})$.

It is tempting to ask whether there is a further connection for these cohomology classes in the case of F_2. In this case, a Brunnian braid in P_3 whose braid closure is the classical Borromean rings represents the classical Hopf map

$$\eta : S^3 \to S^2.$$

Multiples of the Hopf map are distinguished by naively evaluating these representing braids against natural crossed homomorphisms corresponding to modular forms of weight 4 in the classical ring of forms associated to the principal congruence subgroup of level 2 in $PSL(2,\mathbb{Z})$.

For example, the elements of $H^1(B_{2g+2}; Sym^k(V_{2g} \otimes \mathbb{R}))$ can be regarded as functions (derivations) on the braid groups with values in $Sym^k(V_{2g} \otimes \mathbb{R})$ which distinguish some, but not all braids.

Q 3.5: The cohomology classes in the previous question arise in the context of modular forms. That is, the classical Eichler-Shimura isomorphism identifies certain rings of modular forms as the cohomology groups of a discrete groups with local coefficients in the symmetric powers of natural representations $Sym^k(V_{2g})$ for $g = 1$. In the special case of subgroups Γ of $SL(2,\mathbb{Z})$, the non-vanishing cohomology groups are concentrated in degree 1, and are given by $H^1(\Gamma; Sym^k(V_2))$. The elements in the ring of classical modular forms are then identified as crossed homomorphisms [33]. These crossed homomorphisms distinguish some braids, but not all braids (and are thus less strong than the Vassiliev invariants of pure braids).

Is there a similar interpretation of Vassiliev invariants of pure braids as crossed homomorphisms $f : P_n \to M$ for some choice of module M over the group ring of the pure braid group P_n? That is, do Vassiliev invariants, and modular forms arise from similar contexts?

Q 3.6: Characterize which braids are distinguished by crossed homomorphisms arising in this context.

Q 3.7: Restricting to P_{2g+2} in B_{2g+2}, as well as reducing modulo 2, it follows directly that

$$H^*(P_{2g+2}; Sym^k(V_{2g} \otimes \mathbb{Z}/2\mathbb{Z}))$$

is isomorphic to

$$H^*(P_{2g+2}; \mathbb{Z}/2\mathbb{Z}) \otimes Sym^k(V_{2g} \otimes \mathbb{Z}/2\mathbb{Z}).$$

Work out the mod-2 Bockstein spectral sequence to obtain the the answers for
$$H^*(P_{2g+2}; Sym^k(V_{2g} \otimes \mathbb{Q})),$$
and
$$H^*(B_{2g+2}; Sym^k(V_{2g} \otimes \mathbb{Q})).$$

Q 3.8: Is there a modification of D. Tamaki's "gravity filtration" [34] of the little 2-cubes which gives a spectral sequence abutting to
$$H^*(B_{2g+2}; Sym^k(V_{2g} \otimes \mathbb{F}))$$
for a field $\mathbb{F}$?

The motivation for this question is to find a practical approach to describe the homology, and cohomology of the braid groups with various natural choices of local coefficients, and which fits with an extension of the classical Eichler-Shimura isomorphism for rings of modular forms.

4. Representation spaces for braid groups and fundamental groups of complements of complex hyperplane arrangements

Given the connections above, it is natural to consider different choices of representations of pure braid groups. Thus, consider the the space of homomorphisms $Hom(G, H)$ where G is the pure braid group P_n, and H is a Lie group. In addition, consider the representation space $Rep(G, H) = Hom(G, H)/Inn(H)$ where the group of inner automorphisms of H, $Inn(H)$, acts in the natural way via conjugation.

Q 4.1: Enumerate the path-components of $Hom(G, H)$. A lower bound is given in [1] in case $G = P_n$, and H is either $U(n)$ or $O(n)$.

Q 4.2: Describe the topology of $Hom(G, H)$. What is the homotopy type of $Hom(G, H)$, or $Rep(G, H)$ where G is the pure braid group, and H is a compact simple Lie group, or H is one of $SL(n, \mathbb{R})$, $SL(n, \mathbb{C})$, or $PSL(n, \mathbb{R})$? For example, if $q \geq 3(n)(n-1)/2$, then $Hom(P_n, O(q))$ has at least L path-components where L is the order of the group $H^1(P_n; \mathbb{Z}/2\mathbb{Z}) \oplus H^2(P_n; \mathbb{Z}/2\mathbb{Z})$ [1].

Q 4.3: A special case is given by $H = O(n)$. Consider $KO^0(BG)$, the group of stable classes of real orthogonal bundles over BG, together with the subgroup generated by the natural images
$$\amalg_{q \geq 1} Hom(G, O(q)) \to KO^0(BG).$$
Call this subgroup
$$KO^0_{rep}(BG).$$

The group $KO^0_{rep}(BP_n)$ was computed in [1]. The answer is a finite abelian 2-group given in terms of the mod-2 cohomology of P_n [1]. Similar, but less specific results follow when P_n is replaced by the fundamental group of a complex hyperplane arrangement.

What is $KO^0_{rep}(BG)$ for the more general case where G is the fundamental group of the complement of any complex hyperplane arrangement?

Q 4.4: Identify conditions on G, and H which imply that the natural map
$$\Theta_{G,H} : \pi_0(Hom(G, H)/Inn(H)) \to [BG, BH]$$

is an isomorphism of sets. In case G is the fundamental group of a closed, orientable surface, and H is a complex, connected semi-simple Lie group, then

$$\Theta_{G,H} : \pi_0(Hom(G,H)/Inn(H)) \to [BG, BH]$$

is an isomorphism of sets [20, 28].

Q 4.5: Compare the topologies $Hom(G, O(n)) \to Hom(G, GL(n, \mathbb{R}))$ for G the fundamental group of the complement of a complex hyperplane arrangement. For example, there are both non-trivial orthogonal, and unitary bundles over $K(\pi, 1)$'s for π a braid group or the mapping class group for genus 2 surfaces which do not arise from $O(n)$ representations, but do arise from $GL(n, \mathbb{R})$ representations [4, 2].

5. Representations and the Lie algebra obtained from the Lie algebras associated to the descending central series of a discrete group

This section, based on joint work with S. Prassidis [14], is founded on the structure of the Lie algebra obtained from their descending central series of P_n. Consider a homomorphism $f : P_n \to G$. A "homological/Lie algebraic" method for checking whether such a homomorphism is faithful arises via T. Kohno's analysis of Vassiliev invariants of pure braids in [25, 26]. To check whether f is a monomorphism by [14], it is sufficient that

  i: $E_0^*(f)$ induce a monomorphism on the "top free Lie algebra" generated by the images of Artin's pure braids $A_{j,n}$ for $1 \leq j \leq n-1$, and
  ii: the element $E_0^*(f)(\Sigma_{1 \leq i < j \leq n} A_{i,j})$ has infinite order.

Attempts to use this process raise the following questions.

  Q 5.1: Identify conditions for a representation of the pure braid group to be faithful, and discrete. For example, if a representation $\rho : P_n \to G$ descends to a faithful representation on the Lie algebra attached to the descending central series, then ρ is faithful. Identify a natural extension of this Lie algebraic criterion which implies that an embedding is discrete.
  Q 5.2: Give practical criteria to identify when the fundamental group of the complement of a complex hyperplane arrangement admits a discrete, and faithful linear representation.
  Q 5.3: Characterize those discrete groups G with the property that the Lie algebra obtained from the descending central series

$$E_0^*(G)$$

is a free Lie algebra modulo quadratic relations. This means that $E_0^*(G)$ is a quotient of a free Lie algebra $L[V] = \oplus_{k \geq 1} L_k[V]$ modulo the Lie ideal generated by some choice of elements γ in $L_2[V]$.

Some basic examples of groups G are listed next with the property that the Lie algebra obtained from the descending central series $E_0^*(G)$ is a free Lie algebra modulo quadratic relations:

  (1) a free group (P. Hall),
  (2) Artin's pure braid groups (T. Kohno, M. Falk-R. Randell, M. Xicoténcatl),
  (3) the fundamental group of certain choices of complements of fibred complex hyperplane arrangements (T. Kohno, M. Falk-R. Randell, M. Xicoténcatl),

(4) the fundamental group of the orbit configuration space associated to a discrete subgroup of $PSL(2,\mathbb{R})$ acting properly discontinuously on the upper 1/2-plane (L. Paris, F. Cohen-T. Kohno-M. Xicoténcatl),
(5) the commutator subgroups (the universal abelian covers) of the previous families of groups (experiments),
(6) the fundamental group of a closed orientable surface (J. Labute),
(7) the Malĉev completion of the Torelli group for genus g surfaces with one boundary, and g sufficiently large (R. Hain), and
(8) the upper triangular basis conjugating automorphisms of a free group, a subgroup of McCool's group (experiments).

Q 5.4: Are there additional examples given by the descending central series for the following groups?
   (1) the basis conjugating isomorphisms of a free group,
   (2) the subgroup of the automorphism group of a free group generated by left translations by commutators of weight at least 2 (noting that some of these translations fail to be automorphisms), or
   (3) the IA-automorphisms of a free group.

6. Embedding spaces, and loop spaces of configuration spaces

Consider the space $\mathcal{K}_{n,1}$ of long knots in $\mathbb{R}^n$, the space of smooth embeddings

$$f: \mathbb{R}^1 \to \mathbb{R}^n$$

which are standard outside of the interval $[-1, +1]$ with $f([-1, +1])$ contained in the unit ball. In case $n = 3$, let $[f]$ denote the isotopy class of f in $\mathcal{K}_{3,1}$ with $\mathcal{K}_{3,1}(f)$ the path-component of a 'long embedding' f.

The homotopy type of the space of long knots in $\mathbb{R}^3$, $\mathcal{K}_{3,1}$, was analyzed in work of Hatcher, and Budney [22, 23, 7, 8]. Let

$$X_{\mathcal{P}rime}$$

denote the union of path-components of f in $\mathcal{K}_{3,1}$ for which $[f]$ runs over non-trivial prime long knots in $\mathbb{R}^3$, and where 'prime' has the classical meaning as given in [32, 7]: knots which are not prime are either the unknot or a connected sum of two or more non-trivial knots. One result of Budney is recorded next [7].

THEOREM 6.1. *The space $\mathcal{K}_{3,1}$ is homotopy equivalent to*

$$C(\mathbb{R}^2, X_{\mathcal{P}rime} \amalg \{*\})$$

that is the labelled configuration space of points in the plane with labels in $X_{\mathcal{P}rime} \amalg \{*\}$. Furthermore, the following hold:
   (1) *Each path-component of $\mathcal{K}_{3,1}$ is a $\mathcal{K}(\pi, 1)$.*
   (2) *The path-components of $Conf(\mathbb{R}^2, n) \times_{\Sigma_n} (X_{\mathcal{P}rime})^n$ for all n, and thus the path-components of $\mathcal{K}_{3,1}$ are given by either the path-component of the unknot, or*

$$Conf(\mathbb{R}^2, n) \times_{\Sigma_f} \prod_{i=1}^n \mathcal{K}_{3,1}(f_i)$$

for certain choices of $f_1, \cdots, f_n \in X_{\mathcal{P}rime}$, and Young subgroups Σ_f.

Since $\mathcal{K}_{3,1}$ is a homotopy abelian H-space with multiplication induced by concatenation. Thus there is a bracket operation in homology with any coefficients

$$H_s(\mathcal{K}_{3,1}) \otimes H_t(\mathcal{K}_{3,1}) \to H_{1+s+t}(\mathcal{K}_{3,1}).$$

By results in [13], this pairing for the homology of $\mathcal{K}_{3,1}$ gives $H_*(\mathcal{K}_{3,1})$ the structure of a Poisson algebra.

The following properties of the homology of $\mathcal{K}_{3,1}$ were proven in [9].

(1) The rational homology of $\mathcal{K}_{3,1}$ is a free Poisson algebra generated by

$$V = H_*(X_{\mathcal{P}rime}; \mathbb{Q}).$$

(2) The homology of $\mathcal{K}_{3,1}$ with $\mathbb{F}_p$ coefficients is a free Poisson algebra generated by

$$V = H_*(X_{\mathcal{P}rime}; \mathbb{F}_p)$$

as described in [13, 9].

(3) The integer homology of $\mathcal{K}_{3,1}$ has p-torsion of arbitrarily large order. The order of the torsion in the homology of $\mathcal{K}_{3,1}(f)$ reflects the the 'depth' of its JSJ-tree [9].

In addition, work of V. Tourtchine exhibited the structure of a Poisson algebra on the E^2-term of the Vassiliev spectral sequence [35]. Similar consequences for the space of smooth embeddings of S^1 in S^3 are noted in [9] via the homotopy equivalence

$$SO(n+1) \times_{SO(n-1)} \mathcal{K}_{n,1} \to Emb(S^1, S^n).$$

Furthermore, there is a "capping construction" which gives a continuous map

$$\Phi(M, q) : \Omega Conf(M, q) \to Emb^{Top}(\mathbb{R}^1, M \times \mathbb{R}^1),$$

with one version described in [11] where $Emb^{Top}(\mathbb{R}^1, M \times \mathbb{R}^1)$ denotes the space of continuous embeddings. These maps specialize to maps

$$\Phi(\mathbb{R}^n, q) : \Omega Conf(M, q) \to Emb^{Top}(\mathbb{R}^1, \mathbb{R}^{n+1})$$

in case $n+1 \geq 3$ as well as a smooth version

$$\Phi^{smooth}(\mathbb{R}^n, q) : \Omega^{smooth} Conf(M, q) \to Emb(\mathbb{R}^1, \mathbb{R}^{n+1}).$$

Q 6.1: Identify the path-components of $\mathcal{K}_{3,1}$ given by $\mathcal{K}_{3,1}(f)$ which have torsion in $H_1(\mathcal{K}_{3,1}(f))$. Are there path-components $\mathcal{K}_{3,1}(f)$ which have odd torsion in their first homology group?

Q 6.2: If $n \geq 3$, is $\mathcal{K}_{n,1}$, or the "framed analogue" $\mathcal{K}_{n,1}^{fr}$, a retract of

$$\Omega^2 \Sigma^2 (\amalg_{q \geq 2} \Omega Conf(\mathbb{R}^{n-1}, q)),$$

at least rationally?

Q 6.3: Describe the p-torsion in the homology of $\mathcal{K}_{n,1}$ or $\mathcal{K}_{n,1}^{fr}$ for $n \geq 4$. Is there p-torsion of arbitrarily large order in the homology of $\mathcal{K}_{n,1}^{fr}$ for $n \geq 4$ as is the case for $n = 3$?

Q 6.4: Describe homology classes for $\mathcal{K}_{n,1}$ arising from a capping construction.

Q 6.5: If $n \geq 3$, is the rational homology of $\mathcal{K}_{n,1}^{fr}$ a free Poisson algebra?

References

[1] A. Adem, D. Cohen, and F. R. Cohen, *On representations and K-theory of the braid groups*, Math. Ann. 326 (2003), no. 3, 515–542.

[2] A. Adem, and F. R. Cohen, *Commuting elements and spaces of homomorphisms*, arXiv:math.AT/0603197.

[3] R. Alperin, B. Farb, and G. Noskov *A strong Schottky Lemma for nonpositively curved singular spaces*, Geometriae Dedicata, **92**(2002), 235–243.

[4] D. J. Benson, F. R. Cohen, *Mapping class groups of low genus and their cohomology*, Memoirs of the American Mathematical Society, **443**(1991).

[5] J. Berrick, F. R. Cohen, Y. L. Wong, and J. Wu, *Configurations, braids, and homotopy groups*, J. Amer. Math. Soc. **19** (2006), 265-326.

[6] J. Birman, *Braids, Links and Mapping Class Groups*, Ann. of Math. Studies, **82**(1975), Princeton Univ. Press, Princeton, N.J..

[7] R. Budney, *Little cubes and long knots*, to appear in Topology, math.GT/0309427.

[8] R. Budney, *The topology of knotspaces in dimension 3*, preprint, math.GT/0506524.

[9] R. Budney, F. R. Cohen, *On the homology of the space of knots*, math.GT/0504206.

[10] F. R. Cohen, *On genus one mapping class groups, function spaces, and modular forms*, Cont. Math. **279**(2001), 103-128.

[11] F. R. Cohen, S. Gitler, *Loop spaces of configuration spaces, braid-like groups, and knots*, Proceedings of the 1998 Barcelona Conference on Algebraic Topology.

[12] F. R. Cohen, T. Kohno, and M. Xicoténcatl, *Orbit configuration spaces associated to discrete subgroups of $PSL(2,\mathbb{R})$*, preprint on the ArXiv.

[13] F. R. Cohen, T. J. Lada, J. P. May, *The homology of iterated loop spaces*, Springer-Verlag Lecture Notes in Mathematics, **533**(1976), 207–351.

[14] F. R. Cohen, and S. Prassidis, *On injective homomorphisms for pure braid groups, and associated Lie algebras*, math.GR/0404278 .

[15] F. R. Cohen, and J. Wu, *On braid groups, free groups, and the loop space of the 2-sphere*, Progress in Mathematics, **215**(2003), 93-105, Birkhaüser, and *Braid groups, free groups, and the loop space of the 2-sphere*, math.AT/0409307.

[16] E. B. Curtis, *Simplicial homotopy theory*, Adv. in Math., **6**(1971), 107-209.

[17] M. Eichler, *Eine Verallgemeinerung der Abelschen Integrale*, Math. Zeit., **67**(1957), 267-298.

[18] M. Falk, and R. Randell, *The lower central series of a fiber–type arrangement*, Invent. Math., **82** (1985), 77–88.

[19] M. Furusawa, M. Tezuka, and N. Yagita, *On the cohomology of classifying spaces of torus bundles, and automorphic forms*, J. London Math. Soc., (2)**37**(1988), 528-543.

[20] W. M. Goldman, *Topological components of the space of representations*, Invent. Math. **93**(1988), no. 3, 557-607.

[21] P. de la Harpe, *Topics in Geometric Group Theory*, Chicago Lectures in Mathematics, University of Chicagi Press, 2,000.

[22] A. Hatcher, *Homeomorphisms of sufficiently-large P^2-irreducible 3-manifolds*. Topology . **15** (1976)

[23] ———, *Topological Moduli Spaces of Knots*, math.GT/9909095.

[24] Y. Ihara, *Automorphisms of Pure Sphere Braid groups and Galois Representations*, The Grothendieck Festschrift II, Progress in Mathematics, **87**(1990), Birkhäuser, 353-373.

[25] T. Kohno, *Linear representations of braid groups and classical Yang-Baxter equations*, Cont. Math., **78**(1988), 339-363.

[26] ———, *Vassiliev invariants and de Rham complex on the space of knots*, in: Symplectic Geometry and Quantization, Contemp. Math., **179**(1994), Amer. Math. Soc., Providence, RI, 123-138.

[27] P. Landweber, *Elliptic cohomology and modular forms*, Elliptic Curves, in: Modular Forms in Algebraic Topology, Princeton, 1986, Springer-Verlag, New York, 1988, 55-68.

[28] J. Li, *The space of surface group representations*, Manuscript. Math. **78**(1993), no. 3, 223-243.

[29] J. Milnor, *On the construction F[K]*, In: A student's Guide to Algebraic Topology, J.F. Adams, editor, Lecture Notes of the London Mathematical Society, **4**(1972), 119-136.

[30] G. Nishida, *Modular forms and the double transfer for BT^2*, Japan Journal of Mathematics, **17**(1991), 187-201.

[31] T. C. Ratliff, *Congruence subgroups, elliptic cohomology, and the Eichler-Shimura map*, Journal of Pure and Applied Algebra, **109**(1996), 295-322.

[32] H. Schubert, *Die eindeutige Zerlegbarkeit eines Knoten in Primknoten*, Sitzungsber. Akad. Wiss. Heidelberg, math.-nat. KI., **3:57-167**. (1949)

[33] G. Shimura, *Introduction to the arithmetic theory of automorphic forms*, Publications of the Mathematical Society of Japan 11, Iwanami Shoten, Tokyo; University Press, Princeton, 1971.

[34] D. Tamaki, *A dual Rothenberg-Steenrod spectral sequence*, Topology, **33**(4)(1994), 631-662.

[35] V. Tourtchine *On the other side of the bialgebra of chord diagrams*, arXiv QA/0411436.

[36] J. Wu, *On combinatorial descriptions of the homotopy groups of certain spaces*, Math. Proc. Camb. Philos. Soc., **130**(2001), no.3, 489-513.

E-mail address: cohf@math.rochester.edu

DEPARTMENT OF MATHEMATICS, UNIVERSITY OF ROCHESTER, ROCHESTER, NEW YORK 14627

Cohomological Structure of the Mapping Class Group and Beyond

Shigeyuki Morita

ABSTRACT. In this paper, we briefly review some of the known results concerning the cohomological structures of the mapping class group of surfaces, the outer automorphism group of free groups, the diffeomorphism group of surfaces as well as various subgroups of them such as the Torelli group, the IA outer automorphism group of free groups, the symplectomorphism group of surfaces.

Based on these, we present several conjectures and problems concerning the cohomology of these groups. We are particularly interested in the possible interplays between these cohomology groups rather than merely the structures of individual groups. It turns out that, we have to include, in our considerations, two other groups which contain the mapping class group as their core subgroups and whose structures seem to be deeply related to that of the mapping class group. They are the arithmetic mapping class group and the group of homology cobordism classes of homology cylinders.

1. Introduction

We begin by fixing our notations for various groups appearing in this paper. Let Σ_g denote a closed oriented surface of genus g which will be assumed to be greater than or equal to 2 unless otherwise specified. We denote by $\mathrm{Diff}_+\Sigma_g$ the group of orientation preserving diffeomorphisms of Σ_g equipped with the C^∞ topology. The same group equipped with the *discrete* topology is denoted by $\mathrm{Diff}^\delta_+\Sigma_g$. The mapping class group $\mathcal{M}_g$ is the group of path components of $\mathrm{Diff}_+\Sigma_g$. The Torelli group, denoted by $\mathcal{I}_g$, is the subgroup of $\mathcal{M}_g$ consisting of mapping classes which act on the homology group $H_1(\Sigma_g;\mathbb{Z})$ trivially. Thus we have an extension

(1) $$1 \longrightarrow \mathcal{I}_g \longrightarrow \mathcal{M}_g \longrightarrow \mathrm{Sp}(2g,\mathbb{Z}) \longrightarrow 1$$

where $\mathrm{Sp}(2g,\mathbb{Z})$ denotes the Siegel modular group. Choose an embedded disk $D \subset \Sigma_g$ and a base point $* \in D \subset \Sigma_g$. We denote by $\mathcal{M}_{g,1}$ and $\mathcal{I}_{g,1}$ (resp. $\mathcal{M}_{g,*}$ and

---

2000 *Mathematics Subject Classification.* Primary 57R20, 55R40, 32G15, 57N05, 57M99, 20J06; Secondary 57N10, 20F28, 17B40, 17B56, 57R32.

Key words and phrases. Mapping class group, automorphism group of free group, moduli space of curves, outer space, tautological algebra, free Lie algebra, derivation algebra, symplectic group, homology cylinder, homology sphere, symplectomorphism group.

The author is partially supported by JSPS Grants 16204005 and 16654011.

©2006 American Mathematical Society

$\mathcal{I}_{g,*}$) the mapping class group and the Torelli group *relative* to D (resp. the base point $*$).

Next let F_n denote a free group of rank $n \geq 2$. Let $\operatorname{Aut} F_n$ (resp. $\operatorname{Out} F_n$) denote the automorphism group (resp. outer automorphism group) of F_n. Let IAut_n (resp. IOut_n) denote the subgroup of $\operatorname{Aut} F_n$ (resp. $\operatorname{Out} F_n$) consisting of those elements which act on the abelianization $H_1(F_n; \mathbb{Z})$ of F_n trivially. Thus we have an extension

(2) $$1 \longrightarrow \operatorname{IOut}_n \longrightarrow \operatorname{Out} F_n \longrightarrow \operatorname{GL}(n, \mathbb{Z}) \longrightarrow 1.$$

The fundamental group $\pi_1(\Sigma_g \setminus \operatorname{Int} D)$ is a free group of rank $2g$. Fix an isomorphism $\pi_1(\Sigma_g \setminus \operatorname{Int} D) \cong F_{2g}$. By a classical result of Dehn and Nielsen, we can write

$$\mathcal{M}_{g,1} = \{\varphi \in \operatorname{Aut} F_{2g}, \varphi(\gamma) = \gamma\}$$

where the element γ is defined by

$$\gamma = [\alpha_1, \beta_1] \cdots [\alpha_g, \beta_g]$$

in terms of appropriate free generators $\alpha_1, \beta_1, \cdots, \alpha_g, \beta_g$ of F_{2g}. Then we have the following commutative diagram

(3)
$$\begin{array}{ccccccccc}
1 & \longrightarrow & \mathcal{I}_{g,1} & \longrightarrow & \mathcal{M}_{g,1} & \longrightarrow & \operatorname{Sp}(2g, \mathbb{Z}) & \longrightarrow & 1 \\
& & \downarrow & & \downarrow & & \downarrow & & \\
1 & \longrightarrow & \operatorname{IAut}_{2g} & \longrightarrow & \operatorname{Aut} F_{2g} & \longrightarrow & \operatorname{GL}(2g, \mathbb{Z}) & \longrightarrow & 1.
\end{array}$$

Similarly, for the case of the mapping class group with respect to a base point, we have

(4)
$$\begin{array}{ccccccccc}
1 & \longrightarrow & \mathcal{I}_{g,*} & \longrightarrow & \mathcal{M}_{g,*} & \longrightarrow & \operatorname{Sp}(2g, \mathbb{Z}) & \dashrightarrow & 1 \\
& & \downarrow & & \downarrow & & \downarrow & & \\
1 & \longrightarrow & \operatorname{IOut}_{2g} & \longrightarrow & \operatorname{Out} F_{2g} & \longrightarrow & \operatorname{GL}(2g, \mathbb{Z}) & \longrightarrow & 1.
\end{array}$$

Next we fix an area form (or equivalently a symplectic form) ω on Σ_g and we denote by $\operatorname{Symp} \Sigma_g$ the subgroup of $\operatorname{Diff}_+ \Sigma_g$ consisting of those elements which preserve the form ω. Also let $\operatorname{Symp}_0 \Sigma_g$ be the identity component of $\operatorname{Symp} \Sigma_g$. Moser's theorem [**84**] implies that the quotient group $\operatorname{Symp} \Sigma_g / \operatorname{Symp}_0 \Sigma_g$ can be naturally identified with the mapping class group $\mathcal{M}_g$ and we have the following commutatvie diagram

(5)
$$\begin{array}{ccccccccc}
1 & \longrightarrow & \operatorname{Symp}_0 \Sigma_g & \longrightarrow & \operatorname{Symp} \Sigma_g & \longrightarrow & \mathcal{M}_g & \longrightarrow & 1 \\
& & \downarrow & & \downarrow & & \| & & \\
1 & \longrightarrow & \operatorname{Diff}_0 \Sigma_g & \longrightarrow & \operatorname{Diff}_+ \Sigma_g & \longrightarrow & \mathcal{M}_g & \longrightarrow & 1
\end{array}$$

where $\operatorname{Diff}_0 \Sigma_g$ is the identity component of $\operatorname{Diff}_+ \Sigma_g$.

In this paper, we also consider two other groups. Namely the arithmetic mapping class group and the group of homology cobordism classes of homology cylinders. They will be mentioned in §8 and §11 respectively.

2. Tautological algebra of the mapping class group

Let $\mathbf{M}_g$ be the moduli space of smooth projective curves of genus g and let $\mathcal{R}^*(\mathbf{M}_g)$ be its tautological algebra. Namely it is the subalgebra of the Chow algebra $\mathcal{A}^*(\mathbf{M}_g)$ generated by the tautological classes $\kappa_i \in \mathcal{A}^i(\mathbf{M}_g)$ ($i = 1, 2, \cdots$) introduced by Mumford [85]. Faber [15] made a beautiful conjecture about the structure of $\mathcal{R}^*(\mathbf{M}_g)$. There have been done many works related to and inspired by Faber's conjecture (we refer to survey papers [30][53][105] for some of the recent results including enhancements of Faber's original conjecture). However the most difficult part of Faber's conjecture, which claims that $\mathcal{R}^*(\mathbf{M}_g)$ should be a Poincaré duality algebra of dimension $2g - 4$, remains unsettled.

Here we would like to describe a topological approach to Faber's conjecture, in particular this most difficult part. For this, we denote by

$$e_i \in H^{2i}(\mathcal{M}_g; \mathbb{Z}) \quad (i = 1, 2, \cdots)$$

the i-th Mumford-Morita-Miller tautological class which was defined in [72] as follows. For any oriented Σ_g-bundle $\pi : E \to X$, the tangent bundle along the fiber of π, denoted by ξ, is an oriented plane bundle over the total space E. Hence we have its Euler class $e = \chi(\xi) \in H^2(E; \mathbb{Z})$. If we apply the Gysin homomorphism (or the integration along the fibers) $\pi_* : H^*(E; \mathbb{Z}) \to H^{*-2}(X; \mathbb{Z})$ to the power e^{i+1}, we obtain a cohomology class

$$e_i(\pi) = \pi_*(e^{i+1}) \in H^{2i}(X; \mathbb{Z})$$

of the base space X. By the obvious naturality of this construction, we obtain certain cohomology classes

$$e \in H^2(\text{EDiff}_+\Sigma_g; \mathbb{Z}), \quad e_i \in H^{2i}(\text{BDiff}_+\Sigma_g; \mathbb{Z})$$

where $\text{EDiff}_+\Sigma_g \to \text{BDiff}_+\Sigma_g$ denotes the universal oriented Σ_g-bundle. In the cases where $g \geq 2$, a theorem of Earle and Eells [12] implies that the two spaces $\text{EDiff}_+\Sigma_g$ and $\text{BDiff}_+\Sigma_g$ are Eilenberg-MacLane spaces $K(\mathcal{M}_{g,*}, 1)$ and $K(\mathcal{M}_g, 1)$ respectively. Hence we obtain the universal Euler class $e \in H^2(\mathcal{M}_{g,*}; \mathbb{Z})$ and the Mumford-Morita-Miller classes $e_i \in H^{2i}(\mathcal{M}_g; \mathbb{Z})$ as group cohomology classes of the mapping class groups. It follows from the definition that, over the rationals, the class e_i is the image of $(-1)^{i+1}\kappa_i$ under the natural projection $\mathcal{A}^*(\mathbf{M}_g) \to H^*(\mathcal{M}_g; \mathbb{Q})$.

Now we define $\mathcal{R}^*(\mathcal{M}_g)$ (resp. $\mathcal{R}^*(\mathcal{M}_{g,*})$) to be the subalgebra of $H^*(\mathcal{M}_g; \mathbb{Q})$ (resp. $H^*(\mathcal{M}_{g,*}; \mathbb{Q})$) generated by the classes $e_1, e_2, \cdots$ (resp. $e, e_1, e_2, \cdots$) and call them the tautological algebra of the mapping class group $\mathcal{M}_g$ (resp. $\mathcal{M}_{g,*}$). There is a canonical projection $\mathcal{R}^*(\mathbf{M}_g) \to \mathcal{R}^*(\mathcal{M}_g)$.

Let us denote simply by H (resp. $H_\mathbb{Q}$) the homology group $H_1(\Sigma_g; \mathbb{Z})$ (resp. $H_1(\Sigma_g; \mathbb{Q})$). Also we set

$$U = \Lambda^3 H / \omega_0 \wedge H, \quad U_\mathbb{Q} = U \otimes \mathbb{Q}$$

where $\omega_0 \in \Lambda^2 H$ denotes the symplectic class. $U_\mathbb{Q}$ is an irreducible representation of the algebraic group $\text{Sp}(2g, \mathbb{Q})$ corresponding to the Young diagram $[1^3]$ consisting of 3 boxes in a single column. Recall here that, associated to any Young diagram whose number of rows is less than or equal to g, there corresponds an irreducible representation of $\text{Sp}(2g, \mathbb{Q})$ (cf. [20]). In our papers [77][81], we constructed a morphism

(6)
$$\begin{array}{ccc}
\mathcal{M}_{g,*} & \xrightarrow{\rho_2} & \left(([1^2]\oplus[2^2])\widetilde{\times}_{\text{torelli}}\Lambda^3 H_{\mathbb{Q}}\right)\rtimes \mathrm{Sp}(2g,\mathbb{Q}) \\
\downarrow & & \downarrow \\
\mathcal{M}_g & \xrightarrow{\rho_2} & ([2^2]\widetilde{\times}U_{\mathbb{Q}})\rtimes \mathrm{Sp}(2g,\mathbb{Q})
\end{array}$$

where $[2^2]\widetilde{\times}U_{\mathbb{Q}}$ denotes a central extension of $U_{\mathbb{Q}}$ by $[2^2]$ corresponding to the unique copy $[2^2]\in H^2(U_{\mathbb{Q}})$ and $(([1^2]\oplus[2^2])\widetilde{\times}_{\text{torelli}}\Lambda^3 H_{\mathbb{Q}}$ is defined similarly (see [**81**] for details). The diagram (6) induces the following commutative diagram.

(7)
$$\begin{array}{ccc}
\left(\Lambda^*\Lambda^3 H_{\mathbb{Q}}^*/([1^2]^{\text{torelli}}\oplus[2^2])\right)^{Sp} & \xrightarrow{\rho_2^*} & H^*(\mathcal{M}_{g,*};\mathbb{Q}) \\
\uparrow & & \uparrow \\
\left(\Lambda^*U_{\mathbb{Q}}^*/([2^2])\right)^{Sp} & \xrightarrow{\rho_2^*} & H^*(\mathcal{M}_g;\mathbb{Q}).
\end{array}$$

On the other hand, we proved in [**51**][**52**] that the images of the above homomorphisms ρ_2^* are precisely the tautological algebras. Here the concept of the *generalized* Morita-Mumford classes defined by Kawazumi [**48**] played an important role. Then in [**82**], the effect of unstable degenerations of Sp-modules appearing in (7) was analized and in particular a part of Faber's conjecture claiming that $\mathcal{R}^*(\mathcal{M}_g)$ is already generated by the classes $e_1, e_2, \cdots, e_{[g/3]}$ was proved (later Ionel [**39**] proved this fact at the level of $\mathcal{R}^*(\mathbf{M}_g)$). Although the way of degenerations of Sp-modules is by no means easy to be studied, it seems natural to expect the following.

CONJECTURE 1. The natural homomorphisms
$$\left(\Lambda^*\Lambda^3 H_{\mathbb{Q}}\right)^{Sp}\to H^*(\mathcal{M}_{g,*};\mathbb{Q}), \quad (\Lambda^*U_{\mathbb{Q}})^{Sp}\to H^*(\mathcal{M}_g;\mathbb{Q})$$
induce isomorphisms
$$\left(\Lambda^*\Lambda^3 H_{\mathbb{Q}}^*/([1^2]^{\text{torelli}}\oplus[2^2])\right)^{Sp}\cong \mathcal{R}^*(\mathcal{M}_{g,*})$$
$$\left(\Lambda^*U_{\mathbb{Q}}^*/([2^2])\right)^{Sp}\cong \mathcal{R}^*(\mathcal{M}_g).$$

Furthermore, the algebras on the left hand sides are Poincaré duality algebras of dimensions $2g-2$ and $2g-4$ respectively.

Here we mention that for a single Riemann surface X, the cohomology $H^*(\mathrm{Jac}(X);\mathbb{Q})$ is a Poincaré duality algebra of dimension $2g$ while it can be shown that there exists a canonical isomorphism
$$H^*(\mathrm{Jac}(X);\mathbb{Q})/([1^2])\cong H^*(X;\mathbb{Q})$$
which is a Poincaré duality algebra of dimension 2. Here
$$[1^2]\subset H^2(\mathrm{Jac}(X);\mathbb{Q})$$
denotes the kernel $\mathrm{Ker}(\Lambda^2 H_{\mathbb{Q}}^*\to\mathbb{Q})$ of the intersection pairing and $([1^2])$ denotes the ideal generated by it. Observe that we can write $\Lambda^*U_{\mathbb{Q}}^* = H^*(PH^3(\mathrm{Jac}))$ which is a Poincaré duality algebra of dimension $\binom{2g}{3}-2g$, where $PH^3(\mathrm{Jac})$ denotes the *primitive part* of the third cohomology of the Jacobian variety. Hence the above conjecture can be rewritten as
$$\left(H^*(PH^3(\mathrm{Jac}))/([2^2])\right)^{Sp}\cong \mathcal{R}^*(\mathcal{M}_g)$$

so that it could be phrased as the *family version* of the above simple fact for a single Riemann surface.

3. Higher geometry of the mapping class group

Madsen and Weiss [64] recently proved a remarkable result about the homotopy type of the classifying space of the stable mapping class group. As a corollary, they showed that the stable rational cohomology of the mapping class group is isomorphic to the polynomial algebra generated by the Mumford-Morita-Miller classes

$$\lim_{g\to\infty} H^*(\mathcal{M}_g; \mathbb{Q}) \cong \mathbb{Q}[e_1, e_2, \cdots].$$

We also would like to mention fundamental results of Tillmann [104] and Madsen and Tillmann [63].

As was explained in [80], the classes e_i serve as the (orbifold) Chern classes of the tangent bundle of the moduli space $\mathbf{M}_g$ and it may appear that, stably and quantitatively, the moduli space $\mathbf{M}_g$ is similar to the classifying space of the unitary group, namely the complex Grassmannian. However, qualitatively the situation is completely different and the moduli space has much deeper structure than the Grassmannian. Here we would like to present a few problems concerning "*higher geometry*" of the mapping class group where we understand the Mumford-Morita-Miller classes as the primary characteristic classes.

First we recall the following problem, because of its importance, which was already mentioned in [80] (Conjecture 3.4).

PROBLEM 2. Prove (or disprove) that the even Mumford-Morita-Miller classes $e_{2i} \in H^{4i}(\mathcal{I}_g; \mathbb{Q})$ are non-trivial, in a suitable stable range, as cohomology classes of the Torelli group.

The difficulty of the above problem comes from the now classical fact, proved by Johnson [40], that the abelianization of the Torelli group is very big, namely $H_1(\mathcal{I}_g; \mathbb{Q}) \cong U_\mathbb{Q}$ ($g \geq 3$). Observe that if $\mathcal{I}_g$ were perfect, then the above problem would have been easily solved by simply applying the Quillen plus construction to each group of the group extension (1) and then looking at the homotopy exact sequence of the resulting fibration. The work of Igusa [38] (in particular Corollary 8.5.17) shows a close connection between the above problem with another very important problem (see Problem 11 in § 4) of non-triviality of Igusa's higher Franz-Reidemeister torsion classes in $H^{4i}(\mathrm{IOut}_n; \mathbb{R})$ (Igusa uses the notation $\mathrm{Out}^h F_n$ for the group IOut_n). We also refer to a recent work of Sakasai [95] which is related to the above problem.

Next we recall the following two well-known problems about the structure of the Torelli group which are related to a foundational work of Hain [29].

PROBLEM 3. Determine whether the Torelli group $\mathcal{I}_g$ ($g \geq 3$) is finitely presentable or not (note that $\mathcal{I}_g$ ($g \geq 3$) is known to be finitely generated by Johnson [42]).

PROBLEM 4. Let $\mathfrak{u}_g$ denote the graded Lie algebra associated to the prounipotent radical of the relative Malcev completion of $\mathcal{I}_g$ defined by Hain [29] and let $\mathfrak{u}_g \to \mathfrak{h}_g^\mathbb{Q}$ be the natural homomorphism (here $\mathfrak{h}_g^\mathbb{Q}$ denotes the graded Lie algebra consisting of symplectic derivations, with positive degrees, of the Malcev Lie algebra of $\pi_1\Sigma_g$). Determine whether this homomorphism is injective or not.

In [**80**], we defined a series of secondary characteristic classes for the mapping class group. However there was ambiguity coming from possible odd dimensional stable cohomology classes of the mapping class group. Because of the result of Madsen-Weiss cited above, we can now eliminate the ambiguity and give a precise definition as follows. For each i, we constructed in [**51**][**52**] explicit group cocycles $z_i \in Z^{2i}(\mathcal{M}_g; \mathbb{Q})$ which represent the i-th Mumford-Morita-Miller class e_i by making use of the homomorphism $\mathcal{M}_g \to U \rtimes \mathrm{Sp}(2g, \mathbb{Z})$ constructed in [**77**] which extends the (first) Johnson homomorphism $\mathcal{I}_g \to U$. These cocycles are $\mathcal{M}_g$-invariant by the definition. Furthermore we proved that such cocycles are unique up to coboundaries. On the other hand, as is well known, any *odd* class e_{2i-1} comes from the Siegel modular group $\mathrm{Sp}(2g, \mathbb{Z})$ so that there is a cocycle $z'_{2i-1} \in Z^{4i-2}(\mathcal{M}_g; \mathbb{Q})$ which comes from $\mathrm{Sp}(2g, \mathbb{Z})$. This cocycle is uniquely defined up to coboundaries and $\mathcal{M}_g$-invariant. Now consider the difference $z_{2i-1} - z'_{2i-1}$. It is a coboundary so that there exists a cochain $y_i \in C^{4i-3}(\mathcal{M}_g; \mathbb{Q})$ such that $\delta y_i = z_{2i-1} - z'_{2i-1}$. Since $H^{4i-3}(\mathcal{M}_g; \mathbb{Q}) = 0$ by [**64**] (in a suitable stable range), the cochain y_i is well-defined up to coboundaries.

Now let $\mathcal{K}_g$ be the kernel of the Johnson homomorphism so that we have an extension

(8) $$1 \longrightarrow \mathcal{K}_g \longrightarrow \mathcal{I}_g \longrightarrow U \longrightarrow 1.$$

Recall that Johnson [**43**] proved that $\mathcal{K}_g$ is the subgroup of $\mathcal{M}_g$ generated by Dehn twists along separating simple closed curves on Σ_g. The cocycle z'_{2i-1} is trivial on the Torelli group $\mathcal{I}_g$ while the cocycle z_{2i-1} (in fact any z_i) vanishes on $\mathcal{K}_g$. It follows that the restriction of the cochain y_i to $\mathcal{K}_g$ is a cocycle. Hence we obtain a cohomology class

$$d_i = [y_i|_{\mathcal{K}_g}] \in H^{4i-3}(\mathcal{K}_g; \mathbb{Q}).$$

This cohomology class is $\mathcal{M}_g$-invariant where $\mathcal{M}_g$ acts on $H^*(\mathcal{K}_g; \mathbb{Q})$ via outer conjugations. This can be shown as follows. For any element $\varphi \in \mathcal{M}_g$, let $\varphi_*(y_i)$ be the cochain obtained by applying the conjugation by φ on y_i. Since both cocycles z_{2i-1}, z'_{2i-1} are $\mathcal{M}_g$-invariant, we have $\delta \varphi_*(y_i) = \delta y_i$. Hence $\varphi_*(y_i) - y_i$ is a cocycle of $\mathcal{M}_g$. By the result of [**64**] again, we see that $\varphi_*(y_i) - y_i$ is a coboundary. Hence the restrictions of $\varphi_*(y_i)$ and y_i to $\mathcal{K}_g$ give the same cohomology class.

DEFINITION 5. We call the cohomology classes $d_i \in H^{4i-3}(\mathcal{K}_g; \mathbb{Q})^{\mathcal{M}_g}$ ($i = 1, 2 \cdots$) obtained above the *secondary* characteristic classes of the mapping class group.

The secondary classes d_i are stable in the following sense. Namely the pull back of them in $H^{4i-3}(\mathcal{K}_{g,1}; \mathbb{Q})$ are independent of g under natural homomorphisms induced by the inclusions $\mathcal{K}_{g,1} \to \mathcal{K}_{g+1,1}$ where $\mathcal{K}_{g,1}$ denotes the subgroup of $\mathcal{M}_{g,1}$ generated by Dehn twists along separating simple closed curves on $\Sigma_g \setminus D$. This is because the cocycles z_{2i-1}, z'_{2i-1} are stable with respect to g. It follows that the secondary classes d_i are defined for *all* g as elements of $H^{4i-3}(\mathcal{K}_{g,1}; \mathbb{Q})^{\mathcal{M}_{g,1}}$ although we have used the result of [**64**], which is valid only in a stable range. However as elements of $H^{4i-3}(\mathcal{K}_g; \mathbb{Q})^{\mathcal{M}_g}$ the class d_i is defined only for $g \geq 12i - 9$ at present, although it is highly likely that it is defined for all g. It was proved in [**75**] that d_1 is the generator of $H^1(\mathcal{K}_g; \mathbb{Z})^{\mathcal{M}_g} \cong \mathbb{Z}$ for all $g \geq 2$. See [**79**] for another approach to the secondary classes.

PROBLEM 6. Prove that all the secondary classes $d_2, d_3, \cdots$ are non-trivial.

Here is a problem concerning the first class d_1. Let C be a separating simple closed curve on Σ_g which divides Σ_g into two compact surfaces of genera h and $g-h$ and let $\tau_C \in \mathcal{K}_g$ be the Dehn twist along C. Then we know that the value of d_1 on $\tau_C \in \mathcal{K}_g$ is $h(g-h)$ (up to a constant depending on g). This is a very simple formula. However at present there is no known algorithm to calculate the value $d_1(\varphi)$ for a given element $\varphi \in \mathcal{K}_g$, say by analyzing the action of φ on $\pi_1 \Sigma_g$.

PROBLEM 7. Find explicit way of calculating $d_1(\varphi)$ for any given element $\varphi \in \mathcal{K}_g$. In particular, determine whether the Magnus representation $\mathcal{I}_{g,1} \to \mathrm{GL}(2g; \mathbb{Z}[H])$ of the Torelli group detects d_1 or not.

Suzuki [103] proved that the Magnus representation of the Torelli group mentioned above is *not* faithful so that it may happen that the intersection of the kernel of the Magnus representation with $\mathcal{K}_g$ is not contained in the kernel of d_1. We may also ask whether the representation of the hyperelliptic mapping class group given by Jones [45], restricted to the intersection of this group with $\mathcal{K}_g$, detects d_1 or not (cf. Kasahara [46] for a related work for the case $g=2$). There are also various interesting works related to the class d_1 such as Endo [14] and Morifuji [71] treating the hyperelliptic mapping class group, Kitano [54] as well as Hain and Reed [33].

Recently Biss and Farb [5] proved that the group $\mathcal{K}_g$ is not finitely generated for all $g \geq 3$ ($\mathcal{K}_2$ is known to be an infinitely generated free group by Mess [69]). However it is still not yet known whether the abelianization $H_1(\mathcal{K}_g)$ is finitely generated or not (cf. Problem 2.2 of [80]).

Finally we would like to mention that Kawazumi [50] is developing a theory of harmonic Magnus expansions which gives in particular a system of differential forms representing the Mumford-Morita-Miller classes on the universal family of curves over the moduli space $\mathbf{M}_g$.

4. Outer automorphism group of free groups

As already mentioned in § 1, let F_n denote a free group of rank $n \geq 2$ and let $\mathrm{Out}\, F_n = \mathrm{Aut}\, F_n / \mathrm{Inn}\, F_n$ denote the outer automorphism group of F_n. In 1986, Culler and Vogtmann [11] defined a space X_n, called the *Outer Space*, which plays the role of the Teichmüller space where the mapping class group is replaced by $\mathrm{Out}\, F_n$. In particular, they proved that X_n is contractible and $\mathrm{Out}\, F_n$ acts on it properly discontinuously. The quotient space

$$\mathbf{G}_n = X_n / \mathrm{Out}\, F_n$$

is called the moduli space of *graphs* which is the space of all the isomorphism classes of metric graphs with fundamental group F_n. Recently many works have been done on the structure of $\mathrm{Out}\, F_n$ as well as $\mathbf{G}_n$, notably by Vogtmann (see her survey paper [106]), Bestvina (see [2]) and many others.

It is an interesting and important problem to compare similarity as well as difference between the mapping class group and $\mathrm{Out}\, F_n$ which will be discussed at several places in this book. Here we would like to concentrate on the cohomological side of this problem.

Hatcher and Vogtmann [37] (see also Hatcher [35]) proved that the homology of $\mathrm{Out}\, F_n$ stabilizes in a certain stable range. This is an analogue of Harer's stability theorem [34] for the mapping class group. More precisely, they proved that the

natural homomorphisms
$$\mathrm{Aut}\, F_n \to \mathrm{Aut}\, F_{n+1}, \quad \mathrm{Aut}\, F_n \to \mathrm{Out}\, F_n$$
induce isomorphisms on the i-dimensional homology group for $n \geq 2i+2$ and $n \geq 2i+4$, respectively. Thus we can speak of the stable cohomology group
$$\lim_{n\to\infty} \widetilde{H}^*(\mathrm{Out}\, F_n)$$
of $\mathrm{Out}\, F_n$.

In the case of the mapping class group, it was proved in [70][73] that the natural homomorphism $\mathcal{M}_g \to \mathrm{Sp}(2g,\mathbb{Z})$ induces an injection
$$\lim_{g\to\infty} H^*(\mathrm{Sp}(2g,\mathbb{Z});\mathbb{Q}) \cong \mathbb{Q}[c_1, c_3, \cdots] \subset \lim_{g\to\infty} H^*(\mathcal{M}_g; \mathbb{Q})$$
on the stable rational cohomology group where the stable rational cohomology of $\mathrm{Sp}(2g,\mathbb{Z})$ was determined by Borel [6][7]. In the case of $\mathrm{Out}\, F_n$, Igusa proved the following remarkable result which shows a sharp contrast with the case of the mapping class group (see Theorem 8.5.3 and Remark 8.5.4 of [38]).

THEOREM 8 (Igusa[38]). *The homomorphism*
$$\widetilde{H}^k(\mathrm{GL}(n,\mathbb{Z});\mathbb{Q}) \longrightarrow \widetilde{H}^k(\mathrm{Out}\, F_n; \mathbb{Q})$$
induced by the natural homomorphism $\mathrm{Out}\, F_n \to \mathrm{GL}(n,\mathbb{Z})$ *is the 0-map in the stable range* $n \geq 2k+1$.

Recall here that the stable cohomology of $\mathrm{GL}(n,\mathbb{Z})$ is given by
$$\lim_{n\to\infty} H^*(\mathrm{GL}(n,\mathbb{Z});\mathbb{Q}) \cong \Lambda_{\mathbb{Q}}(\beta_5, \beta_9, \beta_{13}, \cdots)$$
due to Borel in the above cited papers.

On the other hand, the first non-trivial rational cohomology of the group $\mathrm{Aut}\, F_n$ was given by Hatcher and Vogtmann [36]. They showed that, up to cohomology degree 6, the only non-trivial rational cohomology is
$$H^4(\mathrm{Aut}\, F_4; \mathbb{Q}) \cong \mathbb{Q}.$$
Around the same time, by making use of a remarkable theorem of Kontsevich given in [55][56], the author constructed many homology classes in $H_*(\mathrm{Out}\, F_n; \mathbb{Q})$ (see [80] and §10 below). The simplest one in this construction gave a series of elements
$$\mu_i \in H_{4i}(\mathrm{Out}\, F_{2i+2}; \mathbb{Q}) \quad (i = 1, 2, \cdots)$$
and the first one μ_1 was shown to be non-trivial by a computer calculation. Responding to an inquiry of the author, Vogtmann communicated us that she modified the argument in [36] to obtain an isomorphism $H^4(\mathrm{Out}\, F_4; \mathbb{Q}) \cong \mathbb{Q}$. Thus we could conclude that μ_1 is the generator of this group (see [80][106]). Recently Conant and Vogtmann proved that the second class $\mu_2 \in H_8(\mathrm{Out}\, F_6; \mathbb{Q})$ is also non-trivial in their paper [10] where they call μ_i the Morita classes. Furthermore they constructed many cycles of the moduli space $\mathbf{G}_n$ of graphs by explicit constructions in the Outer Space X_n.

More recently, Ohashi [91] determined the rational cohomology group of $\mathrm{Out}\, F_n$ for all $n \leq 6$ and in particular he showed
$$H_8(\mathrm{Out}\, F_6; \mathbb{Q}) \cong \mathbb{Q}.$$
It follows that μ_2 is the generator of this group. At present, the above two groups (and one more group, $H_7(\mathrm{Aut}\, F_5; \mathbb{Q}) \cong \mathbb{Q}$ proved by Gerlits [25]) are the only known

non-trivial rational homology groups of Out F_n (and Aut F_n). Now we would like to present the following conjecture based on our expectation that the classes μ_i should concern not only the cohomology of Out F_n but also the structure of the arithmetic mapping class group (see §8) as well as homology cobordism invariants of homology 3-spheres as will be explained in §11 below and [**83**].

CONJECTURE 9. The classes μ_i are non-trivial for all $i = 1, 2, \cdots$.

More generally we have the following.

PROBLEM 10. Produce non-trivial rational (co)homology classes of Out F_n.

Next we consider the group IOut$_n$. In [**38**] Igusa defined higher Franz-Reidemeister torsion classes
$$\tau_{2i} \in H^{4i}(\text{IOut}_n; \mathbb{R})$$
as a special case of his general theory. These classes reflect Igusa's result mentioned above (Theorem 8) that the pull back of the Borel classes $\beta_{4i+1} \in H^{4i+1}(\text{GL}(n, \mathbb{Z}); \mathbb{R})$ in $H^{4i+1}(\text{Out } F_n; \mathbb{R})$ vanish. However it seems to be unknown whether his classes are non-trivial or not.

PROBLEM 11 (Igusa). Prove that the higher Franz-Reidemeister torsion classes $\tau_{2i} \in H^{4i}(\text{IOut}_n; \mathbb{R})$ are non-trivial in a suitable stable range.

In the unstable range, where the Borel classes vanish in $H^*(\text{GL}(n, \mathbb{Z}); \mathbb{R})$, there seem to be certain relations between the classes τ_{2i}, (dual of) μ_i and unstable cohomology classes in $H^*(\text{GL}(n, \mathbb{Z}); \mathbb{Q})$. As the first such example, we would like to ask the following specific problem.

PROBLEM 12. Prove (or disprove) that the natural homomorphism
$$H^4(\text{Out } F_4; \mathbb{Q}) \cong \mathbb{Q} \longrightarrow H^4(\text{IOut}_4; \mathbb{Q})^{GL}$$
is an isomorphism where the right hand side is generated by (certain non-zero multiple of) τ_2.

Here is another very specific problem. We know the following groups explicitly by various authors:
$$H^8(\mathcal{M}_{3,*}; \mathbb{Q}) \cong \mathbb{Q}^2 \quad (\text{Looijenga [{\bf 61}]})$$
$$H^8(\text{GL}(6, \mathbb{Z}); \mathbb{Q}) \cong \mathbb{Q} \quad (\text{Elbaz-Vincent, Gangl, Soulé [{\bf 13}]})$$
$$H^8(\text{Out } F_6; \mathbb{Q}) \cong \mathbb{Q} \quad (\text{Ohashi [{\bf 91}]})$$

On the other hand, we have the following natural injection i as well as projection p
$$(9) \qquad \mathcal{M}_{3,*} \xrightarrow{i} \text{Out } F_6 \xrightarrow{p} \text{GL}(6, \mathbb{Z}).$$

PROBLEM 13. Determine the homomorphisms
$$(10) \qquad H^8(\mathcal{M}_{3,*}; \mathbb{Q}) \xleftarrow{i^*} H^8(\text{Out } F_6; \mathbb{Q}) \xleftarrow{p^*} H^8(\text{GL}(6, \mathbb{Z}); \mathbb{Q})$$
induced by the above homomorphisms in (9).

REMARK 14. It seems to be natural to conjecture that the right map in (10) is an isomorphism while the left map is trivial. The former part is based on a consideration of possible geometric meaning of the classes $\mu_i \in H_{4i}(\text{Out } F_{2i+2}; \mathbb{Q})$. For the particular case $i = 2$ here, it was proved in [**13**] that $H^9(\text{GL}(6, \mathbb{Z}); \mathbb{Q}) = 0$. It follows that the Borel class in $H^9(\text{GL}(6, \mathbb{Z}); \mathbb{R})$ vanishes. Because of this, it is likely that the Igusa class $\tau_4 \in H^8(\mathrm{IOut}_6; \mathbb{R})$ would vanish as well and the class μ_2 would survive in $H_8(\mathrm{GL}(6, \mathbb{Z}); \mathbb{Q})$. For the latter part, see Remark 19 below.

PROBLEM 15. Define unstable (co)homology classes of $\mathrm{GL}(n, \mathbb{Z})$. In particular, what can be said about the image of $\mu_i \in H_{4i}(\mathrm{Out}\, F_{2i+2}; \mathbb{Q})$ in $H_{4i}(\mathrm{GL}(2i+2, \mathbb{Z}); \mathbb{Q})$ under the projection $\mathrm{Out}\, F_{2k+2} \to \mathrm{GL}(2k+2, \mathbb{Z})$?

The above known results as well as explicit computation made so far seem to support the following conjecture (which might be something like a folklore).

CONJECTURE 16. *) The stable rational cohomology of $\mathrm{Out}\, F_n$ is trivial. Namely
$$\lim_{n\to\infty} \widetilde{H}^*(\mathrm{Out}\, F_n; \mathbb{Q}) = 0.$$

We can also ask how the cohomology of $\mathrm{Out}\, F_n$ with *twisted coefficients* look like.

PROBLEM 17. Compute the cohomology of $\mathrm{Aut}\, F_n$ and $\mathrm{Out}\, F_n$ with coefficients in various $\mathrm{GL}(n, \mathbb{Q})$-modules.

For example, we could ask how Looijenga's result [**62**] for the case of the mapping class group can be generalized in these contexts. We refer to the work of Kawazumi [**49**] and also Satoh [**97**] for recent results concerning the above problem.

Finally we recall the following well known problem.

PROBLEM 18. Determine whether the natural homomorphisms
$$\widetilde{H}^*(\mathrm{Aut}\, F_{2g}; \mathbb{Q}) \to \widetilde{H}^*(\mathcal{M}_{g,1}; \mathbb{Q})$$
$$\widetilde{H}^*(\mathrm{Out}\, F_{2g}; \mathbb{Q}) \to \widetilde{H}^*(\mathcal{M}_{g,*}; \mathbb{Q})$$
induced by the inclusions $\mathcal{M}_{g,1} \to \mathrm{Aut}\, F_{2g}$, $\mathcal{M}_{g,*} \to \mathrm{Out}\, F_{2g}$ are trivial or not.

We refer to a result of Wahl [**107**] for a homotopy theoretical property of the homomorphism $\mathcal{M}_{g,1} \to \mathrm{Aut}\, F_{2g}$ where g tends to ∞.

REMARK 19. The known results as well as explicit computations made so far seem to suggest that the above maps are trivial. According to a theorem of Kontsevich [**55**][**56**] (see §9 below), the triviality of the second map above is equivalent to the statement that the natural inclusion
$$\mathfrak{l}_\infty^+ \to \mathfrak{a}_\infty^+$$
between two infinite dimentional Lie algebras (see §9 for the definition) induces the *trivial map*
$$H^*(\mathfrak{a}_\infty^+)^{Sp} \to H^*(\mathfrak{l}_\infty^+)^{Sp}$$
in the Sp-invariant cohomology groups. Here the trivial map means that it is the 0-map except for the bigraded parts which correspond to the 0-dimensional homology groups of $\mathcal{M}_{g,*}$ and $\mathrm{Out}\, F_{2g}$.

REMARK 20. In this paper, we are mainly concerned with the rational cohomology group of the mapping class group, $\mathrm{Out}\, F_n$ and other groups. As for cohomology group with finite coefficients or torsion classes, here we only mention the work of Galatius [**22**] which determines the mod p stable cohomology of the mapping class group and also Hatcher's result [**35**] that the stable homology of $\mathrm{Out}\, F_n$ contains the homology of $\Omega^\infty S^\infty$ as a direct summand.

---
*) This conjecture has been recently proved by S. Galatius.

5. The derivation algebra of free Lie algebras and the traces

As in §1, let F_n be a free group of rank $n \geq 2$ and let us denote the abelianization $H_1(F_n)$ of F_n simply by H_n. Also let $H_n^{\mathbb{Q}} = H_n \otimes \mathbb{Q}$. Sometimes we omit n and we simply write H and $H_{\mathbb{Q}}$ instead of H_n and $H_n^{\mathbb{Q}}$. Let

$$\mathcal{L}_n = \oplus_{k=1}^{\infty} \mathcal{L}_n(k)$$

be the free graded Lie algebra generated by H_n. Also let $\mathcal{L}_n^{\mathbb{Q}} = \mathcal{L}_n \otimes \mathbb{Q}$. We set

$$\mathrm{Der}^+(\mathcal{L}_n) = \{\text{derivation } D \text{ of } \mathcal{L}_n \text{ with positive degree}\}$$

which has a natural structure of a graded Lie algebra over $\mathbb{Z}$. The degree k part of this graded Lie algebra can be expressed as

$$\mathrm{Der}^+(\mathcal{L}_n)(k) = \mathrm{Hom}(H_n, \mathcal{L}_n(k+1))$$

and we have

$$\mathrm{Der}^+(\mathcal{L}_n) = \bigoplus_{k=1}^{\infty} \mathrm{Der}^+(\mathcal{L}_n)(k).$$

Similarly we consider $\mathrm{Der}^+(\mathcal{L}_n^{\mathbb{Q}})$ which is a graded Lie algebra over $\mathbb{Q}$.

In the case where we are given an identification $\pi_1(\Sigma_g \setminus \mathrm{Int} D) \cong F_{2g}$, we have the symplectic class $\omega_0 \in \mathcal{L}_{2g}(2) = \Lambda^2 H_{2g}$ and we can consider the following graded Lie subalgebra

$$\mathfrak{h}_{g,1} = \{D \in \mathrm{Der}^+(\mathcal{L}_{2g}); D(\omega_0) = 0\}$$
$$= \bigoplus_{k=1}^{\infty} \mathfrak{h}_{g,1}(k).$$

Similarly we have $\mathfrak{h}_{g,1}^{\mathbb{Q}} = \mathfrak{h}_{g,1} \otimes \mathbb{Q}$.

In our paper [**76**], for each k, we defined a certain homomorphism

$$\mathrm{trace}(k) : \mathrm{Der}^+(\mathcal{L}_n)(k) \longrightarrow S^k H_n$$

where $S^k H_n$ denotes the k-th symmetric power of H_n. We call this "trace" because it is defined as the usual trace of the abelianized *non-commutative Jacobian matrix* of each homogeneous derivation. Here we recall the definition briefly from the above cited paper. Choose a basis $x_1, \cdots, x_n$ of $H_n = H_1(F_n; \mathbb{Z})$. We can consider $\mathcal{L}_n(k+1)$ as a natural submodule of $H_n^{\otimes(k+1)}$ consisting of all the Lie polynomials of degree $k+1$. For example $[x_1, x_2] \in \mathcal{L}_n(2)$ corresponds to the element $x_1 \otimes x_2 - x_2 \otimes x_1 \in H^{\otimes 2}$. By using the concept of the Fox free differential, we can also embed $\mathcal{L}_n(k+1)$ into the set $(H_n^{\otimes k})^n$ of all the n-dimensional column vectors with entries in $H_n^{\otimes k}$ by the following correspondence

$$\mathcal{L}_n(k+1) \ni \eta \longmapsto \left(\frac{\partial \eta}{\partial x_i}\right) \in (H_n^{\otimes k})^n.$$

Here for each monomial $\eta \in \mathcal{L}_n(k+1) \subset H_n^{\otimes(k+1)}$ which is uniquely expressed as

$$\eta = \eta_1 \otimes x_1 + \cdots + \eta_n \otimes x_n \quad (\eta_i \in H_n^{\otimes k}),$$

we have

$$\frac{\partial \eta}{\partial x_i} = \eta_i.$$

DEFINITION 21. In the above terminologies, the k-th trace $\mathrm{trace}(k)\colon \mathrm{Der}^+(\mathcal{L}_n)(k) \longrightarrow S^k H_n$ is defined by

$$\mathrm{trace}(k)(f) = \left(\sum_{i=1}^n \frac{\partial f(x_i)}{\partial x_i}\right)^{\mathrm{ab}}$$

where $f \in \mathrm{Der}^+(\mathcal{L}_n)(k) = \mathrm{Hom}(H_n, \mathcal{L}_n(k+1))$ and the superscript ab denotes the natural projection $H_n^{\otimes k} \to S^k H_n$.

REMARK 22. If we identify the target $\mathrm{Hom}(H_n, \mathcal{L}_n(k+1))$ of $\mathrm{trace}(k)$ with

$$H_n^* \otimes \mathcal{L}_n(k+1) \subset H_n^* \otimes H_n^{\otimes(k+1)}$$

where $H_n^* = \mathrm{Hom}(H_n, \mathbb{Z})$ denotes the dual space of H_n, then it follows immediately from the definition that $\mathrm{trace}(k)$ is equal to the restriction of the contraction

$$C_{k+1}\colon H_n^* \otimes H_n^{\otimes(k+1)} \longrightarrow H_n^{\otimes k}$$

followed by the abelianization $H_n^{\otimes k} \to S^k H_n$. Here

$$C_{k+1}(f \otimes u_1 \otimes \cdots \otimes u_{k+1}) = f(u_{k+1}) u_1 \otimes \cdots \otimes u_k$$

for $f \in H_n^*, u_i \in H_n$. Also it is easy to see that, if we replace C_{k+1} with C_1 defined by

$$C_1(f \otimes u_1 \otimes \cdots \otimes u_{k+1}) = f(u_1) u_2 \otimes \cdots \otimes u_{k+1}$$

in the above discussion, then we obtain $(-1)^k \mathrm{trace}(k)$.

EXAMPLE 23. Let $\mathrm{ad}_{x_2}(x_1)^k \in \mathrm{Der}^+(\mathcal{L}_n)(k)$ be the element defined by

$$\mathrm{ad}_{x_2}(x_1)^k(x_2) = [x_1, [x_1, [\cdots, [x_1, x_2]\cdots] \quad (k\text{-times } x_1)$$
$$\mathrm{ad}_{x_2}(x_1)^k(x_i) = 0 \quad (i \neq 2).$$

Then a direct computation shows that

$$\mathrm{trace}(k)(\mathrm{ad}_{x_2}(x_1)^k) = x_1^k.$$

As was mentioned in [76], the traces are $GL(H_n)$-equivariant in an obvious sense. Since x_1^k generates $S^k H_n$ as a $GL(H_n)$-module, the above example implies that the mapping $\mathrm{trace}(k) \colon \mathrm{Der}^+(\mathcal{L}_n)(k) \longrightarrow S^k H_n$ is surjective for any k. Another very important property of the traces proved in the above cited paper is that they vanish identically on the commutator ideal $[\mathrm{Der}^+(\mathcal{L}_n), \mathrm{Der}^+(\mathcal{L}_n)]$. Hence we have the following *surjective* homomorphism of graded Lie algebras

$$(\tau_1, \oplus_k \mathrm{trace}(k))\colon \mathrm{Der}^+(\mathcal{L}_n) \longrightarrow \mathrm{Hom}(H_n, \Lambda^2 H_n) \oplus \bigoplus_{k=2}^\infty S^k H_n$$

where the target is understood to be an *abelian* Lie algebra. We have also proved that, for any k, $\mathrm{trace}(2k)$ vanishes identically on $\mathfrak{h}_{g,1}$ and that $\mathrm{trace}(2k+1)\colon \mathfrak{h}^{\mathbb{Q}}_{g,1}(2k+1) \to S^{2k+1} H^{\mathbb{Q}}_{2g}$ is surjective. Thus we have a surjective homomorphism

(11) $$(\tau_1, \oplus_k \mathrm{trace}(2k+1))\colon \mathfrak{h}^{\mathbb{Q}}_{g,1} \longrightarrow \Lambda^3 H^{\mathbb{Q}}_{2g} \oplus \bigoplus_{k=1}^\infty S^{2k+1} H^{\mathbb{Q}}_{2g}$$

of graded Lie algebras which we conjectured to give the *abelianization* of the Lie algebra $\mathfrak{h}^{\mathbb{Q}}_{g,1}$ (see Conjecture 6.10 of [80]).

Recently Kassabov [47] (Theorem 1.4.11) proved the following remarkable result. Let $x_1, \cdots, x_n$ be a basis of H_n as before.

THEOREM 24 (Kassabov). *Up to degree $n(n-1)$, the graded Lie algebra $\mathrm{Der}^+(\mathcal{L}_n^{\mathbb{Q}})$ is generated as a Lie algebra and $\mathfrak{sl}(n,\mathbb{Q})$-module by the elements $\mathrm{ad}(x_1)^k$ ($k=1,2,\cdots$) and the element D which sends x_1 to $[x_2,x_3]$ and $x_i (i\neq 1)$ to 0.*

If we combine this theorem with the concept of the traces, we obtain the following.

THEOREM 25. *The surjective Lie algebra homomorphism*

$$(\tau_1, \oplus_k \mathrm{trace}(k)) : \mathrm{Der}^+(\mathcal{L}_n^{\mathbb{Q}}) \longrightarrow \mathrm{Hom}(H_n^{\mathbb{Q}}, \Lambda^2 H_n^{\mathbb{Q}}) \oplus \bigoplus_{k=2}^{\infty} S^k H_n^{\mathbb{Q}},$$

induced by the degree 1 part and the traces, gives the abelianization of the graded Lie algebra $\mathrm{Der}^+(\mathcal{L}_n^{\mathbb{Q}})$ up to degree $n(n-1)$ so that any element of degree $2\leq d \leq n(n-1)$ with vanishing trace belongs to the commutator ideal $[\mathrm{Der}^+(\mathcal{L}_n^{\mathbb{Q}}), \mathrm{Der}^+(\mathcal{L}_n^{\mathbb{Q}})]$. Furthermore, any $\mathfrak{sl}(n,\mathbb{Q})$-equivariant splitting to this abelianization generates $\mathrm{Der}^+(\mathcal{L}_n^{\mathbb{Q}})$ in this range. Hence stably there exists an isomorphism

$$H_1\left(\mathrm{Der}^+(\mathcal{L}_\infty^{\mathbb{Q}})\right) \cong \mathrm{Hom}(H_\infty^{\mathbb{Q}}, \Lambda^2 H_\infty^{\mathbb{Q}}) \oplus \bigoplus_{k=2}^{\infty} S^k H_\infty^{\mathbb{Q}}$$

and the degree 1 part and (any $\mathfrak{sl}(n,\mathbb{Q})$-equivariant splittings of) the traces generate $\mathrm{Der}^+(\mathcal{L}_\infty^{\mathbb{Q}})$.

Although the structure of $\mathfrak{h}_{g,1}^{\mathbb{Q}}$ is much more complicated than that of $\mathrm{Der}^+(\mathcal{L}_n^{\mathbb{Q}})$, Kassabov's argument adapted to this case together with some additional idea will produce enough information about the generation as well as the abelianization of $\mathfrak{h}_{g,1}^{\mathbb{Q}}$ in a certain *stable range*. Details will be given in our forthcoming paper [83]. It follows that any element in the Lie algebra $\mathfrak{h}_{\infty,1}^{\mathbb{Q}}$ can be expressed in terms of the degree 1 part and the traces.

6. The second cohomology of $\mathfrak{h}_{g,1}^{\mathbb{Q}}$

In this section, we define a series of elements in $H^2(\mathfrak{h}_{g,1}^{\mathbb{Q}})^{Sp}$ which denotes the *Sp-invariant part* of the second cohomology of the graded Lie algebra $\mathfrak{h}_{g,1}^{\mathbb{Q}}$.

As is well known, $U_{\mathbb{Q}} = \Lambda^3 H_{\mathbb{Q}}/\omega_0 \wedge H_{\mathbb{Q}}$ and $S^{2k+1} H_{\mathbb{Q}}$ ($k=1,2,\cdots$) are all irreducible representations of $\mathrm{Sp}(2g,\mathbb{Q})$. It is well known in the representation theory that, for any irreducible representation V of the algebraic group $\mathrm{Sp}(2g,\mathbb{Q})$, the tensor product $V\otimes V$ contains a unique trivial summand $\mathbb{Q}\subset V\otimes V$ (cf. [20] for generalities of the representations of the algebraic group $\mathrm{Sp}(2g,\mathbb{Q})$). In our case where V is any of the above irreducible representations, it is easy to see that the trivial summand appears in the second exterior power part $\Lambda^2 V \subset V\otimes V$. It follows that each of

$$\Lambda^2 U_{\mathbb{Q}}, \ \Lambda^2 S^3 H_{\mathbb{Q}}, \ \Lambda^2 S^5 H_{\mathbb{Q}}, \ \cdots$$

contains a unique trivial summand $\mathbb{Q}$. Let

$$\iota_1: \Lambda^2 U_{\mathbb{Q}} \to \mathbb{Q}, \quad \iota_{2k+1}: \Lambda^2 S^{2k+1} H_{\mathbb{Q}} \to \mathbb{Q} \quad (k=1,2,\cdots)$$

be the unique (up to scalars) *Sp*-equivariant homomorphism. We would like to call them *higher intersection pairing* on surfaces which generalize the usual one $\Lambda^2 H_{\mathbb{Q}} \to \mathbb{Q}$. We can write

$$\iota_1 \in H^2(U_{\mathbb{Q}})^{Sp}, \quad \iota_{2k+1} \in H^2(S^{2k+1} H_{\mathbb{Q}})^{Sp}.$$

DEFINITION 26. We define the cohomology classes
$$e_1, t_3, t_5, \cdots \in H^2(\mathfrak{h}_{g,1}^{\mathbb{Q}})^{Sp}$$
by setting
$$e_1 = \bar{\tau}_1^*(\iota_1), \quad t_{2k+1} = \mathrm{trace}(2k+1)^*(\iota_{2k+1})$$
where $\bar{\tau}_1$ denotes the composition $\mathfrak{h}_{g,1}^{\mathbb{Q}} \to \Lambda^3 H_{\mathbb{Q}} \to U_{\mathbb{Q}}$.

CONJECTURE 27. The classes $e_1, t_3, t_5, \cdots$ are all non-trivial. Furthermore they are linearly independent and form a basis of $H^2(\mathfrak{h}_{g,1}^{\mathbb{Q}})^{Sp}$.

REMARK 28. The element e_1 is the *Lie algebra version* of the first Mumford-Morita-Miller class (we use the same notation).

REMARK 29. The Lie algebra $\mathfrak{h}_{g,1}^{\mathbb{Q}}$ is graded so that the cohomology group $H^2(\mathfrak{h}_{g,1}^{\mathbb{Q}})^{Sp}$ is *bigraded*. Let $H^2(\mathfrak{h}_{g,1}^{\mathbb{Q}})_n^{Sp}$ denote the weight n part of $H^2(\mathfrak{h}_{g,1}^{\mathbb{Q}})^{Sp}$ (see §9 for more details). Then by definition we have
$$e_1 \in H^2(\mathfrak{h}_{g,1}^{\mathbb{Q}})_2^{Sp}, \quad t_{2k+1} \in H^2(\mathfrak{h}_{g,1}^{\mathbb{Q}})_{4k+2}^{Sp}.$$
Hence if the elements $e_1, t_3, t_5, \cdots$ are non-trivial, then they are automatically linearly independent. Thus the above conjecture can be rewritten as
$$H^2(\mathfrak{h}_{g,1}^{\mathbb{Q}})_n^{Sp} \cong \begin{cases} \mathbb{Q} & (n = 2, 6, 10, 14, \cdots) \\ 0 & (\text{otherwise}) \end{cases}$$
where the summands $\mathbb{Q}$ in degrees $2, 6, 10, \cdots$ are generated by the above classes.

As for the non-triviality, all we know at present is the non-triviality of e_1, t_3, t_5. The non-triviality of the class t_{2k+1} is the same as that of the class μ_k because of the theorem of Kontsevich described in §9. This will be explained in that section.

7. Constructing cohomology classes of $\mathfrak{h}_{g,1}^{\mathbb{Q}}$

In this section, we describe a general method of constructing Sp-invariant cohomology classes of the Lie algebra $\mathfrak{h}_{g,1}^{\mathbb{Q}}$ which generalize the procedure given in the previous section. As was already mentioned in our paper [80], the homomorphism (11) induces the following homomorphim in the Sp-invariant part of the cohomology

(12) $$H^*\left(\Lambda^3 H_{\mathbb{Q}} \oplus \bigoplus_{k=1}^{\infty} S^{2k+1} H_{2g}^{\mathbb{Q}}\right)^{Sp} \longrightarrow H^*(\mathfrak{h}_{g,1}^{\mathbb{Q}})^{Sp}.$$

By the same way as in [78][51], the left hand side can be computed by certain polynomial algebra
$$\mathbb{Q}[\Gamma; \Gamma \in \mathcal{G}^{odd}]$$
generated by graphs belonging to $\mathcal{G}^{odd}$ which denotes the set of all isomorphism classes of *connected graphs* with valencies in the set
$$3, 3, 5, 7, \cdots$$
of odd integers. Here we write two copies of 3 because of different roles: the first one is related to the target $\Lambda^3 H_{\mathbb{Q}}$ of τ_1 (alternating) while the second one is related to the target $S^3 H_{\mathbb{Q}}$ of $\mathrm{trace}(3)$ (symmetric). The other $2k+1$ $(k = 2, 3, \cdots)$ are related to the target $S^{2k+1} H_{\mathbb{Q}}$ of $\mathrm{trace}(2k+1)$. Thus we obtain a homomorphism

(13) $$\Phi : \mathbb{Q}[\Gamma; \Gamma \in \mathcal{G}^{odd}] \longrightarrow H^*(\mathfrak{h}_{g,1}^{\mathbb{Q}})^{Sp}.$$

The elements $e_1, t_3, t_5, \cdots$ defined in Definition 26 arise as the images, under Φ, of those graphs with exactly two vertices which are connected by $3, 3, 5, 7, \cdots$ edges.

REMARK 30. As was mentioned already in the previous section §6, the cohomology of $\mathfrak{h}_{g,1}^{\mathbb{Q}}$ is bigraded. Let $\Gamma \in \mathcal{G}^{odd}$ be a connected graph whose valencies are v_3^a times 3 (alternating), v_3^s times 3 (symmetric) and v_{2k+1} times $2k+1$ ($2k+1 > 3$). Then
$$\Phi(\Gamma) \in H^d(\mathfrak{h}_{g,1}^{\mathbb{Q}})_n^{Sp} \tag{14}$$
where
$$d = v_3^a + v_3^s + v_5 + v_7 + \cdots$$
$$n = v_3^a + 3v_3^s + 5v_5 + 7v_7 + \cdots.$$
Observe that $n + 2v_3^a$ is equal to twice of the number of edges of Γ. It follows that n (and hence d) is always an even integer.

PROBLEM 31. Find explicit graphs $\Gamma \in \mathcal{G}^{odd}$ such that the corresponding homology classes $\Phi(\Gamma)$ are non-trivial.

8. Three groups beyond the mapping class group

In view of the definition of the Lie algebra $\mathfrak{h}_{g,1}$ (see §5), we may say that it is the "Lie algebra version" of the mapping class group $\mathcal{M}_{g,1}$. However the result of the author [76] showed that it is too big to be considered so and the following question arose: what is the algebraic and/or geometric meaning of the complement of the image of $\mathcal{M}_{g,1}$ in $\mathfrak{h}_{g,1}$? Two groups came into play in this framework in the 1990's. One is the arithmetic mapping class group through the works of number theorists, notably Oda, Nakamura and Matsumoto, and the other is Out F_n via the theorem of Kontsevich described in the next section §9. In this section, we would like to consider the former group briefly from a very limited point of view (see [87][66] and references therein for details). The latter group was already introduced in §4 and will be further discussed in §10.

More recently, it turned out that we have to treat one more group in the above setting and that is the group of homology cobordism classes of homology cylinders. This will be discussed in §11 below. We strongly expect that the traces will give rise to meaningful invariants in each of these three groups beyond the mapping classs group.

Now we consider the first group above. The action of $\mathcal{M}_{g,1}$ on the lower central series of $\Gamma = \pi_1(\Sigma_g \setminus \mathrm{Int}\, D)$ induces a filtration $\{\mathcal{M}_{g,1}(k)\}_k$ on $\mathcal{M}_{g,1}$ as follows. Let $\Gamma_1 = [\Gamma, \Gamma]$ be the commutator subgroup of Γ and inductively define $\Gamma_{k+1} = [\Gamma, \Gamma_k]$ ($k = 1, 2, \cdots$). The quotient group $N_k = \Gamma/\Gamma_k$ is called the k-th nilpotent quotient of Γ. Note that N_1 is canonically isomorphic to $H_{2g} = H_1(\Sigma_g; \mathbb{Z})$. Now we set
$$\mathcal{M}_{g,1}(k) = \{\varphi \in \mathcal{M}_{g,1}; \varphi \text{ acts on } N_k \text{ trivially}\}.$$
Thus the first group $\mathcal{M}_{g,1}(1)$ in this filtration is nothing but the Torelli group $\mathcal{I}_{g,1}$. As is well known, the quotient group Γ_k/Γ_{k+1} can be identified with the $(k+1)$-st term $\mathcal{L}_{2g}(k+1)$ of the free graded Lie algebra generated by H_{2g} (see § 5). It can be checked that the correspondence
$$\mathcal{M}_{g,1}(k) \ni \varphi \longmapsto$$
$$\Gamma \ni \alpha \mapsto \varphi_*(\alpha)\alpha^{-1} \in \Gamma_k/\Gamma_{k+1} \cong \mathcal{L}_{2g}(k+1)$$

descends to a homomorphism

$$\tau_k : \mathcal{M}_{g,1}(k) \longrightarrow \mathrm{Hom}(H_{2g}, \mathcal{L}_{2g}(k+1))$$

which is now called the k-th Johnson homomorphism because it was introduced by Johnson (see [40][41]). Furthermore it turns out that the totality $\{\tau_k\}_k$ of these homomorphims induces an injective homomorphism of graded Lie algebras

(15) $$\mathrm{Gr}^+(\mathcal{M}_{g,1}) = \bigoplus_{k=1}^{\infty} \mathcal{M}_{g,1}(k)/\mathcal{M}_{g,1}(k+1) \longrightarrow \mathfrak{h}_{g,1}$$

(see [76][80] for details). Although there have been obtained many important results concerning the image of the above homomorphism, the following is still open.

PROBLEM 32. Determine the image as well as the cokernel of the homomorphism (15) explicitly.

Note that Hain [29] proved that the image of (15), after tensored with $\mathbb{Q}$, is precisely the Lie subalgebra generated by the degree 1 part. However it is unclear which part of $\mathfrak{h}_{g,1}^{\mathbb{Q}}$ belongs to this Lie subalgebra.

In relation to this problem, Oda predicted, in the late 1980's, that there should arise "arithmetic obstructions" to the surjectivity of Johnson homomorphism. More precisely, based on the theory of Ihara in number theory which treated mainly the case $g=0, n=3$, he expected that the absolute Galois group $\mathrm{Gal}(\overline{\mathbb{Q}}/\mathbb{Q})$ should "appear" in $\mathfrak{h}_{g,1} \otimes \mathbb{Z}_\ell$ outside of the geometric part and which should be Sp-invariant for any genus g and for any prime ℓ. In 1994, Nakamura [86] proved, among other results, that this is in fact the case (see also Matsumoto [65]). This was the second obstruction to the surjectivity of Johnson homomorphism, the first one being the traces in [76]. This raised the following problem.

PROBLEM 33. Describe the Galois images in $\mathfrak{h}_{g,1} \otimes \mathbb{Z}_\ell$.

The above result was proved by analyzing the number theoretical enhancement of the Johnson homomorphism where the geometric mapping class group is replaced by the arithmetic mapping class group which is expressed as an extension

$$1 \longrightarrow \hat{\mathcal{M}}_g^n \longrightarrow \pi_1^{alg} \mathbf{M}_g^n/\mathbb{Q} \longrightarrow \mathrm{Gal}(\overline{\mathbb{Q}}/\mathbb{Q}) \longrightarrow 1$$

and studied by Grothendieck, Deligne, Ihara, Drinfel'd and many number theorists. Nakamura continued to study the structure of the mapping class group from the point of view of number theory extensively (see e.g. [88][89]). On the other hand, Hain and Matsumoto recently proved remarkable results concerning this subject (see [31][32]). In view of deep theories in number theory, as well as the above explicit results, it seems to be conjectured that there should exist an embedding

$$\mathrm{FreeLie}_{\mathbb{Z}}(\sigma_3, \sigma_5, \cdots) \subset \mathfrak{h}_{g,1}$$

of certain free graded Lie algebra over $\mathbb{Z}$ generated by certain elements $\sigma_3, \sigma_5, \cdots$, corresponding to the Soulé elements, into $\mathfrak{h}_{g,1}$ such that the tensor product of it with $\mathbb{Z}_\ell$ coincides with the image of $\mathrm{Gal}(\overline{\mathbb{Q}}/\mathbb{Q})$ for any prime ℓ.

We expect that the above conjectured free graded Lie algebra (the motivic Lie algebra) can be realized inside $\mathfrak{h}_{g,1}$ (in fact inside the commutator ideal $[\mathfrak{h}_{g,1}, \mathfrak{h}_{g,1}]$) explicitly in terms of the traces. In some sense, the elements σ_{2k+1} should be *decomposable* in higher genera. Here we omit the precise form of the expected formula which will be given in a forthcoming paper.

Finally we mention the analogue of the Johnson homomorphisms for the group $\operatorname{Aut} F_n$ very briefly. Prior to the work of Johnson, Andreadakis [1] introduced and studied the filtration on $\{\operatorname{Aut} F_n(k)\}_k$ which is induced from the action of $\operatorname{Aut} F_n$ on the lower central series of F_n. The first group $\operatorname{Aut} F_n(1)$ in this filtration is nothing but the group IAut_n. It can be checked that an analogous procedure as in the case of the mapping class group gives rise to certain homomorphisms

$$\tau_k : \operatorname{Aut} F_n(k) \longrightarrow \operatorname{Hom}(H_n, \mathcal{L}_n(k+1))$$

and the totality $\{\tau_k\}_k$ of these homomorphims induces an injective homomorphism of graded Lie algebras

$$(16) \qquad \operatorname{Gr}^+(\operatorname{Aut} F_n) = \bigoplus_{k=1}^{\infty} \operatorname{Aut} F_n(k)/\operatorname{Aut} F_n(k+1) \longrightarrow \operatorname{Der}^+(\mathcal{L}_n).$$

We refer to [49][93][98] for some of the recent works related to the above homomorphism.

9. A theorem of Kontsevich

In this section, we recall a theorem of Kontsevich described in [55][56] which is the key result for the argument given in the next section. See also the paper [9] by Conant and Vogtmann for a detailed proof as well as discussion of this theorem in the context of cyclic operads. In the above cited papers, Kontsevich considered three kinds of infinite dimensional Lie algebras denoted by $\mathfrak{c}_g, \mathfrak{a}_g, \mathfrak{l}_g$ (commutative, associative, and lie version, respectively). The latter two Lie algebras are defined by

$$\mathfrak{a}_g = \{\text{derivation } D \text{ of the tensor algebra } T^*(H_{\mathbb{Q}})$$
$$\text{such that } D(\omega_0) = 0\}$$

$$\mathfrak{l}_g = \{\text{derivation } D \text{ of the free Lie algebra } \mathcal{L}_{2g}^{\mathbb{Q}} \subset T^*(H_{\mathbb{Q}})$$
$$\text{such that } D(\omega_0) = 0\}.$$

There is a natural injective Lie algebra homomorphism $\mathfrak{l}_g \to \mathfrak{a}_g$. The degree 0 part of both of $\mathfrak{a}_g, \mathfrak{l}_g$ is the Lie algebra $\mathfrak{sp}(2g, \mathbb{Q})$ of $\operatorname{Sp}(2g, \mathbb{Q})$. Let $\mathfrak{a}_g^+$ (resp. $\mathfrak{l}_g^+$) denote the Lie subalgebra of $\mathfrak{a}_g$ (resp. $\mathfrak{l}_g$) consisting of derivations with *positive* degrees. Then the latter one $\mathfrak{l}_g^+$ is nothing other than the Lie algebra $\mathfrak{h}_{g,1}^{\mathbb{Q}}$ considered in §5. Now Kontsevich described the *stable* homology groups of the above three Lie algebras (where g tends to ∞) in terms of cohomology groups of graph complexes, moduli spaces $\mathbf{M}_g^m$ of Riemann surfaces and the outer automorphism groups $\operatorname{Out} F_n$ of free groups (or the moduli space of graphs), respectively. Here is the statement for the cases of $\mathfrak{a}_\infty, \mathfrak{l}_\infty$.

THEOREM 34 (Kontsevich). *There are isomorphisms*

$$PH_*(\mathfrak{a}_\infty) \cong PH_*(\mathfrak{sp}(\infty, \mathbb{Q})) \oplus \bigoplus_{g \geq 0, m \geq 1, 2g-2+m > 0} H^*(\mathbf{M}_g^m; \mathbb{Q})^{\mathfrak{S}_m},$$

$$PH_*(\mathfrak{l}_\infty) \cong PH_*(\mathfrak{sp}(\infty, \mathbb{Q})) \oplus \bigoplus_{n \geq 2} H^*(\operatorname{Out} F_n; \mathbb{Q}).$$

Here P denotes the *primitive parts* of $H_*(\mathfrak{a}_\infty), H_*(\mathfrak{l}_\infty)$ which have natural structures of Hopf algebras and $\mathbf{M}_g^m$ denotes the moduli space of genus g smooth curves with m punctures.

Here is a very short outline of the proof of the above theorem. Using natural cell structure of the Riemann moduli space $\mathbf{M}_g^m$ ($m \geq 1$) and the moduli space $\mathbf{G}_n$ of graphs, which serves as the (rational) classifying space of $\operatorname{Out} F_n$ by [**11**], Kontsevich introduced a natural filtration on the cellular cochain complex of these moduli spaces. Then he proved that the associated spectral sequence degenerates at the E_2-term and only the diagonal terms remain to be non-trivial. On the other hand, by making use of classical representation theory for the group $\operatorname{Sp}(2g, \mathbb{Q})$, he constructed a quasi isomorphism between the E_1-term and the chain complexes of the relevant Lie algebras ($\mathfrak{a}_\infty$ or $\mathfrak{l}_\infty$). For details, see the original papers cited above as well as [**9**].

There is also the dual version of the above theorem which connects the primitive cohomology of the relevant Lie algebras with the homology groups of the moduli space or $\operatorname{Out} F_n$. We would like to describe it in a detailed form because this version is most suitable for our purpose. The Lie algebras $\mathfrak{l}_\infty^+, \mathfrak{a}_\infty^+$ are graded. Hence their Sp-invariant cohomology groups are *bigraded*. Let $H^k(\mathfrak{l}_\infty^+)_n^{Sp}$ and $H^k(\mathfrak{a}_\infty^+)_n^{Sp}$ denote the weight n part of $H^k(\mathfrak{l}_\infty^+)^{Sp}$ and $H^k(\mathfrak{a}_\infty^+)^{Sp}$ respectively. Then we have the following isomorphisms.

(17) $$PH^k(\mathfrak{l}_\infty^+)_{2n}^{Sp} \cong H_{2n-k}(\operatorname{Out} F_{n+1}; \mathbb{Q})$$

(18) $$PH^k(\mathfrak{a}_\infty^+)_{2n}^{Sp} \cong \bigoplus_{2g-2+m=n} H_{2n-k}(\mathbf{M}_g^m; \mathbb{Q})_{\mathfrak{S}_m}.$$

10. Constructing homology classes of $\operatorname{Out} F_n$

In this section, we combine our construction of many cohomology classes in $H^*(\mathfrak{h}_{g,1}^\mathbb{Q})^{Sp}$ given in §6, §7 with Kontsevich's theorem given in §9 to produce homology classes of the group $\operatorname{Out} F_n$.

First we see that the homomorphism given in (13) is *stable* with respect to g. More precisely, the following diagram is commutative

(19)
$$\begin{array}{ccc} \mathbb{Q}[\Gamma; \Gamma \in \mathcal{G}^{odd}] & \xrightarrow{\Phi} & H^*(\mathfrak{h}_{g+1,1}^\mathbb{Q})^{Sp} \\ \| & & \downarrow \\ \mathbb{Q}[\Gamma; \Gamma \in \mathcal{G}^{odd}] & \xrightarrow{\Phi} & H^*(\mathfrak{h}_{g,1}^\mathbb{Q})^{Sp}, \end{array}$$

where the right vertical map is induced by the inclusion $\mathfrak{h}_{g,1} \to \mathfrak{h}_{g+1,1}$. This follows from the fact that the traces $\operatorname{trace}(2k+1)$ as well as τ_1 are all stable with respect to g in an obvious way.

Next, we see that the cohomology class $\Phi(\Gamma) \in H^*(\mathfrak{h}_{\infty,1}^\mathbb{Q})^{Sp}$ obtained in this way is primitive if and only if Γ is connected. Keeping in mind the fact $\mathfrak{h}_{\infty,1}^\mathbb{Q} = \mathfrak{l}_\infty^+$, the property (14), as well as the version of Kontsevich's theorem given in (17) we now obtain the following theorem.

THEOREM 35. *Associated to each connected graph* $\Gamma \in \mathcal{G}^{odd}$ *whose valencies are* v_3^a *times* 3 *(alternating),* v_3^s *times* 3 *(symmetric) and* v_{2k+1} *times* $2k+1$ *($2k+1 > 3$),*

we have a homology class
$$\Phi(\Gamma) \in H_{2n-d}(\operatorname{Out} F_{n+1}; \mathbb{Q})$$
where
$$d = v_3^a + v_3^s + v_5 + v_7 + \cdots, \quad 2n = v_3^a + 3v_3^s + 5v_5 + 7v_7 + \cdots.$$

REMARK 36. Let Γ_{2k+1} be the connected graph with two vertices both of which have valency $2k+1$. Then $d = 2, 2n = 4k+2$ and $\Phi(\Gamma_{2k+1}) \in H_{4k}(\operatorname{Out} F_{2k+2}; \mathbb{Q})$ is the class μ_k already mentioned in §4.

The following problem is an enhancement of Problem 31 in the context of the homology of the moduli space of graphs rather than the cohomology of the Lie algebra $\mathfrak{h}_{g,1}^{\mathbb{Q}}$.

PROBLEM 37. Give examples of odd valent graphs Γ whose associated homology classes $\Phi(\Gamma) \in H_*(\operatorname{Out} F_n; \mathbb{Q})$ are non-trivial as many as possible. Also compare these classes with the homology classes constructed by Conant and Vogtmann [10] as explicit cycles in the moduli space of graphs. Furthermore investigate whether these classes survive in $H_*(\operatorname{GL}(n, \mathbb{Z}); \mathbb{Q})$, or else come from $H_*(\operatorname{IOut}_n; \mathbb{Q})$, or not.

REMARK 38. It seems that the geometric meaning of Kontsevich's theorem is not very well understood yet. In particular, there is almost no known relations between the classes $\Phi(\Gamma) \in H_*(\operatorname{Out} F_n; \mathbb{Q})$ where the rank n varies. However, there should be some unknown structures here. For example, there are graphs Γ whose associated classes $\Phi(\Gamma)$ lie in $H_8(\operatorname{Out} F_n; \mathbb{Q})$ for $n = 7, 8$ which might be closely related to the class $\mu_2 \in H_8(\operatorname{Out} F_6; \mathbb{Q})$.

11. Group of homology cobordism classes of homology cylinders

In his theory developed in [27], Habiro introduced the concept of *homology cobordism of surfaces* and proposed interesting problems concerning it. Goussarov [26] also studied the same thing in his theory. It played an important role in the classification theory of 3-manifolds now called the Goussarov-Habiro theory. Later Garoufalidis and Levine [24] and Levine [60] used this concept to define a group $\mathcal{H}_{g,1}$ which consists of homology cobordism classes of *homology cylinders* on $\Sigma_g \setminus \operatorname{Int} D$. We refer to the above papers for the definition (we use the terminology *homology cylinder* following them) as well as many interesting questions concerning the structure of $\mathcal{H}_{g,1}$. It seems that the importance of this group is growing recently.

Here we summarize some of the results of [24][60] which will be necessary for our purpose here. Consider $\Gamma = \pi_1(\Sigma_g \setminus \operatorname{Int} D)$, which is isomorphic to F_{2g}, and let $\{\Gamma_k\}_k$ be its lower central series as before. Note that Γ contains a particular element $\gamma \in \Gamma$ which corresponds to the unique relation in $\pi_1 \Sigma_g$. They define

$$\operatorname{Aut}_0(\Gamma/\Gamma_k) = \{f \in \operatorname{Aut}(\Gamma/\Gamma_k);$$
$$f \text{ lifts to an endomorphism of } \Gamma \text{ which fixes } \gamma \bmod \Gamma_{k+1}\}.$$

By making a crucial use of a theorem of Stallings [102], for each k they obtain a homomorphism
$$\sigma_k : \mathcal{H}_{g,1} \longrightarrow \operatorname{Aut}_0(\Gamma/\Gamma_k).$$
The following theorem given in [24] is a basic result for the study of the structure of the group $\mathcal{H}_{g,1}$.

THEOREM 39 (Garoufalidis-Levine). *The above homomorphism σ_k is surjective for any k.*

They use the homomorphisms $\{\sigma_k\}_k$ to define a certain filtration $\{\mathcal{H}_{g,1}(k)\}_k$ of $\mathcal{H}_{g,1}$ and show that the Johnson homomorphisms are defined also on this group. Furthermore they are *surjective* so that there is an isomorphism

$$\mathrm{Gr}^+(\mathcal{H}_{g,1}) = \bigoplus_{k=1}^{\infty} \mathcal{H}_{g,1}(k)/\mathcal{H}_{g,1}(k+1) \cong \mathfrak{h}_{g,1}.$$

They concluded from this fact that $\mathcal{H}_{g,1}$ contains $\mathcal{M}_{g,1}$ as a subgroup. The homomorphisms σ_k fit together to define a homomorphism

$$\sigma : \mathcal{H}_{g,1} \longrightarrow \varprojlim \mathrm{Aut}_0(\Gamma/\Gamma_k).$$

They show that this homomorphism is not surjective by using the argument of Levine [**59**]. We mention a recent paper [**96**] of Sakasai for a related work. Also they point out that, although the restriction of σ to $\mathcal{M}_{g,1}$ is injective, σ has a rather big kernel because $\mathrm{Ker}\,\sigma$ at least contains the group

$$\Theta_{\mathbb{Z}}^3 = \{\text{oriented homology 3-sphere}\}/\text{homology cobordism}.$$

It is easy to see that $\Theta_{\mathbb{Z}}^3$ is contained in the center of $\mathcal{H}_{g,1}$ so that we have a central extension

(20) $$0 \longrightarrow \Theta_{\mathbb{Z}}^3 \longrightarrow \mathcal{H}_{g,1} \longrightarrow \overline{\mathcal{H}}_{g,1} \longrightarrow 1$$

where $\overline{\mathcal{H}}_{g,1} = \mathcal{H}_{g,1}/\Theta_{\mathbb{Z}}^3$.

The group $\Theta_{\mathbb{Z}}^3$ is a very important abelian group in low dimensional topology. In [**21**], Furuta first proved that this group is an infinitely generated group by making use of gauge theory. See also the paper [**16**] by Fintushel and Stern for another proof. However, only a few additive invariants are known on this group at present besides the classical surjective homomorphism

(21) $$\mu : \Theta_{\mathbb{Z}}^3 \longrightarrow \mathbb{Z}/2$$

defined by the Rokhlin invariant. One is a non-trivial homomorphism $\Theta_{\mathbb{Z}}^3 \longrightarrow \mathbb{Z}$ constructed by Frøyshov [**17**] and the other is given by Ozsváth and Szabó [**92**] as an application of their Heegaard Floer homology theory. Neumann [**90**] and Siebenmann [**101**] defined an invariant $\bar{\mu}$ for plumbed homology 3-spheres and Saveliev [**99**] introduced his ν-invariant for any homology sphere by making use of the Floer homology. On the other hand, Fukumoto and Furuta (see [**18**][**19**]) defined certain invariants for plumbed homology 3-spheres using gauge theory. According to a recent result of Saveliev [**100**], these invariants fit together to give a candidate of another homomorphism on $\Theta_{\mathbb{Z}}^3$.

The situation being like this, $\Theta_{\mathbb{Z}}^3$ remains to be a rather mysterious group. Thus we have the following important problem.

PROBLEM 40. Study the central extension (20) from the point of view of group cohomology as well as geometric topology. In particular determine the Euler class of this central extension which is an element of the group

$$H^2(\overline{\mathcal{H}}_{g,1}; \Theta_{\mathbb{Z}}^3) \cong \mathrm{Hom}(H_2(\overline{\mathcal{H}}_{g,1}), \Theta_{\mathbb{Z}}^3) \oplus \mathrm{Ext}_{\mathbb{Z}}(H_1(\overline{\mathcal{H}}_{g,1}), \Theta_{\mathbb{Z}}^3).$$

This should be an extremely difficult problem. Here we would like to indicate a possible method of attacking it, and in particular a possible way of obtaining additive invariants for the group $\Theta_{\mathbb{Z}}^3$, very briefly. Details will be given in a forthcoming paper.

Using the traces, we can define a series of certain cohomology classes
$$\tilde{t}_{2k+1} \in H^2(\varprojlim \mathrm{Aut}_0(\Gamma/\Gamma_m)) \quad (k=1,2,\cdots).$$
These are the "group version" of the elements $t_{2k+1} \in H^2(\mathfrak{h}_{g,1}^{\mathbb{Q}})^{Sp}$ defined in §6 (Definition 26). The homomorphism σ is trivial on $\Theta_{\mathbb{Z}}^3$ so that we have the induced homomorphism $\bar{\sigma}: \overline{\mathcal{H}}_{g,1} \to \varprojlim \mathrm{Aut}_0(\Gamma/\Gamma_m)$.

CONJECTURE 41.
1. $\bar{\sigma}^*(\tilde{t}_{2k+1})$ is non-trivial in $H^2(\overline{\mathcal{H}}_{g,1})$ for any k
2. $\sigma^*(\tilde{t}_{2k+1})$ is trivial in $H^2(\mathcal{H}_{g,1})$ for any k.

The first part of the above conjecture is the "group version" of Conjecture 27 and it should be even more difficult to prove. On the other hand, if the classes $\sigma^*(\tilde{t}_{2k+1})$ were non-trivial, then they would serve as invariants for certain 4-manifolds (2-dimensional family of homology cylinders). This seems unlikely to be the case. Thus the second part is related to the following problem.

PROBLEM 42. Determine the abelianization of the group $\mathcal{H}_{g,1}$. Is it trivial? Also determine the second homology group $H_2(\mathcal{H}_{g,1};\mathbb{Z})$. Is the rank of it equal to 1 given by the signature?

If everything will be as expected, we would obtain non-trivial homomorphisms
$$\hat{t}_{2k+1}: \Theta_{\mathbb{Z}}^3 \longrightarrow \mathbb{Z}$$
as *secondary* invariants associated to the cohomology classes $\tilde{t}_{2k+1}$. There should be both similarity and difference between these cases and the situation where we interpreted the Casson invariant as the secondary invariant associated to the first Mumford-Morita-Miller class e_1 (see [74]). More precisely, they are similar because they are all related to some cohomology classes in $H^2(\mathfrak{h}_{g,1}^{\mathbb{Q}})$. They are different because e_1 is non-trivial in $H^2(\mathcal{H}_{g,1})$ whereas we expect that the other classes $\sigma^*(\tilde{t}_{2k+1})$ would be all trivial in the same group.

Also recall here that Matumoto [67] and Galewsky and Stern [23] proved that every topological manifold (of dimension $n \geq 7$) is simplicially triangulable if and only if the homomorphism (21) splits. In view of this result, it should be important to investigate the mod 2 structure of the extension (20) keeping in mind the works of Birman and Craggs [3] as well as Johnson [44] for the case of the mapping class group.

However we have come too far and surely many things have to be clarified before we would understand the structure of the group $\mathcal{H}_{g,1}$.

Finally we would like to propose a problem.

PROBLEM 43. Generalize the infinitesimal presentation of the Torelli Lie algebra given by Hain [29] to the case of the group of homology cobordism classes of homology cylinders.

12. Diffeomorphism groups of surfaces

Let us recall the important problem of constructing (and then detecting) characteristic classes of smooth fiber bundles as well as those of foliated fiber bundles whose fibers are diffeomorphic to a general closed C^∞ manifold M. This is equivalent to the problem of computing the cohomology groups $H^*(\mathrm{BDiff}\, M)$ (resp. $H^*(\mathrm{BDiff}^\delta M)$) of the classifying space of the diffeomorphim group $\mathrm{Diff}\, M$ (resp. the same group $\mathrm{Diff}^\delta M$ equipped with the discrete topology) of M. Although the theory of higher torsion invariants of fiber bundles developed by Igusa [38] on the one hand and by Bismut, Lott [4] and Goette on the other and also the Gel'fand-Fuks cohomology $H^*_{GF}(M)$ of M (see e.g. [28]) produce characteristic classes for the above two types of M-bundles, there seem to be only a few known results concerning explicit computations for specific manifolds. The same problems are also important for various subgroups of $\mathrm{Diff}\, M$, in particular the symplectomorphism group $\mathrm{Symp}(M,\omega)$ (in the case where there is given a symplectic form ω on M) as well as the volume preserving diffeomorphism group.

Here we would like to propose two problems for the special, but at the same time very important case of surfaces. Note that we have two characteristic classes

$$(22) \qquad \int_{fiber} u_1 c_1^2, \int_{fiber} u_1 c_2 \in H^3(\mathrm{BDiff}^\delta_+ \Sigma_g; \mathbb{R})$$

which are defined to be the fiber integral of the characteristic classes $u_1 c_1^2, u_1 c_2$ of codimention two foliations. In the case of $g = 0$ (namely S^2), Thurston and then Rasmussen [94] (see also Boullay [8]) proved that these classes are linearly independent and vary continuously. Although it is highly likely that their results can be extended to the cases of surfaces of higher genera, explicit construction seems to be open.

PROBLEM 44. Prove that the above characteristic classes induce surjective homomorphism

$$H_3(\mathrm{BDiff}^\delta_+ \Sigma_g; \mathbb{Z}) \longrightarrow \mathbb{R}^2$$

for any g.

The cohomology classes in (22) are stable with respect to g. On the other hand, in [57][58] we found an interesting interaction between the twisted cohomology group of the mapping class group and some well known concepts in symplectic topology such as the flux homomorphism as well as the Calabi homomorphism (see [68] for generalities of the symplectic topology). By making use of this, we defined certain cohomology classes of $\mathrm{BSymp}^\delta \Sigma_g$ and proved non-triviality of them.

In view of the fact that all the known cohomology classes of $\mathrm{BDiff}^\delta_+ \Sigma_g$ as well as $\mathrm{BSymp}^\delta \Sigma_g$ are stable with respect to the genus, we would like to ask the following problem.

PROBLEM 45. Study whether the homology groups of $\mathrm{BDiff}^\delta_+ \Sigma_g$ stabilize with respect to g or not. The same problem for the group $\mathrm{Symp}^\delta \Sigma_g$.

Acknowledgments. The author would like to express his hearty thanks to R. Hain, N. Kawazumi, D. Kotschick, M. Matsumoto, H. Nakamura, T. Sakasai for enlightening discussions as well as useful informations concerning the problems treated in this paper.

References

1. S. Andreadakis, *On the automorphisms of free groups and free nilpotent groups*, Proc. London Math. Soc. 15 (1965), 239–268.
2. M. Bestvina, *The topology of* Out(F_n), Proceedings of the International Congress of Mathematicians, Beijing 2002, Higher Ed. Press, Beijing 2002, 373–384.
3. J. Birman and R. Craggs, *The μ-invariant of 3-manifolds and certain structural properties of the group of homeomorphisms of a closed oriented 2-manifold*, Trans. Amer. Math. Soc. 237 (1978), 283–309.
4. J.-M. Bismut and J. Lott, *Flat vector bundles, direct images and higher real analytic torsion*, J. Amer. Math. Soc. 8 (1995), 291–363.
5. D. Biss and B. Farb, *$\mathcal{K}_g$ is not finitely generated*, Invent. Math. 163 (2006), 213-226.
6. A. Borel, *Stable real cohomology of arithmetic groups*, Ann. Sci. École Norm. Sup. 7 (1974), 235–272.
7. A. Borel, *Stable real cohomology of arithmetic groups II*, in *Manifolds and Groups, Papers in Honor of Yozo Matsushima*, Progress in Math. 14, Birkhäuser, Boston, 1981, 21–55.
8. P. Boullay, $H_3(Diff^+(S^2); \mathbb{Z})$ *contains an uncountable $\mathbb{Q}$-vector space*, Topology 35 (1996), 509–520.
9. J. Conant and K. Vogtmann, *On a theorem of Kontsevich*, Alg. Geom. Topology 3 (2003), 1167–1224.
10. J. Conant and K. Vogtmann, *Morita classes in the homology of automorphism group of free groups*, Geometry and Topology 8 (2004), 1471–1499.
11. M. Culler and K. Vogtmann, *Moduli of graphs and automorphisms of free groups*, Invent. Math. 84 (1986), 91–119.
12. C.J. Earle and J. Eells, *The diffeomorphism group of a compact Riemann surface*, Bull. Amer. Math. Soc. 73 (1967), 557–559.
13. P. Elbaz-Vincent, H. Gangl and C. Soulé, *Quelques calculs de la cohomologie de $GL_N(\mathbb{Z})$ et la K-théorie de $\mathbb{Z}$*, C. R. Math. Acad. Sci. Paris 335 (2002), 321–324.
14. H. Endo, *Meyer's signature cocycle and hyperelliptic fibrations*, Math. Ann. 316 (2000), 237–257.
15. C. Faber, *A conjectural description of the tautological ring of the moduli space of curves*, in *Moduli of Curves and Abelian Varieties*, Carel Faber and Eduard Looijenga, editors, Vieweg 1999.
16. R. Fintushel and R. Stern, *Instanton homology of Seifert fibred homology three spheres*, Proc. London Math. Soc. 61 (1990), 109–137.
17. K. Frøyshov, *Equivariant aspects of Yang-Mills Floer theory*, Topology 41 (2002), 525–552.
18. Y. Fukumoto and M. Furuta, *Homology 3-spheres bounding acyclic 4-manifolds*, Math. Res. Letters 7 (2000), 757–766.
19. Y. Fukumoto, M. Furuta and M. Ue, *W-invariants and Neumann-Siebenmann invariants for Seifert homology 3-spheres*, Topology Appl. 116 (2001), 333–369.
20. W. Fulton and J. Harris, *Representation Theory*, Graduate Texts in Math. 129, Springer Verlag 1991.
21. M. Furuta, *Homology cobordism group of homology 3-spheres*, Invent. Math. 100 (1990), 339–355.
22. S. Galatius, *Mod p homology of the stable mapping class group*, Topology 43 (2004), 1105–1132.
23. D. Galewski and R. Stern, *Classification of simplicial triangulations of topological maifolds*, Ann. of Math. 111 (1980), 1–34.
24. S. Garoufalidis and J. Levine, *Tree-level invariants of three-manifolds, Massey products and the Johnson homomorphism*, in *Graphs and Patterns in Mathematics and Theoretical Physics*, Proc. Sympos. Pure Math. 73 (2005), 173–205.
25. F. Gerlits, *Invariants in chain complexes of graphs*, preprint.
26. M. Goussarov, *Finite type invariants and n-equivalence of 3-manifolds*, C. R. Math. Acad. Sci. Paris 329 (1999), 517–522.
27. K. Habiro, *Claspers and finite type invariants of links*, Geometry and Topology 4 (2000), 34–43.
28. A. Haefliger, *Sur les classes caractéristiques des feuilletages*, Séminaire Bourbaki, 1971/72, Lecture Notes in Mathematics, vol. 317, Springer Verlag, 1973, 239–260.

29. R. Hain, *Infinitesimal presentations of the Torelli groups*, J. Amer. Math. Soc. 10 (1997), 597–651.
30. R. Hain and E. Looijenga, *Mapping class groups and moduli space of curves*, Proc. Symp. Pure Math. 62.2 (1997), 97–142.
31. R. Hain and M. Matsumoto, *Weighted completion of the Galois groups and Galois actions on the fundamental group of* $\mathbf{P}^1 - \{0,1,\infty\}$, Compositio Math. 139 (2003), 119–167.
32. R. Hain and M. Matsumoto, *Galois actions on fundamental groups of curves and the cycle* $C - C_-$, J. Inst. Math. Jussieu 4 (2005), 363-403.
33. R. Hain and D. Reed, *On the Arakelov geometry of the moduli spaces of curves*, Journal of Differential Geometry 67 (2004), 195–228.
34. J. Harer, *Stability of the homology of the mapping class group of an orientable surface*, Ann. of Math. 121 (1985), 215–249.
35. A. Hatcher, *Homological stability for automorphism groups of free groups*, Comment. Math. Helv. 70 (1995), 129–137.
36. A. Hatcher and K. Vogtmann, *Rational homology of* $Aut(F_n)$, Math. Res. Letters 5 (1998), 759–780.
37. A. Hatcher and K. Vogtmann, *Homology stability for outer automorphism groups of free groups*, Alg. Geom. Topology 4 (2004), 1253–1272.
38. K. Igusa, *Higher Franz-Reidemeister Torsion*, AMS/IP Studies in Advanced Mathematics, American Mathematical Society, 2002.
39. E-N. Ionel, *Relations in the tautological ring of* $\mathcal{M}_g$, Duke Math. J. 129 (2005), 157-186.
40. D. Johnson, *An abelian quotient of the mapping class group* $\mathcal{I}_g$, Math. Ann. 249 (1980), 225–242.
41. D. Johnson, *A survey of the Torelli group*, Contemporary Math. 20 (1983), 165–179.
42. D. Johnson, *The structure of the Torelli group I: A finite set of generators for* $\mathcal{I}_g$, Ann. of Math. 118 (1983), 423–442.
43. D. Johnson, *The structure of the Torelli group II: A characterization of the group generated by twists on bounding simple closed curves*, Topology 24 (1985), 113–126.
44. D. Johnson, *The structure of the Torelli group III: The abelianization of* $\mathcal{I}_g$, Topology 24 (1985), 127–144.
45. V. Jones, *Hecke algebra representations of braid groups and link polynomials*, Ann. of Math. 126 (1987), 335–388.
46. Y. Kasahara, *An expansion of the Jones representation of genus 2 and the Torelli group*, Alg. Geom. Topology 1, (2001), 39–55.
47. M. Kassabov, *On the automorphism tower of free nilpotent groups*, thesis, Yale University, 2003.
48. N. Kawazumi, *A generalization of the Morita-Mumford classes to extended mapping class groups for surfaces*, Invent. Math. 131 (1998), 137–149.
49. N. Kawazumi, *Cohomological aspects of the Magnus expansions*, preprint.
50. N. Kawazumi, *Harmonic Magnus expansion on the universal family of Riemann surfaces*, preprint.
51. N. Kawazumi and S. Morita, *The primary approximation to the cohomology of the moduli space of curves and cocycles for the stable characteristic classes*, Math. Res. Letters 3 (1996), 629–641.
52. N. Kawazumi and S. Morita, *The primary approximation to the cohomology of the moduli space of curves and cocycles for the Mumford-Morita-Miller classes*, preprint.
53. F. Kirwan, *Cohomology of moduli spaces*, Proceedings of the International Congress of Mathematicians, Beijing 2002, Higher Ed. Press, Beijing 2002, 363–382.
54. T. Kitano, *On the first Morita-Mumford class of surface bundles over* S^1 *and the Rochlin invariant*, J. Knot Theory Ramifications 9 (2000), 179–186.
55. M. Kontsevich, *Formal (non-)commutative symplectic geometry*, in *The Gelfand Mathematical Seminars 1990-1992*, Birkhäuser Verlag 1993, 173–188.
56. M. Kontsevich, *Feynman diagrams and low-dimensional topology*, in *Proceedings of the First European Congress of Mathematicians*, Vol II, Paris 1992, Progress in Math. 120, Birkhäuser Verlag 1994, 97–121.
57. D. Kotschick and S. Morita, *Signature of foliated surface bundles and the group of symplectomorphisms of surfaces*, Topology 44 (2005), 131–149.

58. D. Kotschick and S. Morita, *Characteristic classes of foliated surface bundles with area-preserving total holonomy*, to appear in Journal of Differential Geometry.
59. J. Levine, *Link concordance and algebraic closure*, Comment. Math. Helv. 64 (1989), 236–255.
60. J. Levine, *Homology cylinders: an enlargement of the mapping class group*, Alg. Geom. Topology 1 (2001), 243–270.
61. E. Looijenga, *Cohomology of $\mathcal{M}_3$ and $\mathcal{M}_3^1$*, in *Mapping Class Groups and Moduli Spaces of Riemann Surfaces*, C.-F. Bödigheimer and R. Hain, editors, Contemporary Math. 150 (1993), 205–228.
62. E. Looijenga, *Stable cohomology of the mapping class group with symplectic coefficients and of the universal Abel-Jacobi map*, J. Algebraic Geometry 5 (1996), 135–150.
63. I. Madsen and U. Tillmann, *The stable mapping class group and $Q(\mathbb{C}P^\infty)$*, Invent. Math. 145 (2001), 509–544.
64. I. Madsen and M. Weiss, *The stable moduli space of Riemann surfaces: Mumford's conjecture*, to appear in Ann. of Math.
65. M. Matsumoto, *Galois representations on profinite braid groups on curves*, J. Reine. Angew. Math. 474 (1996), 169–219.
66. M. Matsumoto, *Arithmetic fundamental groups and the moduli of curves*, School on Algebraic Geometry (Trieste 1999), ICTP Lecture Notes 1, Abdus Salam Int. Cent. Theoret. Phys., Trieste 2000, 355–383.
67. T. Matumoto, *Triangulation of manifolds*, Algebraic and Geometric Topology, Stanford 1976, Proc. Sympos. Pure Math. 32 (1978), Part 2, 3–6.
68. D. McDuff and D.A. Salamon, *Introduction to Symplectic Topology*, second edition, Oxford University Press, 1998.
69. G. Mess, *The Torelli groups for genus 2 and 3 surfaces*, Topology 31 (1992), 775–790.
70. E.Y. Miller, *The homology of the mapping class group*, Journal of Differential Geometry 24 (1986), 1–14.
71. T. Morifuji, *On Meyer's function of hyperelliptic mapping class group*, J. Math. Soc. Japan 55 (2003), 117–129.
72. S. Morita, *Characteristic classes of surface bundles*, Bull. Amer. Math. Soc. 11 (1984), 386–388.
73. S. Morita, *Characteristic classes of surface bundles*, Invent. Math. 90 (1987), 551-577.
74. S. Morita, *Casson's invariant for homology 3-spheres and characteristic classes of surface bundles*, Topology 28 (1989), 305–323.
75. S. Morita, *On the structure of the Torelli group and the Casson invariant*, Topology 30 (1991), 603–621.
76. S. Morita, *Abelian quotients of subgroups of the mapping class group of surfaces*, Duke Math. J. 70 (1993), 699–726.
77. S. Morita, *The extension of Johnson's homomorphism from the Torelli group to the mapping class group*, Invent. Math. 111 (1993), 197–224.
78. S. Morita, *A linear representation of the mapping class group of orientable surfaces and characteristic classes of surface bundles*, in Proceedings of the Taniguchi Symposium on Topology and Teichmüller Spaces held in Finland, July 1995, World Scientific 1996, 159–186.
79. S. Morita, *Casson invariant, signature defect of framed manifolds and the secondary characteristic classes of surface bundles*, Journal of Differential Geometry 47 (1997), 560–599.
80. S. Morita, *Structure of the mapping class groups of surfaces: a survey and a prospect*, Geometry and Topology Monographs 2 (1999), Proceedings of the Kirbyfest, 349–406.
81. S. Morita, *Structure of the mapping class group and symplectic representation theory*, l'Enseignement Math. Monographs 38 (2001), 577–596.
82. S. Morita, *Generators for the tautological algebra of the moduli space of curves*, Topology 42 (2003), 787–819.
83. S. Morita, *Structure of derivation algebras of surfaces and its applications*, in preparation.
84. J. Moser, *On the volume forms*, Trans. Amer. Math. Soc. 120 (1965), 286–294.
85. D. Mumford, *Towards an enumerative geometry of the moduli space of curves*, in *Arithmetic and Geometry*, Progress in Math. 36 (1983), 271–328.
86. H. Nakamura, *Coupling of universal monodromy representations of Galois- Teichmüller modular group*, Math. Ann. 304 (1996), 99–119.

87. H. Nakamura, *Galois rigidity of profinite fundamental groups*, Sugaku Expositions 10 (1997), 195–215.
88. H. Nakamura, *Limits of Galois representations in fundamental groups along maximal degeneration of marked curves, I*, Amer. Jour. Math. 121 (1999), 315–358.
89. H. Nakamura, *Limits of Galois representations in fundamental groups along maximal degeneration of marked curves, II*, Arithmetic fundamental groups and noncommutative algebra, Berkeley 1999, Proc. Sympos. Pure Math. 70 (2002), Amer. Math. Soc., 43–78.
90. W. Neumann, *An invariant of plumbed homology spheres*, Topology Symposium, Siegen 1979, Lect. Notes in Math. 788, Springer, 1980, 125–144.
91. R. Ohashi, in preparation.
92. P. Ozsváth and Z. Szabó, *Absolutely graded Floer homologies and intersection forms for four-manifolds with boundary*, Advances in Math. 173 (2003), 179–261.
93. A. Pettet, *The Johnson homomorphism and the second cohomology of IA_n*, Algebraic and Geometric Topology 5 (2005), 725-740.
94. O. H. Rasmussen, *Continuous variation of foliations in codimension two*, Topology 19 (1980), 335–349.
95. T. Sakasai, *The Johnson homomorphism and the third rational cohomology group of the Torelli group*, Topology and its Applications 148 (2005), 83–111.
96. T. Sakasai, *Homology cylinders and the acyclic closure of a free group*, Algebraic and Geometric Topology 6 (2006), 603-631.
97. T. Satoh, *Twisted first homology groups of the automorphism group of a free group*, J. Pure Appl. Alg. 204 (2006), 334-348.
98. T. Satoh, *New obstructions to the surjectivity of the Johnson homomorphism of the automorphism group of a free group*, preprint.
99. N. Saveliev, *Floer homology of Brieskorn homology spheres*, Journal of Differential Geometry 53 (1999), 15–87.
100. N. Saveliev, *Fukumoto-Furuta invariants of plumbed homology 3-spheres*, Pacific J. Math. 205 (2002), 465–490.
101. L. Siebenmann, *On vanishing of the Rohlin invariant and nonfinitely amphicheiral homology 3-spheres*, Topology Symposium, Siegen 1979, Lect. Notes in Math. 788, Springer, 1980, 172–222.
102. J. Stallings, *Homology and central series of groups*, Journal of Algebra 2 (1965), 170–181.
103. M. Suzuki, *The Magnus representation of the Torelli group $\mathcal{I}_{g,1}$ is not faithful for $g \geq 2$*, Proc. Amer. Math. Soc. 130 (2002), 909–914.
104. U. Tillmann, *On the homotopy of the stable mapping class group*, Invent. Math. 130 (1997), 157–175.
105. R. Vakil, *The moduli space of curves and its tautological ring*, Notices of the Amer. Math. Soc. 50 (2003), 647–658.
106. K. Vogtmann, *Automorphisms of free groups and Outer Space*, Geometricae Dedicata 94 (2002), 1–31.
107. N. Wahl, *From mapping class groups to automorphism groups of free groups*, J. London Math. Soc. (2) 72 (2005), 510-524.

DEPARTMENT OF MATHEMATICAL SCIENCES, UNIVERSITY OF TOKYO, KOMABA, TOKYO 153-8914, JAPAN

E-mail address: morita@ms.u-tokyo.ac.jp

From Braid Groups to Mapping Class Groups

Luis Paris

ABSTRACT. This paper is a survey of some properties of the braid groups and related groups that lead to questions on mapping class groups.

1. Introduction

I have to confess that I am not a leader in the area of mapping class groups and Teichmüller geometry. Nevertheless, I can ask a single question which, I believe, should be of interest for experts in the field.

Question. *Which properties of the braid groups can be extended to the mapping class groups?*

The present paper is a sort of survey of some (not always well-known) properties of the braid groups and related groups which lead to questions on mapping class groups. The reader will find throughout the text several more or less explicit questions, but none of them are stated as "official" problems or conjectures. This choice is based on the idea that the proof often comes together with the question (or the theorem), and, moreover, the young researchers should keep in mind that the surroundings and the applications of a given problem should be the main motivations for its study, and not the question itself (or its author).

Throughout the paper, Σ will denote a compact, connected, and oriented surface. The boundary $\partial \Sigma$, if non empty, is a finite collection of simple closed curves. Consider a finite set $\mathcal{P} = \{P_1, \ldots, P_n\}$ of n distinct points, called *punctures*, in the interior of Σ. Define $\mathcal{H}(\Sigma, \mathcal{P})$ to be the group of orientation-preserving homeomorphisms $h : \Sigma \to \Sigma$ such that h is the identity on each boundary component of Σ and $h(\mathcal{P}) = \mathcal{P}$. The *mapping class group* $\mathcal{M}(\Sigma, \mathcal{P}) = \pi_0(\mathcal{H}(\Sigma, \mathcal{P}))$ *of* Σ *relative to* $\mathcal{P}$ is defined to be the set of isotopy classes of mappings in $\mathcal{H}(\Sigma, \mathcal{P})$, with composition as group operation. We emphasize that, throughout an isotopy, the boundary components and the punctures of $\mathcal{P}$ remain fixed. It is clear that, up to isomorphism, this group depends only on the genus g of Σ, on the number p of components of $\partial \Sigma$, and on the cardinality n of $\mathcal{P}$. So, we may write $\mathcal{M}(g, p, n)$ in place of $\mathcal{M}(\Sigma, \mathcal{P})$, and simply $\mathcal{M}(\Sigma)$ for $\mathcal{M}(\Sigma, \emptyset)$.

2000 *Mathematics Subject Classification.* Primary 20F36; Secondary 57M99, 57N05.

©2006 American Mathematical Society

Let $\mathbb{D} = \{z \in \mathbb{C}; |z| \leq 1\}$ be the *standard disk*, and let $\mathcal{P} = \{P_1, \ldots, P_n\}$ be a finite collection of n punctures in the interior of $\mathbb{D}$. Define a *n-braid* to be a n-tuple $\beta = (b_1, \ldots, b_n)$ of disjoint smooth paths in $\mathbb{D} \times [0,1]$, called the *strings* of β, such that:

- the projection of $b_i(t)$ on the second coordinate is t, for all $t \in [0,1]$ and all $i \in \{1, \ldots, n\}$;
- $b_i(0) = (P_i, 0)$ and $b_i(1) = (P_{\chi(i)}, 1)$, where χ is a permutation of $\{1, \ldots, n\}$, for all $i \in \{1, \ldots, n\}$.

An isotopy in this context is a deformation through braids which fixes the ends. Multiplication of braids is defined by concatenation. The isotopy classes of braids with this multiplication form a group, called the *braid group on n strings*, and denoted by $B_n(\mathcal{P})$. This group does not depend, up to isomorphism, on the set $\mathcal{P}$, but only on the cardinality n of $\mathcal{P}$, thus we may write B_n in place of $B_n(\mathcal{P})$.

The group B_n has a well-known presentation with generators $\sigma_1, \ldots, \sigma_{n-1}$ and relations
$$\begin{aligned} \sigma_i \sigma_j &= \sigma_j \sigma_i & &\text{if } |j - i| > 1, \\ \sigma_i \sigma_j \sigma_i &= \sigma_j \sigma_i \sigma_j & &\text{if } |j - i| = 1. \end{aligned}$$

There are other equivalent descriptions of B_n, as a group of automorphisms of the free group F_n, as the fundamental group of a configuration space, or as the mapping class group of the n-punctured disk. This explains the importance of the braid groups in many disciplines.

Now, the equality
$$(1) \qquad B_n(\mathcal{P}) = \mathcal{M}(\mathbb{D}, \mathcal{P})$$
is the source of the several questions that the reader will find in the paper.

2. Garside groups and Artin groups

We start this section with a brief presentation of Garside groups. Our presentation of the subject draws in many ways from the work of Dehornoy [16] as well as [19], and, like all treatments of Garside groups, is inspired ultimately by the seminal papers of Garside [26], on braid groups, and Brieskorn and Saito [8], on Artin groups.

Let M be an arbitrary monoid. We say that M is *atomic* if there exists a function $\nu : M \to \mathbb{N}$ such that

- $\nu(a) = 0$ if and only if $a = 1$;
- $\nu(ab) \geq \nu(a) + \nu(b)$ for all $a, b \in M$.

Such a function on M is called a *norm* on M.

An element $a \in M$ is called an *atom* if it is indecomposable, namely, if $a = bc$ then either $b = 1$ or $c = 1$. This definition of atomicity is taken from [19]. We refer to [19] for a list of further properties all equivalent to atomicity. In the same paper, it is shown that a subset of M generates M if and only if it contains the set of all atoms. In particular, M is finitely generated if and only if it has finitely many atoms.

Given that a monoid M is atomic, we may define left and right invariant partial orders $\leq_L$ and $\leq_R$ on M as follows.
- Set $a \leq_L b$ if there exists $c \in M$ such that $ac = b$.
- Set $a \leq_R b$ if there exists $c \in M$ such that $ca = b$.

We shall call these *left* and *right divisibility orders* on M.

A *Garside monoid* is a monoid M such that
- M is atomic and finitely generated;
- M is cancellative;
- $(M, \leq_L)$ and $(M, \leq_R)$ are lattices;
- there exists an element $\Delta \in M$, called *Garside element*, such that the set $L(\Delta) = \{x \in M; x \leq_L \Delta\}$ generates M and is equal to $R(\Delta) = \{x \in M; x \leq_R \Delta\}$.

For any monoid M, we can define the group $G(M)$ which is presented by the generating set M and the relations $ab = c$ whenever $ab = c$ in M. There is an obvious canonical homomorphism $M \to G(M)$. This homomorphism is not injective in general. The group $G(M)$ is known as the *group of fractions* of M. Define a *Garside group* to be the group of fractions of a Garside monoid.

Remarks.
(1) Recall that a monoid M satisfies *Öre's conditions* if M is cancellative and if, for all $a, b \in M$, there exist $c, d \in M$ such that $ac = bd$. If M satisfies Öre's conditions then the canonical homomorphism $M \to G(M)$ is injective. A Garside monoid obviously satisfies Öre's conditions, thus any Garside monoid embeds in its group of fractions.
(2) A Garside element is never unique. For example, if Δ is a Garside element, then Δ^k is also a Garside element for all $k \geq 1$ (see [**16**]).

Let M be a Garside monoid. The lattice operations of $(M, \leq_L)$ are denoted by $\vee_L$ and $\wedge_L$. These are defined as follows. For $a, b, c \in M$, $a \wedge_L b \leq_L a$, $a \wedge_L b \leq_L b$, and if $c \leq_L a$ and $c \leq_L b$, then $c \leq_L a \wedge_L b$. For $a, b, c \in M$, $a \leq_L a \vee_L b$, $b \leq_L a \vee_L b$ and, if $a \leq_L c$ and $b \leq_L c$, then $a \vee_L b \leq_L c$. Similarly, the lattice operations of $(M, \leq_R)$ are denoted by $\vee_R$ and $\wedge_R$.

Now, we briefly explain how to define a biautomatic structure on a given Garside group. By [**22**], such a structure furnishes solutions to the word problem and to the conjugacy problem, and it implies that the group has quadratic isoperimetric inequalities. We refer to [**22**] for definitions and properties of automatic groups, and to [**16**] for more details on the biautomatic structures on Garside groups.

Let M be a Garside monoid, and let Δ be a Garside element of M. For $a \in M$, we write $\pi_L(a) = \Delta \wedge_L a$ and denote by $\partial_L(a)$ the unique element of M such that $a = \pi_L(a) \cdot \partial_L(a)$. Using the fact that M is atomic and that $L(\Delta) = \{x \in M; x \leq_L \Delta\}$ generates M, one can easily show that $\pi_L(a) \neq 1$ if $a \neq 1$, and that there exists a positive integer k such that $\partial_L^k(a) = 1$. Let k be the lowest integer satisfying $\partial_L^k(a) = 1$. Then the expression

$$a = \pi_L(a) \cdot \pi_L(\partial_L(a)) \cdot \ldots \cdot \pi_L(\partial_L^{k-1}(a))$$

is called the *normal form* of a.

Let $G = G(M)$ be the group of fractions of M. Let $c \in G$. Since G is a lattice with positive cone M, the element c can be written $c = a^{-1}b$ with $a, b \in M$. Obviously, a and b can be chosen so that $a \wedge_L b = 1$ and, with this extra condition, are unique. Let $a = a_1 a_2 \ldots a_p$ and $b = b_1 b_2 \ldots b_q$ be the normal forms of a and b, respectively. Then the expression
$$c = a_p^{-1} \ldots a_2^{-1} a_1^{-1} b_1 b_2 \ldots b_q$$
is called the *normal form* of c.

Theorem 2.1 (Dehornoy [16]). *Let M be a Garside monoid, and let G be the group of fractions of M. Then the normal forms of the elements of G form a symmetric rational language on the finite set $L(\Delta)$ which has the fellow traveler property. In particular, G is biautomatic.*

The notion of a Garside group was introduced by Dehornoy and the author [19] in a slightly restricted sense, and, later, by Dehornoy [16] in the larger sense which is now generally used. As pointed out before, the theory of Garside groups is largely inspired by the papers of Garside [26], which treated the case of the braid groups, and Brieskorn and Saito [8], which generalized Garside's work to Artin groups. The Artin groups of spherical type which include, notably, the braid groups, are motivating examples. Other interesting examples include all torus link groups (see [47]) and some generalized braid groups associated to finite complex reflection groups (see [2]).

Garside groups have many attractive properties: solutions to the word and conjugacy problems are known (see [16], [46], [25]), they are torsion-free (see [15]), they are biautomatic (see [16]), and they admit finite dimensional classifying spaces (see [18], [10]). There also exist criteria in terms of presentations which detect Garside groups (see [19], [16]).

Let S be a finite set. A *Coxeter matrix* over S is a square matrix $M = (m_{st})_{s,t \in S}$ indexed by the elements of S and such that $m_{ss} = 1$ for all $s \in S$, and $m_{st} = m_{ts} \in \{2, 3, 4, \ldots, +\infty\}$ for all $s, t \in S$, $s \neq t$. A Coxeter matrix $M = (m_{st})_{s,t \in S}$ is usually represented by its *Coxeter graph*, Γ, which is defined as follows. The set of vertices of Γ is S, two vertices s, t are joined by an edge if $m_{st} \geq 3$, and this edge is labeled by m_{st} if $m_{st} \geq 4$.

Let Γ be a Coxeter graph with set of vertices S. For two objects a, b and $m \in \mathbb{N}$, define the word
$$\omega(a, b : m) = \begin{cases} (ab)^{\frac{m}{2}} & \text{if } m \text{ is even}, \\ (ab)^{\frac{m-1}{2}} a & \text{if } m \text{ is odd}. \end{cases}$$
Take an abstract set $\mathcal{S} = \{\sigma_s; s \in S\}$ in one-to-one correspondence with S. The *Artin group of type* Γ is the group $A = A_\Gamma$ generated by $\mathcal{S}$ and subject to the relations
$$\omega(\sigma_s, \sigma_t : m_{st}) = \omega(\sigma_t, \sigma_s : m_{st}) \quad \text{for } s, t \in S, \ s \neq t, \text{ and } m_{st} < +\infty,$$

where $M = (m_{st})_{s,t \in S}$ is the Coxeter matrix of Γ. The *Coxeter group of type* Γ is the group $W = W_\Gamma$ generated by S and subject to the relations

$$s^2 = 1 \quad \text{for } s \in S,$$
$$(st)^{m_{st}} = 1 \quad \text{for } s, t \in S, \ s \neq t, \text{ and } m_{st} < +\infty.$$

Note that W is the quotient of A by the relations $\sigma_s^2 = 1$, $s \in S$, and s is the image of σ_s under the quotient epimorphism.

The number $n = |S|$ of generators is called the *rank* of the Artin group (and of the Coxeter group). We say that A is *irreducible* if Γ is connected, and that A is of *spherical type* if W is finite.

Coxeter groups have been widely studied. Basic references for them are [6] and [30]. In contrast, Artin groups are poorly understood in general. Beside the spherical type ones, which are Garside groups, the Artin groups which are better understood include right–angled Artin groups, 2-dimensional Artin groups, and FC type Artin groups. Right-angled Artin groups (also known as graph groups or free partially commutative groups) have been widely studied, and their applications extend to various domains such as parallel computation, random walks, and cohomology of groups. 2-dimensional Artin groups and FC type Artin groups have been introduced by Charney and Davis [9] in 1995 in their study of the $K(\pi, 1)$-problem for complements of infinite hyperplane arrangements associated to reflection groups.

Let G be an Artin group (resp. Garside group). Define a *geometric representation* of G to be a homomorphism from G to some mapping class group. Although I do not believe that mapping class groups are either Artin groups or Garside groups, the theory of geometric representations should be of interest from the perspective of both mapping class groups and Artin (resp. Garside) groups. For instance, it is not known which Artin groups (resp. Garside groups) can be embedded into a mapping class group.

A first example of such representations are the (maximal) free abelian subgroups of the mapping class groups whose study was initiated by Birman, Lubotzky, and McCarthy [5]. These representations are crucial in the study of the automomorphism groups and abstract commensurators of the mapping class groups (see [31], [33], [34], [41]). Another example are the presentations of the mapping class groups as quotients of Artin groups by some (not too bad) relations (see [39], [37]). However, the most attractive examples of geometric representations are the so-called *monodromy representations* introduced by several people (Sullivan, Arnold, A'Campo) for the Artin groups of type A_n, D_n, E_6, E_7, E_8, and extended by Perron, Vannier [45], Crisp and the author [11] to all small type Artin groups as follows.

First, recall that an Artin group A associated to a Coxeter graph Γ is said to be of *small type* if $m_{st} \in \{2, 3\}$ for all $s, t \in S$, $s \neq t$, where $M = (m_{st})_{s,t \in S}$ is the Coxeter matrix of Γ.

An *essential circle* in Σ is an oriented embedding $a : \mathbb{S}^1 \to \Sigma$ such that $a(\mathbb{S}^1) \cap \partial \Sigma = \emptyset$, and $a(\mathbb{S}^1)$ does not bound any disk. We shall use the description of the circle $\mathbb{S}^1$ as $\mathbb{R}/2\pi\mathbb{Z}$. Take an (oriented) embedding $\mathbb{A}_a : \mathbb{S}^1 \times [0, 1] \to \Sigma$ of the annulus such

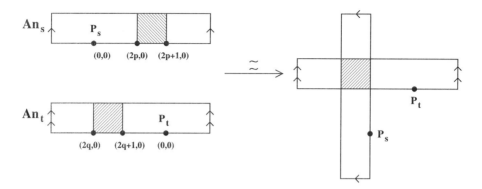

FIGURE 1. Identification of annuli.

that $\mathbb{A}_a(\theta, \frac{1}{2}) = a(\theta)$ for all $\theta \in \mathbb{S}^1$, and define a homomorphism $T_a \in \mathcal{H}(\Sigma)$, which restricts to the identity outside the interior of the image of $\mathbb{A}_a$, and is given by

$$(T_a \circ \mathbb{A}_a)(\theta, x) = \mathbb{A}_a(\theta + 2\pi x, x) \quad \text{for } (\theta, x) \in \mathbb{S}^1 \times [0, 1],$$

inside the image of $\mathbb{A}_a$. The *Dehn twist* along a is the element of $\mathcal{M}(\Sigma)$ represented by T_a. The following result is easily checked and may be found, for instance, in [**4**].

Proposition 2.2. *Let $a_1, a_2 : \mathbb{S}^1 \to \Sigma$ be two essential circles, and, for $i = 1, 2$, let τ_i denote the Dehn twist along a_i. Then*

$$\begin{aligned} \tau_1 \tau_2 &= \tau_2 \tau_1 && \text{if } a_1 \cap a_2 = \emptyset, \\ \tau_1 \tau_2 \tau_1 &= \tau_2 \tau_1 \tau_2 && \text{if } |a_1 \cap a_2| = 1. \end{aligned}$$

We assume now that Γ is a small type Coxeter graph, and we associate to Γ a surface $\Sigma = \Sigma(\Gamma)$ as follows.

Let $M = (m_{s\,t})_{s,t \in S}$ be the Coxeter matrix of Γ. Let $<$ be a total order on S which can be chosen arbitrarily. For each $s \in S$, we define the set

$$\text{St}_s = \{t \in S;\, m_{s\,t} = 3\} \cup \{s\}.$$

We write $\text{St}_s = \{t_1, t_2, \ldots, t_k\}$ such that $t_1 < t_2 < \cdots < t_k$, and suppose $s = t_j$. The difference $i - j$ is called the *relative position* of t_i with respect to s, and is denoted by $\text{pos}(t_i : s)$. In particular, $\text{pos}(s : s) = 0$.

Let $s \in S$. Put $k = |St_s|$. We denote by An_s the annulus defined by

$$\text{An}_s = \mathbb{R}/2k\mathbb{Z} \times [0, 1].$$

For each $s \in S$, write P_s for the point $(0,0)$ of An_s. The surface $\Sigma = \Sigma(\Gamma)$ is defined by

$$\Sigma = \left(\coprod_{s \in S} \text{An}_s \right) / \approx,$$

where $\approx$ is the relation defined as follows. Let $s, t \in S$ such that $m_{s\,t} = 3$ and $s < t$. Put $p = \text{pos}(t : s) > 0$ and $q = \text{pos}(s : t) < 0$. For each $(x, y) \in [0, 1] \times [0, 1]$, the relation $\approx$ identifies the point $(2p + x, y)$ of An_s with the point $(2q + 1 - y, x)$ of An_t (see Figure 1). We identify each annulus An_s and the point P_s with their image in Σ, respectively.

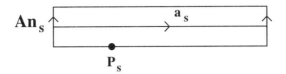

FIGURE 2. The essential circle a_s in An_s.

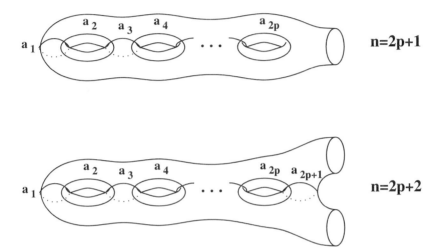

FIGURE 3. Monodromy representation for B_n.

We now define the monodromy representation $\rho : A_\Gamma \to \mathcal{M}(\Sigma)$. Let $s \in S$, and put $k = |\mathrm{St}_s|$. We denote by $a_s : \mathbb{S}^1 \to \Sigma$ the essential circle of Σ such that $a_s(\theta)$ is the point $(\frac{k\theta}{\pi}, \frac{1}{2})$ of An_s (see Figure 2). We let τ_s denote the Dehn twist along a_s. One has $a_s \cap a_t = \emptyset$ if $m_{st} = 2$, and $|a_s \cap a_t| = 1$ if $m_{st} = 3$. Therefore, by Proposition 2.2, we have the following.

Proposition 2.3. *Let Γ be a small type Coxeter graph. There exists a well-defined group homomorphism $\rho : A_\Gamma \to \mathcal{M}(\Sigma)$ which sends σ_s to τ_s for each $s \in S$.*

Example. Consider the braid group B_n, $n \geq 3$. If n is odd, then the associated surface, Σ, is a surface of genus $\frac{n-1}{2}$ with one boundary component. If n is even, then Σ is a surface of genus $\frac{n-2}{2}$ with two boundary components (see Figure 3). For each $i = 1, \ldots, n-1$, let $a_i : \mathbb{S}^1 \to \Sigma$ denote the essential circle pictured in Figure 3, and let τ_i denote the Dehn twist along a_i. Then the monodromy representation $\rho : B_n \to \mathcal{M}(\Sigma)$ of the braid group sends σ_i to τ_i for all $i = 1, \ldots, n-1$.

Let $\mathcal{O}_2$ denote the ring of germs of functions at 0 of $\mathbb{C}^2$. Let $f \in \mathcal{O}_2$ with an isolated singularity at 0, and let μ be the Milnor number of f. To the germ f one can associate a surface Σ with boundary, called the *Milnor fiber* of f, an analytic subvariety $\mathcal{D}$ of $\mathbb{D}(\eta) = \{t \in \mathbb{C}^\mu; \|t\| < \eta\}$, where $\eta > 0$ is a small number, and a representation $\rho : \pi_1(\mathbb{D}(\eta) \setminus \mathcal{D}) \to \mathcal{M}(\Sigma)$, called the *geometric monodromy* of f.

Recall that the simple singularities are defined by the equations

$$\begin{aligned}
A_n &: & f(x,y) &= x^2 + y^{n+1}, & n &\geq 1, \\
D_n &: & f(x,y) &= x(x^{n-2} + y^2), & n &\geq 4, \\
E_6 &: & f(x,y) &= x^3 + y^4, \\
E_7 &: & f(x,y) &= x(x^2 + y^3), \\
E_8 &: & f(x,y) &= x^3 + y^5.
\end{aligned}$$

By [1] and [7], for each $\Gamma \in \{A_n; n \geq 1\} \cup \{D_n; n \geq 4\} \cup \{E_6, E_7, E_8\}$, the group $\pi_1(\mathbb{D}(\eta) \setminus \mathcal{D})$ is isomorphic to the Artin group of type Γ, and $\rho : A_\Gamma \to \mathcal{M}(\Sigma)$ is the monodromy representation of the group A_Γ.

The monodromy representations are known to be faithful for $\Gamma = A_n$ and $\Gamma = D_n$ (see [45]), but are not faithful for $\Gamma \in \{E_6, E_7, E_8\}$ (see [51]). Actually, the monodromy representations are not faithful in general (see [36]).

3. Dehornoy's ordering

Call a group or a monoid G *left-orderable* if there exists a strict linear ordering $<$ of the elements of G which is left invariant, namely, for $x, y, z \in G$, $x < y$ implies $zx < zy$. If, in addition, the ordering is a well-ordering, then we say that G is *left-well-orderable*. If there is an ordering which is invariant under multiplication on both sides, we say that G is *biorderable*.

Recall that the braid group B_n has a presentation with generators $\sigma_1, \ldots, \sigma_{n-1}$ and relations

(2) $$\sigma_i \sigma_j = \sigma_j \sigma_i \quad \text{for } |i - j| > 1,$$

(3) $$\sigma_i \sigma_j \sigma_i = \sigma_j \sigma_i \sigma_j \quad \text{for } |i - j| = 1.$$

The submonoid generated by $\sigma_1, \ldots, \sigma_{n-1}$ is called the *positive braid monoid* on n strings, and is denoted by B_n^+. It has a monoid presentation with generators $\sigma_1, \ldots, \sigma_{n-1}$ and relations (2) and (3), it is a Garside monoid, and the group of fractions of B_n^+ is B_n.

The starting point of the present section is the following result due to Dehornoy [13].

Theorem 3.1 (Dehornoy [13]). *The Artin braid group B_n is left-orderable, by an ordering which is a well-ordering when restricted to B_n^+.*

Left-invariant orderings are important because of a long-standing conjecture in group theory: that the group ring of a torsion-free group has no zero divisors. This conjecture is true for left-orderable groups, thus Theorem 3.1 implies that $\mathbb{Z}[B_n]$ has no zero divisors. However, the main point of interest of Theorem 3.1 is not only the mere existence of orderings on the braid groups, but the particular nature of the construction. This has been already used to prove that some representations of the braid groups by automorphisms of groups are faithful (see [50] and [12]), and to study the so-called "weak faithfulness" of the Burau representation (see [14]), and I am sure that these ideas will be exploited again in the future.

Let G be a left-orderable group, and let $<$ be a left invariant linear ordering on G. Call an element $g \in G$ *positive* if $g > 1$, and denote by $\mathcal{P}$ the set of positive elements. Then $\mathcal{P}$ is a subsemigroup (i.e. $\mathcal{P} \cdot \mathcal{P} \subset \mathcal{P}$), and we have the disjoint union $G = \mathcal{P} \sqcup \mathcal{P}^{-1} \sqcup \{1\}$, where $\mathcal{P}^{-1} = \{g^{-1}; g \in \mathcal{P}\}$. Conversely, if $\mathcal{P}$ is a subsemigroup of G such that $G = \mathcal{P} \sqcup \mathcal{P}^{-1} \sqcup \{1\}$, then the relation $<$ defined by $f < g$ if $f^{-1}g \in \mathcal{P}$ is a left invariant linear ordering on G.

The description of the positive elements in Dehornoy's ordering is based on the following result which, in my view, is more interesting than the mere existence of the ordering, because it leads to new techniques to study the braid groups.

Let B_{n-1} be the subgroup of B_n generated by $\sigma_1, \ldots, \sigma_{n-2}$. Let $\beta \in B_n$. We say that β is σ_{n-1}-*positive* if it can be written
$$\beta = \alpha_0 \sigma_{n-1} \alpha_1 \sigma_{n-1} \alpha_2 \ldots \sigma_{n-1} \alpha_l,$$
where $l \geq 1$ and $\alpha_0, \alpha_1, \ldots, \alpha_l \in B_{n-1}$. We say that β is σ_{n-1}-*negative* if it can be written
$$\beta = \alpha_0 \sigma_{n-1}^{-1} \alpha_1 \sigma_{n-1}^{-1} \alpha_2 \ldots \sigma_{n-1}^{-1} \alpha_l,$$
where $l \geq 1$ and $\alpha_0, \alpha_1, \ldots, \alpha_l \in B_{n-1}$. We denote by $\mathcal{P}_{n-1}$ the set of σ_{n-1}-positive elements, and by $\mathcal{P}_{n-1}^{-}$ the set of σ_{n-1}-negative elements. Note that $\mathcal{P}_{n-1}^{-} = \mathcal{P}_{n-1}^{-1} = \{\beta^{-1}; \beta \in \mathcal{P}_{n-1}\}$.

Theorem 3.2 (Dehornoy [13]). *We have the disjoint union* $B_n = \mathcal{P}_{n-1} \sqcup \mathcal{P}_{n-1}^{-} \sqcup B_{n-1}$.

Now, the set $\mathcal{P}$ of positive elements in Dehornoy's ordering can be described as follows. For $k = 1, \ldots, n-1$, let $\mathcal{P}_k$ denote the set of σ_k-positive elements in B_{k+1}, where B_{k+1} is viewed as the subgroup of B_n generated by $\sigma_1, \ldots, \sigma_k$. Then $\mathcal{P} = \mathcal{P}_1 \sqcup \mathcal{P}_2 \sqcup \cdots \sqcup \mathcal{P}_{n-1}$. It is easily checked using Theorem 3.2 that $\mathcal{P}$ is a subsemigroup of B_n and that $B_n = \mathcal{P} \sqcup \mathcal{P}^{-1} \sqcup \{1\}$.

Dehornoy's proof of Theorem 3.2 involves rather delicate combinatorial and algebraic constructions which were partly motivated by questions in set theory. A topological construction of Dehornoy's ordering of the braid group B_n, viewed as the mapping class group of the n-punctured disk, is given in [24]. This last construction has been extended to the mapping class groups of the punctured surfaces with non-empty boundary by Rourke and Wiest [48]. This extension involves some choices, and, therefore, is not canonical. However, it is strongly connected to Mosher's normal forms [42] and, as with Dehornoy's ordering, is sufficiently explicit to be used for other purposes (to study faithfulness properties of "some" representations of the mapping class group, for example. This is just an idea...). Another approach (see [49]) based on Nielsen's ideas [43] gives rise to many other linear orderings of the braid groups, and, more generally, of the mapping class groups of punctured surfaces with non-empty boundary, but I am not convinced that these orderings are explicit enough to be used for other purposes.

The mapping class group of a (punctured) closed surface is definitively not left-orderable because it has torsion. However, it is an open question to determine whether some special torsion-free subgroups of $\mathcal{M}(\Sigma)$ such as Ivanov's group $\Gamma_m(\Sigma)$ (see [32]) are orderable. The Torelli group is residually torsion-free nilpotent (see [27]), therefore is biorderable. We refer to [17] for a detailed account on Dehornoy's ordering and related questions.

4. Thurston classification

Recall that an *essential circle* in $\Sigma \setminus \mathcal{P}$ is an oriented embedding $a : \mathbb{S}^1 \to \Sigma \setminus \mathcal{P}$ such that $a(\mathbb{S}^1) \cap \partial \Sigma = \emptyset$, and $a(\mathbb{S}^1)$ does not bound a disk in $\Sigma \setminus \mathcal{P}$. An essential circle $a : \mathbb{S}^1 \to \Sigma \setminus \mathcal{P}$ is called *generic* if it does not bound a disk in Σ containing one puncture, and if it is not isotopic to a boundary component of Σ. Recall that two circles $a, b : \mathbb{S}^1 \to \Sigma \setminus \mathcal{P}$ are *isotopic* if there exists a continuous family $a_t : \mathbb{S}^1 \to \Sigma \setminus \mathcal{P}$, $t \in [0, 1]$, such that $a_0 = a$ and $a_1 = b$. Isotopy of circles is an equivalence relation that we denote by $a \sim b$. Observe that $h(a) \sim h'(a')$ if $a \sim a'$ and $h, h' \in \mathcal{H}(\Sigma, \mathcal{P})$ are isotopic. So, the mapping class group $\mathcal{M}(\Sigma, \mathcal{P})$ acts on the set $\mathcal{C}(\Sigma, \mathcal{P})$ of isotopy classes of generic circles.

Let $f \in \mathcal{M}(\Sigma, \mathcal{P})$. We say that f is a *pseudo-Anosov element* if it has no finite orbit in $\mathcal{C}(\Sigma, \mathcal{P})$, we say that f is *periodic* if f^m acts trivially on $\mathcal{C}(\Sigma, \mathcal{P})$ for some $m \geq 1$, and we say that f is *reducible* otherwise.

The use of the action of $\mathcal{M}(\Sigma, \mathcal{P})$ on $\mathcal{C}(\Sigma, \mathcal{P})$ and the "Thurston classification" of the elements of $\mathcal{M}(\Sigma, \mathcal{P})$ play a prominent role in the study of the algebraic properties of the mapping class groups. Here are some applications.

Theorem 4.1.

(1) (**Ivanov [32]**). *Let f be a pseudo-Anosov element of $\mathcal{M}(\Sigma, \mathcal{P})$. Then f has infinite order, $\langle f \rangle \cap Z(\mathcal{M}(\Sigma, \mathcal{P})) = \{1\}$, where $\langle f \rangle$ denotes the subgroup generated by f, and $\langle f \rangle \times Z(\mathcal{M}(\Sigma, \mathcal{P}))$ is a subgroup of finite index of the centralizer of f in $\mathcal{M}(\Sigma, \mathcal{P})$.*

(2) (**Birman, Lubotzky, McCarthy [5]**). *If a subgroup of $\mathcal{M}(\Sigma, \mathcal{P})$ contains a solvable subgroup of finite index, then it contains an abelian subgroup of finite index.*

(3) (**Ivanov [32], McCarthy [40]**). *Let G be a subgroup of $\mathcal{M}(\Sigma, \mathcal{P})$. Then either G contains a free group of rank 2, or G contains an abelian subgroup of finite index.*

The use of the action of $\mathcal{M}(\Sigma, \mathcal{P})$ on $\mathcal{C}(\Sigma, \mathcal{P})$ is also essential in the determination of the (maximal) free abelian subgroups of $\mathcal{M}(\Sigma, \mathcal{P})$ (see [5]), and in the calculation of the automorphism group and abstract commensurator group of $\mathcal{M}(\Sigma, \mathcal{P})$ (see [31], [33], [41], [34]).

The present section concerns the question of how to translate these techniques to other groups such as the (spherical type) Artin groups, or the Coxeter groups.

Let Γ be a Coxeter graph, let $M = (m_{st})_{s,t \in S}$ be the Coxeter matrix of Γ, and let $A = A_\Gamma$ be the Artin group of type Γ. For $X \subset S$, we denote by Γ_X the full subgraph of Γ generated by X, and by A_X the subgroup of A generated by $\mathcal{S}_X = \{\sigma_x; x \in X\}$. By [38] (see also [44]), the group A_X is the Artin group of type Γ_X. A subgroup of the form A_X is called a *standard parabolic subgroup*, and a subgroup which is conjugate to a standard parabolic subgroup is simply called a *parabolic subgroup*. We shall denote by $\mathcal{AC} = \mathcal{AC}(\Gamma)$ the set of parabolic subgroups different from A and from $\{1\}$.

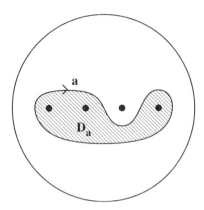

FIGURE 4. The disk $\mathbb{D}_a$.

Let $f \in A$. We say that f is an *(algebraic) pseudo-Anosov element* if it has no finite orbit in $\mathcal{AC}$, we say that f is an *(algebraic) periodic element* if f^m acts trivially on $\mathcal{AC}$ for some $m > 0$, and we say that f is an *(algebraic) reducible element* otherwise.

The above definitions are motivated by the following result.

Proposition 4.2. *Let f be an element of the braid group B_n (viewed as the mapping class group $\mathcal{M}(\mathbb{D},\mathcal{P})$ as well as the Artin group associated to the Dynkin graph A_{n-1}).*

- *f is a pseudo-Anosov element if and only if f is an algebraic pseudo-Anosov element.*
- *f is a periodic element if and only if f is an algebraic periodic element.*
- *f is a reducible element if and only if f is an algebraic reducible element.*

Proof. Recall that $\mathbb{D} = \{z \in \mathbb{C}; |z| \leq 1\}$, $\mathcal{P} = \{P_1, \ldots, P_n\}$ is a set of n punctures in the interior of $\mathbb{D}$, and $B_n = \mathcal{M}(\mathbb{D}, \mathcal{P})$. Let $\sigma_1, \ldots, \sigma_{n-1}$ be the standard generators of B_n. We can assume that each σ_i acts trivially outside a disk containing P_i and P_{i+1}, and permutes P_i and P_{i+1}.

Let $a : \mathbb{S}^1 \to \mathbb{D} \setminus \mathcal{P}$ be a generic circle. Then $a(\mathbb{S}^1)$ separates $\mathbb{D}$ into two components: a disk $\mathbb{D}_a$ bounded by $a(\mathbb{S}^1)$, and an annulus $\mathbb{A}_a$ bounded by $a(\mathbb{S}^1) \sqcup \mathbb{S}^1$, where $\mathbb{S}^1 = \{z \in \mathbb{C}; |z| = 1\}$ (see Figure 4). Let $m = |\mathbb{D}_a \cap \mathcal{P}|$. Then the hypothesis "$a$ is generic" implies that $2 \leq m \leq n - 1$. Furthermore, the group $\mathcal{M}(\mathbb{D}_a, \mathbb{D}_a \cap \mathcal{P})$ is isomorphic to B_m.

Let τ_a denote the Dehn twist along a. If $m = |\mathbb{D}_a \cap \mathcal{P}| \geq 3$, then the center $Z(\mathcal{M}(\mathbb{D}_a, \mathbb{D}_a \cap \mathcal{P}))$ of $\mathcal{M}(\mathbb{D}_a, \mathbb{D}_a \cap \mathcal{P})$ is the (infinite) cyclic subgroup generated by τ_a. If $m = 2$, then $\mathcal{M}(\mathbb{D}_a, \mathbb{D}_a \cap \mathcal{P})$ is an infinite cyclic group, and τ_a generates the (unique) index 2 subgroup of $\mathcal{M}(\mathbb{D}_a, \mathbb{D}_a \cap \mathcal{P})$. Recall that, if τ_{a_1} and τ_{a_2} are two Dehn twists along generic circles a_1 and a_2, respectively, and if $\tau_{a_1}^{k_1} = \tau_{a_2}^{k_2}$ for some $k_1, k_2 \in \mathbb{Z} \setminus \{0\}$, then $k_1 = k_2$ and $a_1 = a_2$. Recall also that, if τ_a is the Dehn twist along a generic circle a, then $g\tau_a g^{-1}$ is the Dehn twist along $g(a)$, for any $g \in \mathcal{M}(\mathbb{D}, \mathcal{P})$. We conclude from the above observations that, if

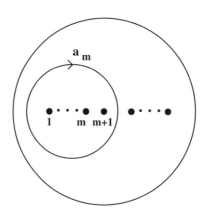

FIGURE 5. The curve a_m.

$g \cdot \mathcal{M}(\mathbb{D}_a, \mathbb{D}_a \cap \mathcal{P}) \cdot g^{-1} = \mathcal{M}(\mathbb{D}_a, \mathbb{D}_a \cap \mathcal{P})$, then $g(a) = a$. Conversely, if $g(a) = a$, then $g(\mathbb{D}_a, \mathbb{D}_a \cap \mathcal{P}) = (\mathbb{D}_a, \mathbb{D}_a \cap \mathcal{P})$, thus $g \cdot \mathcal{M}(\mathbb{D}_a, \mathbb{D}_a \cap \mathcal{P}) \cdot g^{-1} = \mathcal{M}(\mathbb{D}_a, \mathbb{D}_a \cap \mathcal{P})$.

Write $S = \{1, 2, \ldots, n-1\}$. For $X \subset S$, we set $\mathcal{S}_X = \{\sigma_x ; x \in X\}$, and denote by $(B_n)_X$ the subgroup of B_n generated by $\mathcal{S}_X$.

Let $1 \leq m \leq n-2$, and $X(m) = \{1, 2, \ldots, m\}$. We can choose a generic circle $a_m : \mathbb{S}^1 \to \mathbb{D} \setminus \mathcal{P}$ such that $\mathbb{D}_{a_m} \cap \mathcal{P} = \{P_1, \ldots, P_m, P_{m+1}\}$ and $\mathcal{M}(\mathbb{D}_{a_m}, \mathbb{D}_{a_m} \cap \mathcal{P}) = (B_n)_{X(m)}$ (see Figure 5). Observe that, if $a : \mathbb{S}^1 \to \mathbb{D} \setminus \mathcal{P}$ is a generic circle such that $|\mathbb{D}_a \cap \mathcal{P}| = m+1$, then there exists $g \in \mathcal{M}(\mathbb{D}, \mathcal{P}) = B_n$ such that $g\mathcal{M}(\mathbb{D}_a, \mathbb{D}_a \cap \mathcal{P})g^{-1} = \mathcal{M}(\mathbb{D}_{a_m}, \mathbb{D}_{a_m} \cap \mathcal{P}) = (B_n)_{X(m)}$. In particular, $\mathcal{M}(\mathbb{D}_a, \mathbb{D}_a \cap \mathcal{P})$ is a parabolic subgroup of B_n.

Take $X \subset S$, $X \neq S$. Decompose X as a disjoint union $X = X_1 \sqcup X_2 \sqcup \cdots \sqcup X_q$, where $X_k = \{i_k, i_k + 1, \ldots, i_k + t_k\}$ for some $i_k \in \{1, \ldots, n-1\}$ and some $t_k \in \mathbb{N}$, and $i_k + t_k + 1 < i_{k+1}$, for all $k = 1, \ldots, q$. We can choose generic circles $b_1, \ldots, b_q : \mathbb{S}^1 \to \mathbb{D} \setminus \mathcal{P}$ such that

$$\mathbb{D}_{b_k} \cap \mathcal{P} = \{P_{i_k}, \ldots, P_{i_k+t_k}, P_{i_k+t_k+1}\}, \quad \text{for } 1 \leq k \leq q,$$
$$\mathbb{D}_{b_k} \cap \mathbb{D}_{b_l} = \emptyset, \quad \text{for } 1 \leq k \neq l \leq q,$$
$$\mathcal{M}(\mathbb{D}_{b_k}, \mathbb{D}_{b_k} \cap \mathcal{P}) = (B_n)_{X_k} \quad \text{for } 1 \leq k \leq q.$$

(See Figure 6.) It is easily checked that

$$(B_n)_X = (B_n)_{X_1} \times \cdots \times (B_n)_{X_q} = \mathcal{M}(\mathbb{D}_{b_1}, \mathbb{D}_{b_1} \cap \mathcal{P}) \times \cdots \times \mathcal{M}(\mathbb{D}_{b_q}, \mathbb{D}_{b_q} \cap \mathcal{P}),$$

and the only Dehn twists which lie in the center of $(B_n)_X$ are $\tau_{b_1}, \tau_{b_2}, \ldots, \tau_{b_q}$.

Let $g \in \mathcal{M}(\mathbb{D}, \mathcal{P})$ which fixes some $a \in \mathcal{C}(\mathbb{D}, \mathcal{P})$. Then $g\mathcal{M}(\mathbb{D}_a, \mathbb{D}_a \cap \mathcal{P})g^{-1} = \mathcal{M}(\mathbb{D}_a, \mathbb{D}_a \cap \mathcal{P})$ and $\mathcal{M}(\mathbb{D}_a, \mathbb{D}_a \cap \mathcal{P})$ is a parabolic subgroup, thus g fixes some element in $\mathcal{AC}$.

Let $g \in \mathcal{M}(\mathbb{D}, \mathcal{P})$ which fixes some element in $\mathcal{AC}$. Up to conjugation, we may suppose that there exists $X \subset S$ such that $g(B_n)_X g^{-1} = (B_n)_X$. We take again the notations introduced above. Recall that the only Dehn twists which lie in the

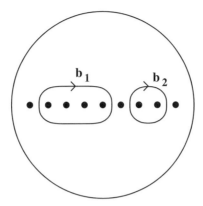

FIGURE 6. The curves b_1, b_2 for $X = \{2, 3, 4, 7\}$.

center of $(B_n)_X$ are $\tau_{b_1}, \tau_{b_2}, \ldots, \tau_{b_q}$, thus g must permute the set $\{\tau_{b_1}, \tau_{b_2}, \ldots, \tau_{b_q}\}$. So, there exists $m \geq 1$ such that $g^m \tau_{b_1} g^{-m} = \tau_{b_1}$, that is, $g^m(b_1) = b_1$.

Let $g \in \mathcal{M}(\mathbb{D}, \mathcal{P})$ which fixes all the elements of $\mathcal{C}(\mathbb{D}, \mathcal{P})$. Let $A \in \mathcal{AC}$. There exist $h \in B_n$ and $X \subset S$ such that $A = h(B_n)_X h^{-1}$. We keep the notations used above. The element $h^{-1}gh$ also fixes all the elements of $\mathcal{C}(\mathbb{D}, \mathcal{P})$. In particular, we have $(h^{-1}gh)(b_k) = b_k$ for all $k = 1, \ldots, q$. This implies that $h^{-1}gh$ fixes $\mathcal{M}(\mathbb{D}_{b_k}, \mathbb{D}_{b_k} \cap \mathcal{P})$ for all $k = 1, \ldots, q$, thus $h^{-1}gh$ fixes $(B_n)_X = \mathcal{M}(\mathbb{D}_{b_1}, \mathbb{D}_{b_1} \cap \mathcal{P}) \times \cdots \times \mathcal{M}(\mathbb{D}_{b_q}, \mathbb{D}_{b_q} \cap \mathcal{P})$, therefore g fixes $A = h(B_n)_X h^{-1}$.

Let $g \in \mathcal{M}(\mathbb{D}, \mathcal{P})$ which fixes all the elements of $\mathcal{AC}$. Let $a \in \mathcal{C}(\mathbb{D}, \mathcal{P})$. Since $\mathcal{M}(\mathbb{D}_a, \mathbb{D}_a \cap \mathcal{P})$ is a parabolic subgroup, we conclude that $g\mathcal{M}(\mathbb{D}_a, \mathbb{D}_a \cap \mathcal{P})g^{-1} = \mathcal{M}(D_a, \mathbb{D}_a \cap \mathcal{P})$, thus $g(a) = a$. □

In spite of the fact that the normalizers of the parabolic subgroups of the spherical type Artin groups are quite well understood (see [**44**], [**28**]), it is not known, for instance, whether the centralizer of an algebraic pseudo-Anosov element g in a spherical type Artin group A has $\langle g \rangle \times Z(A)$ as finite index subgroup. This question might be the starting point for understanding the abelian subgroups of the spherical type Artin groups.

We turn now to a part of Krammer's Ph.D. thesis which describes a phenomenom for Coxeter groups similar to the "Thurston classification" for the mapping class groups: in a non-affine irreducible Coxeter group W, there are certain elements, called *essential* elements, which have infinite order, and have the property that the cyclic subgroups generated by them have finite index in their centralizer. Although almost all elements of W are essential, it is very hard to determine whether a given element is essential.

Let Γ be a Coxeter graph with set of vertices S, and let $W = W_\Gamma$ be the Coxeter group of type Γ. Let $\Pi = \{\alpha_s; s \in S\}$ be an abstract set in one-to-one correspondence with S. The elements of Π are called *simple roots*. Let $U = \oplus_{s \in S} \mathbb{R}\alpha_s$ denote the (abstract) real vector space having Π as a basis. Define the *canonical form* as the symmetric bilinear form $\langle,\rangle : U \times U \to \mathbb{R}$ determined by

$$\langle \alpha_s, \alpha_t \rangle = \begin{cases} -\cos(\pi/m_{st}) & \text{if } m_{st} < +\infty, \\ -1 & \text{if } m_{st} = +\infty. \end{cases}$$

For $s \in S$, define the linear transformation $\rho_s : U \to U$ by

$$\rho_s(x) = x - 2\langle x, \alpha_s \rangle \alpha_s, \quad \text{for } x \in U.$$

Then the map $s \mapsto \rho_s$, $s \in S$, determines a well-defined representation $W \to GL(U)$, called *canonical representation*. This representation is faithful and preserves the canonical form.

The set $\Phi = \{w\alpha_s ; w \in W \text{ and } s \in S\}$ is called the *root system* of W. The set $\Phi^+ = \{\beta = \sum_{s \in S} \lambda_s \alpha_s \in \Phi; \lambda_s \geq 0 \text{ for all } s \in S\}$ is the set of *positive roots*, and $\Phi^- = -\Phi^+$ is the set of *negative roots*. For $w \in W$, we set

$$\Phi_w = \{\beta \in \Phi^+ \ ; \ w\beta \in \Phi^-\}.$$

The following proposition is a mixture of several well-known facts on Φ. The proofs can be found in [29].

Proposition 4.3.
  (1) We have the disjoint union $\Phi = \Phi^+ \sqcup \Phi^-$.
  (2) Let $\lg : W \to \mathbb{N}$ denote the word length of W with respect to S. Then $\lg(w) = |\Phi_w|$ for all $w \in W$. In particular, Φ_w is finite.
  (3) Let $w \in W$ and $s \in S$. Then
  $$\lg(ws) = \begin{cases} \lg(w) + 1 & \text{if } w\alpha_s \in \Phi^+, \\ \lg(w) - 1 & \text{if } w\alpha_s \in \Phi^-. \end{cases}$$
  (4) Let $w \in W$ and $s \in S$. Write $\beta = w\alpha_s \in \Phi$, and $r_\beta = wsw^{-1} \in W$. Then r_β acts on U by
  $$r_\beta(x) = x - 2\langle x, \beta \rangle \beta.$$

Let $u, v \in W$ and $\alpha \in \Phi$. We say that α *separates* u and v if there exists $\varepsilon \in \{\pm 1\}$ such that $u\alpha \in \Phi^\varepsilon$ and $v\alpha \in \Phi^{-\varepsilon}$. Let $w \in W$ and $\alpha \in \Phi$. We say that α is *w-periodic* if there exists some $m \geq 1$ such that $w^m \alpha = \alpha$.

Proposition 4.4. *Let $w \in W$ and $\alpha \in \Phi$. Then precisely one of the following holds.*
  (1) *α is w-periodic.*
  (2) *α is not w-periodic, and the set $\{m \in \mathbb{Z}; \alpha \text{ separates } w^m \text{ and } w^{m+1}\}$ is finite and even.*
  (3) *α is not w-periodic, and the set $\{m \in \mathbb{Z}; \alpha \text{ separates } w^m \text{ and } w^{m+1}\}$ is finite and odd.*

Call α *w-even* in Case 2, and *w-odd* in Case 3.

Proof. Assume that α is not w-periodic, and put

$$N_w(\alpha) = \{m \in \mathbb{Z}; \alpha \text{ separates } w^m \text{ and } w^{m+1}\}.$$

We have to show that $N_w(\alpha)$ is finite.

If $m \in N_w(\alpha)$, then $w^m \alpha \in \Phi_w \cup -\Phi_w$. On the other hand, if $w^{m_1}\alpha = w^{m_2}\alpha$, then $w^{m_1-m_2}\alpha = \alpha$, thus $m_1 - m_2 = 0$, since α is not w-periodic. Since $\Phi_w \cup -\Phi_w$ is finite (see Proposition 4.3), we conclude that $N_w(\alpha)$ is finite. $\square$

For $X \subset S$, we denote by W_X the subgroup of W generated by X. Such a subgroup is called a *standard parabolic subgroup*. A subgroup of W conjugated to a standard parabolic subgroup is simply called a *parabolic subgroup*. An element $w \in W$ is called *essential* if it does not lie in any proper parabolic subgroup. Finally, recall that a Coxeter group W is said to be of *affine type* if the canonical form $\langle , \rangle$ is non-negative (namely, $\langle x, x \rangle \geq 0$ for all $x \in U$).

Now, we can state Krammer's result.

Theorem 4.5 (Krammer [35]). *Assume W to be an irreducible non-affine Coxeter group. Let $w \in W$.*

(1) *w is essential if and only if W is generated by the set $\{r_\beta; \beta \in \Phi^+$ and β w-odd$\}$.*

(2) *Let $m \in \mathbb{N}$, $m \geq 1$. w is essential if and only if w^m is essential.*

(3) *Suppose w is essential. Then w has infinite order, and $\langle w \rangle = \{w^m; m \in \mathbb{Z}\}$ is a finite index subgroup of the centralizer of w in W.*

References

[1] V.I. Arnold, *Normal forms of functions near degenerate critical points, the Weyl groups A_k, D_k, E_k and Lagrangian singularities*, Funkcional. Anal. i Prilov zen. **6** (1972), no. 4, 3–25.

[2] D. Bessis, R. Corran, *Garside structure for the braid group of $G(e,e,r)$*, preprint.

[3] J.S. Birman, *Braids, links, and mapping class groups*, Annals of Mathematics Studies, No. 82, Princeton University Press, Princeton, N.J., 1974.

[4] J.S. Birman, *Mapping class groups of surfaces*, Braids (Santa Cruz, CA, 1986), 13–43, Contemp. Math., 78, Amer. Math. Soc., Providence, RI, 1988.

[5] J.S. Birman, A. Lubotzky, J. McCarthy, *Abelian and solvable subgroups of the mapping class groups*, Duke Math. J. **50** (1983), 1107–1120.

[6] N. Bourbaki, *Groupes et algèbres de Lie, chapitres 4, 5 et 6*, Hermann, Paris, 1968.

[7] E. Brieskorn, *Die Fundamentalgruppe des Raumes der regulären Orbits einer endlichen komplexen Spiegelungsgruppe*, Invent. Math. **12** (1971), 57–61.

[8] E. Brieskorn, K. Saito, *Artin-Gruppen und Coxeter-Gruppen*, Invent. Math. **17** (1972), 245–271.

[9] R. Charney, M.W. Davis, *The $K(\pi,1)$-problem for hyperplane complements associated to infinite reflection groups*, J. Amer. Math. Soc. **8** (1995), 597–627.

[10] R. Charney, J. Meier, K. Whittlesey, *Bestvina's normal form complex and the homology of Garside groups*, Geom. Dedicata **105** (2004), 171–188.

[11] J. Crisp, L. Paris, *The solution to a conjecture of Tits on the subgroup generated by the squares of the generators of an Artrin group*, Invent. Math. **145** (2001), 19–36.

[12] J. Crisp, L. Paris, *Representations of the braid group by automorphisms of groups, invariants of links, and Garside groups*, Pacific J. Math. **221** (2005), no. 1, 1–27.

[13] P. Dehornoy, *Braid groups and left distributive operations*, Trans. Amer. Math. Soc. **343** (1994), 115–150.

[14] P. Dehornoy, *Weak faithfulness properties for the Burau representation*, Topology Appl. **69** (1996), 121–143.

[15] P. Dehornoy, *Gaussian groups are torsion free*, J. Algebra **210** (1998), 291–297.

[16] P. Dehornoy, *Groupes de Garside*, Ann. Sci. Ecole Norm. Sup. (4) **35** (2002), 267–306.

[17] P. Dehornoy, I. Dynnikov, D. Rolfsen, B. Wiest, *Why are braids orderable?* Panoramas et Synthèses, 14, Société Mathématique de France, Paris, 2002.

[18] P. Dehornoy, Y. Lafont, *Homology of Gaussian groups*, Ann. Inst. Fourier (Grenoble) **53** (2003), 489–540.

[19] P. Dehornoy, L. Paris, *Gaussian groups and Garside groups, two generalisations of Artin groups*, Proc. London Math. Soc. (3) **79** (1999), 569–604.

[20] P. Deligne, *Les immeubles des groupes de tresses généralisés*, Invent. Math. **17** (1972), 273–302.

[21] G. Duchamp, D. Krob, *The lower central series of the free partially commutative group*, Semigroup Forum **45** (1992), 385–394.

[22] D.B.A. Epstein, J.W. Cannon, D.F. Holt, S.V.F. Levy, M.S. Paterson, W.P. Thurston, *Word processing in groups*, Jones and Bartlett Publishers, Boston, MA, 1992.

[23] E. Fadell, L. Neuwirth, *Configuration spaces*, Math. Scand. **10** (1962), 111–118.

[24] R. Fenn, M.T. Greene, D. Rolfsen, C. Rourke, B. Wiest, *Ordering the braid groups*, Pacific J. Math. **191** (1999), 49–74.

[25] N. Franco, J. González-Meneses, *Conjugacy problem for braid groups and Garside groups*, J. Algebra **266** (2003), 112–132.

[26] F.A. Garside, *The braid group and other groups*, Quart. J. Math. Oxford Ser. (2) **20** (1969), 235–254.

[27] R. Hain, *Infinitesimal presentations of the Torelli groups*, J. Amer. Math. Soc. **10** (1997), no. 3, 597–651.

[28] E. Godelle, *Normalisateur et groupe d'Artin de type sphérique*, J. Algebra **269** (2003), 263–274.

[29] H. Hiller, *Geometry of Coxeter groups*, Research Notes in Mathematics, 54, Pitman (Advanced Publishing Program), Boston, Mass.–London, 1982.

[30] J.E. Humphreys, *Reflection groups and Coxeter groups*, Cambridge Studies in Advanced Mathematics, 29, Cambridge University Press, Cambridge, 1990.

[31] N.V. Ivanov, *Automorphisms of Teichmüller modular groups*, Topology and geometry - Rohlin Seminar, 199–270, Lecture Notes in Math., 1346, Springer, Berlin, 1988.

[32] N.V. Ivanov, *Subgroups of Teichmüller modular groups*, Translations of Mathematical Monographs, 115, American Mathematical Society, Providence, RI, 1992.

[33] N.V. Ivanov, *Automorphism of complexes of curves and of Teichmüller spaces*, Internat. Math. Res. Notices **14** (1997), 651–666.

[34] N.V. Ivanov, J.D. McCarthy, *On injective homomorphisms between Teichmüller modular groups. I*, Invent. Math. **135** (1999), 425–486.

[35] D. Krammer, *The conjugacy problem for Coxeter groups*, Ph.D. Thesis, Universiteit Utrecht, 1995.

[36] C. Labruère, *Generalized braid groups and mapping class groups*, J. Knot Theory Ramifications **6** (1997), 715–726.

[37] C. Labruère, L. Paris, *Presentations for the punctured mapping class groups in terms of Artin groups*, Algebr. Geom. Topol. **1** (2001), 73–114.

[38] H. Van der Lek, *The homotopy type of complex hyperplane complements*, Ph.D. Thesis, Nijmegen, 1983.

[39] M. Matsumoto, *A presentation of mapping class groups in terms of Artin groups and geometric monodromy of singularities*, Math. Ann. **316** (2000), 401–418.

[40] J. McCarthy, *A "Tits-alternative" for subgroups of surface mapping class groups*, Trans. Amer. Math. Soc. **291** (1985), 583–612.

[41] J. McCarthy, *Automorphisms of surface mapping class groups. A recent theorem of N. Ivanov*, Invent. Math. **84** (1986), 49–71.

[42] L. Mosher, *Mapping class groups are automatic*, Ann. of Math. (2) **142** (1995), 303–384.

[43] J. Nielsen, *Untersuchungen zur Topologie des geschlossenen zweiseitigen Flächen*, Acta Math. **50** (1927), 189–358.

[44] L. Paris, *Parabolic subgroups of Artin groups*, J. Algebra **196** (1997), 369–399.

[45] B. Perron, J.P. Vannier, *Groupe de monodromie géométrique des singularités simples*, Math. Ann. **306** (1996), 231–245.

[46] M. Picantin, *The conjugacy problem in small Gaussian groups*, Comm. Algebra **29** (2001), 1021–1039.

[47] M. Picantin, *Automatic structures for torus link groups*, J. Knot Theory Ramifications **12** (2003), 833–866.

[48] C. Rourke, B. Wiest, *Order automatic mapping class groups*, Pacific J. Math. **194** (2000), 209–227.

[49] H. Short, B. Wiest, *Orderings of mapping class groups after Thurston*, Enseign. Math. (2) **46** (2000), 279–312.

[50] V. Shpilrain, *Representing braids by automorphisms*, Internat. J. Algebra Comput. **11** (2001), 773–777.

[51] B. Wajnryb, *Artin groups and geometric monodromy*, Invent. Math. **138** (1999), 563–571.

INSTITUT DE MATHÉMATIQUES DE BOURGOGNE, UMR 5584 DU CNRS, UNIVERSITÉ DE BOURGOGNE, B.P. 47870, 21078 DIJON CEDEX, FRANCE

E-mail address: lparis@u-bourgogne.fr

Titles in This Series

74 **Benson Farb, Editor,** Problems on mapping class groups and related topics, 2006

73 **Mikhail Lyubich and Leon Takhtajan, Editors,** Graphs and patterns in mathematics and theoretical physics (Stony Brook University, Stony Brook, NY, June 14–21, 2001)

72 **Michel L. Lapidus and Machiel van Frankenhuijsen, Editors,** Fractal geometry and applications: A jubilee of Benoît Mandelbrot, Parts 1 and 2 (San Diego, California, 2002 and École Normale Supérieure de Lyon, 2001)

71 **Gordana Matić and Clint McCrory, Editors,** Topology and Geometry of Manifolds (University of Georgia, Athens, Georgia, 2001)

70 **Michael D. Fried and Yasutaka Ihara, Editors,** Arithmetic fundamental groups and noncommutative algebra (Mathematical Sciences Research Institute, Berkeley, California, 1999)

69 **Anatole Katok, Rafael de la Llave, Yakov Pesin, and Howard Weiss, Editors,** Smooth ergodic theory and its applications (University of Washington, Seattle, 1999)

68 **Robert S. Doran and V. S. Varadarajan, Editors,** The mathematical legacy of Harish-Chandra: A celebration of representation theory and harmonic analysis (Baltimore, Maryland, 1998)

67 **Wayne Raskind and Charles Weibel, Editors,** Algebraic K-theory (University of Washington, Seattle, 1997)

66 **Robert S. Doran, Ze-Li Dou, and George T. Gilbert, Editors,** Automorphic forms, automorphic representations, and arithmetic (Texas Christian University, Fort Worth, 1996)

65 **M. Giaquinta, J. Shatah, and S. R. S. Varadhan, Editors,** Differential equations: La Pietra 1996 (Villa La Pietra, Florence, Italy, 1996)

64 **G. Ferreyra, R. Gardner, H. Hermes, and H. Sussmann, Editors,** Differential geometry and control (University of Colorado, Boulder, 1997)

63 **Alejandro Adem, Jon Carlson, Stewart Priddy, and Peter Webb, Editors,** Group representations: Cohomology, group actions and topology (University of Washington, Seattle, 1996)

62 **János Kollár, Robert Lazarsfeld, and David R. Morrison, Editors,** Algebraic geometry—Santa Cruz 1995 (University of California, Santa Cruz, July 1995)

61 **T. N. Bailey and A. W. Knapp, Editors,** Representation theory and automorphic forms (International Centre for Mathematical Sciences, Edinburgh, Scotland, March 1996)

60 **David Jerison, I. M. Singer, and Daniel W. Stroock, Editors,** The legacy of Norbert Wiener: A centennial symposium (Massachusetts Institute of Technology, Cambridge, October 1994)

59 **William Arveson, Thomas Branson, and Irving Segal, Editors,** Quantization, nonlinear partial differential equations, and operator algebra (Massachusetts Institute of Technology, Cambridge, June 1994)

58 **Bill Jacob and Alex Rosenberg, Editors,** K-theory and algebraic geometry: Connections with quadratic forms and division algebras (University of California, Santa Barbara, July 1992)

57 **Michael C. Cranston and Mark A. Pinsky, Editors,** Stochastic analysis (Cornell University, Ithaca, July 1993)

56 **William J. Haboush and Brian J. Parshall, Editors,** Algebraic groups and their generalizations (Pennsylvania State University, University Park, July 1991)

55 **Uwe Jannsen, Steven L. Kleiman, and Jean-Pierre Serre, Editors,** Motives (University of Washington, Seattle, July/August 1991)

For a complete list of titles in this series, visit the
AMS Bookstore at **www.ams.org/bookstore/**.

PSPUM/74